Walter Larcher   Physiological Plant Ecology

Springer

Berlin
Heidelberg
New York
Barcelona
Budapest
Hong Kong
London
Milan
Paris
Tokyo

Walter Larcher

Physiological Plant Ecology

Ecophysiology and Stress Physiology
of Functional Groups

Third Edition

With 348 Figures

 Springer

Professor Dr. Walter Larcher

Institut für Allgemeine Botanik
der Universität Innsbruck
Sternwartestraße 15
A-6020 Innsbruck
Austria

Translated by:

Joy Wieser

Madleinweg 8
A-6063 Rum bei Innsbruck
Austria

Translated from the German edition "Walter Larcher, Ökophysiologie der Pflanzen, 5. Auflage"
published 1994 by Verlag Eugen Ulmer, Stuttgart

First edition 1975
Second edition 1980
Corrected printing 1983
Reprint 1991
Third edition 1995

1 9 8 7 6 5 2 1

ISBN 3-540-58116-2 3. Aufl. Springer-Verlag Berlin Heidelberg New York
ISBN 0-387-58116-2 3rd ed. Springer-Verlag New York Berlin Heidelberg

ISBN 3-540-09795-3 2. Aufl. Springer-Verlag Berlin Heidelberg New York
ISBN 0-387-09795-3 2nd ed. Springer-Verlag New York Berlin Heidelberg

Library of Congress Cataloging-in-Publication Data. Larcher, W. (Walter), 1929– [Ökophysiologie
der Pflanzen. English] Physiological plant ecology: ecophysiology and stress physiology of function
groups / Walter Larcher. – 3rd ed. p. cm. Translation of: Ökophysiologie der Pflanzen. 5. Aufl.
Includes bibliographical references (p.   ) and index. ISBN 3-540-58116-2 (Berlin: alk. paper). –
ISBN 0-387-58116-2 (New York: alk. paper) 1. Plant ecophysiology. I. Title. QK905.L3713 1995
581.5–dc20 94-24405

The use of general descriptive names, registered names, trademarks, etc. in this publication does
not imply, even in the absence of a specific statement, that such names are exempt from the relevant
protective laws and regulations and therefore free for general use.

Cover photograph © Gurche/West Stock.
Typesetting: K + V Fotosatz GmbH, Beerfelden
SPIN 10560997   31/3111-5 4 3 2 - Printed on acid-free paper

Preface to the Third Edition

Plant ecophysiology is concerned fundamentally with the physiology of plants as modified by fluctuating external influences. According to a definition agreed by the Société d'Écophysiologie, ecophysiology "involves both the descriptive study of the responses of organisms to ambient conditions and the causal analysis of the corresponding ecologically dependent physiological mechanisms, at every level of organization. The ecophysiological approach must take into account polymorphism" (that is, structural and functional diversity) "in individual responses, which are largely responsible for the adaptive capacity of any given population. In this respect, ecophysiological study yields information which is fundamental for an understanding of the mechanisms underlying adaptive strategies" (Vannier 1994).

The aim of the author is to convey the conceptual framework upon which this discipline is based, to offer insights into the basic mechanisms and interactions within the system "plant and environment", and to present examples of current problems. Although physiological ecology is currently an exciting and expanding branch of science it is by no means a recent field of research. I have tried to portray its rich historical background in the choice of illustrations and tabular material; the results presented reflect the broadness of vision, the struggles and successes of the pioneering experimental ecophysiologists, as well as the most recent advances in knowledge. Moreover, one must bear in mind the particular characteristics of different localities; I have tried to include a broad selection of examples illustrating the ecophysiological behavior of plants in the greatest possible variety of habitats.

The present edition of Physiological Plant Ecology has been completely redesigned, revised and updated. The concept of functional diversity has been retained throughout, and amplified by further examples.

Chapter 1, in addition to dealing with the abiotic environment of plants, includes a survey of chemical communication and interactions between organisms. Chapters 2, 3 and 4, respectively, are concerned with carbon metabolism and the production of dry matter, with mineral nutrition and responses to particular soils, and with water relations. Chapter 5 is concerned with processes of development in connection with environmental factors. Lastly, an extensive chapter on stress physiology is devoted to the increasing importance of climatic, edaphic and anthropogenic disturbances to plant life.

The book has been expanded by the addition of many new tables and figures; lists of data have been revised and where possible the values have been converted to SI units. The reference numbers affixed to definitions and data in the text indicate sources; references cited in the tables and the legends of figures indicate literature for more detailed studies. It is my hope that these references will stimulate the reader to a more thorough perusal of the original publications.

My sincere thanks are due to Dr. D. Czeschlik and the team at Springer-Verlag for their highly efficient and friendly cooperation in the production of this new English edition. Also, it is a pleasure to express my appreciation to Ms. Joy Wieser B.Sc. (Physiol. Lond.) who translated the German manuscript into English. Her painstaking engagement with the subject matter has produced a translation that conveys in every way my original intention. I am greatly indebted to the many helpful colleagues who contributed information and advice, or made available to me the originals of figures. Finally, my warmest thanks are due to Professor Dr. R. B. Walker, Seattle, whose comprehensive knowledge of the field has been the source of many invaluable suggestions from which the book has profited greatly.

WALTER LARCHER

Contents

Units and Conversion

Prefixes for SI units

G	giga	(10^9)
M	mega	(10^6)
k	kilo	(10^3)
h	hecto	(10^2)
d	deci	(10^{-1})
c	centi	(10^{-2})
m	milli	(10^{-3})
μ	micro	(10^{-6})
n	nano	(10^{-9})

Energy

$1\ J = 1\ N \cdot m = 1\ kg \cdot m^2 \cdot s^{-2} = 1\ W \cdot s = 0.239\ cal = 10^7\ erg$

$1\ W \cdot h = 3.6\ kW \cdot s = 3.6\ kJ = 0.86\ kcal$

$1\ MJ = 0.278\ kWh$

$1\ cal = 4.1868\ J$

$1\ kcal = 1.163\ W \cdot h$

Pressure

$1\ MPa = 10^6\ Pa = 10\ bar$

$1\ bar = 10^5\ N \cdot m^{-2} = 10^5\ Pa = 100\ J \cdot kg^{-1} = 10^6\ erg \cdot cm^{-3}$

$1\ bar = 750\ Torr = 0.9869\ atm$

$1\ atm = 1.0132\ bar = 760\ Torr$

Amount, concentration

Molarity $= mol \cdot kg^{-1}$ solution

Molality $= mol \cdot kg^{-1}$ solvent

$1\ ppm = 10^{-6}\ mol \cdot mol^{-1}$; $1\ \mu g \cdot g^{-1}$; $1\ \mu l \cdot l^{-1}$

$1\ ppb = 10^{-9}\ mol \cdot mol^{-1}$; $1\ ng \cdot g^{-1}$; $1\ nl \cdot l^{-1}$

$1\ ppm\ CO_2 = 1.83\ mg \cdot m^{-3} = 41.6\ \mu mol \cdot m^{-3} = 0.101\ Pa$

(at 20 °C and 101.3 kPa atmospheric pressure)

$1\ dalton = 1.6605 \cdot 10^{-27}\ kg$

Water potential

Ψ [MPa] $= 0.462 \cdot T_{abs} \cdot \ln a_w = -1.06\, T_{abs} \cdot \log_{10}(100/RH)$

$\pi_{20}^* = 310.7 \log_{10} a_w$ [MPa]

$a_w = p_w/p_w^* = 10^{-2} RH$ [%]

a_w = relative activity of water

p_w = water vapour pressure

p_w^* = saturation water vapour pressure

RH = relative air humidity

$1\, osmol \cdot kg^{-1} = 0.00832 \cdot T_{abs}$ [MPa]

Electrolyte (electrical) conductivity

$1\, S = 1\, \Omega^{-1}$

$1\, S \cdot m^{-1} = 10\, mS \cdot cm^{-1}$

$1\, mmhos \cdot cm^{-1} = 1\, mS \cdot cm^{-1}$

Gas exchange

$1\, g\ CO_2$-gas exchange $\approx 0.73\, g\ O_2$-gas exchange $(RQ[CO_2/O_2] = 1)$

$1\, g\ O_2$-gas exchange $\approx 1.38\, g\ CO_2$-gas exchange

Diffusion coefficients $D_{CO_2} = 0.64\, D_{H_2O}$

Diffusion coefficients $D_{H_2O} = 1.56\, D_{CO_2}$

$0.03\%_{vol}\ CO_2 = 300\,\mu l \cdot l^{-1} = 282\,\mu bar = 28\, Pa\ CO_2$ partial pressure

$1\,\mu l \cdot l^{-1} = 1.963\,\mu g\ CO_2 \cdot l^{-1}$ (at 1013 mbar air pressure and 0 °C)

$1\, mg\ CO_2 \cdot dm^{-2} \cdot h^{-1} = 0.028\, mg\ CO_2 \cdot m^{-2} \cdot s^{-1} = 0.63\,\mu mol\ CO_2 \cdot m^{-2}\, s^{-1}$

$1\, mg\ CO_2 \cdot m^{-2} \cdot s^{-1} = 36\, mg\ CO_2 \cdot dm^{-2} \cdot h^{-1} = 22.7\,\mu mol\ CO_2 \cdot m^{-2} \cdot s^{-1}$

$1\,\mu mol\ CO_2 \cdot m^{-2} \cdot s^{-1} = 0.044\, mg\ CO_2 \cdot m^{-2} \cdot s^{-1} = 1.58\, mg\ CO_2 \cdot dm^{-2} \cdot h^{-1}$

$1\, mg\ H_2O \cdot dm^{-2} \cdot h^{-1} = 1.54\,\mu mol\ H_2O \cdot m^{-2} \cdot s^{-1}$

Conductances (at 20 °C and 101.3 kPa):

$1\, cm \cdot s^{-1} \approx 0.416\, mol \cdot m^{-2} \cdot s^{-1}$

$1\, mol \cdot m^{-2} \cdot s^{-1} \approx 0.024\, mm \cdot s^{-1}$

$1\, mol \cdot m^{-2} \cdot s^{-1} = 0.446\, T_0/T_{abs} \cdot P/P_0\, cm \cdot s^{-1}$

$T_0 = 273\, K$; T_{abs} = absolute temperature

$P_0 = 101.3\, kPa$; P = ambient air pressure [kPa]

Phytomass

$1\, g\ DM \cdot m^{-2} = 10^{-2}\, t \cdot ha^{-1}$

$1\, g$ org. DM $\approx 0.42 - 0.51\, g\ C \approx 1.5 - 1.7\, g\ CO_2$

$1\, g\ C \approx 2 - 2.2\, g$ org. DM $\approx 3.1 - 3.4\, g\ CO_2$

$1\, g\ CO_2 \approx 0.59 - 0.66\, g$ org. DM $\approx 0.27 - 0.30\, g\ C$

Radiation

$1\, W \cdot m^{-2} = 1\, J \cdot m^{-2}\, s^{-1} = 31.53\, MJ \cdot m^{-2} \cdot a^{-1}$

$1\, mol$ photons $= 1.8 \cdot 10^5\, J$ (at λ 650 nm) to $2.7 \cdot 10^5\, J$ (at λ 450 nm)

$1\, cal \cdot cm^{-2} \cdot min^{-1} = 6.98 \cdot 10^2\, W \cdot m^{-2} = 6.98 \cdot 10^5\, erg \cdot cm^{-2} \cdot s^{-1}$

$1\, erg \cdot cm^{-2} \cdot s^{-1} = 1.43 \cdot 10^{-6}\, cal \cdot cm^{-2} \cdot min^{-1} = 10^{-3}\, W \cdot m^{-2}$

Conversions* (multiply by)	$W \cdot m^{-2}$ (PhAR) in µmol photons $m^{-2} \cdot s^{-1}$	klux (400–700 nm) in µmol photons $m^{-2} \cdot s^{-1}$
Daylight (sunny)	4.6	18
Daylight (diffuse)	4.2	19
Metal halide lamp	4.6	14
Fluorescent tube (white)	4.6	12
Incandescent lamp	5.0	20

* After McCree (1981); see also Thimijan and Heins (1983)

Temperature

Mean temperature of daylight hours:

$T_D = 0.5 (T_{max} + T_{min}) + (T_{max} - T_{min})/3\pi$ (Milthorpe and Moorby 1979)

$T_D = 0.29 T_{min} + 0.71 T_{max}$ (Goudriaan and Laar 1994)

Mean temperature nighttime hours: $T_N = 0.25 (T_{max} + 3 \cdot T_{min})$

For further information on units and conversion factors see Šesták et al. (1971), Altman and Dittmer (1972), Slavík (1974), Incoll et al. (1977), Savage (1979), Krizek (1982), Pearcy et al. (1989), Nobel (1991a), Salisbury (1991), Hall et al. (1993)

Abbreviations

A	area
ABA	abscisic acid
ATP	adenosine triphosphate
B	biomass, phytomass
C	concentration
CAM	Crassulacean acid metabolism
CGR	crop growth rate
chl	chlorophyll
CK	cytokinin
CUE	carbon use efficiency
DIC	dissolved inorganic carbon
DM	dry matter
DW	dry weight
Ep	potential evaporation
ET	evapotranspiration
FC	field capacity
FW	fresh weight
g	diffusive conductance
GA	gibberellic acid
I	irradiance
IAA	indoleacetic acid (auxin)
IR	infrared radiation
J	fluxes
JA	jasmonic acid
L	litter
L_{50}	50% lethality
LAI	leaf area index
LHPC	light-harvesting pigment complex
NIR	near infrared radiation
NUE	nitrogen use efficiency
P	pressure
PEP	phosphoenolpyruvate
Ph	photosynthesis
PhAR	photosynthetically active radiation
Phy	phytochrome
PP	primary production
PPFD	photosynthetic photon flux density

Pr	precipitation
PR	production rate
PS	photosystem
PWP	permanent wilting percentage
r	transfer resistance
R	respiration
RDI	relative drought index
RH	relative humidity
RuBP	ribulose-1,5-bisphosphate
RUE	radiation use efficiency
RWC	relative water content
SLA	specific leaf area
t	time
T	temperature
Tr	transpiration
ULR	unit leaf rate
UV	ultraviolet radiation
V	volume
VPD	vapour pressure deficit
WSD	water saturation deficit
WUE	water use efficiency
Y	yield
Φ	photon yield
Ψ	water potential

1 The Environment of Plants

In the far geological past when the first plants were evolving, they were confronted with an environment of water, air and rock. Later, with the assistance of microorganisms and animals, their principal substrate, the soil or pedosphere, was gradually formed. Hydrosphere, atmosphere and pedosphere together make up the *spatial environment* of plants; but the environment of a plant is more than this. It is also the physical and chemical factors in its habitat, and the influences exerted by the other organisms in it, whether favourable or unfavourable to the plant's survival and success. *Environment* is thus the totality of the external conditions acting upon a living organism or community of organisms (biocenosis) in its biotope.

1.1 The Surroundings of Plants

1.1.1 The Atmosphere

The most sensitive region of the global environment is the atmosphere, the thin layer of air enveloping the earth (Fig. 1.1). Its innermost layer is the *tropo-*

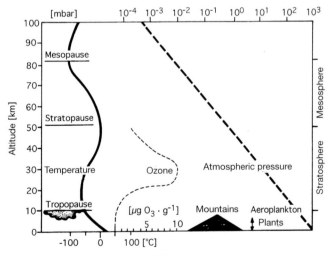

Fig. 1.1. Structure of the atmosphere. (After Häckel 1990)

sphere, the weather zone of the atmosphere. In this layer, the air pressure drops steeply in an outward direction (decrease in air pressure: $10-12$ Pa m^{-1}), and at an altitude of 5500 m is only half that at sea level. At the same time, the temperature of the air also decreases until it reaches a minimum, at the so-called tropopause, at altitudes ranging from 10 km (at middle latitudes) to 18 km (over subtropical and equatorial regions). The extreme stability of the temperature inversion immediately above the tropopause presents an obstacle to the exchange of air with the regions above it, i.e., with the *stratosphere*, so that it is chiefly the troposphere that contains foreign substances and the water vapour resulting from evaporation. In the stratosphere, which extends from the tropopause up to the stratopause, at an altitude of $25-50$ km, the air pressure drops to less than one-tenth of the value at sea level. In this rarefied atmosphere the intense UV radiation brings about the formation of ozone, itself a powerful absorber of ultraviolet radiation. It was only after the formation of this ozone filter that the lethal short-wave UV radiation reaching the earth dropped to a level at which life on land became possible.

The **air enveloping the globe** provides carbon dioxide for plants, and oxygen for all living organisms. The primaeval atmosphere contained large quantities of carbon dioxide, ammonia and methane. Today, the main components of the air of the troposphere are 78 vol.% nitrogen, 21 vol.% oxygen, rare gases (0.95 vol.%) and carbon dioxide (0.035 vol.%). Other constituents are water vapour, trace gases such as methane, sulphur dioxide, halides, volatile nitrogen compounds, ozone and photo-oxidants, as well as aerosols, ash, dust and soot.

The atmosphere contains roughly 1200×10^{12} t of oxygen, the greater part formed by autotrophic organisms and accumulated over aeons of time (see Fig. 2.1). Oxygen losses are continuously replaced by the photosynthesis of phytoplankton in the oceans, and by terrestrial plants, chiefly those of the forests. However, the oxygen emitted by the terrestrial vegetation does not constitute a long-term net gain for the global atmospheric oxygen budget because this photosynthetically released oxygen is consumed again in roughly the same amounts in the course of microbial breakdown of organic matter. In the short term, however, and on a smaller scale, respiration by plants and animals is important. The oxygen consumed by respiring living organisms and by combustion processes can be replaced only by the activity of phytoplankton. In aquatic environments organic detritus sinks to the bottom, where its decomposition is largely achieved by anaerobic processes, i.e. without using up oxygen. The atmosphere contains 721×10^9 t carbon in the form of CO_2 (see Fig. 6.103). In the lower atmosphere at the present time, the average concentration of carbon dioxide is $350 \ \mu l \ l^{-1}$ (equivalent to 35 Pa partial pressure), although this figure is increasing continuously (see Chap. 6.3.3.2).

Substances entering the atmosphere are dispersed very rapidly over great distances; within only a few weeks or even days, emissions are conveyed by air currents over entire continents and oceans. Striking examples of this are seen after severe volcanic eruptions or following the escape of radioactive substances.

1.1.2 The Hydrosphere

The hydrosphere comprises the oceans, groundwater, inland waters, polar ice, glaciers, and the water of the atmosphere. The oceans cover 71% of the surface of the globe and contain 74% of the world's **total water reserves**. These gigantic volumes of water store enormous quantities of energy and material, and as a consequence are the most important stabilizer of geophysical and geochemical processes. The second largest reservoir of water is the continental water, most of it *groundwater*, of which only 1% is near enough to the surface to be within reach of plant roots; the rest seeps down to depths of hundreds of metres. Although *surface waters*, i.e. lakes and rivers, account for only a small proportion of the hydrosphere, their importance as plant environments is very considerable. The water floating above the continents and oceans as *clouds, fog and water vapour*, although amounting to only 0.001% of the earth's water, is of immense significance for the global water and heat balance on account of its high rate of turnover (average residence time of water vapour in the atmosphere = 10 days). Quantitatively, the water cycle is the earth's largest flow of matter, and at the same time the most important flow of energy because the majority of the solar radiation absorbed by the earth's surface is dissipated in the evaporation of water. Thus, the hydrosphere is a determinant component of the climate system of the earth.

The various categories of water bodies differ considerably in their **chemical composition**. The principal constituent of *seawater* [3] is NaCl ($10.8 \, g \, l^{-1}$ Na^+, $9.4 \, g \, l^{-1}$ Cl^-, at a mean density of 1.027); also important are the cations Mg^{2+} ($1.3 \, g \, l^{-1}$), Ca^{2+} and K^+ (roughly $0.4 \, g \, l^{-1}$ of each), and the anions SO_4^{2-} ($2.7 \, g \, l^{-1}$), HCO_3^- ($0.14 \, g \, l^{-1}$) and Br^- ($0.009 \, g \, l^{-1}$). Numerous other elements are present in low concentrations (a few $mg \, l^{-1}$ to $\mu g \, l^{-1}$), either in the ionic form or as compounds (e.g. silicates, organically bound nitrogen). *Freshwater* usually contains large quantities of Ca^{2+} and HCO^{3-}. Where mineral solutes have found their way into the surface water, either from inflows or from the surrounding land, eutrophic zones of water are created. An excess of nutrients can build up in rivers and lakes as a result of inflowing nitrate- and phosphate-loaded waste water. The concentrations of solid material and dissolved mineral substances in a eutrophic lake (from 0.03 to $>5 \, mg \, P \, l^{-1}$ and 0.5 to $>15 \, mg \, N \, l^{-1}$) are up to 300 times those in an oligotrophic lake [142].

The solubility of *gases in water* rises with increasing pressure and decreasing temperature but falls with increasing salt concentration. Carbon dioxide is readily soluble in water (Table 1.1). The proportion of free CO_2 depends on the pH of the water; it is high in the acid range, whereas only bicarbonate and carbonate ions are present at pH values above 9 (Fig. 1.3). The dissolved HCO_3^- can be bound by cations, chiefly Ca^{2+} and Mg^{2+}. The water of oceans and lakes contains 0.14% of the entire carbon of the earth in the form of bicarbonate and carbonate ions or as dissolved inorganic carbon (DIC), the majority in the deep water of the oceans (about 38000×10^9 t). The surface water contains a mere 0.2% of the organic carbon of the hydrosphere; the par-

BIOSPHERE

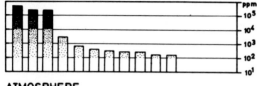

ATMOSPHERE

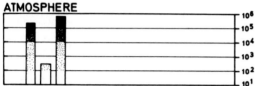

HYDROSPHERE

LITHOSPHERE

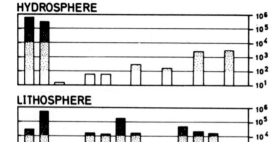

H O C N Ca K Si Mg P S Al Na Fe Cl

Fig. 1.2. Composition of the biomass, atmosphere, hydrosphere and lithosphere, in terms of the relative numbers of atoms (atoms per million atoms, not the proportion by weight) of the various chemical elements. The composition of living organisms is clearly distinct from that of the three components of their environment; they select from the available elements, according to their needs. The scale of the ordinate is logarithmic (different shading of the bars). For example, in the biomass H, O, C and N are present in the greatest proportions: 4.98×10^5 atoms per million (i.e. about 50% of all atoms) are hydrogen atoms; oxygen and carbon atoms each comprise 2.49×10^5 atoms per million (about 25%), and 2.7×10^3 (about 0.3%) are nitrogen atoms. (Deevey 1970)

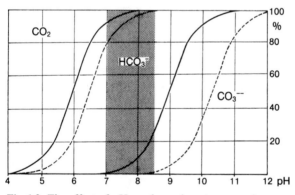

Fig. 1.3. The effect of pH on the carbonate system in seawater (*solid line*) and freshwater (*broken line*). *Shaded area*: Average pH values of seawater. (Ott 1988). The ordinate indicates the percent of the total CO_2 or CO_3^{2-} present at particular pH values

Table 1.1. Solubility (μM) and volume concentration (%) of CO_2 and O_2 in water at different temperatures, in equilibrium with the air. Partial pressure in the air: 35 Pa CO_2, 21 kPa O_2. (Šesták et al. 1971)

Temperature (°C)	Carbon (μM)	Dioxide (%)	Oxygen		Mole ratio O_2/CO_2
			(μM)	(%)	
0	26	0.059	458	1.02	17.6
10	18	0.041	356	0.79	19.8
15	15	0.035	318	0.72	21.2
20	13	0.030	291	0.66	22.4
25	11	0.026	263	0.59	23.9
30	9	0.023	245	0.55	27.2

tial pressure of oxygen in surface water is on average in equilibrium with that of the atmosphere (see Table 1.1).

The carbon dioxide and oxygen contents of the uppermost and best illuminated layers of water (euphotic zone) fluctuate diurnally and seasonally due to the photosynthetic rhythms of the photoautotrophic organisms (Fig. 1.4). The oceans in particular play an important role as a buffering system on account of the intensive exchange of O_2 and CO_2 between hydrosphere and atmosphere.

Water currents bring about mixing of the surface water and serve to equilibrate surface water and water of the deeper levels. Below the euphotic zone in poorly mixed waters, the oxygen concentration drops steadily with increasing depth due to *oxygen consumption* by animals, microorganisms and reducing substances in the detritus. Hence, where circulation is poor, as is frequently the case in lake basins and certain regions of seas, for example, the

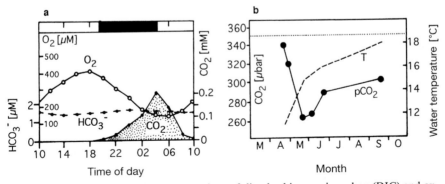

Fig. 1.4 a, b. Fluctuations in the concentrations of dissolved inorganic carbon (DIC) and oxygen in different waters. **a** Changes in the concentrations of dissolved CO_2, HCO_3^- and O_2 at 20–22 °C and pH 7–8 through 24 h in a fish pond in central Europe on a sunny July day. (After Ondok et al. 1984; Pokorný and Ondok 1991). **b** Seasonal alterations in the partial pressure of CO_2 (pCO_2) and temperature (T) of surface water in the NE Atlantic. (After Watson et al., from Williamson and Platt 1991)

Baltic or the Black Sea, there may be a complete absence of oxygen in the deeper water. Vertical convection currents in the *upwelling regions* of the continental shelves carry mineral substances to the upper layers of water. The return of inorganic breakdown products to the surface water ensures that the circulation of material is maintained.

Water currents also play a role in *temperature equilibration*. This is clearly illustrated by the development of temperature-induced density stratification in their absence. Because the incoming radiation is absorbed by the uppermost water layers, only these and none of the deeper layers are warmed up. Warm water being less dense than cold, and because the movement of water by wind and tides can eliminate the density gradient only down to a limited depth, two distinct bodies of water are formed. The narrow transitional zone between the two is known as the *thermocline*. A permanent thermocline exists deep down below the euphotic zone in oceans; an additional seasonal thermocline develops only in middle latitudes. Near the poles and in equatorial regions, however, the water temperature remains constant throughout the year. This explains why no mixing takes place in tropical oceans and why the water of the uppermost, light-absorbing layers is permanently deficient in mineral substances. An exception to this is a considerable upwelling of nutrient-rich cold water on the west coast of S. America near the equator. Density stratification also occurs in lakes, a warmer surface layer floating above the cooler deeper water. Thorough mixing of the body of water takes place only when the surface water cools down sufficiently, i.e. in fall and spring, or in winter at middle and higher latitudes, or diurnally in equatorial regions.

1.1.3 The Lithosphere and the Soil

The **earth's crust** is the storehouse for the variety of chemical elements needed by living organisms for building up their body substance. By far the most commonly occurring element of the earth's crust is minerally bound oxygen (see Fig. 1.2). Second on the list is silicon, followed by aluminium, sodium, iron, calcium, magnesium, potassium and phosphorus in descending order. All other chemical elements are present in concentrations (w/w) of less than 0.1%. Exchange of material takes place between lithosphere and hydrosphere (by transport of weathering products and by the formation of sediments on the bottom of all bodies of water) and with the atmosphere (drifting of dust, volcanic eruptions).

The outer crust of the earth is composed of igneous, metamorphic and sedimentary rocks. The igneous rocks were formed as the molten rock or magma cooled down and crystallized. Slow cooling resulted in coarse-grained *plutonic rocks*, with a high content of silicon and aluminium (acidic granite, syenite and diorite) or of silicon, magnesium and iron (basic gabbro and peridotite). The so-called *extrusive rocks*, on the other hand, the result of rapid cooling, are fine-grained or amorphous, and may also be either acidic or basic (basalt, porphyry, lavas and tuff). *Metamorphic rocks*, as their name implies, originate from volcanic and sedimentary rocks, as a result of the enormous pressure and heat created by the subsidence of the earth's crust following tectonic movements. Slate, phyllite, quartzite and marble are examples of metamorphic rocks, most of which are easily weathered. *Sedimentary rocks* are formed by the compaction and cementation of sediments, either from fragments of rock or deposits of erosion products (conglomerates, sandstone, marl, shales), as well as from deposits of biogenic origin, such as limestone from reef rock or chalk, or from silicate slates.

The principal constituents of the rocks of the earth's crust are silicate minerals (such as quartz, feldspars, augite, hornblende, mica) and carbonate minerals (calcite, dolomite). Ores, phosphates, sulphates, alkali salts and alkali earth salts, fossil carbon beds, as well as accumulations of individual elements such as sulphur, are found either as larger deposits or scattered.

Although the basic material from which soil is formed is provided by the lithosphere, **soil** is more than simply the loosened surface of rocks. Transformations and the mingling of organic and mineral substances have resulted in the production of a complex system known as the *pedosphere.* The picture presented by a landscape bare of soil, e.g. after a volcanic eruption, is one of lunar sterility. Only a few pioneer plants such as cyanobacteria, aerial algae, lichens and mosses are capable of thriving on bare rock or sand.

Soils are formed by weathering, and by humification of plant detritus and dead soil organisms. Outcropping rock is broken up and roughened by mechanical *weathering* caused by water, wind and ice, and by energy released in connection with glacial movements, or when rubble on steep slopes is displaced by landslides and avalanches. As a result of chemical weathering the soluble salts are leached out; the carbonates, feldspars, hornblende, augite and mica are hydrolyzed and minerals rich in iron or manganese are oxidized. In soils already inhabited by living organisms, weathering is promoted by penetration of rock clefts by roots, and by the excretions of plants and microorganisms.

The combined effect of the above processes leads to the production of strata of *secondary soil minerals*, characterized by a large internal surface, predominantly negative surface charges, and high adsorption capacity (oxides, clay minerals such as kaolin, montmorillonite, vermiculite and illite; Table 1.2). In the warm, humid climates of the tropics and subtropics, the combined action of frequently alternating swelling and shrinking and intensive chemical weathering leads to the degradation of soil minerals to iron hydroxides and *amorphous soil colloids* (aluminium hydroxide, silica colloids). In soils of this kind plant nutrients are not easily accumulated; the soil becomes sterile and its surface quickly forms a crust when deprived of vegetation (laterization).

Humus formation begins with the breakdown of detritus (litter) and results in chemically stable humic substances. These are dark, finely particulate, high-molecular substances with an average composition of $44-58\%$ C, $0.5-4\%$ N, $42-46\%$ O and $6-8\%$ H. Litter is converted via a series of intermediate steps into substances that are not easily broken down, such as lignin, cutin, humic substances (strongly acidic, water-soluble fulvic acid, colloidal humic acids and highly polymerized stable humins).

Table 1.2. Properties of ion exchangers in the soil. (Jeffrey 1987; Kuntze et al. 1988)

Exchanger	Specific surface area $(m^2 g^{-1})$	Mean CEC[1] $(eq \, kg^{-1})$
Hydrated oxides of Al and Fe	25 – 40	0.03 – 0.05
Kaolinite	10 – 20	0.03 – 0.15
Illites	100 – 300	0.2 – 0.5
Vermiculite	600 – 800	1 – 2 (8)
Montmorillonite	700 – 1200	0.8 – 1.2 (10)
Humic substances	800	1.5 – 5

[1] Cation exchange capacity.

How quickly litter is broken down depends on the ease of decomposition of the organic matter, the chemical constitution and pH of the weathered substrate, and the climatic conditions. Cellulosic material decomposes in one-third of the time required by woody plant fragments containing tannin (Fig. 1.5). The decomposition of organic matter in the soil is promoted by a weakly acid to weakly alkaline pH, accessibility of oxyen, by moisture and heat. The annual rate of decay, k, can be calculated [106] from the ratio of the quantity of litter produced annually, L, to the total amount of litter, A.

$$k = L/L + A \; [\%] \tag{1.1}$$

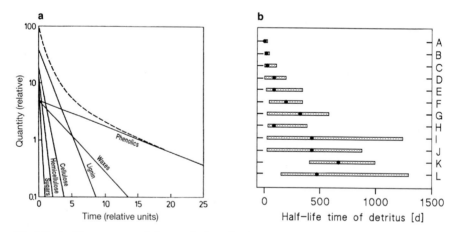

Fig. 1.5 a, b. Time courses of decay of plant detritus. **a** Breakdown of the various organic components of litter. (Minderman, as cited in Chapman 1976). **b** Decomposition rates (half-life time) of different components of detritus. *A* Microalgae; *B* freshwater plants; *C* marsh plants; *D* macroalgae; *E* sea grasses; *F* grasses; *G* sedges; *H* mangroves; *I* leaves of deciduous trees; *J* litter from shrubs; *K* conifer needles; *L* leaves of evergreen trees. *Bars* extend to the 95% confidence limits; *points* represent mean values. (After Enriquez et al. 1993)

The *rate of decomposition* indicates the percent of the annual litter production decomposed each year. In tropical rainforests 95% of the litter is decomposed each year, whereas in deciduous forests of the temperature zone decomposition takes 2−4 years, in coniferous forests of the boreal zone up to 14 years, in montane forests and tundra even decades [1]. Decomposition in steppe regions takes, on an average, 2 years, and proceeds rapidly in spring and summer, but then slows down, which is why the litter layer is thickest in winter.

The *soil texture* is determined by particle size and humus content. Tilled soils containing less than 1% organic matter are considered to be poor in humus, those with 1−4% are termed humic, those with more than 4% strongly humic and those with more than 8% are rated as humus-rich. Humin substances bind with calcium carbonate to form weakly soluble calcium humates, and with clay minerals to form clay-humus complexes which adhere to one another, forming particles between which are small spaces. The system of pores formed by these spaces is filled partly with air and partly with water, and is the environment in which the soil organisms live (the edaphon). The size of the channels is determined in part by the feeding tunnels of the soil animals and the root canals, but primarily by the particle size of the minerals. Sandy soils with coarse, angular particles measuring 0.06 to 2 mm in diameter contain many and large pores. Clay soils with flake-like particles smaller than 0.002 mm are dense and because their pores are fine the pore volume is large, but they are nevertheless poorly aerated. Loamy soils can be classified according to their clay content, ranging from sandy loam to heavy loam.

In undisturbed soils, a profile of *horizons* develops; between the upper layers of litter and humus and the lowermost horizon of mineral weathering products are zones of mixed composition, containing varying proportions of humus. The arrangement of the layers in the *soil profile* is characteristic for the particular type of soil, and is governed by the climate, vegetation and underlying rock (see Table 1.7). Under its natural vegetation, each climate zone produces its own characteristic soil type. In some regions, however, a mature soil cannot develop and the resulting soil is of an extreme type, such as sandy or rocky soils, salt-, soda- and chalky soils in dry regions, or mires and swampy soils in terrain of glacial origin. Cultivation and over-exploitation cause drastic changes in the soil, and even result in extensive *soil erosion*. But even without human intervention soils are continuously in a state of flux: they grow and mature to a maximum climax, and they can age and deteriorate due to acidification, hardening, and depletion or even exhaustion of their nutrients. *Soil development* and deterioration are exceptionally fast processes in the tropics. Tropical soils are severely degraded if the dynamic equilibrium between the decomposition of organic matter in the soil and replacement of litter by the vegetation is disrupted. In cold, wet regions, on the other hand, the formation of soil from parent rock through the stage of raw humus up to full maturity progresses extremely slowly. In certain regions affected by advancing ice during the glacial periods, soil formation is still incomplete.

Soil is a *multiphase system* in which solid, liquid and gaseous phases occur. It presents a vast receiving space, especially efficient in buffering physical and chemical influences. The soil colloids influence the water-holding capacity, by both attracting and exchanging electrolytes. Thus the soil acts as reservoir and filter in the exchange of material between lithosphere, soil water, soil gases and the organisms living and rooting in the soil. Mechanical filtration, sorption and biological breakdown mechanisms can effectively bind pollutants entering the soil via precipitations or inflows. This not only means that toxic substances cannot seep immediately into the ground water, but also that their uptake by plants can be delayed or even prevented. A disadvantage of the high storage capacity of soil, and particularly of forest soils, is the long period of time required for recovery and normalization following heavy loading with pollutants.

1.1.4 The Phytosphere − a Part of the Ecosphere

The part of the earth that supports life, including every living organism within it, constitutes the **biosphere**. Plants grow in every region of the earth: in the oceans and inland waters, and on land − where plants are even encountered in such inhospitable regions as deserts and icefields. Of all living organisms, plants contribute by far the greatest mass: roughly 99% of the total mass represented by all living organisms on the earth (*biomass*) is plant material (*phytomass*). Consequently, the vegetation is a stabilizing factor in the cycling of matter and has a decisive influence on both climate and soil.

The immediate surroundings of plants, the **phytosphere**, is the realm in which the life cycle of the plant proceeds under the influence of the specific conditions imposed at the habitat, and is also the scene of the morphogenetic and evolutionary processes involved in selection and adaptation. The region occupied by roots, the *rhizosphere*, represents a special ecological compartment within the phytosphere.

Within this inorganic and organic environment, plants in turn influence their surroundings. The ecological interplays between biosphere and inorganic environment, in the form of energy flows, cycling of matter and interactions between ecological systems make up the **ecosphere**. The organisms coexisting in an *ecosystem*, which is a spatially and functionally defined part of the ecosphere (such as a forest, a lake, grassland, or a region shaped by human habitation), are connected by a diversity of relationships and interactions.

The **trophic** relationships between the various partners in an ecosystem involve the exchange of materials between autotrophic plants (primary producers) and heterotrophic links in the food chain (phytophages and other consumers, microbial decomposers). This is achieved in a great variety of ways, such as symbiosis (lichenization, mycorrhizal and root nodule symbiosis), parasitization, plant carnivory (digestion of trapped animal organisms) and saprophily (nutrients obtained by breakdown of dead organic substance).

Highly complex interactions regulate the coexistence of the individuals within a population, of the different species of plants in the vegetation, and of the plants, microorganisms and animals. Examples of **ecological interaction** are: *cooperation* of mature plants in shielding juvenile forms from strong irradiation, overheating or excessive cooling; *competition* due to lack of space, competition for light, nutrients and water; *chemical communication*, i.e. release of signal substances by plants, microorganisms and animals, influencing the development and appearance of other partners in the ecosystem (release of auxins by plant roots, promoting growth of soil fungi and bacteria, hormone-like substances from parasitic fungi and insects inducing plant galls and deformations, estrogens produced by plants influencing the population development of mammals, ecdysones of plant origin influencing the larval development of insects); interactions via *secondary plant substances* (see Sect. 1.1.4.2). Such interplay between organisms, which can extend beyond the limits of the individual ecosystem, ensures the stability of the ecosphere as an entity by promoting or inhibiting the development, spread and stability of a species.

In the following section two important interactions of plants and their biotic environment will be discussed: these are, firstly, the biogeochemical exchange processes in the root region and, secondly, the effect of bioactive plant metabolites on the plants, microorganisms and animals in the vicinity.

1.1.4.1 Biogeochemical Exchange Processes in the Rhizosphere

A lively exchange of material and gases takes place between the roots of vascular plants and soil organisms (algae, fungi, actinomycetes, bacteria, animals). Plant litter dropping to the ground is mineralized via the chain of decomposers, and the products are released in part in gaseous form (CO_2, CH_4, NH_3, N_2). In an intact ecosystem, mineralization and the production of organic substance are geared to one another. At high rates of mineralization, the minerals bound in the organic matter are quickly set free and returned to circulation. This means that the plants have a better supply of nutrient ions and are consequently able to produce a greater phytomass. If litter decomposition and breakdown of humus are too slow, the plants adjust to a slower rate of growth.

The **rate of mineralization** may vary over short distances, depending on the composition of the microflora, the aeration, temperature, moisture and pH of the soil, as well as on the quantity and nature of the decomposable substrate. It is highest in moist, neutral or weakly basic humus-rich soils, and is usually low in wet soils and in acid soils containing little calcium. Dryness of the soil and lack of heat (temperatures below $4-5\,°C$) inhibit the mineralizing activities of soil organisms. One of the factors responsible for the remarkable productivity of tropical rainforests is the rapid recycling of mineral substances. The favourable temperature and constantly moist conditions promote the activity of decomposer microorganisms to such a degree that the nutrient elements remain organically bound for only a brief period before they are again available to the plants in an inorganic form.

The activity of decomposer microorganisms is an especially important factor in the regulation of the **nitrogen flow** in an ecosystem. Nitrogen that has been removed as NO_3^- or NH_4^+ from the soil by autotrophic plants is first in-corporated into the phytomass and then returned to the soil. Thus far, the nitrogen is to a large degree protected from leaching. Nitrogen-containing organic substances are attacked by *ammonifying* microorganisms and other decomposers. Strictly speaking, the transition from the organic to an inorganic form is already achieved by the ammonifying bacteria. As a rule, however, the ammonia arising from microbial activity undergoes further conversion. Many microorganisms are able to exploit the different valency and oxidation levels of nitrogen between $N(-3)$ and $N(+5)$ to gain energy. *Nitrifying* (chemoautotrophic) bacteria, for example, oxidize NH_3 and NH_4^+ via nitrite to nitrate, and denitrifying bacteria liberate molecular nitrogen as the end product. Denitrification is particularly high in wet or flooded soils (in rice paddies, for example).

Not only is the rate of *nitrogen mineralization* (see Table 3.12) influenced by the above-mentioned climatic and edaphic factors, it also depends on the suitability of the litter for microbial decomposition, an indication of which is provided by the ratio of the carbon and nitrogen contents (C/N) of the substrate. Substances with a very high C/N ratio (about $100:1$, e.g. straw and lignified litter) are not readily utilizable by microorganisms unless additional nitrogen sources are available. The most favourable ratio for microbial decom-position is between $10:1$ and $30:1$ (e.g. leaf litter and humus).

In soils with a small pore volume, the oxygen consumption by roots and soil organisms results in a gradual decrease in the oxygen concentration in the soil air with increasing depth (Fig. 1.6). The constantly warm tropical soils are poor in oxygen on account of their high respiratory activity. In compacted, wet soils or soils covered by ice, inadequate gas exchange can produce a state of severe oxygen depletion, with detrimental or even lethal consequences for plant life. At the same time, the CO_2 content of the soil air rises to several times the concentration in the atmosphere. The extent of CO_2 accumulation depends, in turn, on the biological activity in the soil and the speed of gaseous diffu-sion, and is accordingly higher during the growing season, and in deeper horizons. The CO_2 content increases successively in the following soils, ar-

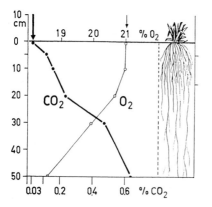

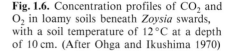

Fig. 1.6. Concentration profiles of CO_2 and O_2 in loamy soils beneath *Zoysia* swards, with a soil temperature of 12 °C at a depth of 10 cm. (After Ohga and Ikushima 1970)

ranged in order of decreasing pore volume: sand, loam, clay, up to marshy soils.

The CO_2 released into the layer of air immediately above the ground is termed *soil respiration*. Gaseous exchange between soil and atmosphere takes the direction of the partial pressure gradient of CO_2. Within the soil, the major mechanism by which gas concentrations are equilibrated is diffusion; on the soil surface, air movements also play a part in this equilibration. Fluctuation in the groundwater level and filling of the soil pores with water also results in CO_2 being driven out of the soil. Soil respiration is a valuable indicator of the intensity of the decomposition processes in the soil. With rising temperature, soil respiration also increases, and at a given temperature it is at its highest when soil moisture is optimal. In temperate zones the soils of forest, heath and grass swards give off an average of $0.1-0.5$ g CO_2 m^{-2}h^{-1}, and those of meadows up to 1 g CO_2 m^{-2}h^{-1}. In nutrient-rich tropical forests $1-1.2$ g CO_2 m^{-2}h^{-1} are recorded in the wet season, whereas on sandy soils, in dry regions and in the tundra the values are as low as $0.05-0.2$ g CO_2 m^{-2}h^{-1} [225].

The most intensive **exchange of substances** between plants, microorganisms and inorganic soil particles takes place at the root surface. In the soil-root contact layer the fine roots take up water and nutrient ions, and release organic compounds (diffusates, exudates and decay products from dead rhizodermis cells). *The absorbing root surface* of cultivated herbaceous plants amounts to roughly 1 cm^2 per cm^3 of soil. For purposes of comparison the root area can also be described by means of the *root area index* (RAI: m^2 root area per m^2 of ground); the values for saplings of deciduous tree species of the temperate zone are between 0.6 and 2.4, for conifer seedlings around $0.05-0.1$, for dense conifer stands about 10, and for semiarid shrubland 2 to 4 [104, 125]. Distribution and density of roots depend primarily on the specific morphology of the root system of the individual species (root types: Figs. 1.7 and 1.8) and vary over the course of the year, spreading in the spring or rainy season, dying off at the end of the growth season. Soil properties exert a very strong influence on the way in which roots spread; the root system develops in a characteristic manner (Fig. 1.9), depending on the pH, nutrient and water contents (see Fig. 4.9), the aeration of the soil, its depth, and the presence of obstructive or compacted horizons.

In intimate contact with the fine roots and separated by only a few μm is an actively metabolizing and multiplying community of bacteria and actinomycetes, known as the **rhizoplane**. This association with a specific microflora is encouraged by the soluble carbohydrates, oligosaccharides, organic acids, enzymes, vitamins and growth substances exuded by the roots. The mucus covering of the root caps, a colloidal carbohydrate (mucigel), provides additional substrate for the microflora. The plant, too, profits from this exchange of material with the rhizoplane. In addition to provision of the plant with nutrients produced by the bacteria and passed on via the mycorrhizal fungi, its roots are protected from pathogenic organisms by the antibiotic excretions of the microflora.

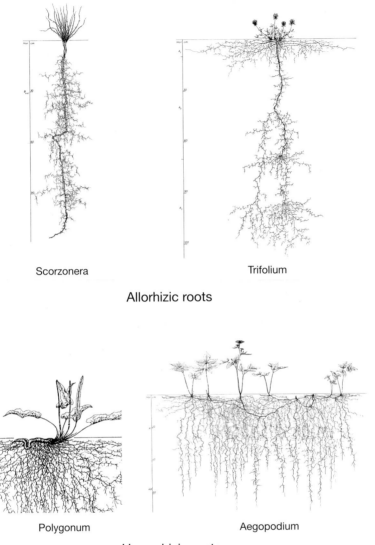

Scorzonera   Trifolium

Allorhizic roots

Polygonum   Aegopodium

Homorhizic roots

Fig. 1.7. Basic types of root systems. In plants with *allorhizic roots*, the root system develops from the primary root which produces one or more cylindrical main roots (e.g. *Scorzonera villosa*). Lateral roots may spread near the soil surface as in *Trifolium trichocephalum*. In plants with a typical *homorhizic* root system the roots arise from basal shoots or from rhizomes (*Polygonum bistorta*) or runners (*Aegopodium podagraria*). (From Kutschera and Lichtenegger 1992)

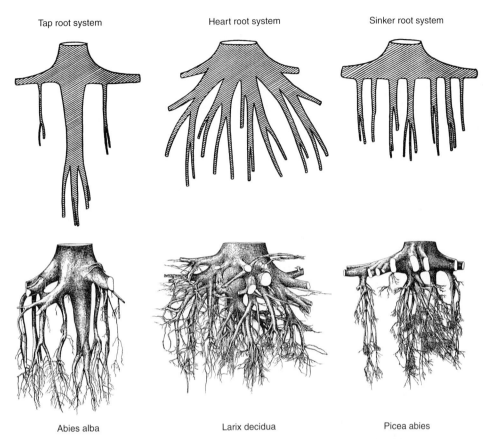

Tap root system   Heart root system   Sinker root system

Abies alba   Larix decidua   Picea abies

Fig. 1.8. Root systems of conifer trees of the temperate zone. *Abies alba* has a *tap root* system, *Larix decidua* a *heart root* system, *Picea abies* a *sinker root* system. (After Köstler et al. 1968). In tropical forests and on oxygen-deficient soils, trees may develop special types of roots, such as *stilt roots, tabular roots* and *buttress roots.* (Richards 1979; Longman and Jeník 1987)

Vascular plants obtain mineral substances which would otherwise be difficult for them to take up, with the help of the greater biochemical capacities of their symbiotic **mycorrhizal fungi**. The fungi increase the absorbing surface of the roots hundred- to thousandfold. The far-spreading mycelial network in the soil collects, and conducts to the host, substances and water from the soil. Fungal growth is supported by copious supplies of carbohydrate from the host, which in return is provided with nutrients, trace elements, amino acids and growth substances by the excretions of the fungi or by disintegration of the hyphae. The great majority of plants are hosts to these symbiotic root fungi. Various forms of mycosymbiotrophy are recognized (Fig. 1.10):

Of worldwide distribution and the most important form of mycotrophy is the *vesicular arbuscular (VA) type of mycorrhiza,* produced by genera of the

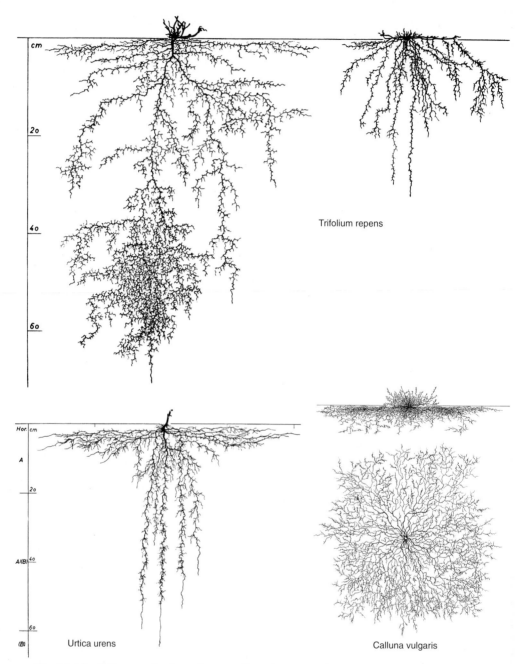

Fig. 1.9. The influence of soil on the spreading of roots: *Trifolium repens* on sandy loam above porous gravel (*left*) and on compacted clay soil (*right*). *Urtica urens* on loamy soil with a nitrogen-rich layer at the surface. *Calluna vulgaris* roots (above, section view; below, plan view) with mycorrhiza, extending into organic soil layers. (From Kutschera 1960; Kutschera and Lichtenegger 1992)

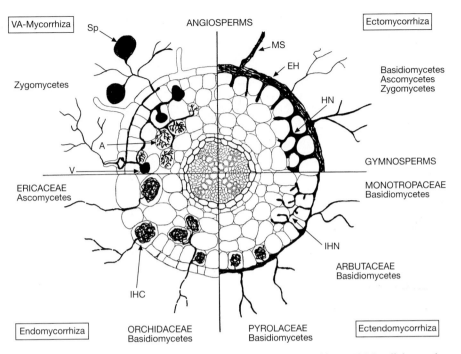

Fig. 1.10. A schematic overview of the different forms of mycorrhiza. *MS* Mycelial strands; *EH* external hyphal mantle; *HN* Hartig net; *IHN* intercellular hyphal net; *IHC* intracellular hyphal complexes; *V* fungal vesicle; *A* arbuscule; *Sp* spore. (After Gianinazzi and Gianinazzi-Pearson 1988)

Zygomycetes (e.g. *Glomus*). This type occurs in most herbaceous plants and many woody plants, as well as in ferns and mosses. Infection of the young roots is followed by arboreal branching of the hyphae (arbuscules) between and within the cells of the root cortex. In this way the internal surface of the symbiont is enlarged and material exchange between the partners is intensified. In older parts of roots the hyphae develop terminal vesicles which store lipids when mature. Exchange of material between plant and fungus takes place in the root cortex; enzymatic dissolution of the arbuscules provides the plant mainly with polyphosphate. Certain Fabaceae establish a three-way symbiosis with VA mycorrhiza and nodule bacteria (e.g. *Glycine max* with *Glomus fasciculatum* and *Bradyrhizobium japonicum*), which ensures a better supply of phosphorus and nitrogen for the host plant even under unfavourable conditions, such as water deficiency [208].

Host-specific fungal symbiosis in root tissues of Ericaceae is usually formed with Ascomycetes (*ericoid* and *arbutoid* mycorrhiza); the saprophytic genus *Monotropa* acts as host to Basidiomycetes. The seedlings and roots of Orchideae form associations with *Rhizoctonia* species.

Particularly in the provision of forest trees (conifers, Fagaceae) with mineral substances, a greater role is played by *ectomycorrhiza*. In this form a dense mycelial network envelopes the young, non-suberized root tips, which consequently assume a swollen, club-like appearance and no longer produce root hairs. The hyphae penetrate between the cells of the root cortex where they form the so-called Hartig net. The fungal mantle puts out hyphae into the soil which take over the function of the missing root hairs.

Forest trees fulfill most of their phosphorus and nitrogen requirements via mycorrhizal roots. The improvement in their nutrient supplies enables the plants to produce more biomass on poor soils and also makes possible their existence on dry, cold and mineral-deficient sites. On raw humus soils the plants with ericoid mycorrhiza profit from the fact that these are able to break down lignin efficiently.

An excess of nitrogen leads to a decline in the mycorrhizae. With respect to heavy metals, uptake is not dependent on the concentration of the metals alone, but also differs from species to species of plant and fungus. The presence of mycorrhiza can increase the uptake of heavy metals by the host plant, and this can mean better supplies of iron and trace elements, but also facilitates the entry of toxic elements such as cadmium and lead. On the other hand, the mycorrhizal fungi can bind heavy metals in their cell wall complex and so are able to buffer the plants against toxic effects. The root-fungal symbiosis is sensitive to acid immissions, and conifers in particular lose their mycobionts under the influence of heavy SO_2 loads (Fig. 1.11). In addition to this the species spectrum of the mycorrhiza-forming fungi may change in polluted soils.

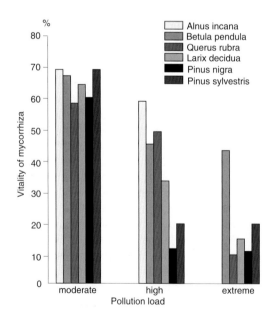

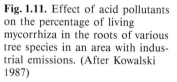

Fig. 1.11. Effect of acid pollutants on the percentage of living mycorrhiza in the roots of various tree species in an area with industrial emissions. (After Kowalski 1987)

1.1.4.2 Chemical Interactions Mediated by Bioactive Plant Substances

Plants produce a great variety of substances that exert a regulatory effect on other plants, microorganisms or animals, rather than being utilized as nutrients. Naturally occurring bioactive substances of this kind, also known as **ecomones** or **infochemicals** [45], play important roles in the ecosystem as signal, recognition, defence substances, and as inhibitors and poisons. Some of these agents act *intraspecifically*, as autotoxins, inhibitors (e.g. of germination) and pheromones (attractants and trail markers), whereas others, termed *allelochemicals*, affect other species of plants, microorganisms and animals. Infochemicals are termed *allomones* if they are of advantage only to the organisms producing them, *kairomones* if beneficial only to the consumer, or *synomones* if both partners benefit. In nature, interrelationships are often obscure and complex, so that the effects of bioactive substances cannot be strictly classified. Nevertheless, a thorough knowledge of these chemoecological relationships is essential for an understanding of the connections and interactions existing within an ecosystem, and for the planning of ecologically sound plant protection measures.

Bioactive plant substances are intermediary and end-products of secondary metabolism and are therefore also referred to as **secondary plant substances**. They are biosynthesized from precursors arising from primary metabolism (Fig. 1.12), the most important synthetic pathways being those leading from carbohydrate- and fat metabolism via acetyl-coenzyme-A, mevalonic acid and isopentenylpyrophosphate to *terpenoids* and *steroids*, and from sugar- and amino-acid metabolism via shikimic acid and the acetate polyketide pathway to *phenol bodies* and their derivatives (e.g. phenylpropanes, flavonoids, tannins, numerous lichen substances) and from amino acids to the *alkaloids*. It is estimated that there are 10^5 ecobiochemically active natural products; Table 1.3 gives an overview of the classes of substances and their taxonomic distribution.

This diversity of secondary plant substances developed during the course of phylogenesis, coevolving in interplay with microbial parasites and especially with animal consumer organisms. Since specialists have repeatedly succeeded in overcoming every kind of defence substance and every poison produced by plants, new modifications and syntheses have been continually necessary. That adaptations of this nature take place within a relatively short period of time can be deduced from the widespread differentiation of *chemoecotypes* and chemotaxonomically distinguishable races (even on the basis of smell, as in some Lamiacae species). The earliest chemical defence substances (Fig. 1.13) were non-specific phenylpropane compounds (resins, lignin; Fig. 1.14), phenol derivatives (condensed tannins) and flavonoids. With the evolution of the angiosperms and the appearance of herbaceous forms, the spectrum of natural substances was greatly enlarged by products of the acetate-mevalonate pathway.

The quantity of infochemicals present differs in the individual organs, tissues and even cells (e.g. secretory elements, tannin idioblasts) of a plant; they

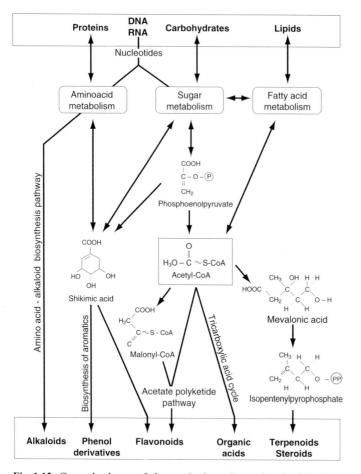

Fig. 1.12. General scheme of the synthetic pathways involved in the secondary metabolism of plants. (After Schlee 1992, expanded)

also alter during the course of individual development and ageing, as well as from season to season. Within a few days, or even from one hour to the next, the concentration of biogenic plant substances can fluctuate considerably. Nutrient conditions and particularly the action of many different kinds of stressors cause modifications in the quantity and composition (Table 1.4) of infochemicals, which in turn influences the behaviour of the other ecosystem partner. Plants under stress and plants with plentiful supplies of nitrogen are preferentially attacked by insects. The plant responds by stepping up its production of the appropriate defence substance.

Innumerable interrelationships are based on chemical communication between organisms; some of the most widespread and particularly important effects are described in the following:

Table 1.3. Occurrence and effects of bioactive secondary plant products. (Harborne 1988; Schlee 1992; supplemented by data from Rundel 1978; Crawford 1989; Lüning 1990; Henssen and Jahns 1991). For the role of secondary metabolites in plant defence against herbivores and pathogens: Bennett and Wallsgrove (1994)

Class	Occurrence	Effects
Terpenoids		
Monoterpenes	Frequent, in essential oils	Attractive taste or odour
Sesquiterpenes	In angiosperms, especially in Asteraceae, in essential oils and resins	Bitter and toxic
Diterpenoids	Frequent, in resins and latex	Some toxic, sticky
Saponins	Especially in Liliiflorae, Solanaceae, Scrophulariaceae	Toxic, antimicrobial
Cardenolides	Especially in Apocynaceae, Asclepiadaceae	Toxic, bitter
Sterols	Specific distribution	Signal substances
Carotenoids	Universal distribution	Coloration
Polyterpenes	In laticifers	Deterrent
Halogenated terpenes	Marine algae	Toxic
Phenolics		
Simple phenols	Universal in leaf and cortex, often in lichens and seaweeds	Anti-microbial, allelopathic, deterrent
Halogenated phenols	Marine algae	Bitter
Flavonoids	Universal distribution	Coloration
Tannins, depsides	In parenchyma cells and tannin idioblasts, and in lichens	Bitter, anti-microbial
Dibenzofuran	Lichens	Anti-microbial, toxic
Nitrogen compounds		
Alkaloids	Widely in angiosperms, especially in root, leaf, and fruit	Toxic, bitter
Amines and peptides	Widely in angiosperms, often in flowers	Attractive odour
Cyclic polypeptides	Poisonous algae (especially Cyanobacteria, Dinophyceae, Chrysophyceae)	Toxic, antibiotic
Non-protein-forming amino acids	Frequent, especially in Fabales	Many toxic
Cyanogenic glycosides	Occasional, e.g. in fruits and leaves	Toxic
Glucosinolates	Occasional, e.g. in Brassicaceae	Acrid, bitter
Other secondary compounds		
Coumarins	Frequent, especially in Fabales, Rubiaceae, Poaceae	Allelopathic, anti-microbial
Polyacetylenes	Chiefly in Apiaceae and Asteraceae	Some toxic

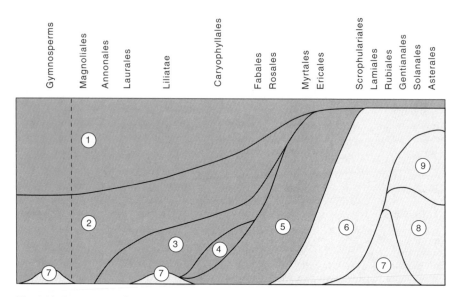

Fig. 1.13. Occurrence of bioactive products of secondary metabolism in *primitive* and *derived* taxa. In the course of evolution, products of the shikimate pathway (*dark shading*) became increasingly replaced by those of the mevalonate-acetate pathway (*pale shading*). *1* Lignin; *2* condensed tannins; *3* isoquinoline alkaloids; *4* betalains; *5* gallotannins; *6* indole alkaloids and iridoids; *7* steroids; *8* sesquiterpenes; *9* polyacetylene. (After Kubitzky, from Frohne and Jensen 1992, supplemented by data from Harborne 1988)

Attraction of pollinators; zoochory: The presence of colour pigments and aromatic substances in flowers and fruits attracts the insects, birds and other animals, by which pollination or fruit dispersal (*zoochory*) are effected. Since both partners, plant and animal, profit from this arrangement, the attractive substances function as synomones.

Table 1.4. Effects of mineral nutrient supply and water deficit on the secondary metabolism of vascular plants. (Gershenzon 1984)

Class of secondary metabolite	Nitrogen		Deficiency of			Water deficiency
	Fertilized	Deficient	P	K	S	
Terpenoids						
Herbs		×	+	×		+
Trees		−				−
Phenol derivatives		+	+	+	+	×
Alkaloids	+	−	×	+		+
Glucosinolates		+			−	+
Cyanogenic glycosides		−				+

+ Increase; − decrease; × no clear trend.

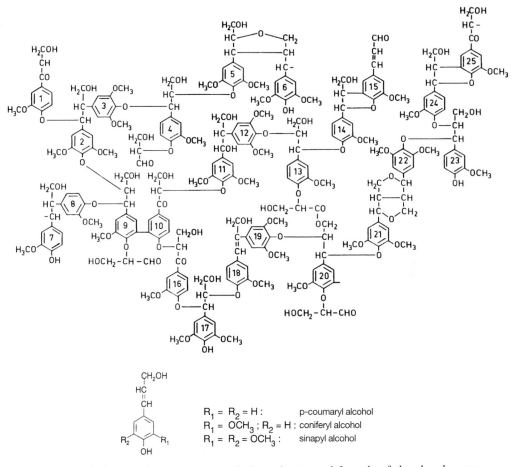

Fig. 1.14. Structural model of beech lignin and structural formula of the phenylpropane units (cinnamyl alcohols) from which the three-dimensional polymer is constructed. The lignin of dicotyledons contains mainly sinapyl- and coniferyl alcohols. Gymnosperm lignin is composed principally of coniferyl alcohols. Lignin of graminoids contains, in addition to sinapyl- and conferyl alcohols, up to 30% p-coumaryl alcohol and small amounts of cinnamic acids. (After Nimz 1974, from Lin and Dence 1992)

The *colour* of flowers (petals, anthers, pollen) and fruits results partly from deposits of coloured compounds in biomembranes (carotenoids) and cell walls (brown phlobaphenes, black melanins), but mainly from glycosidic-bound coloured substances dissolved in the cell sap. The colour itself depends on the type of chromophore structure (Fig. 1.15; flavone structure in flavonoids, betalain structure in betacyanins and betaxanthines) as well as substitutions in the basic skeleton, on possible complexes with Fe^{3+} and Al^{3+}, and on the pH of the cell sap. In this way a finely graduated range of colour shades has developed, ranging from cream, to yellow, pink, red, blue and deep violet

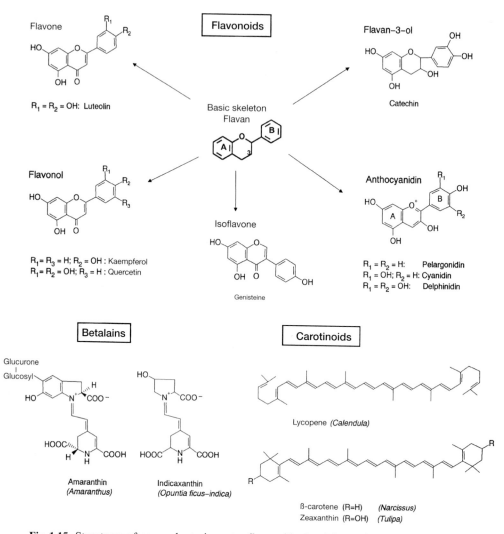

Fig. 1.15. Structures of some plant pigments: flavonoids, betalains and carotenoids

(Table 1.5). Ultraviolet "landing or honey guides" on blossoms are recognizable to Hymenoptera and some Diptera. Birds are attracted by bright colours,
and also by the ultraviolet reflected from the surface layers of wax on certain
fruits.

Aromatic substances attract insects, birds and mammals to flowers and
fruits at the most favourable times, which are when the flowers are opening
or the fruits ripening. Often, the emission of scent is attuned to the diurnal
activity of the pollinating animals (moths, bats) or of the fruit consumers
(Fig. 1.16). The scents emitted by plants are almost invariably mixtures of

Table 1.5. Flower pigments and pollinators. (After Harborne 1988)

Colour	Pigment	Occurrence (examples)	Pollinators (examples)
White, cream	Leucoanthocyanidin, quercetin	Frequent	Chiefly bees
Yellow	Carotenoids, flavonols, chalcons	Frequent	Bees, butterflies, birds
Yellow/purple	Betalains	Caryophyllales	Bees, butterflies
Orange	Carotenoids, pelargonidin + aurone	*Lilium* *Antirrhinum*	Bees, butterflies, birds
Pink	Peonidin	*Paeonia,* *Rosa rugosa*	(Bees[a]), dipterans, butterflies, birds
Red/purple	Pelargonidin, cyanidin (also with carotenoids)	Frequent	(Bees[a]), dipterans, butterflies, birds
Blue	Cyanidin, delphinidin (co-pigment Al/Fe^-)	*Centaurea,* *Gentiana*	Bees, butterflies, birds
Mauve	Delphinidin	Frequent	Bees, butterflies
Green	Chlorophylls	*Hellborus,* *Dorstenia*	Dipterans, bats

[a] Bees and other Hymenoptera are insensitive to red; they are attracted to red flowers by UV-nectar guides and yellow or blue stamens.

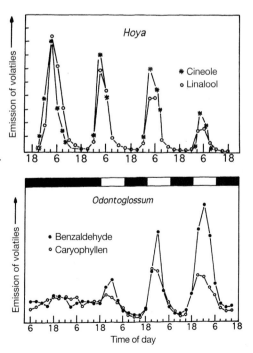

Fig. 1.16. Timing of emission of flower scents. In *Hoya carnosa* the production of volatiles follows a circadian rhythm (endogenous "biological clock"). The flowers of *Odontoglossum constrictum* also emit volatiles in a diurnal rhythm, but in this case the timing is determined by the light/dark changeover (*unshaded* and *shaded bars*). The release of the scent substances is a cytologically regulated excretory process. (From Matile and Altenburger 1988; Altenburger and Matile 1990)

volatile secondary plant metabolites, especially of monoterpenes (essential oils), aliphatic alcohols, ketones and esters, fatty acids, aromatics (e.g. vanillin, eugenol), amines and indole derivatives. An unusual case is presented by the highly specific attraction of pollinators by *Ophrys* species [124]. Cyclic terpenoids emanating from the orchid blossoms simulate the sexual pheromones of wild bees; the copulatory movements of the duped males successfully bring about the complicated transfer of pollen to the stigma.

Stimulation by means of signal substances: plant exudates attract a variety of animals (nematodes, insects), and may also elicit an attack by phytoparasites. The chemical messengers put out by the plant thus come under the heading of kairomones. Chemical signals from their prospective host plant are necessary to initiate germination of seeds of parasitic Scrophulariacae, Oro-banchacae, Balanophoracae, Rafflesiacae and Hydnoracae. For example, the tiny seeds of the *Striga* species (Scrophulariaceae) that cause serious economic damage in the tropics and subtropics by parasitizing the roots of millet, corn, sugar cane, legumes and root crops, can lie for decades in the soil before they are specifically stimulated by a root exudate from a host plant, after which they germinate and form haustoria. The processes of germination and adhesion are initiated by a variety of highly effective substances (strigol, benzoquinone derivatives). Similarly, signal substances are active between other host-parasite partners, and also for directing *Rhizobium leguminosarum* to the roots of Fabaceae.

Allelopathy: plants may give off substances that are injurious to other plants, or even prevent their becoming established in the vicinity. This effect is termed *allelopathy* [154]. In most cases these substances (known collectively as allomones) are short-chain fatty acids, essential oils, phenolic compounds, alkaloids, steroids and coumarin derivatives. They may be released into the air, excreted by the roots, or washed out of the shoot and into the soil by rain. Certain floating plants (e.g. *Nymphaea odorata* and *Brasenia schreberi*) release their allelopathic substances into the water, suppressing duckweed (*Lemna minor*) [53].

Although it is not easy to demonstrate chemical effects of this kind in nature, there is unquestionable evidence that they do take place (Fig. 1.17). For instance, it is well known that the compound juglone can prevent germination in many plant species. The leaves and the root exudates of *Juglans regia* contain a naphthalene glucoside that is itself not allelopathic. Only after its hydrolysis and oxidation by soil microorganisms to hydrojuglone is the active inhibitor juglone finally formed. *Calluna* and *Arctostaphylos* species release carboxyphenolic acids and hydroxycinnamic acids, which inhibit the growth of grasses and herbs. Especially in steppes and arid shrub formations many Asteracea (*Parthenium, Encelia, Artemisia*), Lamiaceae, Myrtaceae, Rutaceae and Rosaceae give off allelopathic terpenes and water-soluble phenolic substances. In such places the inhibited areas around the older plants are recolonized only when the vegetation and the inhibitory substances in the

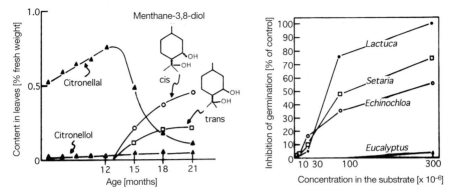

Fig. 1.17. Allelopathic substances. *Left* Content of ethereal oils in the leaves of *Eucalyptus citriodora* during the course of development and ageing; *right* allelopathic inhibition of the germination of seeds of *Lactuca sativa, Setaria viridis* and *Echinochloa crus-galli* by menthane-3,8-diol from eucalyptus leaves. The germination of eucalyptus seeds is not impaired. (From Mizutani 1989). Allelopathic effects occur chiefly in plant communities into which foreign species have been introduced. For example, the suppression of a herbaceous undergrowth in *Eucalyptus* stands occurs only outside Australia. The bioactive compound juglone from walnut trees only exerts an allelopathic effect on plant species *not* occurring in the original distribution area of *Juglans* spp.: the rich herbaceous layer in walnut forests in Kyrgyztan is the result of coevolutive development within this phytocoenosis. (Rabotnov 1995)

organic horizons of the soil have been decomposed by microorganisms or destroyed by fire. The lack of undergrowth in some tropical forests and in bamboo stands is probably also due to the action of phytotoxins. Similarly, lichen substances such as orcinol depsides (see Fig. 1.18) and usnic acid have an allelopathic effect on conifer seedlings and an antibiotic effect on fungi.

Defence against infection by means of phytotoxins and phytoalexins: plants can defend themselves in a number of ways against infection and invasion of their tissues by phytopathic bacteria and fungi. Measures of this kind are, for example, a hypersensitive reaction to attack, thus provoking rapid death of the affected cells and so depriving the invaders of nutrition; by encapsulating the intruding organisms with polysaccharides; and by means of antimicrobial agents which are either present in the plant tissue in appropriate amounts before infection (preformed or *preinfectional agents*), or are only produced after attack by viruses, bacteria or fungi has taken place (induced or postinfectional agents). Preformed fungotoxic and fungistatic substances are present in many plants. Common examples of these are allyl sulfides (in *Brassica* and *Allium* spp.), lactones (in tulips and *Ranunculus*), saponins (in ivy and oats), quinones (in apple trees), salicylic acid [192], flavonoids, tannins and terpenoids (Fig. 1.18).

Post-infectionally, the affected plant tissue synthesizes *phytoalexins* upon contact with the invading organism or its products. At the present time, more than 200 phytoalexins have been identified, of which particular chemical struc-

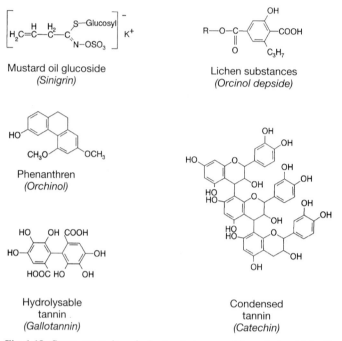

Mustard oil glucoside
(Sinigrin)

Lichen substances
(Orcinol depside)

Phenanthren
(Orchinol)

Hydrolysable
tannin
(Gallotannin)

Condensed
tannin
(Catechin)

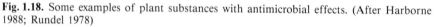

Fig. 1.18. Some examples of plant substances with antimicrobial effects. (After Harborne 1988; Rundel 1978)

tures predominate in certain families, such as isoflavonoids in Fabaceae, sesquiterpenes in Solanaceae, polyacetylenes in Asteraceae and phenanthrenes (orchinol) in Orchidaceae, but quinone, coumarin and stilbene derivatives also occur (Fig. 1.19). Furthermore, one and the same plant species may produce several different phytoalexins. Synthesis of phytoalexins is also triggered by non-pathogenic organisms (mycorrhizal fungi, rhizoplane bacteria), which improves the plant's general powers of defence.

Chemical defence against herbivory: plants are protected from herbivores by a variety of means, such as thorns, spines, hairs, lignified and silicified cell walls, and especially by protective chemical substances, which may be repellent, or unpleasant to the taste (irritant or bitter), sticky, or toxic. Selection and coevolution in response to grazing have resulted in the production of a wide range of defence substances in plants. In sufficient concentrations, some of these compounds are repellent to every type of herbivore, as are, for example, polyphenols and tannins, which precipitate protein, thus making them uneconomical for consumers; or cyanogenic glycosides, from which cyanic acid is liberated by hydrolytic cytoplasmic enzymes following cell disruption (Fig. 1.20). The majority of bioactive plant products, especially those with a specific protective function, are biochemically characteristic for the plant species.

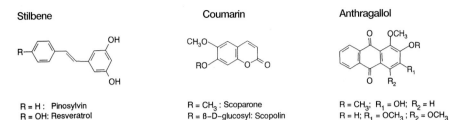

Stilbene

R = H : Pinosylvin
R = OH: Resveratrol

Coumarin

R = CH₃ : Scoparone
R = β–D–glucosyl: Scopolin

Anthragallol

R = CH₃; R₁ = OH; R₂ = H
R = H; R₁ = OCH₃; R₂ = OCH₃

Fig. 1.19. Phytoalexins. Derivatives of *stilbene* (diphenylethylene), such as pinosylvin in the heartwood of species of pine and resveratrol in many plant species, have a fungistatic effect; in the leaves of grape vines resveratrol is formed after the plant is attacked by *Botrytis cinerea*, and also as a result of UV irradiation. *Coumarins* are encountered throughout the plant kingdom before and after infection: scoparon and scopolin inhibit the postinfectional spread of certain fungi (*Phytophthora, Verticillium, Penicillium* etc.) in the shoot cortex and the pericarp of citrus fruits. *Anthragallols* are derivatives of anthraquinones; the methyl ethers shown here exert a fungitoxic effect on *Phytophora cinnamomi* in the cortex of *Cinchona*. (From Gottstein und Gross 1992)

Certain classes of plant substances are especially effective against *mammalian herbivores*, above all the alkaloids (Fig. 1.21) but also lichenous substances (vulpinic acid and atranorin in *Letharia vulpina*; various depsides in *Cladonia* spp.). An example of highly specific defence is seen in the proteins (phytohaemoagglutinins or lectins) of Fabaceae and Euphorbiaceae, which cause agglutination of erythrocytes. Root exudates (e.g. from *Tagetes* plants) repel *nematodes*. Plants with a high content of cyanogenic glycosides are unpalatable to *snails*, and in regions with mild winters which allow snails to multiply rapidly, biochemical races of Fabaceae (*Trifolium repens, Lotus corniculatus*) capable of producing these repellents have developed. The ability to synthesize cyanogenic glycosides is controlled by two genes.

Many secondary plant compounds affect phytophagous *insects* by acting as repellents or metabolic inhibitors, by disturbing reproduction, or simply by virtue of their toxicity (Fig. 1.22). Good examples of such substances are

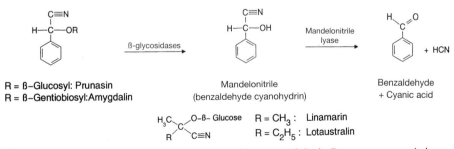

R = β–Glucosyl: Prunasin
R = β–Gentiobiosyl: Amygdalin

β-glycosidases →

Mandelonitrile
(benzaldehyde cyanohydrin)

Mandelonitrile lyase →

Benzaldehyde
+ Cyanic acid

+ HCN

R = CH₃ : Linamarin
R = C₂H₅ : Lotaustralin

Fig. 1.20. Commonly occurring cyanogenic glycosides: *amygdalin* in Rosaceae; *prunasin* in e.g. Rosaceae, Myrtaceae, Caprifoliaceae and ferns; *linamarin* and lotaustralin in Fabales (Fabaceae, *Acacia*), Euphorbiaceae, Linaceae and Asteraceae. If, due to cell disruption, the glycosides compartmentalized in the vacuoles come into contact with β-glucosidases, free hydrocyanic acid is formed (*top*)

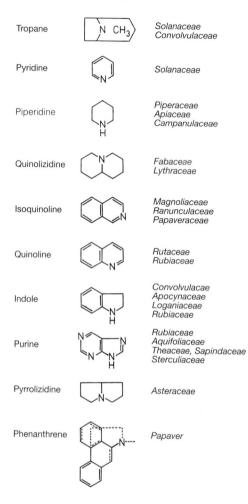

Fig. 1.21. Ring systems of alkaloids and examples of families in which such alkaloids occur

triterpenes, mustard oil glycosides, saponins, alkaloids, protease inhibitors, non-proteinogenic amino acids (which have an inhibitory effect on the protein metabolism of insects) and steroids. Ferns, Cycadaceae, Taxaceae, Podocarpaceae and certain angiosperms contain phytoecdysones which, as moulting hormones, regulate the development of insect larvae. Plants of arid regions are well provided with glandular hairs and surface resins, with latex and other terpenoids in glands and ducts, as well as flavonoid glycosides and phototoxic secondary substances.

Often, biosynthesis is induced and enhanced in plants by herbivory. This is borne out by the observation that the leaves of intensively grazed grasses contain more silica than those on ungrazed areas. Damage by herbivores causes an increase in the content of polyphenols, tannins and terpenes in birches [27] and poplars, and the same applies to damage caused by insects and mollusks. Clearly, the greater investment of metabolites in the complicated

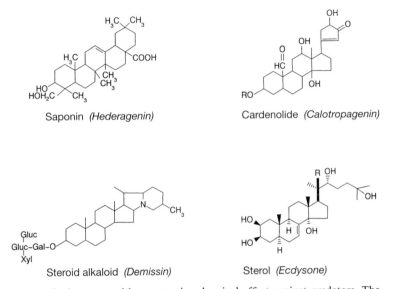

Fig. 1.22. Examples of triterpenes with a protective chemical effect against predators. The saponin *hederagenin* is present as a glycoside in the poisonous leaves and berries of *Hedera helix*. The cardiac glycoside *calotropagenin* from *Asclepias curassavica* is toxic to almost all herbivores; only the larvae of the monarch butterfly have coevolved and adapted to the poison, so that it not only does them no harm, it even protects them (and the butterfly and its eggs) from enemies. The steroid alcohol *demissin* occurring in the wild potato *Solanum demissum* protects it from attack by the potato beetle. *Ecdysones* occur principally in phylogenetically primitive plant groups and may have played an important role in the producer-consumer relationship early in an evolution. (After Crawford 1989; Schlee 1992)

synthesis of defence substances takes place at the expense of the energy budget and dry matter production of the plants exposed to such stress (see Table 6.1).

1.2 Radiation and Climate

All life on the earth is maintained by the *flow of energy* radiated by the sun and entering the biosphere. By means of photosynthesis this radiation energy is fixed in plants in the form of latent chemical energy, from which all links in the food chain derive the energy needed for carrying on vital processes. Radiation is also the primary energy source for the turnover of organic materials, and by regulating the heat and water balance of the earth it provides the energy conditions essential for living organisms. For plants, however, radiation is not only a source of energy (photoenergetic effect), it is also a stimulus governing development (photocybernetic effect), and occasionally also a stress factor (photodestructive effect). Each of these effects is elicited by the uptake of light quanta and every process dependent upon radiation is mediated by highly specific photoreceptors whose absorption spectra correspond to the respective photobiological event (see Table 1.6). Important factors here are the timing, duration, direction of incidence and spectral composition of the incoming radiation.

The radiant energy reaching different places on the earth's surface depends upon their orientation with respect to the sun. Hence, as a result of the rotation and revolution of the earth, the input of solar energy is a periodically fluctuating environmental factor. It imposes a climatic rhythmicity upon all terrestrial phenomena. The periodic alternation of night and day is the astronomical *trigger* responsible for regulating diurnal and seasonal biological rhythms. Moreover, solar radiation, by acting as a signal, controls many developmental processes, such as germination, directional growth and the shaping of external form.

1.2.1 Radiation

1.2.1.1 The Radiation Environment

The biosphere receives solar radiation at wavelengths ranging from 290 nm to about 3000 nm. Radiation at shorter wavelengths is absorbed in the upper atmosphere by ozone and by oxygen in the air; the longer-wavelength limit is determined by the water-vapour and carbon-dioxide content of the atmosphere. An average of 45% of the incoming solar energy falls within the spectral range 380–710 nm, which is the range utilized for photosynthesis by plants (*photosynthetically active radiation*, PhAR; often defined as the 400–700 nm range). Adjoining the lower end of this range is the *ultraviolet*

Table 1.6. Effect of radiation on plant life. (Ross 1981)

Spectral region	Wavelength (nm)	Percent of solar radiant energy	Photosynthetic	Effects of radiation		
				Photomorphogenetic	Photodestructive	Thermal
Ultraviolet	290–380	0–4	Insignificant	Slight	Significant	Insignificant
Photosynthetically active range (PhAR)	380–710	21–46[a]	Significant	Significant	Slight	Significant
Infrared	750–4000	50–79[a]	Insignificant	Significant	Insignificant	Significant
Longwave radiation	4000–100000		Insignificant	Insignificant	Insignificant	Significant

[a] Depending on position of sun and degree of cloud cover.

radiation (UV-A 315–380 nm and UV-B 280–315 nm), the upper end is followed by *infrared radiation* (IR 750–4000 nm; see Table 1.6). Plants also receive *thermoradiation* (long wave IR 4000 to 10^5 nm), and themselves emit this type of radiation.

At the outer limits of the earth's atmosphere the *intensity of radiation* is 1360 W m^{-2} (the *solar constant*). Of this, only an average of 47% reaches the earth's surface. More than half is lost, being cast back into space as a result of refraction and diffraction in the high atmosphere, or scattered or absorbed by particles suspended in the air. The totality of incoming radiation reaching a horizontal surface is called the *global radiation*; it is composed both of the direct solar radiation and the diffuse sky light. At sea level, the global radiation attains maximal values of about 1 kW m^{-2} and intensities in the PhAR range of 400 W m^{-2} (equivalent of photosynthetic photon flux density, PPFD, of approximately 1800 µmol photons m^{-2} s^{-1}; cf. conversion, p. XII).

The *irradiance* at a particular locality depends in the first place on the latitude. Thus the tropics, especially the high pressure regions where clouds are few, receive an above-average quantity of solar radiation; an average of 70% of the incoming radiation penetrates the clear air above dry regions near the equator (Fig. 1.23). Towards the poles, the annual total global radiation gradually decreases. At greater altitudes, mainly because of the greater clarity of the atmosphere, more radiation reaches the ground than on valley floors. Moreover, local topography and time are also important factors determining the available light since this is dependent on the shape of the horizon, the angle of incidence of radiation, and the atmospheric conditions (clouds, haziness).

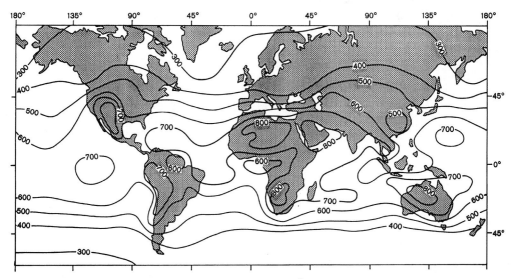

Fig. 1.23. Annual totals of global radiation in GJ km^{-2}. (After DeJong as cited in Schultz 1988)

In **water**, radiation is more strongly attenuated than in the atmosphere. Long-wave thermoradiation is absorbed in the upper few millimetres, infrared radiation in the uppermost centimetres, whereas UV penetrates to a metre. Visible radiation reaches greater depths, where blue-green twilight prevails (Fig. 1.24). The intensity of radiation in bodies of water depends on the magnitude and nature of illumination above their surface, on the amount of reflection and back scattering of light at or near the surface, and on attenuation as the rays pass through the water. With increasing depth, the radiation intensity decreases exponentially, since radiation is absorbed and scattered by the water itself as well as by dissolved materials, suspended particles of soil and detritus, and plankton. The layer of water above the level at which lack of light prevents the existence of autotrophic plants is called the *euphotic* zone. In the open ocean 1% of the radiation penetrates down to about 150 m, whereas near the coast, depending on the purity of the water, the corresponding depth does not exceed 50 m. In clear lakes the depth to which light penetrates in sufficient amounts to support vascular plants is only about 5 m, although sessile algae occur down to 20–30 m. In flowing water the light intensity can decrease drastically within a few centimetres or decimetres. As a consequence of the very rapid attenuation of radiation in water, the daily duration of photosynthesis decreases with increasing depth.

Light scarcely penetrates soil at all; in sandy and clay soils only 1%, at most, of the incident radiation reaches a depth of 2–5 mm below the surface [231]. Red light penetrates farthest into sandy soil, blue light the least. However, even in such small amounts, light exerts physiological effects on the soil flora, and on the seeds in the uppermost layers.

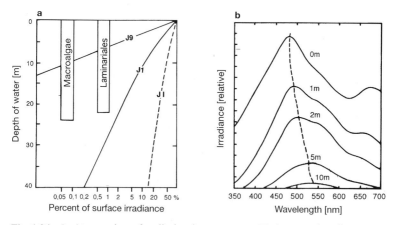

Fig. 1.24a, b. Attenuation of radiation in seawater. **a** Underwater irradiance as a percentage of surface irradiance in the spectral range 350–700 nm with increasing depth, for different Jerlov water types: JI clear water of tropical oceans; J1 clear water in coastal regions; *J9* coastal water with maximum transmission in green wavelengths. *Bars* maximum depths of algal distribution in coastal waters. **b** Spectral distribution of underwater daylight in coastal waters (Jerlov type 5). (After Jerlov 1976; Lüning and Dring, as cited in Lüning 1990)

1.2.1.2 The Distribution of Radiation in the Plant Cover

Closed plant stands build up a system in which successive layers of leaves partially overlap and shade one another. The incident light is absorbed progressively in its passage through the layers, so that in a deep canopy it is almost completely utilized (Fig. 1.25). The arrangement of leaves during shoot growth and leaf development results in a structure which, by virtue of its finely tuned adjustments to the steep light gradient within the crown, exerts a *photohomeostatic effect*.

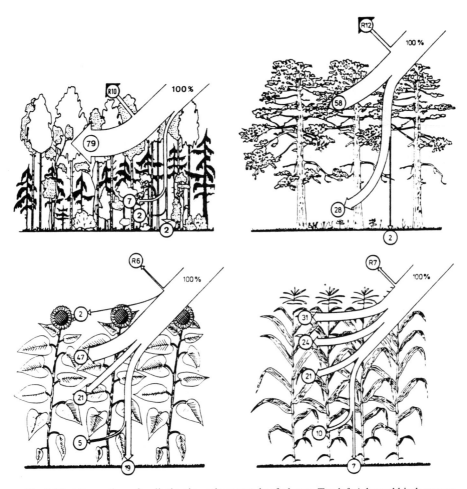

Fig. 1.25. Attenuation of radiation in various stands of plants. *Top left* A boreal birch-spruce mixed forest (Kairiukshtis 1967); *top right* a thin pine forest (Cernusca 1977); *below left* a sunflower field (Hiroi and Monsi 1966); *below right* a maize field (Allen et al. 1964). *R* canopy reflectance. Most of the radiation penetrating dense, flat-leaved stands is absorbed and scattered in the upper third, whereas in stands with narrow, erect leaves the radiation is more evenly distributed

Incident radiation reaches the interior of a plant stand in a variety of ways. In the first place, it can enter as direct radiation through gaps in the stand as well as at its margins (Fig. 1.26), then also as scattered light after reflection by leaves and the soil surface, and finally by passage through the leaf blades. The **attenuation of radiation** in a stand of plants depends mainly on the density of the foliage, on the arrangement of the leaves within the stand, and on the inclination of the leaves to incoming radiation. The degree of average can be expressed quantitatively by the *leaf-area index* (LAI). The cumulative LAI indicates the total area of leaves above a certain area of ground [271]:

$$LAI = \frac{\text{total leaf area}}{\text{ground area}} \qquad (1.2)$$

Leaf area and ground area are both expressed in m^2, so that LAI is a dimensionless measure of the amount of cover. Total leaf area refers to the area of only one side of the leaves. With a LAI of 4, a given area of ground would be covered, theoretically, by four times that area of leaves, arranged, of course, in several layers. The radiation entering forests is also attenuated by tree trunks and branches. This component is included in the measurements obtained by optical procedures (plant area index = PAI) [32].

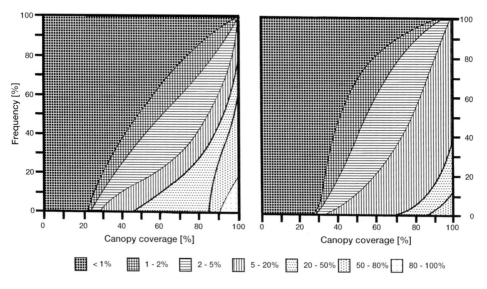

Relative transmittance

Fig. 1.26. Diagram of PhAR penetration through gaps in the canopy of mixed conifer-deciduous forests on a sunny day (*left*) and an overcast day (*right*) in July. *Relative transmittance:* light intensity below canopy in percent of maximum irradiance in the open field during a clear summer day. (After Tselniker 1968)

On its way through the plant canopy the radiation must pass through successive layers of leaves. In the process, its intensity decreases almost exponentially with increasing amount of cover, in accordance with the *Lambert-Beer Extinction Law*. If the layering of the foliage is taken to be homogeneous, the decrease in radiation can be computed from Monsi and Saeki's modification [155] of the extinction equation,

$$I_Z = I_0 \cdot e^{-k \cdot LAI} \tag{1.3}$$

where I_Z is the intensity of radiation at a certain depth from the top of the plant canopy; I_0 is the radiation incident at the top of the canopy; k is the extinction coefficient for this particular plant community; LAI is the total leaf area above the level at which I_Z is estimated, per unit ground area (the cumulative LAI).

The *attenuation coefficient* indicates the degree of decrease of light due to absorption and scatter within the canopy. In grain fields, meadows, reeds, and in sea-grass mats (Zostera communities), where the leaves are mostly upright (more than three-quarters of the leaves are erectophilous, at an angle of more than 45° above the horizontal), the attenuation coefficient is usually between 0.3 and 0.5, and in the middle of the stand the light intensity is still at least half that of the external light (Fig. 1.27: graminaceous type). In contrast, in plant communities with planophilous leaves, like forbs, or beneath a cover of water lilies, in the understorey of forests or in fields of clover or sunflower, the attenuation coefficient is greater than 0.7, and halfway down two-thirds to three-quarters of the incident light has been absorbed (Fig. 1.27: dicotyledon

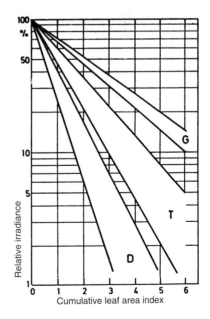

Fig. 1.27. The exponential decrease of light intensity in different stands of plants as a function of leaf area index. The cumulative LAI is derived by summation of the index values for the individual horizontal layers of leaves in the stand. In broadleaved dicotyledonous communities (*D*) the attenuation of light is considerable even with a low LAI, whereas in grass communities (*G*) attenuation occurs more gradually; stands of trees (*T*) represent an intermediate position. (After Monsi and Saeki 1953; Kira et al. 1969). For different aquatic stands see Pokorný and Ondok (1991)

type). In forests with closely packed crowns and dense foliage, so much light is absorbed in the upper layers that very little is left at the trunk level and on the forest floor. In such forests the attenuation of light is similar to, or even more abrupt than that under dicotyledonous herbs. Forests composed of tree species with loose crowns (birch, eucalyptus) attenuate the light as gradually as do grass communities. In agricultural and horticultural plantations, attenuation depends on plant density and on the distance between the rows, so that by careful spacing the best possible absorption of light can be obtained.

The interception of radiation in a stand can also be characterized in terms of the space occupied by foliage. The *foliage density* is calculated from the total leaf area and the volume of the stand, and is expressed as m^2 leaf area per m^3 stand volume. In a field of maize in summer, with a LAI of 4.2, the foliage density was $2.6 \, m^2 \, m^{-3}$ [103], whereas the leaf area density for subtropical broadleaved evergreen forests with a LAI of about 7 is only $2.5-3.5 \, m^2 \, m^{-3}$ [279] because although the attenuation coefficient in the crown is high (roughly 0.8), it drops steeply around the trunks where there are few leaves. Foliage density indices for tree canopies are especially informative; values so far obtained for solitary shrubs and for natural hedges lie between 1.5 and $3.2 \, m^2 \, m^{-3}$ [126].

For comparative purposes, a useful measure of the irradiance in poorly illuminated habitats, such as the forest undergrowth or beneath dense herbaceous plant communities (rosette plants and mosses growing flat on the ground), or in caves and crevices, is the *relative irradiance*. This expresses the amount of the outside light (as a ratio I_z/I_0, or as percent of I_0) reaching a plant in its habitat. In broadleaved forests and conifer forests of the temperate zone, an average of $3-10\%$ of the outside light penetrates the tree canopy during the growing season, whereas during winter, when deciduous trees are bare, penetration can be as high as $50-70\%$. On the floor of dense conifer forests, in the evergreen broadleaved forests of warm-temperate regions, and in species-rich tropical forests the *minimal relative irradiance* may drop to less than 1%. Even in forests with a closed canopy there are gaps through which sun rays frequently reach the forest floor. When the sun is at its zenith, these fluctuating patches attain $40-80\%$ of the intensity of the outside light, but as the incoming rays gradually approach the horizontal, the spots of light (sunflecks) on the ground become smaller and disappear [2].

Plants whose leaves during the majority of daylight hours do not receive the minimum amount of light necessary for photosynthetic compensation (see Chap. 2.2.5.1) soon deteriorate and eventually die. The minimal relative irradiance for vascular plants lies as a rule between 0.5 and 1.0%. Mosses, which have no non-green parts to maintain, can survive with even 0.5% irradiance, aerial algae with as little as 0.1%.

In deciduous forests the amount of light that can penetrate changes with the seasonal changes in foliation. Development of the undergrowth is attuned to this light regime (Fig. 1.28), with the herbaceous flora often developing in spring before the trees come into leaf, or persisting in autumn after leaf fall. Especially the spring flowers (spring geophytes such as *Galanthus, Scilla,*

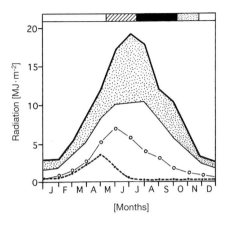

Fig. 1.28. Irradiance above and inside a deciduous forest in England, and the effect of the state of foliation. *Stippled area* Direct solar radiation; *upper line* monthly totals of global radiation above the canopy; *lower limit of stippled area* diffuse radiation; *circles* irradiance in a large clearing; *broken line* irradiance beneath the tree canopy; *horizontal shading at the top* leaf unfolding (*hatched*), full foliation (*black*), leaf fall (*stippled*). (After Anderson, from Walter and Breckle 1986)

Anemone), go through a large part of their life cycle in the period between snow melt and maximum foliation of the tree layer. Winter green species of the forest understorey (e.g. *Hepatica, Pachysandra*) utilize the greater supply of light for carbon assimilation and growth, as long as winter temperatures permit.

1.2.1.3 The Radiation Environment of the Individual Plant

From all sides, the aboveground parts of a plant receive radiation in a variety of forms: as direct sunlight, scattered sky light, diffuse radiation in overcast weather, and reflected radiation from the ground. Growth form (mode of branching) and leaf arrangement determine the individual field of light in which a plant lives. Most plants orient their *assimilatory surfaces* so that as few leaves as possible are continuously exposed to direct radiation and so that most leaves are in semi-shade, where they receive diffuse light. Erect leaves (e.g. of macchia or chapparal shrubs, of spherical cushion plants and of many monocotyledons), leaves positioned in profile (e.g. of *Iris* and *Lactuca* species), pendulous leaves (e.g. of *Eucalyptus*) and those with curved surfaces, (cylindrical leaves, scale-like leaves, needles and assimilatory shoot axes) intercept incoming radiation at an acute angle, thus avoiding injuries from strong irradiation and overheating.

In the crowns of solitary trees and shrubs, a *light gradient* develops from the margins of the crown to its interior. The attenuation of radiation within the crown depends on the characteristic architecture and stage of development of the axial system, on the type and state of foliation, and on age of the plant. A distinction is made between *sun crowns* (pine, larch, birch, many savanna trees) and *shade crowns* (many conifers, beech, evergreen broadleaved trees), depending on the ability of the individual species to develop specialized shade leaves. The average irradiance of the inner foliage of the crown provides an indication of the specific light requirements of the shade leaves, and of their phenotypic plasticity. The innermost leaves in sun crowns receive, on an average, 10–20% of the outside light whereas in shade crowns leaves persist where the relative irradiance is only 1–3% (Fig. 1.29).

1.2.1.4 Reception of Radiation by Leaves

Of the radiation falling on a leaf, some is reflected, some is absorbed, and the remainder is transmitted (Fig. 1.30).

Reflectance

The light leaving a leaf is partly light reflected by the surface and partly **scattered** radiation from inside the leaf. The *reflectance* of a leaf depends on the nature of its surface; for example, it can be greatly increased by a dense covering of hairs. Reflectance from the surface of lichens depends on the moisture present; when lichens dry out, the air replacing the water gives the thalli a brighter appearance than they have when saturated.

In the *visible* range, leaves reflect an average of 6–10% of the incoming radiation [74]. Shiny leaves of certain tree species of the warm temperate and tropical rainforests can reflect up to 12–15% of visible light; this reflected and

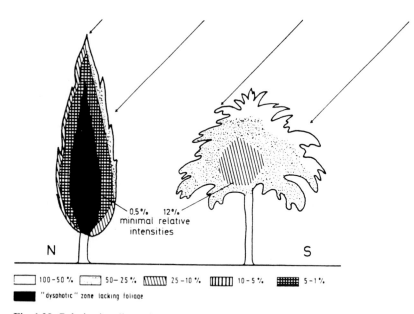

Fig. 1.29. Relative irradiance in dense and more open crowns of trees at noon on clear days in July and August. In the dense cypress crown (*left*) there is a sharp decline in irradiance even in the outermost regions of the crown. When the irradiance becomes less than 0.5% of that in the open, the green twigs inside the crown turn yellow and dry up because, as a rule, the negative carbon balance of unproductive branches is not compensated by imports from other parts of the tree. The more open olive crown (*right*) with its small, strongly light-scattering leaves is penetrated by diffuse light, so that even in the darkest parts of the crown there are leaf-bearing branches. (Unpubl. data of the author)

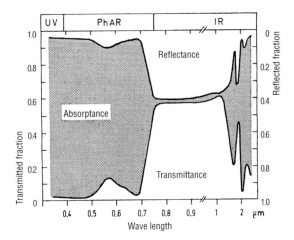

Fig. 1.30. Relative reflectance, transmittance and absorptance of a poplar leaf (*Populus deltoides*) as a function of wavelength. (After Gates 1965)

scattered light is responsible for the relative brightness in the interior of these "lucidiphyllous" forests [118]. Green light is strongly reflected (10–20%), orange and red light least so (3–10%). Reflection of red light plays an important role in communication between neighbouring plants. Because the red/far-red ratio in the reflected light increases with increasing distance from the plant reflecting it, the receiving plant, via its phytochrome system (Phy_r/Phy_{fr}; red/far-red equilibrium), is able to detect the presence and distance of its neighbours [232].

Little *ultraviolet* radiation is reflected by leaves (3% or less), whereas in the *infrared* range 70% of the incident perpendicular radiation is reflected. Reflection in the near infrared region (NIR: 750–1350 nm) can be identified by means of spectral measurements or with false colour films. Deviations from the typical species-specific *NIR-reflection pattern* provide an indication of leaf damage caused by climatic conditions, emissions, or fungal disease. Spectral shifts in reflectance can be detected from airplanes or satellites; the inclusion of spectral analysis in measurement of radiation reflected from leaves makes it possible to determine the distribution of biomass and the primary production of continents and oceans, as well as alterations in plant cover (e.g. when the vegetation turns green in spring; leaf discoloration; stress situations).

Absorptance

Most of the radiation penetrating a leaf is absorbed. In its *passage through the leaf* the amount of radiation reaching successive cell layers decreases exponentially (Fig. 1.31). The light in the intercellular spaces is totally reflected. Depending on the leaf structure and the chloroplast content of the mesophyll cells, leaves generally absorb 60–80% of the PhAR (Fig. 1.32a). Idioblasts of certain succulents, and fibre bundles in sclerophylls and palm leaves facilitate the passage of light into deep-lying mesophyll layers [261]. The leaves of some herbaceous species that grow in the deep shade of tropical rainforests possess lens-shaped cells (see Fig. 2.31) which focus the weak light on the clusters of

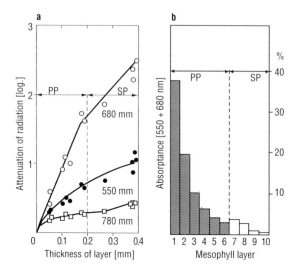

Fig. 1.31 a, b. Attenuation of radiation inside leaves of *Camellia japonica*. a Attenuation (negative logarithm of transmittance) in the palisade parenchyma (*PP*) and in the spongy parenchyma (*SP*) of the green (550 nm), red (680 nm) and far-red (780 nm) regions of the spectrum. b Absorption of radiation in successive layers of the mesophyll. (After Terashima and Saeki 1985)

chloroplasts in the mesophyll. The epidermal cells of the same leaves also contain anthocyanin, which reabsorbs the light leaving the mesophyll.

Absorption of *visible* light depends primarily on the chloroplast pigments. Hence, the spectral absorptance curves of leaves show maxima that coincide with the absorption maxima of chlorophylls and carotenoids. However, this does not mean that the absorption spectrum of the whole leaf is identical with that of chloroplast suspensions or of chlorophyll extracts (Fig. 1.32b). The greater part of the *ultraviolet radiation* is retained by cuticular and suberized outer layers of the epidermis, as well as by phenolic compounds in the cell sap

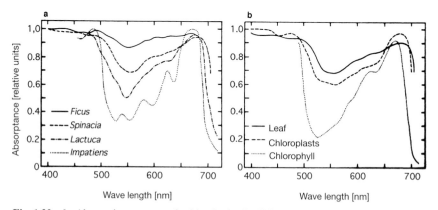

Fig. 1.32 a, b. Absorption spectra of a histologically different leaves and b an intact spinach leaf, a chloroplast suspension and a chlorophyll extract from the same leaf. (After Moss and Loomis, as cited in Kubin 1985). For optical parameters of leaves of many plant species, see Gausman and Allen (1973) and Gausman (1985)

of the uppermost cell layers, so that at most 2–5%, and usually less than 1% of the UV radiation reaches the deeper layers of the leaf. Thus the epidermis, and also hair layers, act as an efficient UV filter for the assimilatory parenchyma; as an example, the trichomes on the leaves of olive trees contain considerable amounts of flavonoids and absorb up to 60% of UV-B [110]. Pigment substances deposited in the outer layers of lichens filter UV and visible light. Up to 2000 nm, little *infrared* is absorbed by the leaves, but in the range of long-wave heat radiation above 7000 nm it is almost completely (97%) absorbed. Accordingly, the plant behaves like a black body with respect to radiation.

Transmittance

Transmittance by leaves depends on their *structure and thickness*. Soft, flexible leaves transmit 10–20% of the solar radiation, very thin leaves transmit as much as 40%, but thick, coarse leaves almost none at all (less than 3%) [112]. Transmission is greatest at the *wavelength bands* where reflection is great – that is, in the green and particularly in the near infrared. Radiation filtered through foliage is therefore particularly rich in wavelengths around 500 nm and over 800 nm. Beneath a canopy of leaves a red-green cast prevails, and in the depths of a dense forest only far red and infrared radiation remain (Fig. 1.33). Therefore in the characterization of the light climate of a particular habitat the red/far-red ratio should always be determined because of its importance for plant development. Only about 0.5–2% of the incident light, mainly long-wave, penetrates through *bud scales* and the *bark* of thinner twigs [188]. The apical meristem of buds receives more far-red and NIR (700–840 nm) than red (600–690 nm), so that the Phy_r/Phy_{fr} ratio changes as the buds de-

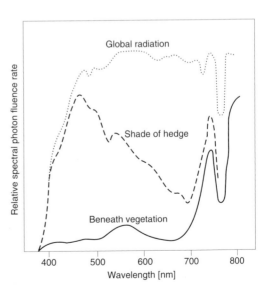

Fig. 1.33. Spectral distribution of direct solar radiation (at noon), on the shaded side of a hedgerow and beneath a closed plant canopy. (After Smith 1981). Spectral photon fluence rate distributions for conifer forests can be found in Alexeev (1975); for tropical forests in Johnson and Atwood (1970)

velop, and according to season. These signals are perceived by the phytochrome system which, via gene activation, triggers the appropriate steps in development and differentiation.

1.2.1.5 Adaptation of Plants to the Local Radiation Climate

Plants exhibit several kinds of adaptation – modulative, modicative and evolutive – to the prevailing quantity and quality of radiation. Adaptation is used here in a broad sense, and includes short-term reactions, acclimation, and evolution of genotypes.

Modulative adaptations take place quickly and are reversible, the original behaviour being quickly resumed once the initial conditions are reestablished. Examples of *photomodulation* are photonastic movements, such as the closure of guard cells; leaf movements, or *solar tracking*, by which the leaf surface is orientated so as to achieve optimal exposure to incident radiation; the diurnal and weather-dependent opening and closing of flowers (photonasty). The repositioning of chloroplasts (photodinesis) is a modulative reaction to fluctuating light intensities: in weak light they are oriented perpendicularly to the incident light, whereas in strong light they move to the cell walls, aided by the contraction of cytoplasmic filaments, and adopt a profile position. Modulative adaptations to radiation with an immediate effect on the photosynthetic processes take place via changes in the chloroplasts (ultrastructural reorganization in thylakoids and light activation of enzyme syntheses).

Modificative responses adjust the plants to the average conditions of radiation during morphogenesis. The phenotypic differentiation of tissues and organs is, as a rule, irreversible. If the light conditions change later on, new shoots are produced and the original, and now ill-adapted, leaves senesce and are shed. Plants that develop in bright light produce a robust axial system; their leaves have several layers of mesophyll and cells rich in chloroplasts, as well as a dense vascular network (Fig. 1.34). As a consequence of the structural adaptations and the more active metabolic processes (see Table 2.9), plants adapted to strong light produce more dry matter, the energy content of which is greater, and their fertility is higher (frequency of flowering, fruit setting and yield). Plants adapted to dim light have long internodes and thin leaves with a larger surface. These characteristics enable them to thrive in locations where only a modest amount of energy is available.

Evolutive adaptations to the available radiation are part of the genotype and determine the habitat preferences of the different plant species and *photoecotypes*. The classification of plants as dim-light plants, shade plants (sciophytes), sun plants (heliophytes) and strong-light plants (occupying open habitats in the high mountains, in deserts, and on sea coasts) reflects an ecological differentiation resulting from selection and adaptability. The genetic basis of ecotypes [255] can be demonstrated by rearing and studying the progeny of different plant populations under identical conditions. The *response norm* of a plant is genetically determined. Thus sun plants can adapt to shade,

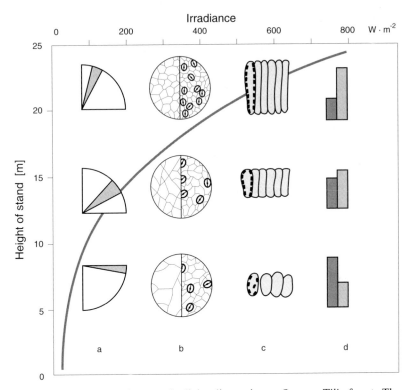

Fig. 1.34a–d. Structural adaptations to the light climate in an *Quercus-Tilia* forest. The *curve* indicates the average irradiance at noon on clear days in July. **a** Predominant leaf inclination. **b** Density of venation on the leaf surface (*left half*) and stomatal density on the lower surface (*right half*). **c** Typical shape of the cells of the palisade parenchyma. **d** Chlorophyll content per cell, referred to fresh weight (*left*) and surface area (*right*). (After Tselniker 1978; Goryshina 1980, 1989). For modificative responses to the microclimate in conifer forests see Aussenac (1973), for structural and functional adaptations of sclerophyllous leaves of mediterranean shrubs see Gratani (1993)

but not to the same extent as the genetic shade plants, which in turn behave analogously in the reverse direction. The remarkable ontogenetic plasticity of trees and climbing plants of dark forests is a genotypic feature. In the juvenile phase their leaves are morphologically and functionally of the shade type, and only with the transition to the adult phase can they acquire heliophytic characteristics; shoots arising from the earliest-formed parts of the stem behave like the juvenile stages (cyclophysis).

Modulative, modificative and evolutive adaptations are not mutually exclusive, but are superimposed so as to permit fine adjustments guaranteeing the greatest possible utilization of the available radiation. The dense filling of the spaces in the plant canopy, particularly striking in tropical forests, is due to the exploitation of the available light by plant species with genetically differing light requirements. The variety of life forms makes it possible for lianas and epiphytes to exploit the ecological niches of light in the multi-layered levels of the crowns of the tree storey. Moreover, *secondary effects* of radiation (heat, effects on water balance) are involved in all adaptations to local light intensity. Thus, sun plants are also adapted to higher temperatures, dry air and intermittent water stress.

Plants adapt not only to the intensity of light but to its spectral composition as well. **Chromatic adaptation** has been demonstrated primarily in blue and red algae, in which the relative amounts of specific accessory pigments change with the spectral shift in light as it penetrates to deeper water. In terrestrial plants, too, the composition of chloroplast pigments can change according to the spectral composition of the available light. The ability of plants to compensate for qualitative alterations in radiation by changes in their pigments is not only of ecological advantage, but is also of agricultural value, as for example in growing under artificial light.

1.2.2 Climate

1.2.2.1 Climate Regions

Climate is an important factor for plant growth; it governs the extent of their area of distribution and sets limits for their survival. This is seen on a large scale in the global distribution of the various types of vegetation according to zonal climate and soil type (Table 1.7), and on a smaller scale in the distribution of plant species and communities according to local conditions (Fig. 1.35). Plants occupy the habitat most suited to their specific requirements [266]. By restriction to habitats where the microclimate is favourable (warm slopes in a cold climate, or moist ravines in arid regions), plant species can succeed even beyond the limits of their principal area of distribution.

The average conditions and the usual weather pattern at a particular place constitute what is termed *climate*. The **macroclimate**, which is observed and recorded by a network of meteorological stations, forms the basis for the characterization of zonal and regional climates. The various *types of zonal climate* result from the different energy balance prevailing at different latitudes. At intermediate and high latitudes, for example, the climate is of the seasonal type: the alternation of positive radiation balance in summer and negative balance in winter produces a warm season with long days, and a cold season with short days. In the equatorial zone, where the number of hours of daylight varies only slightly during the year, the climate is described as thermodiurnal (Fig. 1.36). At low latitudes and in regions with a mediterranean climate, rainy- and dry seasons occur (see Fig. 6.57). *Regional variations in climate* are conditioned by position with relation to the seas (maritime or continental climates), to oceanic currents (e.g. warming by the Gulf Stream and Kuroshio current; cooling by Labrador, Humboldt, and Benguela currents), to

Table 1.7. Climate zones, soil types and predominant vegetation. (Walter and Breckle 1991)

Climate type	Soils	Vegetation
Equatorial diurnal climate, usually permanently humid	Equatorial brown clays, ferralitic soils (ferralsol)	Evergreen tropical rain-forest, seasonal features almost completely missing
Tropical Summer rains and cool dry season (humid to arid)	Red clays or red earths, ferralitic savanna soils	Tropical deciduous forests or savannas
Subtropical arid desert climate, sparse rainfall	Desert soils (arenosols, yermosols, xerosols) also saline soils (solonetz, solonchak)	Subtropical desert vegetation, predominantly stony landscape
Mediterranean Winter rain and summer drought (semiarid to semihumid)	Mediterranean brown earths, often fossil terra rossa	Sclerophyllous woody plants, sensitive to prolonged frost
Warm-temperate often with summer rain maximum, or mild-maritime	Red or yellow forest soils, slightly podzolized	Temperate evergreen forests
Cool-temperate with short cold winter	Forest brown earths and grey forest soils	Broadleaved deciduous forests (bare in winter)
Continental arid-temperate, cold winters	Steppe soils (chernozems, castanozems, sierozems)	Steppe to desert, only hot in summer
Boreal cold-temperate with cool summer and long winter	Podzols or raw humus-bleached earths	Boreal coniferous forests
Polar very short summer	Tundra humus-rich soils with pronounced solifluction	Treeless tundra vegetation, usually on permafrost

prevailing wind (trade winds), and to orographic climatic divides (leeward and windward of mountain ranges).

Within the zonal and regional macroclimates, independent climatic areas of various sizes can form. Climate can be considerably modified locally by the nature of the landscape (*mesoclimate*), and above all by the influence of geomorphology on the directional climatic factors of radiation and wind. Large expanses of water affect the surroundings over a distance of many kilometres and large settlements exhibit a specific *city climate*, characterized by less solar radiation, higher air temperatures, lower air humidity and reduced wind velocities.

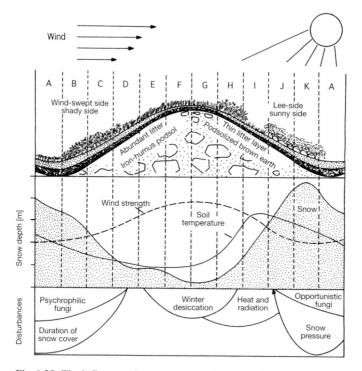

Fig. 1.35. The influence of topography and of climatic factors on the distribution of vegetation types and soil formations above silicate rock in the subalpine belt of the Central Alps. *A* Snow bed, *Soldanella*, mosses; *B Rhododendron ferrugineum* with mosses; *C Rhododendron ferrugineum* with *Vaccinium uliginosum*; *D* dwarf shrub heath dominated by *Vaccinium uliginosum*; *E Loiseleuria procumbens; F* lichen heath, wind-eroded bare sites; *G* open vegetation (rosette- and cushion plants, *Juncus trifidus*); *H Arctostaphylus uva ursi* and *Vaccinium vitis idaea; I* barren patches; *J Juniperus nana. Calluna vulgaris, Vaccinium vitis idaea; K Rhododendron ferrugineum* and *Juniperus nana.* **Stresses** Long periods with snow cover greatly reduce the length of the growing season, and the deprived plants are attacked by psychrophilic fungi; on windy hill tops and on sunny slopes the plants often lack the protection of snow in winter and are thus exposed to low temperatures and winter desiccation; in wind-sheltered situations facing south, strong irradiation in summer can produce high temperatures near the ground. (After Aulitzky 1963)

In *mountains*, in compliance with the adiabatic temperature gradient, with progressively decreasing atmospheric pressure the temperature drops by an average of $6.5 \, K \, km^{-1}$. With increasing altitude, irradiance on clear days increases, the wind blows harder and more often, the air humidity is lower, evaporation is greater and, as a rule, precipitation and duration of snow cover also increase (Fig. 1.37). On tropical mountains a cloud bank forms at greater altitudes, where the air humidity is usually high. Due to the shorter growing season with increasing altitude, and the abrupt changes in climate over relatively short distances, mountains represent selection filters and acclimatization gradients for the flora of the surrounding region. On slopes and in valley

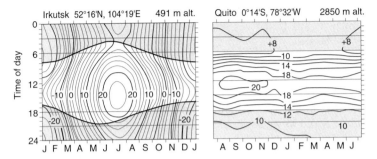

Fig. 1.36. Mean air temperature, as a function of time of day (*ordinate*) and time of the year (*abscissa*). Duration of daylight and night time (*shaded areas*) are also shown. *Irkutsk* is at an intermediate latitude and has an extreme continental climate with marked annual fluctuations of temperature); *Quito* has an equatorial highland climate with marked daily fluctuations of temperature with almost no annual fluctuation. (After Troll 1955, 1964)

basins, microclimatic differences in irradiation, temperature and evaporation occur. In valleys, on account of the topographically restricted horizon, the hours of sunlight are reduced. The annual mean duration of sunshine is reduced by as much as 60% in north-south valleys. On the mountains of temperate regions, slopes facing south to southwest warm up more than flat areas and north-facing slopes. Evaporation is higher on the windy ridges, and in winter more snow accumulates in the sheltered depressions. In tropical and sub-

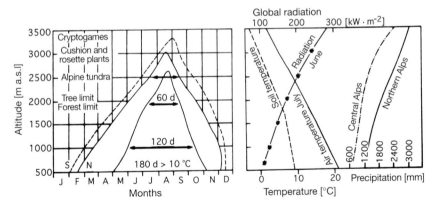

Fig. 1.37. Changes in climatic conditions with increasing altitude, and altitudinal limits of vegetation in the Alps. *Left* Altitudinal belts and plant distribution; duration of snow cover (*gridded area*) on shady north slopes (*N*) and slopes facing south (*S*), and length of the growing season (days with mean temperature above 10 °C). Observations of A. v. Kerner-Marilaun in the Tyrolean Alps between 1863 and 1878; the exact data were published by F. v. Kerner (1888) and provide a data base for studies on changes in climate. *Right* Increase in global radiation on cloudless days (daily totals for June), and in annual precipitation (mm) in the Central and Northern Alps; decrease in mean air temperature in July, and in the mean annual temperature of the soil at a depth of 1 m. (From data of Steinhauser et al. 1960; Turner 1970; Franz 1979)

tropical mountains, gullies and depressions are wetter and also wind-sheltered, which means that trees and shrubs grow to higher altitudes than on the exposed slopes.

1.2.2.2 Bioclimate and Phytosphere

The bioclimate is the microclimate of the plant cover, from the upper surface of the stand down to the deepest roots in the soil. It is a *characteristic climate*, determined by the nature and height of the plant cover. The presence of plants affects the local characteristics so that the vegetation influences the environment of which it is a part. The bioclimate, however, is also continuously subjected to influences from all climatic spheres. The zonal climate, for example, determines the irradiance, which is dependent on the latitude, whereas the nature of precipitation is dependent on regional conditions. The degree of warming, evaporation, and the distribution of precipitation depend largely on local topography.

The Bioclimate of the Individual Plant
The bioclimate surrounding the individual plant is a boundary layer climate, determined mainly by the position of the leaves with respect to incoming radiation and by the effect of wind.

Energy uptake by means of radiation absorption, heat storage, loss of heat by convection and the withdrawal of latent heat by evaporation regulate the **heat budget of plants**. Although plants are poikilothermic organisms, their own temperature is not exactly the same as that of their environment. When exposed to strong irradiation the plant is closely enveloped in a heated mantle of air. The wind carries away the outer layer, down to a few millimeters above the leaf surface (Fig. 1.38), thus promoting the loss of heat. The smaller and the more divided the leaves, and the greater the wind velocity, the more effectively is heat lost to the surrounding air by means of convection. Rosette plants and cushion plants that sit tightly on the ground warm up more than do erect

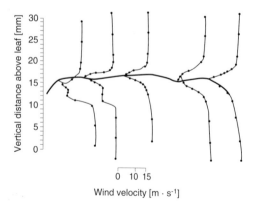

Fig. 1.38. Profiles of mean wind speed in a laminar stream in the boundary layer over a poplar leaf. (Grace 1977)

species (Fig. 1.39), especially if they grow in places affording shelter from wind. Under such conditions leaf *temperatures* of 8–10 K above air temperatures often develop, and it is not unusual for leaves to heat up by as much as 10–15 K above air temperature [73]. Succulent leaves and shoots, fleshy fruits and tree trunks store heat and attain especially high temperatures. Evaporation removes energy, since it is used for vaporization. The *cooling effect of evaporation* corresponds to an energy consumption of 70 W m^{-2} per 1 g H_2O dm^{-2} h^{-1} evaporated (i.e. 0.1 mm H_2O h^{-1}). Due to the physiological control of stomatal transpiration, plants are able to influence energy exchange with the environment. Cooling by evaporation is particularly effective at low air humidities and under windy conditions, provided that the plants take up enough water to maintain a high level of transpiration.

In and around tree crowns, bushes, and in tussock grasses the velocity of the wind is reduced and the *wind profile* is characteristically altered (Fig. 1.40). The air is slowed down in front of solitary trees, on either side the wind speed increases, but behind and beneath the tree the air is calmer. The result is an uneven distribution of rain and snow around individual trees or groups of trees, which especially affects the water balance of the herbaceous layer.

The Bioclimate in Plant Stands
In plant stands a large part of the incident radiant energy is converted into sensible heat in a narrow zone near the surface of the canopy and is used in

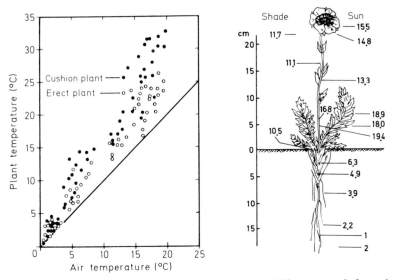

Fig. 1.39. *Left* Leaf temperatures of alpine plants with different growth forms in sunlight. Plants growing close to the ground (*filled circles*) clearly warm up more than upright forms (*open circles*). (Salisbury and Spomer 1964). *Right* Temperature distribution in the arctic species *Novosieversia glacialis* on a sunny July day when the air temperature was 11.7 °C (Tikhomirov 1963)

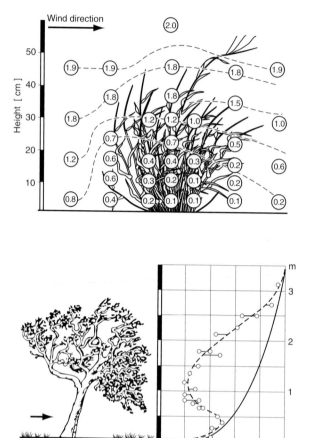

Fig. 1.40. Wind distribution around savanna plants. *Top* Wind velocities (m s^{-1}) inside a tussock grass (*Axonopus*). *Below* Wind profile on the lee side of a solitary tree (*Curatella americana*) measured immediately behind the crown (*broken line*) and over a treeless area (*solid line*). (After Vareschi 1960)

evaporation. The greatest temperature fluctuations take place in this active layer. With increasing depth inside the stand, the wind gradually subsides. *Aerodynamic exchange processes*, such as the transport of water vapour and carbon dioxide, take place within the vegetation principally by diffusion, since air turbulence no longer plays an appreciable role. Hence, vertical gradients develop in the vegetation profile, their direction being reversed with the change from day to night (Fig. 1.41). Beneath closed stands the bioclimate is more stable, warmer and wetter than the climate of the outer air (Fig. 1.42).

On sunny days the temperature at the surface of *forests* is higher than that of the air above it. During the night the surface of the stand cools down more quickly and to a greater degree than the interior; beneath the crowns of the trees nocturnal cooling due to loss of thermal radiation is far smaller, so that young plants beneath are less exposed to danger from frost. Under the canopy

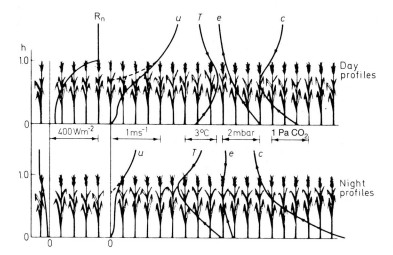

Fig. 1.41. Differences between day and night in the profiles of aerodynamic exchange at different heights (*h* metres in a stand of cereal grain. *S* Radiation balance; *u* wind velocity; *T* temperature; *e* water vapour pressure; *c* CO_2 concentration. (From Monteith 1973)

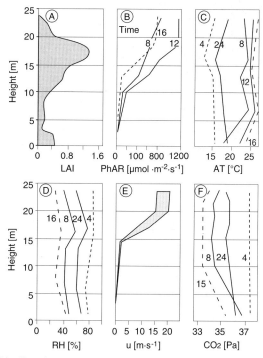

Fig. 1.42 A–F. Vertical profiles of bioclimatic parameters in a central European mixed deciduous *Carpinus-Quercus* forest in midsummer. **A** Vertical distribution of leaf area index (LAI). **B** Photosynthetically active radiation (PhAR). **C** Air temperature. **D** Relative air humidity. **E** Wind velocity. **F** CO_2 concentration. (After Eliáš et al. 1989). For tropical forests see Aoki et al. (1975)

of tropical and subtropical evergreen rainforests the daily temperature amplitude amounts to merely 2–5 K, and the annual fluctuation to less than 10 K.

A dense vegetation has an important influence on the amount and spatial
distribution of the precipitation reaching the ground. Due to interception in
the forest canopy (see Chap. 4.4.1.2) only part of the rain or snow falling in
the open can reach the forest undergrowth. The tree canopy also comes in contact with the passing mists and low clouds, and may filter out noxious dust
particles.

The Bioclimate of the Rhizosphere

Roots accommodate to the microclimate of the soil. The temperature of the
soil changes within very small distances, being dependent upon the type of
ground cover and the heat capacity of the soil. Where vegetation is sparse, the
degree to which the soil warms up depends on its colour, water and air content,
and on its structure. Loose, well aerated soils warm up superficially, whereas
compact, wet soils conduct heat to deeper levels. During the night the soil surface cools down and the heat flow in the soil is reversed. The soil thus buffers
the heat balance of the habitat by taking up a considerable quantity of heat
during the day, and releasing this at night. In regions with a seasonal climate,
a seasonal temperature cycle – measurable even down to greater depths in the
soil (Fig. 1.43) – is superimposed on the diurnal temperature fluctuations.

Under a closed cover of vegetation the soil is protected from strong irradiation and from radiation loss. Snow, too, creates nearly constant temperature
conditions in the soil; under a metre of snow the temperature of the plants and
the upper layers of soil is not far from 0 °C.

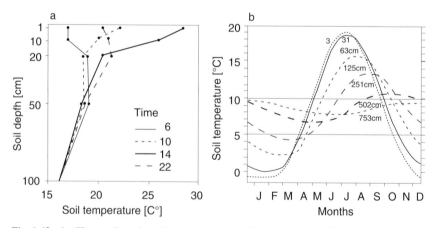

Fig. 1.43 a, b. Fluctuations in soil temperature. **a** Temperature profiles in soil below bare
ground, measured at four different times on a typical summer day in Illinois. (Rosenberg
1974). **b** Annual temperature fluctuations at various soil depths in northern Europe. (Geiger
1961). Temperature profiles for tropical soils are given by Schulz (1960) and Vareschi and
Huber (1971)

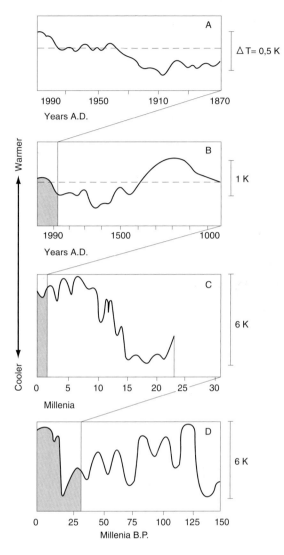

Fig. 1.44 a – d. Reconstruction of global fluctuations in temperature in earlier epochs, based on **a** meteorological measurements, **b** historical records, **c** palynological and glaciological data for Europe and North America, and **d** measurements of oxygen isotopes in deep-sea sediments. The amplitude of temperature fluctuations is given in Kelvin. (After Webb 1986; see also Roeckner 1992; Webb and Bartlein 1992). For secular fluctuations of decennial mean temperatures and extreme annual values from all parts of the world, see Lauscher (1981)

1.2.2.3 The Variability of Climate

Climate and weather are continuously changing. The interplay of the meterological factors of radiation, air temperature, air humidity, wind and precipitation, determines the frequent and short-lasting fluctuations in weather conditions. *Weather fluctuations* lasting for several days or weeks (periods of rain or frost) occur at irregular intervals in connection with global weather. *Long-term* changes in the earth's climate [273] proceed over decades or even millenia (Fig. 1.44). Since the end of the Pleistocene there have been frequent

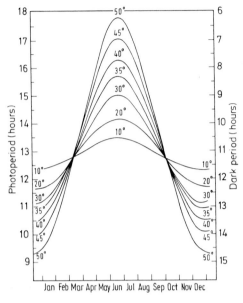

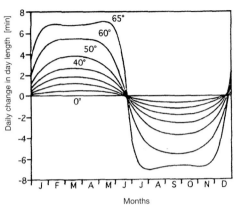

Fig. 1.45. Changes in photoperiod at different latitudes. *Left* The effect of latitude on seasonal photoperiod and *right* the daily change in photoperiod over the course of the year. (From Downs and Hellmers 1975; Salisbury and Ross 1992)

alternations of warm and cold periods; and within the space of only a few decades we have experienced fluctuations in the advance and retreat of glaciers. To the cosmic and geological causes of changes in global climate must now be added the effects of mankind's activities on the energy budget of the atmosphere (see Chap. 6.3.3.2).

The regularly occurring *periodic changes* in the climatic factors of the environment are connected with the diurnal and seasonal changes in solar radiation. The alternation of day and night results from the earth's rotation (*diurnal photoperiodism*). Since the phase of irradiation is a period of warming-up, and night time is a period of cooling, the photoperiodism is accompanied by a diurnal *thermoperiodism,* albeit with a slight lag. This is due to the fact that the air temperature reaches its maximum slightly later than the sun reaches zenith, and its daily minimum not until the end of the night.

With increasing latitude the shifts in the ratio of daylight to darkness in the course of the year (*seasonal photoperiodism*) become more pronounced (Fig. 1.45). At intermediate latitudes the day lasts about 16 h in summer, but is only half as long in winter. Beyond the polar circles alternation of night and day ceases altogether for a period of time centring at the solstices. The seasonal variation in photoperiod is accompanied by seasonal changes in temperature. Differences between the energy budget at low latitudes, where insolation is high, and at higher latitudes with less intense radiation, affect global circulation in the atmosphere and in the oceans, and thus influence the distribution of precipitation. The result is a periodic alternation of rainy and dry seasons (*seasonal hydroperiodism*).

2 Carbon Utilization and Dry Matter Production

2.1 Carbon Metabolism in the Cell

2.1.1 Photosynthesis

In the earliest geological period, as the first prokaryotes (archaebacteria, sulphur bacteria, cyanobacteria) developed photosynthetically active membranes, the environment was strongly anoxic. The primaeval atmosphere was a reducing one, and even the hydrosphere contained very little free oxygen.

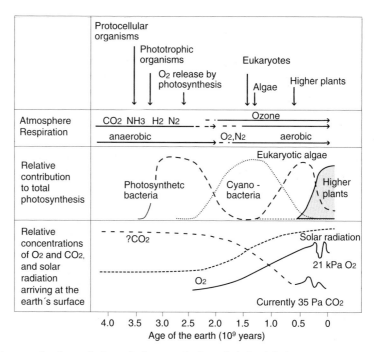

Fig. 2.1. Possible steps in the evolution of photosynthesis and their global consequences. (Lawlor 1993)

Photoautotrophic organisms, through their ability to carry on photosynthesis, created the energetic and material basis for the evolution of life on earth (Fig. 2.1). The two end products of photosynthesis, the oxygen evolved and the carbon assimilated, are equally important to all living organisms. *Oxygen* became the prerequisite for respiration, which is the most efficient form of biological oxidation, providing energy for metabolism and for maintaining cellular structure. *Carbohydrates* became the universal substrate for respiration and the starting point for a wide diversity of biosyntheses. As evolution progressed to the highly differentiated terrestrial vascular plants, so, too, did the productive capacity of plants steadily increase. The vegetation of the earth, due to its photosynthetic activity, represents an immeasurably large, continuously replenishable source of bioenergy.

In photosynthesis, radiant energy is absorbed and transformed into the energy of chemical bonds. For every gram atomic weight of carbon taken up, 479 kJ of potential energy are obtained. Photosynthesis involves light-driven *photochemical* processes, *enzymatic* processes requiring no light (the so-called dark reactions) and processes of *diffusion* which bring about exchange of carbon dioxide and oxygen between the chloroplasts and the outside air. Each one of these processes is dependent on internal and external factors and can thus limit the yield of the overall process. In the following sections the bioenergetic and biochemical aspects of photosynthesis are discussed only insofar as they are of ecological significance.

2.1.1.1 The Photochemical Process: Energy Conversion

The primary requirement for photosynthesis to take place is the absorption of radiation by the chloroplasts. The receptors for this radiation are chlorophylls, with absorption maxima in the red and blue (see Fig. 1.32), and accessory plastid pigments (carotene, xanthophyll) which absorb in the blue and UV. Cyanobacteria, red algae and cryptomonads contain, additionally, biliproteids (phycocyanin and phycoerythrin); photoautotrophic purple bacteria possess bacteriochlorophyll with a principal absorption band in the far red.

Absorption of radiation depends to a large extent on the concentration of photosynthetically active pigments. Particularly under intense light, this can be the limiting factor for the photochemical process (Fig. 2.2). *Chlorophyll deficiency*, which is evident in the yellowish to white appearance of the plant (*chlorosis*), considerably lowers the rate of photosynthesis. This condition occurs when the leaves are beginning to unfold and again in the autumn, when they turn yellow; leaves also become chlorotic when chlorophyll is destroyed as a result of too much or too little light, when the mineral balance is disturbed, by exposure to noxious gases, or as a result of viral infections. Lastly, chlorophyll deficiency can be genetically determined, as in mutants with mottled or yellow leaves.

Photosynthesis takes place in the *chloroplasts*. These are cellular organelles, surrounded by a double membrane and containing in their stroma a system of

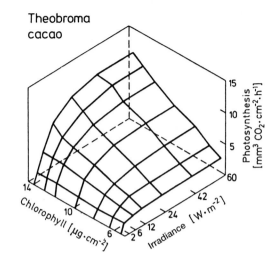

Fig. 2.2. Relationships between potential photosynthesis and the chlorophyll content of maturing leaves of *Theobroma cacao* in response to light intensity. The measurements were made under CO_2-saturated conditions ($2400\,\mu l\,l^{-1}$ in the air) to ensure that the rate of photosynthesis was not limited by secondary processes. (Baker and Hardwick 1976)

thylakoid membranes (with the photosynthetically active pigments), ribosomes, plastid DNA (the genome of the chloroplast) and various other inclusions (e.g. plastoglobules). The conversion of radiation energy to chemically bound energy takes place in the thylakoids.

The *photochemical process* is initiated when photosynthetically utilizable radiation is captured by the chloroplasts. Two pigment systems, arranged in series, are involved in the light-driven reactions. Each of these photosystems is connected with a light-harvesting pigment-protein complex (LHPC; Fig. 2.3). Antennal pigments (chlorophylls, pheophytin, phycobilins as well in cyanobacteria and algae, and carotenoids) absorb the incoming light quanta, which are then passed on to the reaction centre. *Photosystem I* (PS I) is phylogenetically the older photosystem and consists of pigment collectives with a particular structural organization; the predominant component of these is chlorophyll a. The reaction center is a chlorophyll-a-protein complex with an absorption peak at 700 nm – hence the alternative designation of this complex, "pigment 700". The ratio of total chlorophyll to P 700 ranges from 300 : 1 to 600 : 1, depending on species and condition of the plant. In sun leaves the ratio of chlorophyll a to b is higher, and the proportion of PS I is high as compared with shade leaves. *Photosystem II* contains a chlorophyll-a-protein complex with maximal absorption at about 680 nm ("P 680"), and has a higher proportion of chlorophyll b and xanthophylls than photosystem I; the xanthophylls play a special role in removing excess energy (xanthophyll cycle, see Chap. 6.2.1.1).

Light stimulates the active chlorophyll in the reaction centers to release electrons; those released from P 700 are used for the reduction of $NADP^+$. The electrons required for the reduction of chlorophyll are provided by *photolysis* of water or by another suitable electron donor, for example, H_2S in photoautotrophic bacteria. An optimal rate of photosynthesis requires a steady *flow*

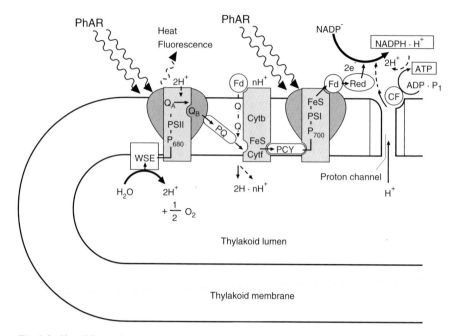

Fig. 2.3. Simplified scheme of the components of photosynthetic electron transport across the thylakoid membrane. *WSE* Water splitting enzyme; *PSII* photosystem II with light-harvesting complex II (*dark stippled*), P_{680} and plastoquinone acceptors Q_A and Q_B; *PQ* plastoquinone pool; cytochrome b_6/f complex containing ferredoxin (*Fd*), cytochrome b_6 (*Cytb*), sulphur-iron centres (*FeS*) and cytochrome f (*Cytf*); *PCY* plastocyanin; PSI photosystem I with light-harvesting complex I, P_{700} and FeS; *Red* ferredoxin-NADP$^+$ reductase; *CF* coupling factor with ATP synthetase. (After Hess 1991)

of electrons between PS II and PS I; this is the rate-determining step in the light reaction. In the course of photolysis protons and oxygen are also released. The protons accumulate inside the thylakoid envelope, so that a proton gradient is established between the inner and outer surface of the thylakoid membrane; this pH gradient provides the energy for the formation of ATP. The photochemical process delivers via this *non cyclic electron transport* reduction equivalents (NADPH$_2$) high-energy phosphate (ATP) and oxygen:

$$2\,H_2O + 2\,NADP^+ + 2\,ADP + 2\,P_i \xrightarrow[P_{600}+P_{680}]{8\,h_\nu} 2\,NADPH_2 + 2\,ATP + O_2 \quad (2.1)$$

The electron released by P 700 after absorption of radiation can also return to the oxidized chlorophyll molecule by way of a number of redox systems. This *cyclic electron transport* is also coupled with formation of ATP (cyclic phosphorylation).

Only part of the absorbed radiation energy is used for the photochemical process; the rest is converted into fluorescent and phosphorescent light and heat. Recordings of the *fluorescence* of green tissues (in vivo chlorophyll-a-fluorescence), especially during the phase of induction in light following a dark phase, gives information about the state of the photosystems, the course of electron transport, and the formation of the proton gradients required for ATP synthesis. In vivo fluorescence analyses are particularly useful in ecophysiological studies concerning light adaptations, and for detecting changes in the photosynthetic apparatus during development and in connection with stress.

The quantum yield of photosynthesis. The yield of photochemical reactions depends on the energy of the light quanta, which in turn depends on the wavelength of the absorbed radiation. The *quantum efficiency* Φ_a of photosynthesis is expressed in mol of oxygen liberated (occasionally in mol of CO_2 converted or of carbohydrate formed) per mol of absorbed photons:

$$\Phi_a \ [O_2] = \frac{\text{photochemical work (mol } O_2)}{\text{absorbed light quanta (mol photons)}} \tag{2.2}$$

Quantum efficiencies of $0.05-0.12$ mol O_2 per mol absorbed photons are achieved by terrestrial vascular plants [19, 146], $0.01-0.07$ by submersed vascular plants, $0.03-0.09$ by macroalgae, and $0.07-0.12$ by microalgae [69].

2.1.1.2 Fixation and Reduction of Carbon Dioxide: Conversion of Matter

The energy and $NADPH_2$ gained in the primary reactions are used for the conversion of carbon dioxide to carbohydrates, which are of higher energy value. This reaction takes place in the stroma of the chloroplasts.

The Pentose-Phosphate Pathway for CO_2 Fixation (Calvin-Benson Cycle)
After its arrival in the chloroplasts, carbon dioxide is first of all bound to an acceptor (Acc), the pentosephosphate *ribulose-1,5-bisphosphate (RuBP)*, which then undergoes carboxylation catalyzed by the enzyme RuBP-carboxylase (Rubisco). This enzyme is present in considerable amounts in the leaves (it accounts for 50% of the soluble proteins in the chloroplasts). The carboxylation product, a 6-carbon molecule, decomposes immediately to produce two molecules of 3-phosphoglyceric acid (PGA). Each of these molecules contains 3 carbon atoms and the process is therefore also called the C_3 *pathway of CO_2 assimilation*. PGA is reduced to glyceraldehyde-3-phosphate (GAP) over several steps involving ATP and $NADPH_2$. The general formula for the reduction of carbon dixoide is:

$$nCO_2 + nAcc + 2nNADPH_2 + 2nATP \xrightarrow{\text{Enzyme}} (CH_2O)_n + 2nNADP^+$$
$$+ 2nADP + 2nP_i + nAcc + nH_2O \tag{2.3}$$

The triose phosphate GAP flows into a pool of carbohydrates of different carbon chain-lengths (C_3-C_7), which provides material for the synthesis of various substances (sugars, starch, carboxylic acids, amino acids) and for the regeneration of the acceptor.

The **efficiency of carboxylation,** or in other words the speed with which CO_2 is processed after being taken up, is limited chiefly by the quantity and the activity of the enzyme and by the availability of the carbon dioxide. Other factors exerting an influence are the acceptor concentration, temperature, the degree of hydration of the protoplasm, the supply of minerals (especially phosphate), and the stage of development and state of activity of the plant. Phytohormones such as abscisic acid, which regulates ion fluxes, also play a role.

The higher the CO_2 concentration in the medium surrounding a plant (water or the external air) the more efficient is the catalytic activity of the RuBP-carboxylase. The CO_2 concentration in the intercellular air of plants growing in swampy habitats may rise to as much as 4% by internal diffusion from the roots [25].

Biochemical concentration mechanisms are especially effective in boosting carboxylation:

1. *Water plants,* which are less well provided with CO_2 than terrestrial species, can utilize HCO_3^- in addition to dissolved CO_2. This ability is based on a number of different mechanisms such as the active transport of HCO_3^- into the cells (e.g. in cyanobacteria and many green algae), conversion into CO_2 in the cell wall by carbonic anhydrase (in microalgae), and by proton-coupled polar transport (in Characeae, in leaves of submersed macrophytes). Carbonic anhydrase, a zinc-containing enzyme, occurs in cell walls, in the cytosol and especially in the chloroplasts of all plants; it accelerates the conversion of HCO_3^- to CO_2 by a factor of 100. Polar transport in the leaves of water plants is a result of increased activity of the H^+-ATPases following illumination; this leads to a drop in pH in the apoplast of the epidermis on the underside of the leaf, with a consequent decrease in the ratio HCO_3^-/CO_2, i.e. more CO_2 is made available. Cyanobacteria, algae and marine vascular plants utilize HCO_3^- to a greater extent than most freshwater macrophytes. However, not all organs of submersed cormophytes are able to obtain dissolved inorganic carbon (DIC) in this way. Green axial shoots with extensive aerenchyma behave like the tissues of terrestrial plants (e.g. *Elodea* and *Egeria* [191]). The phycobionts of lichens also employ CO_2-concentrating mechanisms, which explains why lichens [8] can assimilate even down to very low partial pressures of CO_2.

2. In many plants the *dicarboxylic acid pathway* has developed as a special means of concentrating CO_2 for fixation. This pathway will now be described in detail.

The Dicarboxylic Acid Pathway for CO_2 Assimilation

In roughly 10% of all known plant species the first product of CO_2 fixation is not a 3-carbon molecule but oxaloacetic acid, a dicarboxylic acid with 4 carbon atoms, which is formed by β-carboxylation of *phosphoenol pyruvate* (PEP). Reduction to the carbohydrate, however, always takes place by the pentose phosphate pathway.

The binding of CO_2 to PEP is a widespread biochemical reaction (Wood-Werkman reaction) and is encountered in the metabolism both of microorganisms and higher organisms. From the starting point of this genetically fixed reaction, two variants of the dicarboxylic acid pathway (also called the C_4 pathway on account of the 4 C atoms of oxaloacetic acid) has evolved in two variant forms as part of:

1. a *two-step process* in which fixation of the CO_2 (C_4-process) and formation of carbohydrate (C_3 process) take place in two spatially distinct tissues (*C_4 syndrome*) and

2. a *two-period process,* by which the dicarboxylic acid produced from nocturnally fixed CO_2 is stored overnight in vacuoles and then decarboxylated during the day to provide CO_2 to the chloroplasts of the same cells for processing via the pentosephosphate pathway (*Crassulacean Acid Metabolism,* CAM). An overview of the characteristics of the three different metabolic pathways, C_3, C_4 and CAM is shown in Table 2.1. Plants that use the dicarboxylic acid pathway for CO_2 fixation, employ one of two contrasting production strategies: the achievement of *maximum assimilation* by the highly efficient C_4 grasses, and *flexibility and buffering of carbon assimilation* by CAM plants.

An indicator of the part played by the C_4 process in the overall process of carbon assimilation of a plant is provided by the ratio of the stable isotopes ^{13}C to ^{12}C in the dry matter. The ratio $^{13}C/^{12}C$ of plant matter is referred to a standard and expressed in parts per 1000

$$\delta^{13}C = \frac{(^{13}C/^{12}C)_{sample} - (^{13}C/^{12}C)_{standard} \cdot 1000}{(^{13}C/^{12}C)_{standard}} \qquad (2.4)$$

The present-day content of $^{13}CO_2$ in the total CO_2 in the air is roughly 1.1%. Plant matter contains less ^{13}C because although carboxylases prefer ^{12}C. RuBP carboxylase discriminates much more against ^{13}C ($-28‰$) than PEP-carboxylase ($-9‰$), which uses HCO_3^-. Hence, C_3 terrestrial plants have lower $\delta^{13}C$ values (-23 to -36, mean $-27‰$) than C_4 plants (-10 to -18, mean $-13‰$) [204]; values for CAM plants vary according to functional type (see Fig. 2.10). The values for water plants range more widely, i.e. from -11 to $-50‰$ [114], and thus is not a clear indicator of photosynthetic pathway as in terrestrial plants. Furthermore, $\delta^{13}C$ values are dependent on weather conditions, especially during drought, and also may be affected by the irradiance received by the leaves, uptake of CO_2 from the air in the soil, growing at low temperatures, the decreasing air pressure in mountains, and by ozone stress.

The C_4-Syndrome:
Hatch-Slack-Kortschak Pathway

A striking feature in C_4 terrestrial plants is a wreathlike arrangement of larger chlorenchyma cells around the bundle sheath of their leaves ("*Kranz-type*" anatomy; see Fig. 2.6). In these and the other mesophyll cells CO_2 is bound by the primary acceptor PEP, to form oxaloacetate (see Fig. 2.12, left-side). This is then reduced to malate by a $NADPH_2$-dependent enzyme, malate dehydro-

Table 2.1. Characteristics of plants with different types of CO_2 fixation. (After Black 1973; Kluge and Ting 1978; Osmond 1978; Sestak 1985)

Charactertistic	C_3	C_4	CAM
Leaf structure	Laminar mesophyll, parenchymatic bundle sheaths	Mesophyll arranged radially around chlorenchymatic bundle sheaths ("Kranz"-type-anatomie)[c]	Large vacuoles
Chloroplasts	Granal[a]	Mesophyll: granal; bundle-sheath cells: granal or agranal[b]	Granal
Chlorophyll a/b	ca. 3	ca. 4	≤ 3
Primary CO_2-acceptor	RuBP (substrate: CO_2)	PEP (substrate: HCO_3^-)	In light: RuBP In dark: PEP
First product of photosynthesis	C_3 acids (PGA)	C_4 acids (oxaloacetate, malate, aspartate)	In light: PGA In dark: malate
Carbon-isotope ratio in photosynthates ($\delta^{13}C$)	ca. -20 to -40‰	ca. -10 to -20‰	ca. -10 to -35‰
Photosynthesis depression by O_2	Yes	No	Yes
CO_2 release in light (apparent photorespiration)	Yes	No	No
CO_2-compensation concentration at optimal temperature	$30-50\ \mu l\,l^{-1}$	$<10\ \mu l\,l^{-1}$	In light: $0-200\ \mu l\,l^{-1}$ In dark: $<5\ \mu l\,l^{-1}$
Ratio of mesophyll resistance to minimal stomatal resistance	$4-5$	$0.5-1$	
Net photosynthetic capacity	Slight to high	High to very high	In light: slight In dark: medium
Light saturation of photosynthesis	At intermediate intensities	No saturation, even at highest intensities	At intermediate to high intensities
Redistribution of assimilation products	Slow	Rapid	Variable
Dry matter production	Medium	High	Low

[a] Thylakoids stacked.
[b] Thylakoids lamellar.
[c] Not in aquatic plants.

genase, although in some species aspartate is the transport form produced. Instead of being processed in the mesophyll cells, the dicarboxylic acids are conveyed to the *bundle sheath*. In the chloroplasts of the bundle-sheath cells the malate and aspartate are broken down by specific enzymes into pyruvate and CO_2. The CO_2 thus released is captured by RuBP and processed via the pentose phosphate pathway; the pyruvate is returned to the mesophyll cells and is used for the regeneration of PEP.

The enzyme *PEP carboxylase,* cooperating with the enzyme *carboanhydrase,* which rapidly converts CO_2 into HCO_3^-, is extremely efficient. This system acts as a CO_2 pump, and can also function at very low concentrations of HCO_3^- and CO_2 and at elevated temperatures. The mesophyll cells thus withdraw almost all of the CO_2 from the intercellular air, so that the CO_2 compensation concentration drops to below $10 \, \mu l \, l^{-1}$.

Due to the absence of photorespiration in the mesophyll cells and also because of their ability to carry on photosynthesis even when the concentration of CO_2 inside the leaves is very low (with the stomata to a large extent closed, for example), C_4 plants have an advantage, as compared with C_3 plants, at high temperatures and in times of moderate drought. Particularly under strong irradiation the *high carboxylation efficiency* is a considerable competitive advantage. A further advantage for C_4 plants is that their nitrogen requirements are lower on account of the small amount of protein in the mesophyll chloroplasts. A disadvantage, however, is the sensitivity of many C_4 plants to cold. Low temperature (approximately below 5 to 7 °C) [186] during the period of growth has a negative effect on their development, possibly due to inhibition of transport in the phloem. Accordingly, the abundance of C_4 species

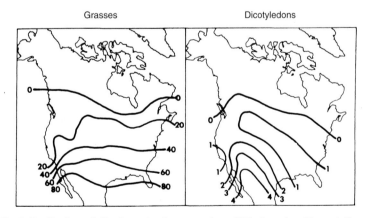

Fig. 2.4. Latitudinal distribution of C_4 plants in the vegetation of N. America. Percent C_4 species in the grass floras *(left)* and among dicotyledonous herbs *(right)*. (After Teeri and Stowe 1976; Stowe and Teeri 1978, from Ehleringer 1979). Similar proportions of C_4 plants are found in Europe. (Collings and Jones 1985). In the arid regions of N. Africa (Batanouny et al. 1988), Asia Minor (Vogel et al. 1986) and Australia (Hattersley 1983) 40–90% of the indigenous species are C_4 plants. In the semi-arid Andes of NW Argentine, 35% of the species in the submontane belt (up to 2300 m a.s.l.) and 20% in the montane belt (up to 3300 m) are C_4 plants. (Ruthsatz and Hofmann 1984)

decreases from regions with a warm, dry climate towards cooler or wetter surroundings, or in other words, with increasing latitude and with increasing altitude (Figs. 2.4 and 2.5). In spite of this, a few C_4 species (e.g. *Paspalum conjugatum, Cyperus* spp., *Spartina angelica, Echinochloa crus-galli* [277]) are found in shady, wet or cool places.

C_4 plants are frequent among annual species, especially summer annuals, and hemicryptophytes, but they are seldom encountered among winter annuals and geophytes. Tall shrubs and trees have apparently not developed the C_4 syndrome. So far, CO_2 incorporation via the C_4 pathway has been shown to exist in roughly 2000 angiosperm species from 18 different families, and in *Anacystis nidulans* [51, 190]. Some aquatic vascular plants (*Eleocharis, Hydrilla*) are able to bind large quantities of CO_2 via PEP-carboxylase in the cytoplasm and to complete carbon assimilation in the chloroplasts of the same cells via RuBP carboxylase. Many C_4 plants are found, especially among the Poaceae (for example, sugar cane, maize, millet, and many savanna grasses), Cyperaceae, Portulaceae, Amaranthaceae, Chenopodiaceae and Euphorbiaceae, although in some genera (*Atriplex, Kochia, Euphorbia, Flaveria, Cyperus, Panicum*) only a few species belong to this type. Crosses between a C_4 species and a closely related C_3 species are intermediate with respect to structure and metabolism (Fig. 2.6). Certain species also have both C_4- and C_3- *ecotypes,* as, for example, the grass *Alloteropsis semialata*. Quite obviously, the evolution of C_3- to C_4-types is in full progress, since *transitional forms* are encountered ($C_3 - C_4$ intermediate) in which the histological and biochemical C_4 characteristics are less well developed. Within the genus *Flaveria* of the Asteraceae there is a gradual transition from *F. cronquistii* (C_3) via *F. ramosissima* and *F. floridana* ($C_3 - C_4$) to *F. trinerva* (C_4) [218]. Other species that can be regarded as $C_3 - C_4$-intermediates are *Panicum milioides* and, in

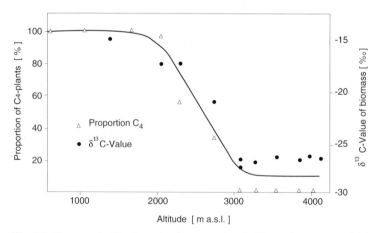

Fig. 2.5. Decrease in C_4 plants in the vegetation in Kenya (as percent of biomass) and increasingly negative isotope ratios $^{13}C/^{12}C$ ($\delta^{13}C$ value) with altitude. (From Tieszen et al. 1979)

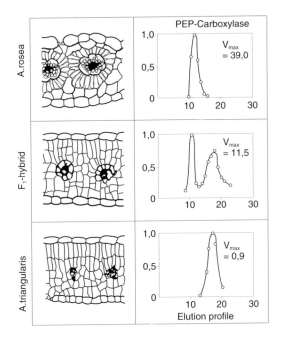

Fig. 2.6. Leaf structure and elution profiles of PEP carboxylase of *Atriplex rosea* (C$_4$), *Altriplex triangularis* (C$_3$) and their F$_1$ hybrids. V_{max} Substrate-saturated activity of PEP carboxylase in µmol CO$_2$ mg^{-1} chl min^{-1}. (Osmond et al. 1980)

the Brassicaceae, *Moricandia arvensis* [26]. In one and the same individual, the mode of carbon assimilation can alter according to the stage of development (in maize leaves [38] the C$_4$ syndrome develops during tissue differentiation), with adaptation to light (*Flaveria brownii* [166]), or with a shift from one habitat to another (*Eleocharis vivipara*; terrestrial form is C$_4$; aquatic, C$_3$).

CO$_2$ Fixation in CAM Plants

The diurnal alternation between CO$_2$ conservation by the formation of acids, and CO$_2$ utilization for the formation of carbohydrate requires that the chloroplast-containing cells have sufficient storage volume in the vacuoles. Hence, the photosynthetic organs of CAM plants consist of chlorenchyma with large, rounded cells, or at least they contain cell layers with a large storage space („degree of succulence": see Fig. 2.10) for carboxylic acids and for water.

CO$_2$ is taken up in the dark by CAM plants through their open stomata and is then bound to phosphoenol pyruvate with the aid of PEP carboxylase. The product of this *dark fixation of CO$_2$* is oxaloacetate, which is subsequently reduced to malate by the NADH$_2$-dependent malate dehydrogenase. The PEP comes from glycolysis so that, as malate is formed, the starch content of the chloroplasts is correspondingly reduced (Fig. 2.7). The first product of fixation enters the vacuoles as malic acid and is accumulated in the cell sap, which becomes progressively more acid during the night (from pH 5 to pH 3). With daylight the next morning, the malate is transported back from the vacuole into the cytosol and the chloroplasts, where it is decarboxylated. The CO$_2$ liberated is taken up by RuBP and reduced to carbohydrate. During the day-

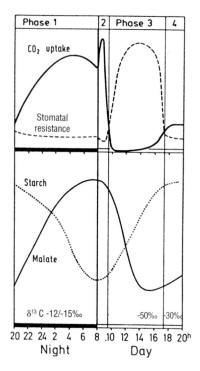

Fig. 2.7. Typical time courses of CO_2 uptake, diurnal acid rhythm, starch storage and stomatal opening in a CAM plant. *Phase 1* primary CO_2 fixation and accumulation of malate in the dark with stomata open (diffusion resistance small), accompanied by starch breakdown; *Phase 2* beginning of the light period, initial uptake of external CO_2 through the wide-open stomata; *Phase 3* movement of malate out of the vacuole and into the cytoplasm and chloroplasts, malate decomposition and release of CO_2, stomatal closure, feedback inhibition of PEP carboxylase, photosynthetic assimilation of internal CO_2, starch synthesis; *Phase 4* reactivation of PEP carboxylase following reduction of the high malate level, opening of the stomata and influx of external CO_2, and the beginning of renewed malate synthesis at the end of the light period. $\delta^{13}C$ changes in $^{13}C/^{12}C$ isotope ratio. (After Kluge and Ting 1978; Griffiths 1991). For different gas exchange patterns of CAM plants see Neales (1975)

time the normal C_3 *cycle* is active. The progressive breakdown of malate is accompanied by a rise in the pH of the cell sap (*diurnal acid rhythm*); the intracellular CO_2 concentration drops, the stomata begin to open, and new CO_2 can flow in from the outside and be fixed by PEP.

The diurnal sequence of events – CO_2 uptake, acid rhythm and assimilation – is recognizable in principle, in all CAM plants, but there are great differences in the amount and duration of CO_2 uptake in dark and light, depending on species, environmental conditions, such as water supply, temperature and day length, and stage of development; e.g. the storage volume of the mesophyll and the enzyme spectrum required for CAM are still underdeveloped in juvenile plants (Fig. 2.8). Such differences can be expressed numerically as the ratio of nocturnal to daytime CO_2 uptake.

In certain CAM plants, such as stem succulent cacti, agaves, aloes, and many Bromeliaceae and Orchidaceae for example, the acid rhythm is employed under all circumstances. Other CAM species, on the other hand, especially among the Aizoaceae and Crassulaceae, only employ CAM in response to stress from drought or salt (Fig. 2.9). In *Mesembryanthemum crystallinum* salt stress elicits de novo synthesis of enzymes of Crassulacean metabolism (see Fig. 6.4).

CAM plants are often found in environments subject to periodic drought, and in habitats with a poor substrate. All of the cacti and most of the Asclepiadaceae and Euphorbiaceae in the succulent-inhabited steppes and de-

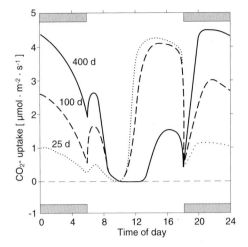

Fig. 2.8. CO_2 uptake over a 24-h period by seedlings of *Agave deserti* of different ages. The youngest, 25-day-old seedlings were 14 mm tall, the 100-day-old seedling were 32 mm, and the 400-day-old individuals were 63 mm tall. The typical CAM pattern develops when the plants are older and larger. (From Nobel 1988)

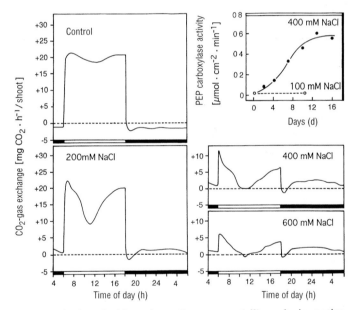

Fig. 2.9. Induction of CAM behaviour in *Mesembryanthemum crystallinum* by increasing salt concentration in the substrate. *Top right* Increase in PEP-carboxylase activity during 2 weeks' treatment with 400 mM NaCl in the root medium. (After Winter and Lüttge 1976; Winter 1985)

serts of the tropics and subtropics are CAM plants; among epiphytes, 50–60% of all Bromeliaceae and Orchidaceae are CAM plants; and Crassulaceae growing on rocks and cliffs, and on shallow soil, behave as CAM plants according to need. In all such habitats it is advantageous for the plants to separate the stages of carbon assimilation in time, fixing the CO_2 by night and processing it on the following day. In this way the carbon supply is ensured, without incurring the risk of water losses. In periods of severe drought, when the stomata remain shut day and night, the CO_2 formed in respiration is recaptured, thus restoring the carbon balance and at the same time conserving water. Although

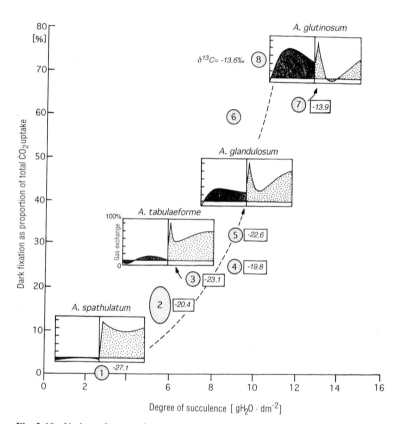

Fig. 2.10. Variant forms of crassulacean acid metabolism in the genus *Aeonium*. *1: A. spathulatum; 2: A. holochrysum; 3: A. tabulaeforme; 4: A. arboreum; 5: A. glandulosum; 6: A. undulatum; 7: A. glutinosum; 8: A. haworthii.* The greater the storage capacity of the leaves (degree of succulence), the greater is the proportion of dark fixation in the total 24 h CO_2 uptake. The daily profiles show the pattern of CO_2 uptake under standardized conditions (10 °C at night, 20 °C during the day). CO_2 uptake during the dark phase is *shaded black,* that during the light phase is *stippled.* At the above-mentioned temperatures *A. spathulatum* behaves like a C_3 plant, *A. tabulaeforme* and *A. glandulosum* take up CO_2 mainly during the day, and *A. glutinosum* chiefly at night. (Lösch 1984, 1987, and pers. comm.). *Italicized values* Isotope ratio ^{13}C/^{12}C in cultivated plants well-supplied with water, in July. (After Pilon-Smits et al. 1991)

this closed CO_2 circuit is "idling", it nevertheless serves to conserve carbon, as well as to protect the photosynthetic apparatus from the damaging effect of excessive radiant energy, since the continuously circulating CO_2 acts as terminal electron acceptor.

More than 20000 plant species of 25 different families use CAM for assimilating carbon [82, 243]. Among them are pteridophytes, such as certain epiphytic Polypodiaceae and aquatic species of *Isoetes;* monocotyledons, including Orchidaceae, Bromeliaceae, Agavaceae and Liliaceae, and representatives of dicotyledonous families, among them Cactaceae, Crassulaceae, Aizoaceae, Portulacaceae, Euphorbiaceae, Piperaceae, Asclepiadaceae, Geraniaceae, Gesneriaceae, Rubiaceae, Asteraceae and others. The CAM behaviour may be restricted to *certain organs* of a plant. The deciduous *Frerea indica* (Asclepiadaceae) [130] binds CO_2 by the C_3 pathway in its leaves, but by way of CAM in the succulent axial parts of the shoot. By so doing, this plant achieves high carbon yields during the wet season while it is in leaf, but when it is bare during the dry season it can take up carbon in a way that is especially economical of water.

The wide distribution of CAM in the plant kingdom, as well as the occurrence of diverse variants of the procedure, suggest that it has developed along a number of different evolutive routes. A characteristic of all CAM plants is that they are undemanding; this allows them the alternative of moving into sparsely inhabited ecological niches, like epiphytic and cliff habitats. Gradual transitions occur from C_3 plants, via intermediate C_3-CAM steps, up to types that are totally dependent on CAM (Fig. 2.10); in addition, there are highly specific differentiations, which lead to the aquatic CAM plants in oligotrophic waters (e.g. *Crassula aquatica* and *Littorella uniflora*).

2.1.2 Photorespiration: the Glycolate Pathway

In light, a metabolic process coupled with photosynthesis takes place in chloroplast-containing plant tissues (with the exception of C_4 mesophyll cells which contain no RuBP-carboxylase-oxygenase) which, like mitochondrial respiration, consumes O_2 and releases CO_2 in the light but, unlike to respiration, ceases in the dark. This O_2/CO_2 gas exchange has been called *light respiration* or photorespiration.

The primary substrate for photorespiratory metabolism is again ribulose bisphosphate, which can be an acceptor not only for CO_2 but also for O_2. By taking up oxygen, RuBP is split into phosphoglyceric acid and phosphoglycolate. The supply of CO_2 and O_2 regulates the relationship between acceptor carboxylation (photosynthesis) and acceptor oxidation (respiration) via the enzyme *RuBP-carboxylase/oxygenase*. A high partial pressure of oxygen favours photorespiration, a large supply of CO_2 favours photosynthesis (Fig. 2.11). As the formation of phosphoglycolate is dependent on the supply of RuBP via the Calvin cycle, photorespiratory O_2 uptake and CO_2 release increase with a rise in photosynthetic activity resulting from an increase in photon flux.

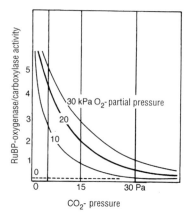

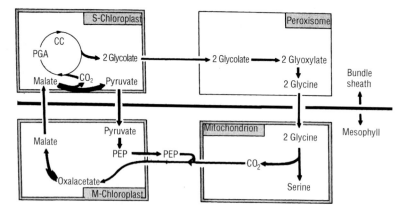

Fig. 2.11. The activity ratio of RuBP oxygenase to RuBP carboxylase in relation to the partial pressures of oxygen and carbon dioxide. (Lawlor 1993)

The phosphoglycolate is split into glycolate and phosphate in the chloroplasts. The glycolate is transported from the chloroplasts into the peroxisomes (Fig. 2.12), cell compartments about the size of mitochondria, which contain glycolate oxidase, catalase and transaminases. In the peroxisomes, when oxygen is taken up, glycolate is oxidized to glyoxylate, which is then either completely degraded to CO_2 via oxalate by further uptake of oxygen, or converted into glycine by transamination. Glycine is transported out of the peroxisomes into the mitochondria (or chloroplasts also), where two molecules of glycine are coupled to one molecule of serine with the release of CO_2. Serine is taken over by the amino acid metabolism or, after deamination, is converted to

Fig. 2.12. Schematic representation of the glycolate pathway of photorespiration in C_4 plants. *CC* Calvin cycle; *PGA* 3-phosphoglycerate; *PEP* phosphoenolpyruvate. CO_2 released from the mitochondrion is immediately fixed in the mesophyll cells. In C_3 plants the glycolate pathway proceeds similarly from the Calvin cycle into the mitochondrion, but the CO_2 escapes. (From Hess 1991)

glycerate via hydroxypyruvate. The glycerate can be phosphorylated in the chloroplasts and returned to the Calvin cycle or used elsewhere.

Under natural conditions (21 kPa pO_2 and 35 Pa pCO_2 in the air, strong irradiation, temperatures between 20 and 30 °C) C_3 plants immediately lose about 20%, or in the extreme case 50% [223], of the photosynthetically acquired CO_2 in the form of *photorespiratory* CO_2. In aquatic plants, which utilize HCO_3^- and guard against scarcity of CO_2 supplies by employing carboanhydratase, photorespiration is low. C_4 plants release no CO_2 in light. Photorespiration in these plants occurs only in the cells of the bundle sheaths, and the CO_2 produced is refixed in the mesophyll cells before it can leave the leaf (see Fig. 2.12). This CO_2 trap prevents loss of substance as a result of photorespiration and makes possible a higher rate of production.

Although carbon is lost in glycolate metabolism, photorespiration by no means represents a necessary evil: it is in fact a branching point for pentose metabolism, connecting with amino acid and fat metabolism. Above all, it diverts the electron flow in excessive light, thus protecting the photosynthetic apparatus and especially the PS II.

2.1.3 Release of Energy by Catabolic Processes

Whereas anabolic processes bring about the synthesis of substances that make up the plant, catabolism or breakdown of these substances provides energy for metabolic processes in the cell. The substrate for these reactions is carbohydrate or fat; in the exergonic decomposition of these substances by *glycolysis*, in the *tricarboxylic acid cycle* (Krebs-Martius cycle) and in the β-*oxidation* of fatty acids, hydrogen is split off stepwise and energy is liberated. The hydrogen is taken up by pyridine- or flavine nucleotides. Most of the energy is obtained when hydrogen is transferred to the final hydrogen acceptor. In **mitochondrial respiration** this acceptor is atmospheric oxygen, which receives the hydrogen in the mitochondria after its passage through an electron transport chain. In the process, ATP is formed by *respiratory chain phosphorylation;* other energy-rich phosphates are produced in the cytosol by glycolysis via *substrate-chain phosphorylation*. The entire process yields 36 mol of ATP, 2 mol of GTP, and a free enthalpy $\Delta G°$ of -2.87 MJ ($= -686$ kcal) per mol of glucose.

In stressed, injured or senescing tissues, terminal electron transport can take a different route, avoiding the HCN (and CO) inhibitable cytochromes. This **cyanide-resistant respiration** converts a large part of the redox energy into heat. An extreme case is the brief activation of this metabolic pathway during flowering of the spadix of some Araceae; the result is an appreciable rise in temperature, which attracts pollinators. Cyanide-resistent respiration is also involved in the climacteric respiration of ripening fruits, and in roots it can be stimulated by an excess of NO_3^- and NH_4^+. Wherever this "overflow respiration" [127] occurs there is a considerable rise in O_2 consumption and in CO_2 liberation.

In **fermentation**, hydrogen is taken up by reducible organic compounds and in **anaerobic respiration** by inorganic ions such as nitrate and sulphate. Because the potential difference between hydrogen and oxygen is greater than that between hydrogen and other oxidizing agents, these catabolic processes yield far less energy than aerobic respiration, which attains an efficiency of $30-40\%$.

In addition to glycolysis and the citric-acid cycle, there is a third way in which glucose can be broken down, particularly in highly differentiated plants. This is by means of the direct oxidation of glucose-6-phosphate, with the reduction of $NADP^+$ instead of NAD^+. In this pathway, decarboxylation results in ribulose-5 phosphate, which is then reconverted to glucose-6-phosphate via the **oxidative pentose phosphate cycle.** The significance of this metabolic pathway lies less in its energy yield than in the provision of *metabolites*. The oxidative pentose phosphate cycle occurs in the cytosol and in the chloroplasts. It is closely related to the reductive pentose phosphate cycle (Calvin cycle) with which it has many reactions in common, and both cycles share in the $NADP^+$ pool.

2.2 Gas Exchange in Plants

2.2.1 The Exchange of Carbon Dioxide and Oxygen

Carbon metabolism in the cell is linked with the atmosphere by **gas exchange.** This implies the exchange of CO_2 and O_2 between the interior of the plant and its surroundings. In *photosynthetic gas exchange* the plant takes up CO_2 and gives off O_2, whereas in *respiratory gas exchange* the direction of transport of the two gases is reversed. Which of the two processes predominates at any particular time can be detected at the leaf surface of assimilating plants. When more CO_2 is consumed in photosynthesis than is freed simultaneously in respiratory processes, this net uptake of CO_2 is known as the *apparent* CO_2 uptake or *net photosynthesis* (Ph_n), in contrast to the *total* CO_2 fixed, which is termed *gross photosynthesis* (Ph_g). Due to the continuous production of CO_2 in catabolic processes, gross photosynthesis can only be estimated. For questions concerning photosynthetic assimilation, it is usually sufficient to know the values for net photosynthesis, which can be obtained by measuring gas exchange.

In the dark, the respiratory CO_2 escapes. Under unfavourable conditions for assimilation, photosynthetic activity may just balance of CO_2 set free simultaneously by respiration. This state of *compensation* is recognizable by the absence of net gas exchange.

The exchange of gas between the cells and the surroundings of the plant takes place by *diffusion* and *mass flow.* The turnover is enormous; for every gram of glucose formed 1.47 g of CO_2 are required, which is roughly the amount contained in 2500 liters of air.

2.2.1.1 The Diffusion Process

The transport of CO_2 and O_2 are described by Fick's law of diffusion:

$$\frac{dm}{dt} = -D \cdot A \cdot \frac{dC}{dx} \tag{2.5}$$

The *diffusion rate* (quantity displaced, dm, in the time interval dt) depends upon the diffusion constant, D, and is greater the steeper the *concentration gradient*, dC/dx, in the direction of diffusion, x, and the greater the exchange area, A. The diffusion constant depends both upon the substance considered and the medium in which diffusion takes place; in air, CO_2 and O_2 can diffuse about 10^4 times as fast as in water. Fick's law can be applied to *gas exchange in plants* in a simplified form [70]:

$$J = \frac{\Delta C}{\Sigma r} \tag{2.6}$$

The diffusional flux (J) is greater the steeper the *concentration gradient* (ΔC) between the ambient air and the reaction site in the cell, and is lowered by a number of factors influencing *resistance to diffusion* (Σr). Σr includes the relevant diffusion constants, the resistances at phase interfaces, and the spatial dimensions involved in the situation.

2.2.1.2 The Concentration Gradient

Assuming that photosynthesis proceeds to the point at which the CO_2 in the chloroplasts is exhausted, and that during respiration in the mitochondria the O_2 concentration there falls to zero, then the concentration gradients of these two gases are determined by their concentration in the surroundings of the plants.

The above-ground organs of terrestrial plants are well supplied with oxygen. The green, photosynthetically active plastids in the cortex of shoots, in the woody parenchyma and pith of slender branches, and in fruits, liberate oxygen into the intercellular system. For such massive organs, this effect of photosynthesis is possibly of greater significance than the rather low yield of carbon. More likely to suffer from oxygen deficiency are roots and underground shoots. However, in plants such as water lilies [39] and alder [214] they do receive oxygen that is moved by processes of thermodiffusion through the intercellular system along thermal gradients existing between leaves and roots and rhizomes. For the normal requirements of mitochondrial respiration, an oxygen concentration of 1–3% is adequate; lower concentrations result in decreases in the intensity of respiration, and impairment of root growth (see Chap. 6.2.3).

Under natural conditions the CO_2 concentration gradient from the surroundings into the leaf is very small. The *CO_2 compensation concentration*, at which gas exchange disappears depends primarily on the mode of carbox-

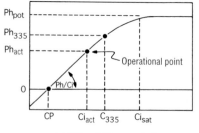

Leaf internal CO_2 concentration

Fig. 2.13. Relation between the intensity of photosynthesis and the CO_2 concentration in the intercellular system of the leaf. At CO_2 saturation of photosynthesis (Ci_{sat}) the potential photosynthetic capacity of the leaf is reached (Ph_{pot}). With sufficient light and open stomata, the actual photosynthesis (Ph_{act}) depends on the supply of CO_2. At a certain Ph_{act} a corresponding actual internal CO_2 concentration (Ci_{act}) is reached; this is termed the "operational point". The difference between the photosynthetic intensity Ph_{335} at equilibrium with the ambient CO_2 concentration (335 µl l^{-1}) and the actual Ph_{act} is due to stomatal diffusion resistance. *CP* is the CO_2 compensation concentration. The slope *Ph/C$_i$* indicates the photosynthetic utilization of CO_2. (After Lange et al. 1987)

ylation (C_3, C_4) and on the intensity of photorespiration; the CO_2 gradient is therefore steeper for C_4 plants than for C_3 plants. The level of the CO_2 compensation concentration for C_3 plants depends on the intensity of irradiation (higher values in weak light), on temperature (higher values with increasing temperature), on conditions with respect to nutrition, water and vigour, and on daily and seasonal adjustments in metabolism [10].

Increasing supplies of CO_2 improve assimilation up to the point of *CO_2 saturation of photosynthesis* (Fig. 2.13). The atmospheric CO_2 is not sufficient for CO_2 saturation to be attained, and hence CO_2 deficiency is the most common yield-limiting factor. The slope Ph/C_i (commonly A/C_i) of the CO_2-effect curve expresses the *photosynthetic utilization of CO_2* and is a measure of the activity of the carboxylation enzymes (corresponding to the efficiency of carboxylation). Therefore with an increasing supply of CO_2 the rate of photosynthesis rises more steeply in C_4 plants than in the C_3 type, whose CO_2 requirements are only satisfied at a CO_2 concentration exceeding 1000 µl l^{-1}. The rate of photosynthesis at the CO_2 saturation point is known as the *potential photosynthetic capacity* (Ph_{pot}).

When the CO_2 content of the air is artificially raised, C_3 plants can bind up to two to three times as much CO_2 as under natural conditions. This is sometimes exploited by growers to increase their yields. By raising the CO_2 in the air in greenhouses (CO_2 fertilization) the growth of tomatoes, cucumbers and leafy vegetables can be doubled.

2.2.1.3 Carbon Dioxide Pathway and Transfer Resistances in the Leaf

On its way to the chloroplasts CO_2 has to overcome a series of barriers. An overview of the transport pathways and the transfer resistances on and within the leaf is shown in Fig. 2.14.

Near the surface of a plant the exchange of gas molecules slows down. This *boundary layer resistance* r_a is particularly high if a stationary layer of air forms on the leaf surface, which is the more likely if the leaves are large, grooved or hairy. The surface resistance is lowered by air movements, and becomes negligible at wind velocities of $3-5 \ m \ s^{-1}$ (Fig. 2.15).

The primary means of entry in gas transport is via the stomatal apparatus. When the stomata are fully open the *stomatal diffusion resistance* (r_s, has a specific minimal value, depending on the size and density of the stomata (see Table 2.2); when the stomata are closed r_s approaches infinity. In terrestrial plants, *cuticular uptake* of CO_2 by the leaf is negligible. In the dark, however, respiratory CO_2 that has accumulated in the intercellular spaces diffuses through the cuticle and peridermal tissues due to the steeper concentration gradients.

The *minimal* stomatal diffusion resistance to CO_2 is particularly low in herbs of sunny habitats and in crop plants, but is relatively high in the leaves of shade plants, in conifer needles and leaves of sclerophyllous evergreen trees. In C_3 plants with fully open pores, the stomatal diffusion resistance is always lower than the sum of the CO_2 transfer resistances inside the leaf, whereas in C_4 plants the internal resistances are so small that the width of the stomatal opening is in any case a limiting factor for CO_2 transfer (Fig. 2.16).

The internal resistances to CO_2 transfer, also termed *residual resistance*, r_r, are composed of the resistance met on entering the liquid phase of the cell, the delay due to assimilation of CO_2 in the chloroplasts, i.e. the "*carboxylation resistance*", r_{cx}, and the limitation of photosynthetic performance as a result of inadequate provision of energy and redox equivalents by the primary reactions, known as the *excitation resistance*", r_{exc}.

During photosynthesis the highest concentration of CO_2, C_a, is at the outside of the layer of air covering the leaf. As a result of the succession of resistances to gas transfer the **CO_2 concentration in the leaf** (C_i) sinks below that of the outside air. The partial pressure of CO_2 in the intercellular air is therefore determined by the CO_2 consumed and the CO_2 entering to replace it. The stomata frequently respond to internal and external conditions in such a way that, for a given partial pressure of CO_2 in the outside air, the partial pressure of the CO_2 inside the leaf remains fairly constant, because r_s changes in proportion to Ph_n. As long as this "feed forward" tendency to optimization is maintained, the ratio C_i/C_a remains constant, usually at about 0.7 [30]. Theoretically, the partial pressure of CO_2 in the chloroplasts (p_c) ought to sink to zero; but in fact in C_3 plants, under natural conditions and even under high irradiance, the intercellular partial pressure of CO_2 (p_i) does not decrease in linear proportion to the chloroplast partial pressure; the average ratio p_c/p_i is 0.7, and p_c/p_a is roughly 0.5 [29].

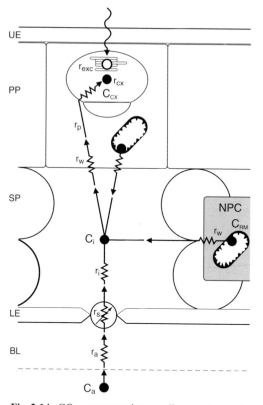

Fig. 2.14. CO_2 concentration gradient and transfer resistances in a leaf during photosynthesis. *UE* Upper epidermis; *PP* palisade parenchyma; *SP* spongy parenchyma; *LE* lower epidermis; *NPC* cells not photosynthetically active; *BL* boundary layer. During photosynthesis a gradient in the CO_2 concentration is established from the outside air (C_a) via the intercellular air (C_i) to the minimal concentration at the site of carboxylation (C_{cx}). In the intercellular system of the leaf, CO_2 arrives not only from outside but also from the cells, as a result of mitochondrial respiration (C_{RM}) and photorespiration. The transfer resistances interposed are the boundary-layer resistance r_a, the stomatal resistance r_s, diffusion resistances in the intercellular system r_i, and resistances associated with the processes of dissolving and transporting CO_2 in the liquid phases of the cell wall (r_w) and in the protoplasm (r_p). r_{cx} indicates the "carboxylation resistance", and r_{exc} the "excitation resistance". For further details see Šesták et al. (1971), Chartier and Bethenod (1977). For a transfer scheme for lichens see Cowan et al. (1992)

2.2.1.4 Stomatal Regulation of Gas Exchange

By varying the width of the stomatal pores the plant is able to control the entry of CO_2 into the leaf. The guard cells are continuously in motion, with the pores in a state of oscillation between narrowing and widening. At any point in time not all the guard cells of a leaf are open to the same extent (Fig. 2.17). *Non-uniform, patchwise stomatal closure* is especially pronounced in stress situations.

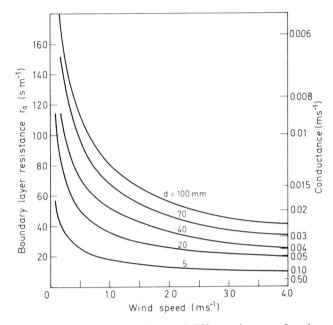

Fig. 2.15. Decrease in boundary-layer resistance over leaves of different sizes as a function of wind velocity. *d* Leaf dimension parallel to the direction of air movement. (Grace 1977). For the effect of flutter on the boundary layer, see Roden and Pearcy (1993)

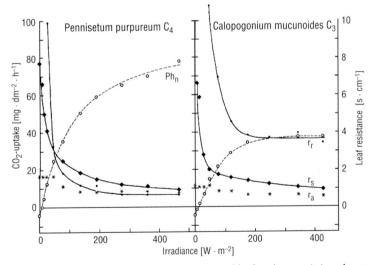

Fig. 2.16. Response of external transfer resistances (r_a, r_s), residual resistance (r_r) and net photosynthesis (Ph_n, *dashed curves*) of leaves of a C_3 plant (*Calopogonium mucunoides*) and a C_4 plant (*Pennisetum purpureum*) to increasing irradiance. In C_3 plants at light saturation r_r is considerably larger than r_s, whereas in C_4 plants it is somewhat smaller. (Ludlow and Wilson 1971)

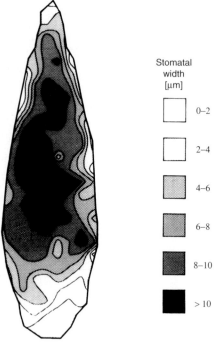

Stomatal
width
[μm]

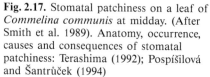

0–2

2–4

4–6

6–8

8–10

> 10

Fig. 2.17. Stomatal patchiness on a leaf of *Commelina communis* at midday. (After Smith et al. 1989). Anatomy, occurrence, causes and consequences of stomatal patchiness: Terashima (1992); Pospíšilová and Šantrůček (1994)

Opening of the Stomata

Number, distribution, size, shape and mobility of the stomata are species-specific characteristics, although they vary with adaptation to the habitat, and even from one individual to another (Table 2.2). In some dicotyledonous species there are also differences between male and female individuals: in *Acer negundo* for example, the stomata on the leaves of female trees are more dense, and are more sensitive to dryness of air and soil than those on male trees (sexual dimorphism).

The critical dimension determining stomatal resistance is the *pore width.* Stomatal diffusion resistance increases exponentially with decreasing pore width, according to a hyperbolic function. Its reciprocal $1/r_s$, is the *stomatal conductance,* g_s, which is directly proportional to pore width. The maximal width to which the pore of a stoma can open, which depends upon the shape and the properties of the walls of the guard cells, determines the upper limit of the rate at which gas can flow through. This is expressed numerically as the maximal stomatal conductance. The capacity for opening is greatest in the leaves of herbaceous dicotyledons, deciduous trees with open crowns and trees of tropical forests. It is particularly low in woody plants with thick, stiff leaves.

The proportion of the shoot surface on which stomatal diffusion is possible is called the *pore area.* The pore area is the product of the pore density (number of stomata per mm^2 of leaf surface) and the maximal pore width (ostiole

Table 2.2. Average ranges of stomatal density, pore width and pore area on the abaxial (lower) side of leaves, with the minimal stomatal diffusion resistance for CO_2. (Compiled from data of numerous authors). For differences in stomatal distribution on upper and lower epidermis see Napp-Zinn (1988), Bolhàr-Nordenkampf and Draxler (1993). Numbers in brackets indicate extreme values

Plants	Stomatal density (number per mm² leaf area)	Pore length (μm)	Maximal pore width (μm)	Pore area (% of leaf area)	Minimal stomatal diffusion resistance to CO_2 (s cm^{-1})	Maximal conductance for CO_2 (mmol m^{-2} s^{-1})
Herbaceous plants of sunny habitats	100 – 200 (300)	10 – 20	4 – 5	0.8 – 1	0.5 – 2.6	150 – 700
Herbaceous plants of shady habitats	(30) 50 – 100 (200)	15 – 30	5 – 6	0.8 – 1.2	2.2 – 6.5	60 – 200
Mountain plants	(100) 150 – 300 (500)	10 – 25	5 – 10	1	1 – 2	50 – 100
Grasses	(30) 50 – 100 (500)	20 – 30	3	0.5 – 0.7	0.9 – 5.2	80 – 460
Palms	150 – 180	15 – 24	2 – 5	0.3 – 1	2 – 6	70 – 100 (400)
Tropical forest trees	200 – 600 (900)	10 – 25	3 – 8	1.5 – 3	2 – 6	(70) 100 – (300)
Evergreen broad-leaved trees	200 – 600					100 – 200
Sclerophylls	100 – 500 (1000)	10 – 15	1 – 2	0.2 – 0.5	2.0 – 6.5	100 – 250
Deciduous trees	100 – 300 (600)	5 – 15	1 – 6	0.5 – 1.2	1.6 – 6.5	50 – 250 (300)
Evergreen conifers	40 – 120	15 – 20		0.3 – 1	2.0 – 6.5	200
Desert shrubs	150 – 300	10 – 15		0.3 – 0.5	1.0 – 7.8	50 – 260
Succulents	15 – 50 (100)	ca. 10	ca. 10	0.1 – 0.4	3.1 – 8	60 – 130

The specific stomatal diffusion resistance can be calculated from the width and length of the pores and the stomatal density (Brown and Escombe 1900). The formula of Parlange and Waggoner (1970) is:

$$r_s = \{[d/\pi \cdot ab] + [\ln (4a/b)/\pi \cdot a]\}/(D \cdot n), \text{ where}$$

a = semilength of pores, b = semiwidth of pores, d = depth of stomata, n = stomatal density, D = diffusivity.

area). In most plants the pore area amounts to $0.5-1.2\%$ of the leaf surface, although in trees of tropical forests it can be as much as 3%. The pore area is especially small in succulents and scleromorphic leaves; on the assimilatory organs of succulents the pore density is low, and the stomata of the leaves of sclerophyllous plants and evergreen dwarf shrubs are able to open only slightly.

The Physiological Mechanism of Stomatal Movement
Opening and closing of the stomata is effected by a *turgor difference* between the guard cells and the adjacent epidermal cells (subsidiary cells). If the turgor of the guard cells becomes greater than that of the epidermal cells the stomata open; when not turgid, they are closed.

Increase in turgor is an *osmoregulatory process* brought about by the active transport of ions. In particular, when K^+ is transported from the subsidiary cells into the guard cells the pores *open*. Charge imbalance is prevented by the movement of inorganic anions (e.g. Cl^-), by the formation of organic anions (e.g. malate) or by the release of protons (the pH value rises as the pores open). An important role in the *closure* of the stomata seems be played by the controlled entry of cytoplasmic Ca^{2+} via calcium channels and its release via calmodulin, acting together with abscisic acid (ABA). Transport of ions is dependent upon a supply of energy (ATP) and is influenced by endogenous signal substances and effectors (phytohormones, photocybernetic sensitizers). Thus the readiness of the stomata to open and close during the course of the day varies according to the states of activity, development, stress, and adaptation. Toxic excretions of plant parasites may either lock the stomata in an open position, as is the case with fusicoccin, a wilting toxin of *Fusicoccum amygdali* which functions as an ABA antagonist, or, like a phytotoxin of *Helminthosporum mayalis,* cause stomatal paralysis.

Control of Stomatal Movement
Although the stomata respond to many influences, their movements appear to be chiefly governed by two control circuits, one involving CO_2 and the other H_2O.

The **CO_2 control circuit** operates in response to the partial pressure of CO_2 in the intercellular space. Its influence is most apparent in the dark, when the epidermis is exposed to air of varying CO_2 *concentrations*. When the partial pressure of CO_2 in the air is higher than $150-250$ Pa the stomata close; when it drops below this level they open. In the *light* the partial pressure of CO_2 in the intercellular space falls as CO_2 is used up during photosynthesis. That the stomata open in light is for the most part an indirect effect of lack of CO_2 (but there is also a direct effect of light). The effect of CO_2 concentration on pore opening is particularly clear in *CAM plants*. Their stomata are open during the night, when the CO_2 partial pressure in the intercellular space falls due to intensive malate formation, and are closed in the light, when CO_2 released during malate breakdown accumulates in the intercellular space before it can be further metabolized (see Fig. 2.7).

Table 2.3. External factors influencing stomatal opening in C_3 and C_4 plants. (Milburn 1979)

External factor	Alteration in factor	Response (increase in stomatal aperture)
PhAR	Increase	Up to light saturation of photosynthesis
Temperature	Toward optimal temperature for photosynthesis	Between 15 and 35 °C
Air humidity	Increase	Above 90% RH
CO_2	Decrease in partial pressure	Below 30 Pa
Atmospheric pollutants	Decrease	Only at lowest concentrations
Fusicoccin	Increase	$0 - 10 \, \mu M$

The **H₂O control circuit** comes into play during water deficiency, when the abscisic acid concentration in the leaves rises, hindering the osmoregulation of the guard cells. As a result, the opening capacity of the pores slowly declines, so that when severe water deficiency threatens, the stomata do not respond to external factors and remain closed.

Factors Influencing Pore Width

The degree of stomatal opening adjusts continuously to changes in the environment (Table 2.3), but with the behaviour of the guard cells dependent on their endogenous state of reactivity.

The *interplay of external factors* usually results in an intermediate pore width. Only seldom, and briefly, are the stomata completely open, since it is uncommon for all those conditions that favor opening to coincide (Fig. 2.18). On the other hand, extreme situations that force prolonged complete pore closure are a common occurrence, especially in drought.

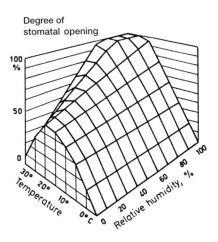

Fig. 2.18. The effect of temperature and air humidity on the degree of stomatal opening in leaves of *Ligustrum japonicum* at an irradiance of 540 W m⁻². (Wilson 1948)

Most important among the *internal factors* are the *phytohormones.* Abscisic acid, phaseic acid, cytokinins and gibberellins are all involved in the adjustment of the stomatal regulatory system to the processes of *growth and development.* Opening behaviour varies with the age and stage of development of a plant, such as during leaf formation, at the beginning of the reproductive phase and during senescence.

2.2.2 Specific Capacity of Net Photosynthesis

The maximal rate of CO_2 uptake under natural conditions of atmospheric CO_2 content and optimal conditions with respect to all other environmental factors is a constitutional characteristic of certain groups and species of plants. This capability for net photosynthesis is termed *photosynthetic capacity,* and can be expressed quantitatively by specific maximal values (Ph_{max}) under natural conditions of CO_2 supply.

Large **differences in photosynthetic capacity** exist within the plant kingdom. A survey of ecophysiologically different constitutional types is given in Table 2.4; maximal photosynthetic rates of selected species are shown in Table 2.5.

On an average, C_4 plants have the highest photosynthetic capacities, especially the subtropical and tropical grasses. Very high intensities of photosynthesis can also be achieved by certain C_3 plants if they contain unusually large amounts of RuBP carboxylase and have very low stomatal resistance values with open stomata. Agricultural crop plants also have very high capacities, largely due to successful breeding. Many C_3 plants, on the other hand, especially the woody plants, achieve only low photosynthetic yields. Nevertheless, certain species of woody plants, such as pioneer trees or shrubs in sunny locations, or fast-growing poplars, have photosynthetic capacities at least equal to that of the average herbaceous C_3 plant.

The photosynthetic yield of *shade plants* is only half to one-third that of plants growing in *sunny* locations. Species with a small assimilatory surface, like grasses with rolled leaves, dwarf ericaceous shrubs with *grooved leaves, conifer needles, shrubs with chlorophyllous stems,* and *succulents,* capture little of the incident light and therefore photosynthesize at only moderate rates. *Shoot axes* are able to fix CO_2 in the cortical parenchyma and even in the chloroplast-containing woody parenchyma, although the yield is very low and at most serves to compensate for some of the respiratory CO_2 losses of the twigs. However, in plants that are bare for longer periods of time, such as deciduous trees and shrubs of the temperate zone and leaf-shedding plants of arid regions, this form of photosynthesis may be significant for their carbon balance. The same holds true for *green fruits;* the photosynthetic yield of the chlorenchyma is small, but is sufficient to replace come of the respiratory losses.

At the lower end of the list are the *ferns, mosses* and *lichens,* with extremely low rates of photosynthesis. This is due, among other factors, to the thinness

Table 2.4. Average maximum values for net photosynthesis under conditions of natural CO_2 availability, saturating light intensity, optimal temperature and adequate water supply. (Summary drawn from data in original publications of many authors)

Plant group	CO_2 uptake	
	$\mu mol\ m^{-2}\ s^{-1}$ [a]	$mg\ g^{-1}\ DM\ h^{-1}$ [b]
Terrestrial plants		
Phanerograms		
Herbaceous plants		
C_4 plants	30 – 60 (70)	60 – 140
C_3 plants		
Winter annual desert plants	20 – 40 (60)	
Crop plants	20 – 40	30 – 60
Mesophytes of sunny habitats	20 – 30 (40)	30 – 60
Plants of dunes and seashore	20 – 30	
Spring geophytes	15 – 20	25 – 40
Mountain plants	15 – 30	25 – 60
Tall forbs	10 – 20	30 – 40
Shade plants	(2) 5 – 10	10 – 30
Plants of dry habitats	15 – 30	15 – 40
Arctic plants	8 – 20	
Grasses and sedges	5 – 15 (20)	8 – 35
Root hemiparasites	(1) 4 – 7	
Shoot hemiparasites	2 – 8	
CAM plants		
in the light	(2) 5 – 12	0.3 – 2
in the dark	6 – 10 (20)	1 – 1.5
Woody plants		
Tropical crop plants	10 – 15	
Tropical species of the first		
stage of succession	12 – 20 (25)	
Tropical lianas (sun leaves)	15 – 20	
Trees of tropical rain forests		
sun leaves	10 – 16	10 – 25
shade leaves	5 – 7	5 – 8
saplings (extreme shade)	1.5 – 3 (5)	
Broadleaved evergreens of the subtropics		
and warm-temperate regions		
sun leaves	6 – 12 (20)	
shade leaves	2 – 4	
Deciduous trees		
sun leaves	10 – 15 (25)	
shade leaves	3 – 6	
Coniferous trees		
deciduous	8 – 10	10 – 20
evergreen	3 – 6 (15)	3 – 18
Mangroves	4 – 8 (12)	
Sclerophylls of periodically dry regions	4 – 10 (16)	3 – 10 (18)
Palms	4 – 10 (20)	
Bamboos	4 – 6	

Table 2.4 (continued)

Plant group	CO$_2$ uptake	
	μmol m^{-2} s^{-1} [a]	mg g^{-1} DM h^{-1} [b]
Desert shrubs (small)	(3) 10 – 15 (30)	(4) 8 – 15 (35)
Dwarf shrubs of heath and tundra		
deciduous	6 – 15	15 – 30
evergreen	3 – 6 (10)	4 – 10
Cryptograms		
Ferns		
in open habitats	8 – 10	
in shade	2 – 5	
Mosses	2 – 3	0.6 – 3.5
Lichens	0.3 – 2 (5)	0.3 – 2.5 (4)
Aquatic plants		
Swamp plants, floating plants	12 – 25 (30)	
Freshwater macrophytes	(5) 7 – 10 (25)	
Seaweeds of the tidal zone	2 – 6	5 – 30 (50)
Planktonic algae	(2) 10 – 30	2 – 3
	mg O$_2$ mg^{-1} Chl h^{-1}	

[a] To allow comparison of photosynthetic capacity of different plant types, the photosynthetic rates are standardized per unit surface area. The surface area is that area receiving radiation, not the total area of upper and lower surface.
[b] The photosynthetic rate per unit dry weight of leaf; this number can be used to calculate the length of time required for a leaf to acquire the carbon necessary to form another leaf of a given mass. This is why the CO$_2$ uptake is also shown in units of weight.

of the assimilatory organs, and in the case of lichens to specific diffusion resistances and the large amount of heterotrophic mycobiont within the thallus.

Aquatic plants form a distinct group. Values for planktonic algae are difficult to compare with those of vascular plants, because free-living unicellular algae are seldom adequately supplied with minerals and thus give very variable and low values. In experiments carried out under optimal conditions, however, cultures of marine algae achieved maximal rates of photosynthesis amounting to $350 - 1000\ \mu$mol CO$_2$ mg^{-1} chlorophyll per hour [115]. Despite the fact that certain aquatic macrophytes attain rates of photosynthesis comparable with those of terrestrial herbaceous shade plants, submersed aquatic plants are, on the whole, characterized by a low capacity to take up CO$_2$. One reason for this is the poor supply of inorganic carbon. Although water may contain roughly 160 times as much CO$_2$ as the air, it takes much longer to reach the leaves of aquatic plants than atmospheric CO$_2$ requires to reach the leaves of terrestrial plants. In addition, the transport mechanisms for HCO$_3^-$ are less well developed in freshwater macrophytes than in marine algae.

Genetically determined differences in photosynthetic capacity are in some cases quite large (Table 2.6). They provide the basis for selective breeding of more productive plants in agriculture, horticulture and forestry. The causes of the specific differences in net photosynthetic capacity lie in the anatomical pe-

Table 2.5. Maximum values for net photosynthesis of plant species with high photosynthetic capacities, under conditions of normal atmospheric CO_2 content. (From data of Björkman et al. 1972; Seeley and Kammereck 1977; Patterson and Duke 1979; Osmond et al. 1982; Nelson 1984; Marek 1988; Ceulemans and Saugier 1991; Nobel 1991 b; Dufrêne and Saugier 1993). All values are in $\mu mol\ CO_2\ m^{-2}\ s^{-1}$

C_4 plants	
Cenchrus ciliaris	68
Pennisetum typhoides	64
Saccharum-hybrids	64
Sorghum sudanese	57
Zea mays	55
Tidestromia oblongifolia	50
C_3 plants (herbaceous species)	
Camissonia claviformis	60
Triticum boeoticum	45
Typha latifolia	43
Oxyza sativa	40
Helianthus anuus	28
Glycine max	27
Eichhornia crassipes	20
CAM plants	
Agave mapisaga	34
Agave fourcroydes	23
Opuntia ficus-indica	20
Woody plants (C_3)	
Salix-species	20 – 35
Populus tristis	30
Populus-species	20 – 25
Hevea brasiliensis	20 – 26
Elais guineensis	20 – 25
Fraxinus pennsylvanica	20 – 25
Ailanthus altissima	20
Prosopis glandulosa	20
Eucalyptus pauciflora	15 – 20
Pinus sylvestris	17

culiarities of the leaf structure, the ease with which air passes through the intercellular system, the shape and distribution of the stomata, and the quantity and activity of the enzymes involved in carboxylation.

2.2.3 Specific Activity of Mitochondrial Respiration

The respiratory activity of a plant depends upon its constitutional type and genotypic characteristics; it differes from organ to organ, and with the availability of substrate, the phase of development, the state of activity, and also the environmental conditions. The *genetic* differences are due to the existence of isoenzymes of the key enzymes involved in catabolic and oxidative processes. A high *sugar level*, such as may occur in connection with high rates of

Table 2.6. Maximal net photosynthesis (Ph_{max} in μmol CO_2 m^{-2} s^{-1}) and specific parameters of gas exchange at light saturation and optimal age of leaves of different *Populus* clones. Stomatal diffusion resistance r_s and internal residual resistance r_r are given in s cm^{-1}. In poplars, the variety-specific differences in net photosynthesis (Beaupré: Nigra Ghoy is as 3:1) are chiefly due to differences in CO_2 transfer inside the leaf (r_r 4:1). (Ceulemans et al. 1980)

Populus clone	Ph_{max}	r_s	r_r
Beaupré	12.0	3.12	8.03
Unal	11.5	2.78	8.45
Trichobel	11.1	3.25	7.08
Ghoy	10.6	1.65	14.08
Italica	9.8	1.77	9.74
Robusta	9.5	2.09	11.36
Columbia River	7.0	0.73	21.28
Nigra Ghoy	3.8	1.50	33.56

CO_2 assimilation, or following breakdown of starch at low temperatures, causes an increase in respiration, whereas during the night and at low carbohydrate levels the respiratory rate gradually becomes lower. Respiration is elevated in reponse to *increased need* of metabolic energy for growth, uptake of ions, or during stress.

Between different plant species differences in respiratory activity in the order of magnitude of 1:10 to 1:20 have been found. An overview of morphological, ecophysiological and distributional types (Table 2.7) shows that under comparable conditions the respiration rate of herbaceous plants, especially rapidly growing species, is twice that of the *foliage* of deciduous trees. The latter, in turn, respires on an average five times as actively as evergreen assimilatory organs. Within one and the same group, heliophytes respire (at 20 °C) more intensely than shade plants, and plants of the Arctic and high mountains at a higher rate than those of warmer climates. Respiration of mosses is carried on at a low rate (about $1-2$ mg g^{-1} h^{-1} [226]), that of lichens varies over a wide range, depending on morphological structure and geographical distribution ($0.5-3$ mg g^{-1} h^{-1} [203]); respiratory intensities in seaweeds range from 0.5 to 5 mg g^{-1} h^{-1} [28, 194].

The *flowers* and *fruits* of an individual plant respire more strongly than its leaves. The respiratory activity of *branches* and *lignified roots* is similar to that of **trunks**, but twigs and thin branches in the crown of a tree respire more intensely (Fig. 2.19). Trunk respiration differs greatly from one species to another (from 2 to 5 mg CO_2 dm^{-2} h^{-1} in summer, for diameters of $10-20$ cm) [209].

In woody shoots and roots it is primarily the cortex, cambium and the outermost cell layers of the wood that respire actively. Expressed with respect to fresh or dry weight, at a given temperature the respiratory activity (CO_2 release per unit mass) of twigs, branches and tree trunks of different thicknesses decreases with increasing diameter, because the ratio of cortex mass to wood mass becomes smaller. Expressed with respect to peridermal surface (CO_2 re-

Table 2.7. Maintenance respiration of mature leaves of vascular plants in summer at $20\,°C$ at the beginning of the night. (Compiled from measurements made by numerous authors). As a rule, respiratory intensity declines in the course of the night

Plant group	CO$_2$ release	
	$\mu mol\ m^{-2}\ s^{-1}$	$mg\ g^{-1}\ DW\ h^{-1}$
Herbaceous crop plants[a]	2 – 6	3 – 8
Herbaceous wild flowers		
sun species	3 – 5	5 – 8
shade species	1 – 3	2 – 5
Hemiparasites	3 – 5 (8)	
Holoparasites	3 – 6	
Tropical forest trees		
sun leaves	0.3 – 0.5 (2)	
shade leaves	0.05 – 0.2	
Deciduous trees		
sun leaves	1 – 2	3 – 4
shade leaves	0.2 – 0.5	1 – 2
Evergreen broadleaved trees of the temperate zone		
sun leaves	0.8 – 1.4	ca. 0.7
shade leaves	0.2 – 0.5	ca. 0.3
Evergreen coniferous trees of the boreal zone		
light-adapted needles	0.5 – 0.7	ca. 1
shade needles		ca. 0.2
Northern and alpine ericaceous dwarf shrubs		
deciduous	0.6 – 1.5	2 – 3
evergreen	0.3 – 1	0.5 – 1.5
Desert shrubs and semishrubs		0.8 – 1.5 (3)
CAM plants		(0.2)
		0.5 – 1 (2.5)
Aquatic vascular plants		$mg\ O_2\ g^{-1}\ h^{-1}$:
		(0.5) 0.8 – 1.2 (1.5)

[a] There is no significant difference between the activities of mitochondrial respiration in C_3 and in C_4 plants. (Byrd et al. 1992).

lease per unit surface area), however, the respiratory activity remains constant over a wide range of size classes. Therefore it is advisable when studying plant organs with a high content of dead supporting material, also to express respiratory activity with respect to the content of protein or protein nitrogen.

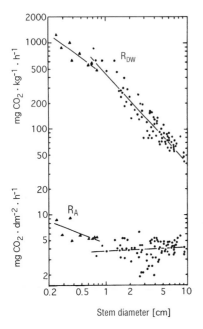

Fig. 2.19. Respiration of *Pinus densiflora* branches and stems of various diameters, referred to dry weight (R_{DW}) and to shoot surface area (R_A). *Triangles* Branches of the same year, in autumn; *dots* 1–11 year-old stems. (Negisi 1974). For *Abies amabilis:* Sprugel (1990)

2.2.4 The Influence of Stage of Development and State of Activity on Respiration and Photosynthesis

Respiratory activity and photosynthetic capacity, although characteristic of a plant species, are not constant values. Not only in the course of individual development, but also in connection with seasonal and even diurnal fluctuations in the plant's activity, gas exchange behaviour can change appreciably.

Respiration and Age
Younger plants respire more strongly than older plants, and growing parts of the plant respire at a particularly high intensity. According to need, the cells are able to adjust the production of respiratory ATP via a feedback mechanism. In seedlings, at the tips of roots, during unfolding of the leaves and in developing fruits, *constructional respiration* amounts to three to ten times the *maintenance respiration*. With differentiation and maturation of the tissues, respiratory activity usually drops to the level of maintenance respiration (Figs. 2.20 and 2.21). At the onset of the breakdown processes connected with senescence there may be a transient *climacteric rise in respiration* in the leaves and fruits of certain plant species. This indicates an alteration in metabolism, resulting in the discoloration of the foliage and the release of gaseous metabolic products (for example ethylene) by fruits.

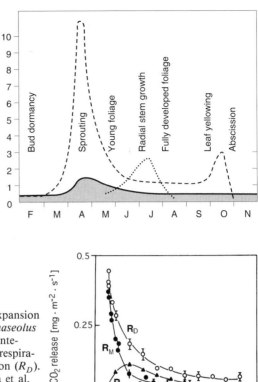

Fig. 2.20. Mitochondrial respiration of leaves of deciduous (*broken line*) and evergreen trees (*solid line*), and stem respiration (*dotted curve*) at different phenological stages. (After Eberhardt 1955; Pisek and Winkler 1958; Negisi 1966; Malkina and Tselniker 1990; Paembonan et al. 1992)

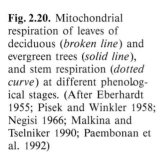

Fig. 2.21. Dark respiration during expansion to maturity of a primary leaf of *Phaseolus vulgaris*, and the proportion of maintenance respiration (R_M) and growth respiration (R_G) in the total dark respiration (R_D). (From Kaše and Čatský, from Tichá et al. 1985)

Photosynthesis and Development

Photosynthetic capacity also changes in the course of development. In the *sprouting phase,* photosynthetic activity is initially so small that it cannot keep pace with the intensive growth respiration. This is because leaves that are in the process of unfolding and have not yet attained their full size take up too little light, their chloroplasts are not fully active and carboxylation has not reached its peak capacity (Fig. 2.22). Fully developed foliage is at the height of its photosynthetic performance. In some herbaceous species the photosynthetic capacity begins to decline after only a few days or weeks and becomes increasingly smaller with advancing age. Shortly before the shoot dies off, chlorophyll breakdown and degeneration of the chloroplasts lead to complete cessation of photosynthesis.

In fast-growing herbaceous plants, individual leaves on the shoot may exhibit considerable differences in photosynthetic capacity, depending on their state of differentiation. The continuous formation and unfolding of new leaves ensures a *constant carbon gain,* and the plant as a whole is able to maintain a fairly stable level of CO_2 fixation throughout the growing period (Fig. 2.23).

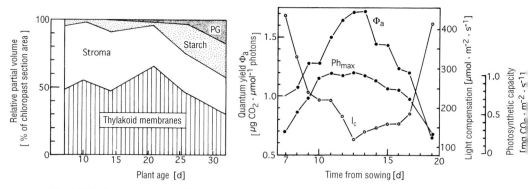

Fig. 2.22. Alterations in the chloroplast structure and in photosynthetic capacity during development and ageing of the primary leaf of *Phaseolus vulgaris*. *Left* Proportions of structural components in the chloroplasts. *PG* Plastoglobules. (From Kutík et al. 1988). *Right* Photosynthetic capacity (Ph_{max}), quantum yield (Φ_a, slope of the light effect curve, referred to absorbed radiation) and light compensation point (I_C). (After Čatský and Tichá 1980). In young leaves of certain tropical rainforest plants the RuBP carboxylase accumulates very slowly. (Kursar and Coley 1992)

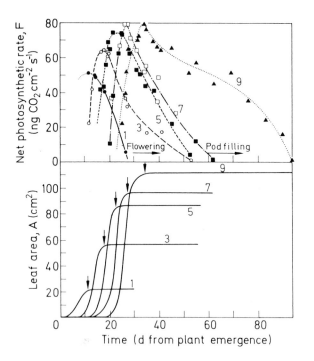

Fig. 2.23. *Above* Net photosynthesis and growth in area of the first, third, fifth, seventh and ninth leaves of soybean plants, in relation to age of the plant. The continued production of new leaves ensures that the plant as a whole can maintain a high photosynthetic capacity over a longer period of time, despite a decrease in the capacity of individual leaves as they reach maturity. *Below* Sequence and sizes of the leaves. (Woodward 1976)

In deciduous trees of temperate zones, photosystems I and II are fully developed within 20 days after unfolding of the leaves, so that photosynthesis attains its peak of activity in early summer. The pattern of photosynthetic activity differs greatly in different tree species, depending on the individual rhythm of leaf ageing and renewal (Fig. 2.24). In pioneer species of the early successional stage, such as birch, alder and poplar, the crowns grow continuously, and their leaves have an extremely high photosynthetic capacity but are short-lived. Forest trees of the climax stage, such as maple, hornbeam and oak, put out new shoots in bursts, the photosynthetic capacity of the leaves is lower, but remains almost unchanged until the leaves turn colour. In most evergreen woody plants, the decline in photosynthetic capacity is, on the whole, determined by the functional longevity of the leaves, and their capacity decreases from year to year.

During the *flowering and fruiting phases* an increase in photosynthetic capacity can be observed in fruit trees, crop plants, and also in wild plants.

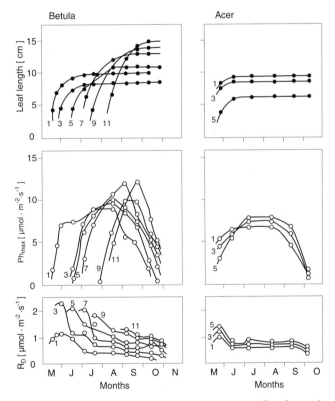

Fig. 2.24. Seasonal pattern of the photosynthetic capacity of a pioneer tree (*Betula maxi-mowicziana*) and a tree species of the late successional stage (*Acer mono*) of the deciduous forest zone in Japan. *Above* Development of leaves during the assimilation period (*numbers* refer to position of leaf); *middle* photosynthetic capacity of the leaves at light saturation and optimal temperature; *below* dark respiration of the leaves at 20 °C. (After Koike 1990)

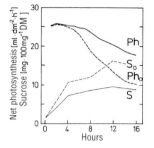

Fig. 2.25. The influence of the presence of fruit on the photosynthetic capacity of the leaves of *Solanum melongena*. *Ph* Photosynthesis of leaves of plants with fruit, and Ph_0 of plants without fruit; *S* sucrose content of leaves of plants with fruit, and S_0 of those without. Under continuous saturating irradiance, photosynthesis declines sooner in plants bearing no fruit due to accumulation of photosynthates in the leaves. (After Claussen and Biller 1977)

If the fruits are removed, the CO_2 uptake drops (Fig. 2.25), largely because carbohydrate no longer is being conveyed to the fruit. Phytohormones play a role in regulations of this nature, which are geared to the assimilatory processes of the whole plant.

There is also a *seasonal* alteration in photosynthetic capacity in connection with adaptation to *environmental conditions*. For example, herbs in the understorey of deciduous forests are acclimated to the fluctuating light climate beneath the tree canopy (Fig. 2.26); perennial plants in regions with alternating rainy and dry seasons produce seasonal leaves with different types of photosynthetic behaviour ("seasonal dimorphism"). At the beginning of winter the photosynthetic capacity of the leaves and needles of evergreen trees of the temperate zone decreases, even in the absence of frost (Fig. 2.27).

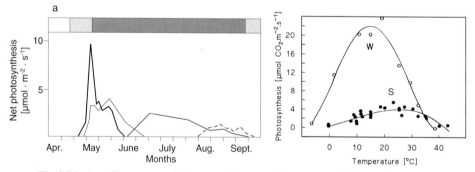

Fig. 2.26a, b. Adjustments of photosynthetic capacity to seasonal changes in environmental conditions. **a** Differences in the photosynthetic capacity of the foliage of *Pulmonaria obscura* in the undergrowth of an oak forest,. The highest rates are seen in spring leaves (*thick line*), before the trees have come into leaf. As soon as the trees are in full leaf and the undergrowth receives too little light, the spring leaves of *Pulmonaria* soon become senescent and are replaced by two flushes of shade-adapted summer leaves (*thin lines*) and autumn leaves (*broken line*).*Pale stippled blocks* Development and senescence of the tree foliage. *Dark stippled* Shade period, during full foliation of tree crowns. (After Goryshina 1969). **b** Photosynthetic capacity of seasonally dimorphous leaves of the mediterranean subshrub *Phlomis fruticosa*, and a shift in the temperature optimum. *W* Winter leaves (cool rainy season) *S* summer leaves (dry season). (After Kyparissis and Manetas 1993)

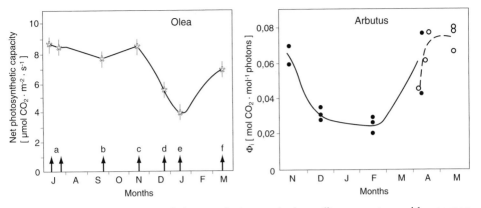

Fig. 2.27. Seasonal course of photosynthetic capacity in mediterranean trees with evergreen leaves. *Left* Photosynthetic capacity of *Olea europaea* from northern Italy, measured under standardized conditions (600 µmol photons m^{-2} s^{-1}, optimal temperature, water saturation, 30 Pa CO_2). *a* Leaves of the current year in early summer; *b* after a short dry period in late summer; *c* during the autumn rainy period; *d* in winter, before the first frost; *e* during a period with night frosts down to $-5\,°C$ (with high level of soluble carbohydrates in mesophyll); *f* leaves of the previous year, in spring. Data from Larcher (1961). *Right* Quantum yields of leaves of *Arbutus unedo* in Portugal. Φ_i Slope of the linear portion of the light effect curve at CO_2 saturation, referred to the incoming PhAR; ● mature leaves; ○ new leaves. (After Beyschlag et al. 1990)

2.2.5 The Effect of External Factors on CO_2 Exchange

CO_2 exchange is influenced by a number of external factors. As a photochemical process, photosynthesis is primarily dependent on the availability of radiation. The biochemical processes in photosynthesis are affected by the availability of CO_2, by temperature, and by the supplies of water and minerals. The entry of CO_2 via the stomata is mainly limited by the consequences of a lowered water potential. Of the environmental factors influencing respiratory gas exchange, temperature is the most important. Toxic pollutants from the environment are detrimental to all processes involved in CO_2 exchange.

Response Patterns of CO_2 Exchange
Under the influence of environmental climatic factors (light, temperature and air humidity) and soil factors (water and mineral supplies), the pattern of CO_2 uptake can take one of two forms [20].

Saturation curves are characteristic for the response to environmental factors that, in increasing supply, promote photosynthesis and above a certain threshold bring about no further rise, but also cause no immediate disturbance over a wide range of high dosage. Excellent examples of this type of behaviour are seen in the saturation curves with elevated supplies of CO_2 (see Fig. 2.13) and with increasing irradiance (see Fig. 2.28). Saturation curves also make it

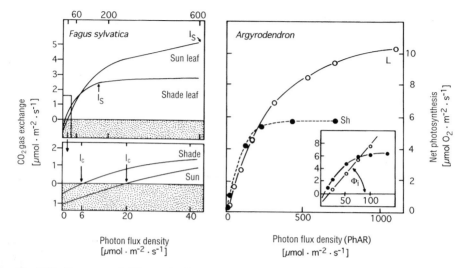

Fig. 2.28. Response of CO_2 gas exchange of sun and shade leaves to light intensity. *Left* Leaves of *Fagus sylvatica* from the periphery of the crown of the tree (*sun leaves*) and from the shaded interior of the crown (*shade leaves*). The effect of weak light is shown in the *lower left diagram*, with the abscissa expanded. Shade-adapted leaves respire at a lower rate than those adapted to light; they reach the light compensation point (I_C) at a lower intensity, and also after this point they utilize weak light better, but they are light-saturated at a lower intensity (I_S). Measurements were made at 30 °C and 30 Pa CO_2. (After Retter 1965). *Right* Leaves from the periphery of the crown (*L*) and the centre of the crown (*Sh*) of *Argyrodendron*, a shade-tolerant tree of tropical rainforests. Measurements were made under conditions of CO_2 saturation. The slope Φ_i of the linear portion of the light effect curve corresponds to the light quantum yield. (After Pearcy, from Anderson and Osmond 1987). For light response of photorespiration, see Chmora (1993)

possible to detect when the changing external factor *alone* is no longer responsible for limiting the speed of the assimilatory processes.

The position of the saturation limit is specific and can be employed to characterize the plant's behaviour. As a rule, the later the onset of saturation, the greater is the photosynthetic gain. At injurious overdosages the saturation curve may reach a *maximum*, then steadily decline. This can be seen at high values of irradiance (see Fig. 2.30) and in response to toxic concentrations of certain mineral substances.

Optimum curves are invariably an expression of sensitivity to under- and overdosage of external factors. The extent of the optimum range is a measure of the ecophysiological flexibility of a species. Photosynthesis responds to temperature changes with a very characteristic optimum curve (see Fig. 2.37). In lichens and mosses, optimum curves for CO_2 uptake occur in connection with water saturation of the thalli (see Fig. 2.45).

The Limitation of Photosynthesis by Environmental Factors
Only rarely do all environmental factors favour intensive uptake of CO_2; more usually, the intensity of one or another factor is unsuitable. If unfavourable conditions are only short-lived the reduction in carbon gain is only small. The consequences are more serious if the external factors become stressful for the plant. *Temporary limitation* of photosynthesis, for example due to short-lived lack of water, cold or hot weather leads to a decrease in carbon gains in proportion to the duration of the unfavourable conditions. If these are too severe, or last too long, it takes longer for the plant to reactivate photosynthesis by repair mechanisms after conditions have improved, if it recovers at all. This situation is referred to as the *after-effect of stress.*

Adaptation of Photosynthesis
The genetically fixed behaviour pattern of a plant includes not only its immediate reaction to changeable external factors, but also the extent of its ability to adapt to the usual environmental conditions and even to stress. The greater the ability to adapt, the better is the photosynthetic yield and the longer it can be maintained under changed conditions. Adaptations result in *optimization* and *harmonization,* both of which are achieved via adjustments that do not result in the highest capacity, but lead, rather, to a compromise between gains and risks. An example of this is the sensitive response of the guard cells, which are able to adjust the CO_2/H_2O exchange in such a way that the plant "neither starves nor thirsts".

2.2.5.1 The Light Response of Photosynthesis

The Light-Effect Curve
If the uptake of radiation by the leaves is very small, the amount of CO_2 set free by respiration may exceed the CO_2 fixed by photosynthesis. This results in an apparent release of CO_2. When photosynthesis fixes exactly as much CO_2 as is set free by respiration, and no gas exchange is detectable, it is said that light compensation is attained.

The *light compensation point* I_C is the irradiance at which photosynthetic CO_2 uptake and respiratory CO_2 release are in equilibrium (Fig. 2.28). Plants that respire intensely require more light for compensation than those with weaker respiration. When I_C has been exceeded, there is at first a linear increase in CO_2 uptake, indicating a strict *proportionality* between the available radiation and the photosynthetic yield. The speed of the light reactions is the limiting factor for the overall process in this range; the greater the *quantum yield* Φ_a the steeper the rise of the response curve in the range of proportionality ("*photosynthetic efficiency*" Φ_i, expressed in mol CO_2 per mol *incident* photons). A steep rise in the light-effect curve is therefore an expression of good utilization of light quanta. At very high irradiance the yield of photosynthesis increases only slightly or not at all. At this level, I_S, the photosynthetic process is *light-saturated.* The rate of CO_2 uptake is no longer limited by photochemical but rather by enzymatic processes, and by the supply of CO_2. The positions of I_C and I_S reflect the light conditions in the natural habitat of the plant and are characteristic for the different plant types (Table 2.8).

A comparison of the light-effect curves of different plant types (Fig. 2.29) clearly shows that the *C_4 plants* are outstanding: even at the highest irradiance millet and maize do not reach light saturation, and at only moderate intensities they assimilate better than C_3 plants. The opposite kind of response is seen in plants in which the photosynthetic apparatus is highly sensitive to light, and whose light-response curve shows a irradiance optimum

Table 2.8. Light response of net photosynthesis of single leaves, under conditions of ambient CO_2 and optimal temperature. (From measurements of numerous authors)

Plant group	Compensation irradiance I_C (μmol photons m^{-2} s^{-1})	Light saturation I_S (μmol photons m^{-2} s^{-1})
Terrestrial plants		
Herbaceous flowering plants		
C_4 plants	20 – 50	> 1500
Desert plants		> 1500
Agricultural C_3 plants	20 – 40	1000 – 1500
Heliophytes	20 – 40	1000 – 1500
Spring geophytes	10 – 20	300 – 1000
Sciophytes	5 – 10	100 – 200 (400)
Woody plants		
Tropical forest trees		
sun leaves	15 – 25	(400) 600 – 1500
shade leaves	5 – 10	200 – 300
young plants	2 – 5	50 – 150
Deciduous broadleaved trees and shrubs		
sun leaves	20 – 50 (100)	600 – > 1000
shade leaves	10 – 15	200 – 500
Non-tropical evergreen broadleaved trees		
sun leaves	10 – 30	600 – 1000
shade leaves	2 – 10	100 – 300
Coniferous trees		
sun shoots	30 – 40	800 – 1100
shade shoots	2 – 10	150 – 200
Ferns		
Sunny habitats	ca. 50	400 – 600 (800)
Shade ferns	1 – 5	50 – 150
Mosses	5 – 20	150 – 300
Lichens	50 – 150	300 – 600
Aquatic plants		
Planktonic algae		200 – 500
Intertidal-zone seaweeds	5 – 8	200 – 500
Deep-water algae	2	150 – 400
Submersed vascular plants	8 – 20 (30)	(60) 100 – 200 (400)

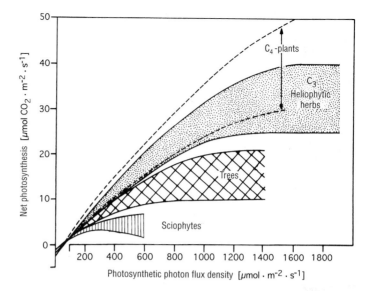

Fig. 2.29. Light-effect curves of net photosynthesis for plants of different functional groups at optimal temperature and with the natural CO_2 supply. Measurements made by numerous authors

(Fig. 2.30). Plants of the understorey of dense forests, aquatic plants and a variety of cryptogams belong to this group.

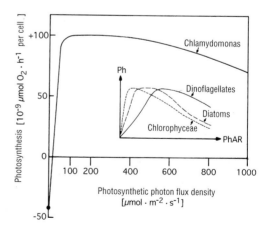

Fig. 2.30. Typical optimum curves showing the effect of light on photosynthesis of planktonic algae. In green algae (*Chlamydomonas reinhardii*) the inhibition of photosynthesis begins early, whereas the optimum range for diatoms is broader, and dinoflagellates require higher intensities of light to achieve optimal photosynthesis. (After Neale 1987 and Ott 1988)

Acclimation to Sunny and Shady Habitats

Leaves adapted to shade respire less than sun leaves and therefore compensate at considerably lower light intensities. In general, the I_K of herbs in shady habitats and tree seedlings in dense forests is about 5 mol photons $m^{-2} s^{-1}$. In addition, shade leaves utilize weak light better than sun leaves, and very soon reach their light saturation point. Heliophytes, on the other hand, are better able to utilize strong light due to the greater capacity of their electron transport system and higher carboxylase activity, and thus achieve considerably larger photosynthetic yields (Table 2.9 and 2.10).

Phenotypical *adaptation to the light climate* of the habitat proceeds mainly during the formation and subsequent differentiation of the leaves, in the course of which the morphological (see Fig. 1.31), histological (Fig. 2.31), ul-

Table 2.9. Differences between leaves adapted to strong light and to low light. (Compiled from data of numerous authors)

Characteristic	Sun leaves	Shade leaves
Leaf characteristics		
Leaf area (surface area/dry weight)	−	+
Internal surface area (A^{mes}/A)[a]	+	−
Thickness of mesophyll	+	−
Thickness of palisade parenchyma	+	−
Stomatal density	+	−
Chloroplasts per leaf area	+	−
Chloroplast characteristics		
Size of chloroplasts	−	+
Stromal/thylakoid volume	+	−
Thylakoids per granum	−	+
Chlorophyll content per chloroplast	−	+
Chlorophyll a/b ratio	+	−
P700/chlorophyll	×	×
P680/chlorophyll	+	−
CF_1/chlorophyll	+	−
α-Carotene/β-carotene	−	+
Lutein/V + A + Z xanthophylls[b]	−	+
Electron transport chain	+	−
Functional characteristics		
Activity of the photosystems	+	−
Speed of electron transport	+	−
ATP-synthetase activity/chlorophyll	+	−
Activity of RuBP carboxylase	+	−
Quantum yield	×	×
Carboxylation efficiency	+	−
Photosynthetic capacity	+	−
Mitochondrial respiration	+	−

[a] Mesophyll cell surface area/leaf area.
[b] Violaxanthin + antheraxanthin + zeaxanthin.
+ Higher, greater; − lower, less; × no clear difference.

Table 2.10. Characteristics of sun and shade leaves of beech and ivy, with values for various features that reflect the extent of adaptation. (Lichtenthaler et al. 1981; Hoflacher and Bauer 1982)

Characteristic	Fagus sylvatica			Hedera helix		
	Sun	Shade	Sun/shade[a]	Sun	Shade	Sun/shade
Leaf surface (cm^2)	28.8	48.9	*0.6*			
Leaf thickness (μm)	185	93	*2*	409	221	*1.85*
Specific leaf area (dm^2 g^{-1} DM)				0.97	2.6	*0.37*
Stomatal density (N mm^{-2})	214	144	*1.5*			
Stomatal conductance (cm s^{-1})				0.65	0.33	*2*
Number of chloroplasts						
per leaf surface (10^9 N dm^{-2})				5.09	2.45	*2.4*
per leaf volume (10^9 N cm^{-3})				1.24	1.11	*1.1*
Chlorophyll concentration (a + b)						
chl/leaf (mg/leaf)	1.6	1.9	*1.2*			
chl/surface area (mg dm^{-2})				8.7	5.5	*1.6*
Chlorophyll a/b	3.9	3.9	*1.3*	3.3	2.8	*1.2*
RuBP carboxylase activity (μmol CO$_2$ dm^{-2} h^{-1})				398	202	*2*
Net photosynthetic capacity (mg CO$_2$ dm^{-2} h^{-1})	3.5	1.3	*2.7*	22.3	9.4	*2.4*
Light compensation point (W m^2)	2.5	1	*2.5*			
Light saturation (W m^{-2})	85	44	*1.9*			
(μmol m^{-2} s^{-1})				600	250	*2.4*
Dark respiration (mg dm^{-2} h^{-1})	0.5	0.16	*3.1*			

[a] Italicized values are ratios.

trastructural and biochemical traits (see Table 2.9) underlying the characteristic features of CO$_2$ exchange in strong and weak light forms are developed. The genotype is responsible for the limits applying to these adaptive processes: for example, that the full expression of the morphological and functional characters typical of heliophytes is possible only under conditions of high irradiance, and that only sciophytes are able to develop extreme shade leaves in low light habitats. However, sciophytes can also partially adapt to a gradually increasing light intensity, so that stronger irradiance causes less damage.

Light Response of Photosynthesis in the Natural Habitat

If gas exchange is not limited by other environmental factors such as water supply and temperature, under natural conditions the response of net photosynthesis to light availability is a near-linear one, up to the saturation range. In the case of C$_4$ plants, this means that they can make full use of the light at noon on clear days (Figs. 2.32 and 2.33). In C$_3$ plants, photosynthetic performance increases up to the time of day when irradiation exceeds I_S. Light passing clouds have little effect on the rate of photosynthesis of heliophytes, but a greater effect is produced by the more pronounced fluctuations in illumina-

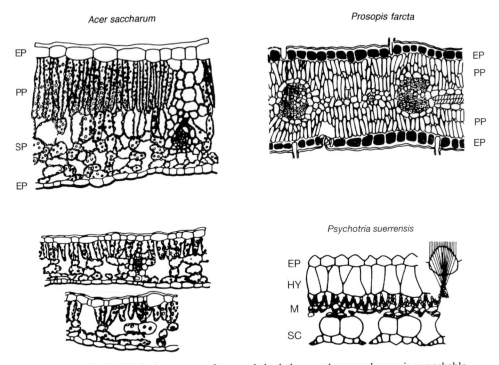

Fig. 2.31. The histological structure of sun and shade leaves. *Acer saccharum* is remarkable for its high degree of phenotypic plasticity. The transverse sections are of leaves taken from the sun-side of the periphery of the crown of a solitary tree (*top*), from the interior of the crown (*middle*) and from the lower branches beneath the dense crown canopy (*below*). The bifacial sun leaves have an upper epidermis (EP) with a thick outer wall, often a multilayered palisade parenchyma (PP) with long, closely packed cells and a thick spongy parenchyma (SP). Leaves from the extreme shade are thin-layered, with short palisade cells and large intercellular spaces. (After Hanson, from Salisbury and Ross 1992). The specific range of adaptation can be expressed quantitatively as the "plasticity index" (sun-leaf thickness minus shade-leaf thickness/sun-leaf thickness; Carpenter and Smith 1981). The equifacial leaf of *Prosopis farcta* is adapted to strong light and illumination from both sides. (After Weiglin and Winter 1991). Leaves of the genotypic shade plant *Psychotria suerrensis* in the undergrowth of dense tropical rainforests develop only a thin mesophyll (*M*), between expanded hypodermal layers (*Hy*). Beneath the stoma is a large substomatal cavity (*SC*). A striking feature is presented by the papillose epidermal cells, which guide light to the interior (enlarged *in the inset*, showing a lens effect). (After Lee 1986)

tion caused by variable cloud cover. The sciophytes on the woodland floor and the shade leaves in the interior of tree crowns are affected by the variations in the sunlight penetrating the foliage. Intermittent sunflecks lasting for some minutes trigger photosynthetic induction processes if their intensity reaches ten times that of the average incident radiation. For minutes after the passing of a beam of sunlight across a leaf, the secondary reactions of photosynthesis

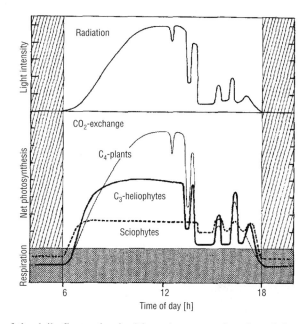

Fig. 2.32. Schematic diagram of the daily fluctuation in CO_2 exchange as a function of the available radiation. C_4 plants can utilize even the most intense illumination for photosynthesis, and their CO_2 uptake follows closely the changes in radiation intensity. In C_3 plants photosynthesis becomes light saturated sooner, so that strong light is not completely utilized. Sciophytes, adapted to utilization of dim light, take up more CO_2 than heliophytes in the early morning and late evening, as well as during periods when they receive little light due to cloud cover or to shade from the tree canopy, but they cannot utilize bright light as efficiently as the heliophytes. (Original)

are engaged in processing the accumulation of $NADPH_2$ and ATP produced by the primary reactions during the sunfleck (post-illumination fixation of CO_2 [185]). In this way even very brief phases of strong light can be exploited by the leaves. Depending on the frequency and sequence of fluctuating light, plants of the under storey in dense forests can achieve higher photosynthetic yields than would be expected with the same quantity of light under constant conditions.

Light Response of Photosynthesis Within a Stand of Plants
Observations on single leaves could lead to the assumption that on sunny days light is available in excess. This is not so, however, either for the whole plant or for stands of plants. During the course of the day, light reaches the individual leaves of a plant at many different angles of incidence; thus the leaves are only rarely exposed to full light. Thus, the highest light intensities occurring in nature can be utilized by plant stands (Fig. 2.34).

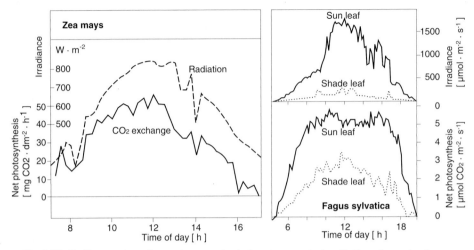

Fig. 2.33. Daily patterns of net photosynthesis in response to irradiation. In maize leaves (C_4), Ph_n follows each daily fluctuation in irradiance. In sun leaves of the beech, Ph_n follows the changing irradiance only up to about 1000 μmol photons $m^{-2}s^{-1}$; in shade leaves Ph_n undergoes short-term fluctuations associated with the brief fluctuations in light intensity. (After Hesketh and Baker 1967; Schulze 1970)

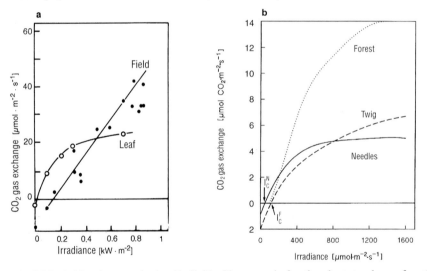

Fig. 2.34 a, b. Net photosynthesis of individual leaves and of entire plant stands as a function of light intensity. **a** Comparison of a wheat leaf and a wheat field (LAI 3.2). The CO_2 up-take of the single leaf is referred to the projected leaf area, and that of the field to the area of the canopy. (After Evans 1973). **b** Light-effect curve for needles of *Picea sitchensis* illuminated from all sides (Ph_n referred to the projected needle area), for whole twigs illuminated bilaterally, with mutual shading of the needles (Ph_n referred to the silhouette area), and for the forest canopy illuminated naturally from above (Ph_n referred to ground area); I_C^N = light compensation of the needles, I_C^F that of the forest stand. (After Jarvis and Leverenz 1983). Average of CO_2 fluxes above forest canopies: conifer stands 15–20, deciduous forest 20–25 tropical rainforest 22, eucalyptus stands 28 $μmol \cdot m^{-2} \cdot s^{-1}$ (Ceulemans and Saugier 1991)

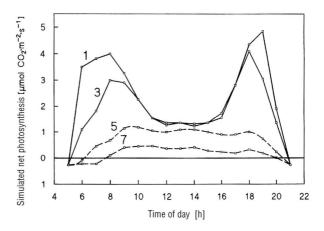

Fig. 2.35. Simulated diurnal courses of CO_2 uptake by individual leaves in a vertical succession of 6-cm-thick layers of foliage in a stand of *Quercus coccifera* on a hot day in July at the beginning of the summer dry period. *1* Sun leaves of the top of the canopy; *3* third layer of sun leaves of the upper crown; *5* shade leaves in the interior of the crown; *7* shade leaves of the lower crown, above which the cumulative LAI is 3.5. (After Tenhunen et al. 1989). Early simulations of canopy photosynthesis including leaf inclination were made by Kuroiwa (1969)

In a stand of plants there are large differences in the contribution of the various layers of foliage to the total photosynthetic yield. In grass communities with mainly erect leaves, photosynthesis is most intense in the middle zone, in which the majority of the penetrating light is absorbed. In forests and dense crowns of trees, light saturation of photosynthesis is attained only sporadically in a narrow outermost layer. Within the crowns and beneath the tree canopy (where diffuse radiation penetrates better than direct solar beams), photosynthetic yield decreases in proportion to the steep decline in light intensity. The shade-adapted leaves in the interior of the crowns play an important role in overcoming this disadvantage, since they are better able to utilize weak light than the non-adapted leaves. This is a photohomeostatic effect, buffering the strong contrasts in light intensity. Because shade leaves are also less exposed to overheating, dry air and wind, they make a modest, but in changeable weather reliable, contribution towards supplying the tree with its basic energy requirements (Fig. 2.35).

Photosynthesis of Phytoplankton Populations
In bodies of water the distribution of phytoplankton follows the light gradients and the density profile. The photosynthesis-depth curve corresponds to the light-response curve for photosynthesis (see Fig. 2.30). In the case of algae this shows an optimum, below which there is too little light for maximum photosynthesis, and a region nearer the water surface, in which when the sun is high there is too much light, and this causes photoinhibition. At this time, therefore, the highest rates of photosynthesis are not measured immediately below the

surface of the water but somewhat deeper, the exact depth depending on the light intensity and the turbidity of the water. On overcast days and during seasons with weaker radiation, when the light does not exceed optimal intensities, the region of maximal photosynthesis is shifted up to the water surface. At greater depths, photosynthetic activity declines until eventually it is just able to compensate for respiration. Compensation usually occurs at the depth at which not more than 1% of the surface radiation penetrates (the *compensation depth*).

2.2.5.2 Responses of Respiration and Photosynthesis to Temperature

Temperature affects metabolic processes by way of its influence on the reaction kinetics of chemical events and on the activities of the various enzymes involved. According to Van't Hoff's *reaction rate-temperature rule* (RRT rule), the reaction rate k rises exponentially with temperature. The increase in reaction rate that results from a temperature increase of 10 °C is expressed by the *temperature coefficient Q_{10}*, given by

$$\ln Q_{10} = \frac{10}{T_2 - T_1} \ln \frac{k_2}{k_1} \tag{2.7}$$

where $T_2 - T_1$ is the temperature interval, k_2 and k_1 the associated rates of reaction. The temperature coefficient is fairly constant over a small range of temperatures: 1.4–2.0 for most enzyme reactions and 1.03–1.3 for physical processes. When the influence of temperature over a large range is of interest, it should be kept in mind that the Q_{10} of a metabolic process varies with temperature. At low temperatures it is large, since as a rule enzyme reactions are rate-limiting, whereas at higher temperatures it becomes smaller because physical processes such as diffusion velocity become limiting.

Temperature Response of Mitochondrial Respiration

Dark respiration increases exponentially with rising temperature. Below 5 °C the activation energy for the various processes involved in respiration is large, and thus the Q_{10} is high. At temperatures higher than 25–30 °C the temperature coefficient for respiration drops to 1.5 or less in the majority of plants. At yet higher temperatures biochemical reactions proceed so rapidly that substrate and metabolites cannot be provided quickly enough and the respiratory intensity soon falls off (Fig. 2.36). At temperatures between 50 and 60 °C heat damage to enzymes and functionally important membrane structures causes respiration to cease.

Temperature Response of Photosynthesis

Temperature chiefly affects photosynthesis by way of electron transport and the secondary processes. At cool, suboptimal temperatures the rates at which carbon is fixed and reduced increase as the temperature rises, until an optimal value is reached. At temperatures above the optimal range the ratio CO_2/O_2

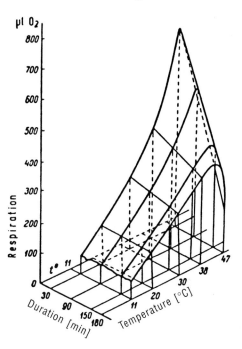

Fig. 2.36. Dark respiration in the leaves of *Podophyllum peltatum* with rising temperature and with increasing duration of these temperatures. (From Semikhatova 1974)

decreases; this results in a decrease in carboxylation efficiency of the enzyme RuBP carboxylase/oxygenase. Very high temperatures impair the interplay between the various reactions involved in carbon metabolism and transport of material. This, together with the inhibition of the membrane-bound photochemical processes, results in a rapid breakdown of photosynthesis.

Very high temperatures result in the complete cessation of CO_2 uptake; if more favourable conditions follow, and if the plant has suffered no structural injury, photosynthetic function slowly recovers.

The temperature response of net photosynthesis can be defined by three parameters: these are the *cold limit* (i.e. the temperature minimum of net photosynthesis), the *temperature optimum,* and the *heat limit* (the temperature maximum of net photosynthesis; Fig. 2.37).

The Optimal Temperature Range

The range of temperature in which net photosynthesis is more than 90% of the maximum attainable can be regarded as optimal. Under natural conditions the temperature optimum for *net photosynthesis of C_3 plants* is lower than the optimum for potential photosynthetic capacity (at CO_2 saturation) and for gross photosynthesis, because at higher temperatures the increased rates of photorespiration and mitochondrial respiration (in non-green tissues) diminish the net photosynthetic yield. The position and span of the temperature optimum is a specific characteristic of a plant. It can, however, be altered by external factors; at low light intensities, for example, the optimum temperature

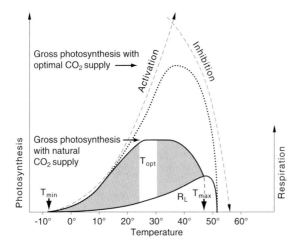

Fig. 2.37. Diagram of the temperature response of photosynthesis and respiration in C_3 plants. Gross photosynthesis increases as a result of temperature activation of the enzymes involved, until inhibitory effects lead to a decline in photosynthetic activity. The difference between gross photosynthesis and light respiration, R_L, gives the net photosynthesis (*hatched area*). The decline in gross photosynthesis due to heat is accompanied by decreased photorespiration. T_{opt} Temperature optimum for net photosynthesis; T_{min} cold limit; T_{max} heat limit. (Original)

range is wider and is displaced towards lower temperatures. A flat optimum has the advantage that fluctuations in temperature over a wide range cause only slight changes in photosynthetic intensity.

The optimal range of temperature for net photosynthesis in most C_3 plants is $15-30\,°C$ (Table 2.11). The lowest optima are seen in lichens of cold regions (the Arctic and especially in the Antarctic). Temperatures between 10 and 20 °C are optimal for sciophytes, and also for species flowering in early spring, when air temperatures are low. Herbs on sunny sites and trees of warm climates, on the other hand, achieve their highest photosynthetic yields between 25 and 35 °C, and desert shrubs even at 40 °C and more. Many *C_4 plants* still assimilate intensively at high temperatures. The C_4 syndrome thus provides the biochemical basis for colonization of hot localities. There are, however, also C_4 plants with low temperature optima for net photosynthesis, such as *Spartina* and *Atriplex* species of cool regions. In *CAM* plants the optimum temperature is high during the light phase, whereas dark-fixation of CO_2 is adjusted to the lower nocturnal temperatures (Fig. 2.38).

The Cold Limit for Net Photosynthesis

In leaves of tropical plants CO_2 uptake comes to a standstill at cool, above-freezing temperatures, chiefly due to the cold sensitivity of their thylakoids (see Chap. 6.2.2.3). Plants of the temperate zone, however, continue to assimilate CO_2 even at temperatures below 0 °C, as long as the cells remain supercooled.

Table 2.11. Temperature response of net photosynthesis under conditions of ambient CO_2 and light saturation. Compiled from data of numerous authors

Plant group	Low-temperature limit for CO_2 uptake (°C)	Temperature optimum of Ph_n (°C)	High-temperature limit for CO_2 uptake (°C)
Herbaceous flowering plants			
C_4 plants of hot habitats	+ 5 to 10	30 – 40 (50)	50 – 60
Agricultural C_3 plants	– 2 to 0	20 – 30 (40)	40 – 50
Heliophytes	– 2 to 0	20 – 30	40 – 50
Sciophytes	– 2 to 0	10 – 20	ca. 40
Desert plants	– 5 to 5	20 – 35 (45)	45 – 50 (60)
CAM plants			
during day	– 2 to 0	(20) 30 – 40	45 – 50
during night	– 2 to 0	10 – 15 (23)	25 – 30
Winter annuals and spring geophytes	– 5 to – 2	10 – 20	30 – 40
Plants of high mountains	– 6 to – 2	15 – 25	38 – 42
Aquatic cormophytes	ca. 0	(15) 20 – 30 (35)	45 – 52
Woody plants			
Evergreen broadleaved trees of the tropics and subtropics	0 to 5	25 – 30	45 – 50
Mangroves	0 to 5	25 – 30	ca. 40
Sclerophyllous trees and shrubs of dry regions	– 5 to – 1	20 – 35	42 – 45
Deciduous trees of the temperate zone	– 3 to – 1	20 – 25	40 – 45
Evergreen coniferous trees	– 5 to – 3	10 – 25	35 – 42
Dwarf shrubs of heath and tundra	ca. – 3	15 – 25	40 – 45
Mosses			
Arctic and subarctic zone	ca. – 8	5 – 12	ca. 30
Temperate regions	ca. – 5	10 – 20	30 – 40
Lichens			
Cold regions	– 10 to – 15	8 – 15 (20)	25 – 30
Deserts	ca. – 10	18 – 20	38 – 45
Tropics	– 2 to 0	ca. 20	25 – 35
Algae			
Snow algae	ca. – 5	0 – 10	30
Thermophilic algae	20 to 30	45 – 55	65

As soon as ice forms in the tissues photosynthesis stops abruptly; this occurs at temperatures around – 5 °C in spring geophytes and plants of the high mountains, and at – 3 to – 5 °C in evergreen leaves and needles [182]. Down to these temperatures photosynthesis functions normally. The resumption of photosynthesis after thawing, however, takes place slowly; the extent of the post-freezing depression of assimilation depends on the severity and duration of the frost. The lower these nocturnal temperatures, the longer the time elaps-

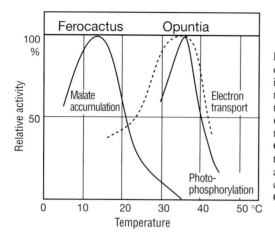

Fig. 2.38. Temperature dependence of nocturnal malate accumulation in *Ferocactus acanthodes* and of the primary processes of photosynthesis in isolated chloroplasts of *Opuntia polyacantha.* The optimal temperatue for the dark fixation of CO_2 by cacti is roughly 20 °C lower than that for the photosynthetic assimilation of the CO_2 released during the day. (After Nobel 1977; Gerwick et al. 1977)

ing next day before CO_2 uptake is resumed, and the lower the maximal values attained (Fig. 2.39). A series of night frosts results in a progessive reduction of the hours of daylight used for the uptake of CO_2.

The Upper Temperature Limit for Net Photosynthesis
At high temperatures the primary processes of photosynthesis are the first to be inhibited (see Table 6.4). The temperature maximum for net photosynthesis can be considered as the compensation point for CO_2 exchange (heat compensation point); the more heat-sensitive the photosynthesis and the more rapid the heat-induced enhancement of respiration, the sooner is this point reached.

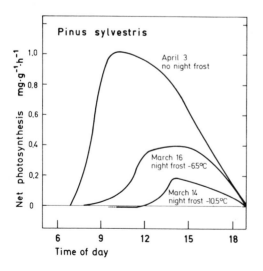

Fig. 2.39. After-effect of night frost on the time course of net photosynthesis in pine twigs during the following day. (After Polster and Fuchs 1963). In contrast to vascular plants, lichens in polar regions are able to photosynthesize at subfreezing temperatures down to −10 °C or lower; therefore they can actively take up CO_2 even under winter conditions (Kappen 1993)

Acclimation to the Prevailing Temperature

The rates of photosynthesis and respiration adapt to the temperature prevailing at a given time. *Modulative temperature adaptation* occurs within a few days, or sometimes a few hours. Possible mechanisms involved are: changes in substrate concentrations, e.g. accumulation of sugar as a response to cold; replacement of certain enzymes by isoenzymes with the same action but different temperature optima; and by chemical and structural alterations in the biomembranes. *Evolutive* temperature acclimation ensures that photosynthesis and respiration are adjusted to the average temperature climate in the habitat of the plant species.

Respiratory activity is enhanced during adaptation to cold and diminished during adaptation to heat: in other words, biochemical regulation is employed to counteract the tendencies in the physicochemically determined rate of reaction. This results in a better supply of energy at low temperatures, while the slower rise in respiratory rate at higher temperatures permits more economical consumption of metabolites. In the ideal case, metabolism is so perfectly adjusted that respiratory intensity is kept at about the same level over a relatively broad range of temperatures (thermohomeostasis). This is illustrated in Fig. 2.40. Rapid modulative adaptation is particularly important to plants in locations where there are wide, abrupt fluctuations in temperature. The activity of maintenance respiration also adjusts to the different temperature conditions prevailing over the course of the year (Fig. 2.41). In evergreen woody plants that continue to assimilate during winter, such as mediterranean sclerophylls or the broadleaved evergreens of warm-temperate regions, respiratory activity is increased during the cool season. In this way respiration rates remain fairly constant throughout the year.

That respiratory activity is adapted to the average temperature of the region over which the species is distributed is clearly shown by the fact that, referred to a standard temperature, the respiration of tropical plants is much weaker than that of plants of temperate zones, whereas plants of the arctic or high

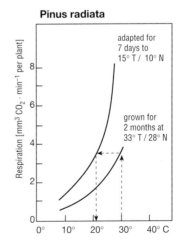

Fig. 2.40. Modulative adjustment of dark respiration of pine seedlings to the temperature at which they were grown. If warm-adapted seedlings (raised at 33 °C by day and 28 °C by night) are transferred to a cool room (15 °C by day, 10 °C by night), activity is doubled within a week. Furthermore, the temperature curve of respiration rises more steeply. As a result the plants now respire just as rapidly at 21 °C as formerly at 30 °C. If the seedlings are subsequently returned to the warm room the respiration gradually returns to its original level (not shown in the diagram). (After Rook 1969)

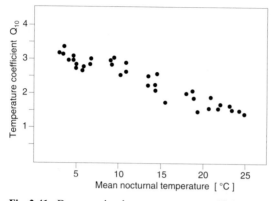

Fig. 2.41. Decrease in the temperature coefficient Q_{10} for dark respiration in *Chamaecyparis obtusa* with increasing nocturnal temperature over the course of the year. (Hagihara and Hozumi 1991)

mountain regions respire at much greater intensities [128, 239]. Leaves of evergreen woody plants of the tropics, subtropics and the mediterranean region respire at 20 °C between 0.1 and 0.2 μmol m^{-2} s^{-1}, needles of cool-temperature conifers between 0.4 and 0.5 μmol m^{-2} s^{-1}, leaves of arctic dwarf shrubs 0.6 – 1.5 μmol m^{-2} s^{-1}; if the respiratory intensity of these plants is referred to the mean temperature of the growth period, then for all localities the average respiratory intensity is found to range only between 0.25 and 0.4 μmol m^{-2} s^{-1} [132]. The same applies for herbaceous plants and for a comparison of plants of valley floors with those of the mountains [220].

The optimum range of temperature and the temperature limits of **net photosynthesis** undergo modulative shifts in acclimating to the thermal conditions in the habitat. Evolutive adaptation is often expressed in a greater readiness for such modulative adaptations. For example, mountain ecotypes of *Oxyria diagyna* react to heat treatment with a greater rise in the temperature optimum and the heat limit for CO_2 uptake than do arctic ecotypes. Conversely, following exposure to cold, photosynthesis of the arctic ecotypes increases at low temperatures sooner and more effectively than that of mountain ecotypes (Fig. 2.42). As a further example: clones of the evergreen shrub *Atriplex lentiformis* taken from the hot desert and from cool coastal localities differ with respect to the temperature curve for photosynthesis only if they have been grown at high temperatures. The heat-resistant desert clone in this case is fully active in photosynthesis, whereas CO_2 uptake in the coastal clone is drastically reduced as a result of heat damage. Ecotype differentation [255] can lead to stepwise shifts in the optimal and limiting temperatures of net photosynthesis; this is obviously connected with the temperatures in the plant's surroundings.

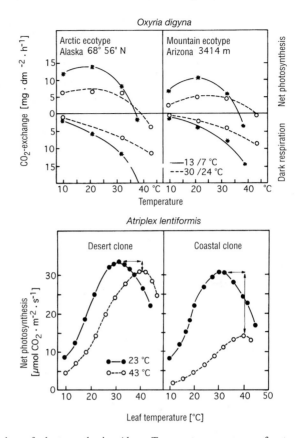

Fig. 2.42. Temperature acclimation of photosynthesis. *Above* Temperature response of net photosynthesis and dark respiration in arctic and mountain ecotypes of *Oxyria digyna* grown at low (day/night: $13/7\,°C$) and high ($30/24\,°C$) temperatures. (After Billings et al. 1971). *Below* Displacement of the temperature-dependence curves for photosynthesis of hot-desert clones and cool-coastal clones of *Atriplex lentiformis* (C_3) grown at daytime temperatures of 23 and $43\,°C$. The *arrows* denote the amplitude of acclimation (horizontal distance) and the difference in photosynthetic capacity (vertical distance). (After Pearcy, from Osmond et al. 1980)

2.2.5.3 CO$_2$ Exchange and Water Supply

Like carbon dioxide, water is a substance necessary for photosynthesis. However, it is not this small amount that makes it a limiting factor, but rather the larger amount of water necessary to maintain a high degree of hydration of the protoplasm. The decrease in cell volume and turgor that result from water shortage are accompanied by a drop in photosynthetic activity (Fig. 2.43). Oxidative processes, however, are less sensitive than photosynthesis, and only when cellular dehydration becomes severe is there a reduction in photorespiration and the mitochondrial respiratory processes. Consequently, water short-

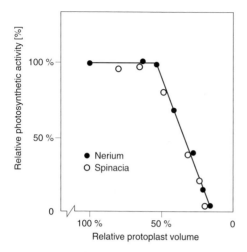

Fig. 2.43. Decline in photosynthetic activity in leaf segments with decreasing protoplast volume due to loss of water. In both plant species (*Nerium oleander*, a xerophyte, and *Spinacia oleracea*, a mesophyte), the photosynthetic activity decreased continuously after a reduction in volume to about 55% of the maximum. Photosynthesis ceased entirely when the volume reached roughly 20% of the maximum. The upper threshold for inactivation corresponds to a water potential of -1 MPa in *Spinacia* and -6 MPa in *Nerium*). (Kaiser 1982)

age results in a gradually decreasing ratio of CO_2 fixation to photorespiration.

In vascular plants, the first effect of water deficiency is a narrowing of the stomata and thus a reduction in CO_2 exchange. Even a low air humidity may cause premature closing of the stomata, the more so if accompanied by soil-moisture depletion. A low water potential in the leaves also causes immediate inhibition of photosynthesis (Fig. 2.44), mainly due to its effect on electron transport and phosphorylation. As a rule both mechanisms are involved: in plants with especially sensitive stomata (e.g. many tree species) the lower uptake of CO_2 is due to stomatal narrowing, but in many herbs and in xerophytic species it is primarily the result of metabolic inhibition.

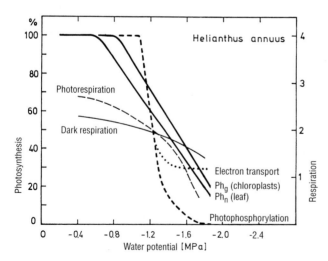

Fig. 2.44. Impairment of photosynthesis and respiration as a result of water deficiency. The intensity of the various processes involved in photosynthesis in sunflower leaves dehydrated to different levels is given as a percentage of that in leaves with adequate water supply. (After Boyer and Bowen 1970; Keck and Boyer 1974; Ortis-Lopez et al. 1991)

CO₂ Exchange and Degree of Hydration of Poikilohydric Plants

In poikilohydric plants such as algae, lichens and mosses the degree of hydration of the protoplasm is matched to the humidity of the surroundings (see Chap. 4.1). These plants rapidly soak up water as soon as it is available, but they can quickly dry out again through evaporation; thus their water content fluctuates over short intervals with the weather conditions, and stays in equilibrium with the water-vapour content of the air.

The dependence of the photosynthesis of poikilohydric species on water content takes the form of an *optimum curve*. The high diffusion resistances to the entry of CO_2 encountered in water-saturated lichens and mosses result in a depression of net photosynthesis (Fig. 2.45). Also, as the plant begins to dry out both photosynthesis and respiration gradually decrease. Characteristic for a species is the relative air humidity (RH) (in equilibrium with the water content of the plant) at which the uptake of CO_2 stops: for aerial algae the limit is in the range of 70–90% RH, for lichens 80–96% RH [129], and for mosses it is usually above 90% RH [187]. Lichens with photobiontic green algae are less sensitive to dehydration than those with blue-green photobionts, since the latter contain lower amounts of osmotically active compounds [131].

The photosynthetic apparatus of thallophytes is well suited to the frequent and pronounced fluctuations in cellular water content. When totally dried-out thalli or leaflets again become wet, it takes only a few hours before the photosynthetic process is *reactivated,* chiefly by neosynthesis of RuBP carboxylase. Heteromeric species of lichens with green algae can take up enough water

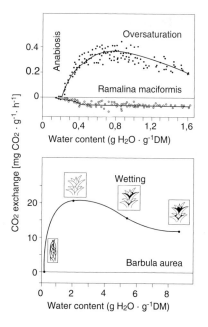

Fig. 2.45. Photosynthesis and dark respiration (R_D) in poikilohydric plants as a function of water content. In a lichen (*Ramalina maciformis*) and a moss (*Barbula aurea*) metabolism is activated during the phase of hydration; when the lichen thallus is oversaturated, and the moss cushion is filled with water, there is a drop in CO_2 uptake due to the high diffusion resistances. (From Lange et al. 1987; Rundel and Lange 1980)

vapour from moist air, even without additional liquid water, to start off their metabolism and to achieve a positive CO_2 balance.

Lichens exploit every available opportunity to take up CO_2 during their periods of hydration. This is illustrated in Fig. 2.46, which shows the CO_2 exchange of a desert lichen, measured in its natural habitat, throughout the day. During the night the thalli take up moisture from the air, and in the early hours of the morning from dew as well. After sunrise, only 3 h remain for the lichen to assimilate carbon before it becomes dry and a rigid until nightfall.

Gas Exchange in Vascular Plants During Water Deficiency

The curve of gas exchange vs. water loss shows two critical points: the threshold between full photosynthetic capacity and *reduced capacity* and the *null point for gas exchange* (see Fig. 2.48). The threshold is reached when water stress is such that the stomata begin to close. In dry air, closure is already elicited by a slight drop in water potential (in moist air somewhat later). The response to low air humidity varies from species to species. If water is supplied after the first critical point has just been passed, recovery of gas exchange is rapid.

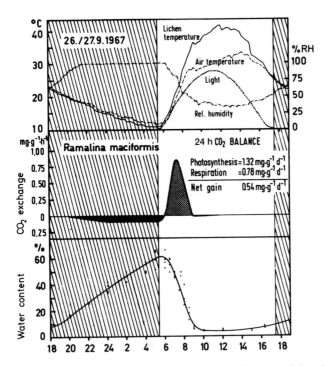

Fig. 2.46. CO_2 exchange and water content of the desert lichen *Ramalina maciformis* during the course of the day. *Upper frame:* irradiance, temperature and air humidity. The nighttime dew supplies the lichen with moisture, but in the morning it rapidly dries out again. *Stippled area* CO_2 uptake; *black area* CO_2 release; *hatched* night-time. (Lange et al. 1970)

The *null point for gas exchange* is determined by the total or near total closure of the stomata and by the direct effect of water deficiency on the protoplasm; it is reached only when dehydration of the leaves is well advanced. However, once this state has been reached, recovery is delayed, and after severe desiccation the original photosynthetic capacity may, under certain conditions, never be achieved again (Fig. 2.47). Repeated drying cycles may also result in hardening, via the modulative adaptation of photosynthetic processes.

The *sensitivity of CO_2 exchange to water shortage,* which is reflected by the positions of the two critical values described above, is to a large extent a characteristic feature of a plant species (Fig. 2.48). Sciophytes and many tree species are extremely sensitive to even slight losses of water. Heliophytes and herbaceous crop plants, on the other hand, can tolerate more or less severe water deficiency, depending on species and conditions of growth; in this re-

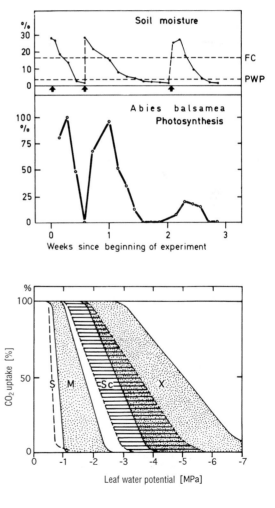

Fig. 2.47. Effect of drought on the net photosynthesis of one-year-old seedlings of *Abies balsamea*. As a result of watering (*arrows*), the soil water was raised above field capacity (*FC*). Thereafter soil moisture was absorbed to below the permanent wilting percentage (*PWP*). After a brief period of drought, net photosynthesis recovered quickly and completely, but if the soil remained below the PWP for several days, recovery was incomplete. (Clark 1961)

Fig. 2.48. Limitation of CO_2 uptake with decreasing water potential in the mesophyll. *S* Succulents; *M* mesophytes (herbaceous dicotyledons, grasses and woody plants of humid regions); *Sc* sclerophyllous shrubs and trees of semi-humid and semiarid regions; *X* xerophytes (herbs, grasses and shrubs of arid regions). (Data of numerous authors)

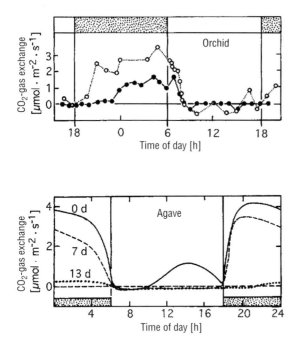

Fig. 2.49. CO_2 exchange of CAM plants during drought. *Above* CO_2 exchange during the course of the day, for the epiphytic orchid *Schomburgkia humboldtiana* during the rainy season (○) and dry season (●) in Venezuela. (From Griffiths et al. 1989). *Below* Gas exchange of *Agave lechuguilla* well supplied with water (control: 0 d) and with increasing drought (7 and 13 days without water). (From Nobel 1988). During water deficiency the stomata remain closed throughout the day and only CO_2 from nocturnal uptake and CO_2 released by respiration is available. During long-lasting periods of drought, little CO_2 is taken up during the night, so that only the CO_2 arising from respiration can be recycled

spect there is little difference between C_4 and C_3 plants. As would be expected, in plants of dry regions and desert plants in particular the limit is at especially low water potentials. At the first sign of water deficiency, CAM plants keep their stomata closed all day, later also increasingly in the night (Fig. 2.49). The respiratory CO_2 can no longer escape and following breakdown of the accumulated malate it is "recycled" in photosynthesis.

The response of net photosynthesis to drought can best be explained in terms of water balance. With an increasingly negative water balance the water potential of a plant also becomes more negative and the stomata remain further and longer closed during the second half of the day (Fig. 2.50). The princi-

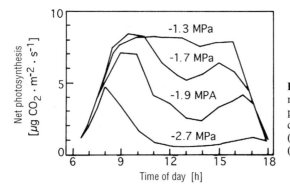

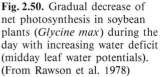

Fig. 2.50. Gradual decrease of net photosynthesis in soybean plants (*Glycine max*) during the day with increasing water deficit (midday leaf water potentials). (From Rawson et al. 1978)

ple exemplified is: the more sensitive a species is to lack of water and the drier the conditions, the earlier in the day are restrictions imposed on assimilative activity (Fig. 2.51).

Carbon Gain and Water Loss

In addition to permitting the uptake of CO_2, the stomata allow the passage of water vapour from the interior of the leaf to the surroundings. In order to take up CO_2, the plant necessarily gives off water, and when it reduces the amount of water lost, the influx of CO_2 is reduced as well. This connection was very early recognized [91, 151] and expressed numerically by the overall ratio of assimilation to water consumption. The relationship between photosynthesis and transpiration is expressed as the **water use efficiency of photosynthesis** (WUE_{Ph}):

$$WUE_{Ph} = \frac{Ph}{Tr} \ (\mu mol \ CO_2 \ m^{-2} \ s^{-1}/mmol \ H_2O \ m^{-2} \ s^{-1}) \tag{2.8}$$

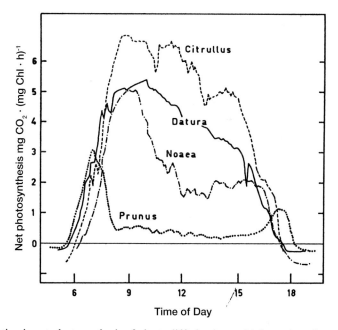

Fig. 2.51. Daily fluctuation in net photosynthesis of plants differing in sensitivity to drought and in the degree of drought stress to which they are exposed, at the end of the dry period in the Negev desert. *Citrullus colocynthis* and *Datura metel* were watered, and the Ph_n values were slightly reduced only during the hottest hours of the day and in the afternoon. The desert plant *Noaea mucronata* after high CO_2 uptake in the morning, vigorously restricted its photosynthesis in the later morning. *Prunus armeniaca* suffered considerably from lack of water toward the end of the dry season, and appreciable CO_2 uptake was only possible in the early morning and late evening. (From Schulze et al. 1972)

The water use efficiency of photosynthesis expresses quantitatively the instantaneous gas exchange of a leaf. In order to exclude the effects of the environmentally dependent concentration gradients for CO_2 and H_2O and so to characterize the *specific* diffusion behaviour of a plant as determined by its structural and functional traits, it is useful to know the *ratio of the transfer resistances* [97] for carbon dioxide and water vapour $\Sigma r^{H_2O}/\Sigma r^{CO_2}$ (or the ratio of the conductances g^{CO_2}/g^{H_2O}). When the stomata are fully open, the ratio of the resistances for most C_3 plants is $0.1-0.4$, for many trees 0.5, and for C_4 plants it can be as much as 0.8 [64, 121].

The conditions for diffusion in the two exchange processes are not identical. The concentration gradient for CO_2 from the outside air to the chloroplasts is much less steep than the H_2O-water vapour gradient between the interior of the leaf and the atmosphere, as long as the outside air is not saturated with water. At a temperature of $20\,°C$ and a relative humidity of 50%, the water-vapour gradient is about 20 times as steep as the CO_2 gradient. For this reason alone, evaporation of water proceeds much more rapidly than the uptake of CO_2. Moreover, the water molecules, being smaller, diffuse 1.5 times as fast as the larger CO_2 molecules, given identical gradients. There is also a fundamental difference with respect to the diffusion pathways: firstly, the route is longer for CO_2, which must enter the chloroplasts, and secondly, the movement of CO_2 in solution is exceedingly slow. The water use efficiency of photosynthesis is therefore always changed whenever any of the factors affecting diffusion is altered.

When the *stomata are fully open*, CO_2 uptake is limited to a greater extent than transpiration by the diffusion resistances inside the leaf. In this respect

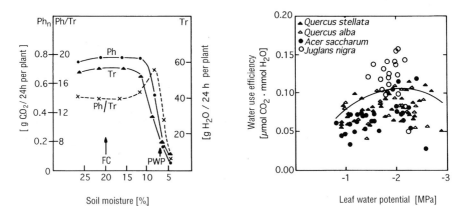

Fig. 2.52. Water use efficiency of photosynthesis under dry conditions. *Left* Daily totals for net photosynthesis (*Ph*) and transpiration (*Tr*) and the effect of decreasing soil moisture on the quotient Ph/Tr (mg CO_2/g H_2O) for 6-week-old wheat plants. *FC* Field capacity; *PWP* permanent wilting percentage. Irradiation: $150\,W\,m^{-2}$ (16 h daily); air temperature: $22\,°C$; air humidity 70%. (After Aho et al. 1979). *Right* More economical water use by leaves of tree seedlings at the commencement of stomatal narrowing in response to water deficit. Here, WUE refers to the leaf conductance for water vapour (instead of transpiration). (From Ni and Pallardy 1991). For different patterns in various tree species see Larcher (1963a, 1965) and Müllerstael (1968)

the C_3 plants, on account of their lower carboxylation efficiency, are at a disadvantage compared with the C_4 species. The most successful compromise between water consumption and CO_2 uptake is achieved when the stomata are *partially open.* This is evident at the beginning of water deficiency, when both exchange processes are already slightly limited and the ratio Ph/Tr is at its highest (Fig. 2.52). With *stomata almost closed,* the ratio Ph/Tr falls sharply because the inflow of CO_2 is more strongly reduced than transpiration, and also because water vapour continues to be lost through the cuticle.

In the field the WUE_{Ph} is affected by the climatic conditions. In the early hours of the morning, WUE_{Ph} is usually at its highest, as long as the air is still moist and photosynthesis has, in bright light, already reached its maximum rate. During the day, as the air is warmed up by the sun, the humidity of the air lessens, evaporation is enhanced by air turbulence, and the Ph/Tr ratio decreases. A reduction in the Ph/Tr ratio also often occurs over the seasons, in connection with ontogenetic development. Because of the decline in photosynthetic capacity in the course of ontogenetic development, WUE_{Ph} also drops, in many species by the time of flowering (e.g. maize [5]), and in any case during fruit development. Stress also reduces the WUE_{Ph}.

For ecological, agricultural and forestry purposes, however, the *relation of dry matter production to water consumption* over the entire period of growth is more informative than instantaneous gas-exchange ratios. The **water use efficiency of productivity** (WUE_P) is defined as follows:

$$WUE_P = \frac{\text{organic dry matter production}}{\text{water consumption}} \text{ g DM kg}^{-1} H_2O \qquad (2.9)$$

Dry matter production and water consumption can be expressed with reference to a single plant or to a *plant stand*; in the latter case (the water use efficiency of *primary productivity*) the production of organic dry-matter is referred to the area of the stand [see Eq. (2.19)], and the value for water consumption is the overall evapotranspiration (see Chap. 4.4.1). The WUE_P is an integral expression of the cumulative increase in dry matter and the water consumption over longer periods, extending from weeks to the entire growth period. Conversion of gas exchange ratios (WUE_{Ph} or g^{CO_2}/g^{H_2O}) by means of conversion factors (CO_2 into dry matter, see p. XII) can be useful in providing estimations but is not quite accurate. The reason for this is that the dry matter production of plants is depends not only on the intensity of gas exchange, but even more on the CO_2 exchange balance (cf. Chap. 2.3.1) and the specific pattern of assimilate allocation.

The water requirement per unit of dry mass produced varies among different species and varieties and is very strongly dependent on the individual state of development, stand density and environmental conditions, particularly the water supply and rate of evaporation. Knowing the WUE_P of crop plants, the grower can select species and varieties appropriate to the situation and, in dry areas, adjust the amounts of water used for irrigation (Table 2.12).

Table 2.12. Average water use efficiencies of productivity WUE_P (g DM kg^{-1} H$_2$O). Data of numerous authors

C$_4$ plants	3 – 5
Herbaceous C$_3$ plants	
Cereals	1.5 – 2
Legumes	1.3 – 1.4
Potatoes and root crops	1.5 – 2.5
Sunflowers, young plants	3.6
Sunflowers, flowering plants	1.5
CAM plants	6 – 15 (30)
Woody plants	
Tropical broadleaved trees (cultivated)	1 – 2
Temperate zone broadleaved trees	3 – 5
Sclerophyllous shrubs	3 – 6
Coniferous trees	3 – 5
Oil palms	3.5

2.2.5.4 CO$_2$ Exchange and Mineral Nutrition

Photosynthesis and respiration are affected in a wide variety of ways by the nutrient supply. In soils not seriously deficient in particular nutrients, the availability of minerals is less critical for photosynthesis than the climatic factors. Nevertheless, the yield of photosynthesis can almost always be enhanced by the artificial provision of nutrients. Mineral nutrients exert their influence on the intensity of carbon metabolism directly, and via growth and morphogenesis.

Biochemical effects on photosynthesis and respiration result from the fact that the minerals either are incorporated in enzymes and pigments, or participate directly as activators in the process of photosynthesis (Table 2.13). *Nitrogen,* as an essential component of proteins and chlorophyll, is necessary for the formation of thylakoids and enzymes. In fact, a close correlation exists between the nitrogen content of leaves and the quantities of chlorophyll (on an average 50 mol thylakoid nitrogen per mol chlorophyll [60]) and of RuBP-carboxylase (Fig. 2.53). *Orthophosphate* is incorporated in energy-rich compounds, such as ATP, triose-, pentose- and hexosephosphates. The supply of inorganic phosphorus plays a key role in the Calvin cycle and in the transport of metabolites and assimilates; phosphate deficiency results in accumulation of assimilates (sucrose and starch) in the chloroplasts and depresses photosynthesis even under otherwise favourable conditions (Fig. 2.54). The *ions* K$^+$, Ca^{2+}, Mg^{2+} and Cl$^-$ are involved in transfer processes (in chloroplasts; K$^+$ and Cl$^-$ also in movements of guard cells) and play a structural role in the photosynthetic apparatus. *Trace elements* are mainly necessary as constituents or cofactors of enzymes involved in metabolism.

Lack of minerals and alterations in the proportions of the elements taken up by the plant can affect the chlorophyll content and the number, size and

Table 2.13. The role of mineral nutrients in photosynthesis. (Nátr 1975; Gerwick 1982; Marschner 1986)

Function	Structural components and stabilizers	Enzyme components and co-factors	Translocation factors, balancing of charges
Differentiation and stabilization of chloroplast structure	N, S, Mg, Fe, Ca	Mg, (Mn), K, Fe, Zn	
Photochemical process	Mn	Mg, Cl, (Mo)	Cl
Electron transport and photophosphorylation	Mg, P, S, Fe, Cu, (Ca)	Mg, K	K, Mg
CO_2 fixation and Calvin cycle	P	Mg, Mn, K, Zn	
Transport of assimilates, synthesis of starch	P	K, Mg, B	
Stomatal movements			K, Cl

ultrastructure of the chloroplasts; this applies even if the elements in question, e.g. iron, are not themselves incorporated into the chlorophyll molecule. Magnesium and iron deficiency are the cause of *chloroses*, which can lead to a drop in CO_2 uptake to less than one-third. The principal result of too little chlorophyll is that the plants are unable to make full use of intense light, i.e. they behave like sciophytes.

Mineral nutrients also influence gas exchange by their effects on *morphogenesis* (i.e. growth, size and structure of leaves, shoots and roots) and on the *course of development* (e.g. life span). Nitrogen deficiency results in the

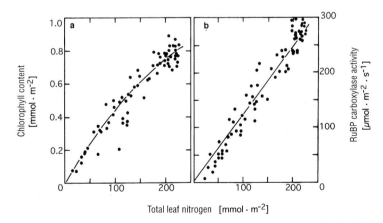

Fig. 2.53 a, b. Correlation between nitrogen content and **a** chlorophyll content and **b** RuBP-carboxylase activity of the flag leaf of *Triticum aestivum*. (From Evans 1983)

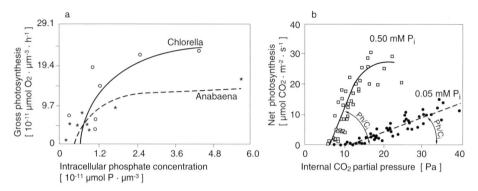

Fig. 2.54a, b. Dependence of photosynthetic capacity on phosphorus supplies. **a** Relationship between intracellular phosphorus content and photosynthesis in a green alga (*Chlorella*) and in a cyanobacterium (*Anabaena*). (Senft 1978). **b** Relationship between photosynthetic CO_2 utilization Ph/C_i (as a measure of the carboxylation efficiency) in soybean plants well supplied with phosphorus (0.50 mM P_i) and under phosphorus deficiency (0.05 mM P_i in nutrient solution). (Lauer et al. 1989)

formation of small leaves with poorly functioning stomata, whereas too much nitrogen causes excessive respiration and thus leads to a reduction in the net yield of photosynthesis.

Carbon Assimilation and Nitrogen Incorporation

The photosynthetic capacity rises with increasing nitrogen content (referred to leaf surface) in linear proportion, until limitation by other factors evokes saturation characteristics (Fig. 2.55). In analogy to water use efficiency, it is possi-

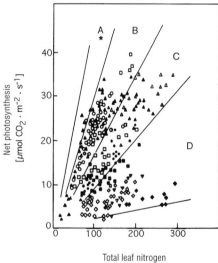

Fig. 2.55. Relationship between photosynthetic capacity and the nitrogen content of leaves. *A* C_4 plants; *B* C_3 crop plants (e.g. high-yield varieties of wheat and rice); *C* winter annuals, mountain plants; *D* sclerophyll shrubs, evergreen trees in semi-arid regions, trees of tropical rainforests. (Measurements of various authors, from Evans 1989 and Körner 1989, supplemented by data in Sage and Pearcy 1987)

ble to calculate nitrogen use efficiency (NUE) [94]. Here, too, a distinction is made between nitrogen use efficiency of photosynthesis (NUE_{Ph}) and a *cumulative* nitrogen use efficiency of productivity (NUE_P) [16]. The *nitrogen use efficiency of photosynthesis* is defined as

$$NUE_{Ph} = \frac{Ph}{N_A} \quad (\mu mol \ CO_2 \ m^{-2} \ s^{-1}/mmol \ N \ m^{-2}) \tag{2.10}$$

where N_A is the nitrogen content per unit leaf surface area (A). For the calculation of the *nitrogen use efficiency of dry matter production* the increase in organic dry matter over a longer period of time (mean residence time, approximately the growth period or functional period of the foliage of 1 year) is related to the nitrogen content of the assimilating organs:

$$NUE_P = \frac{\text{increase in organic dry matter}}{\text{incorporated nitrogen}} \quad (g \ DM \ g^{-1} \ N) \tag{2.11}$$

C_4 plants, due to their highly efficient photosynthesis, and some C_3 crop plants, achieve good yields of CO_2 even if the nitrogen content of their leaves is only moderate. Herbaceous C_3 plants with lower nitrogen use efficiencies compensate for this with higher nitrogen contents, provided that supplies in their habitat are plentiful; with high rates of nitrogen turnover they, also, can maintain a high carbon gain. In dwarf plants, nitrogen is more concentrated in the tissues; plants with leaves surviving longer than one season conserve their nitrogen. In this way, with a moderate photosynthetic capacity and low production, they are able to thrive even on soils poor in nitrogen.

2.2.5.5 Responses of CO_2 Exchange to the Interplay of External Factors

Under natural conditions environmental factors do not exert their effects singly but as a *factor complex*. The photosynthetic response to the interplay of environmental factors differs from the reaction to an isolated factor. An example of this is seen in the shift in the limiting and optimal temperature ranges for net photosynthesis as irradiance increases (Fig. 2.56). The effects of multiple factors as measured under natural conditions thus frequently differ from those recorded in the laboratory. Due to the wide scatter of the data, functional patterns can be represented more realistically by scatter diagrams and boundary lines than by mean values (Fig. 2.57).

The pattern of gas exchange in a plant is determined by the diurnal, seasonal and weather-dependent fluctuations of internal and external conditions. Invariably, one or another *factor is at a minimum* and limits CO_2 uptake for some time. In nature, external factors are only rarely and briefly so favourable and so well harmonized that photosynthesis attains peak values. On an average, the *maximum daily values* for CO_2 uptake only reach 70–80% of the maximal photosynthetic capacity.

A phenomenon of particularly complex origin is the *midday depression* in net photosynthesis, occurring especially often in woody plants. In response to

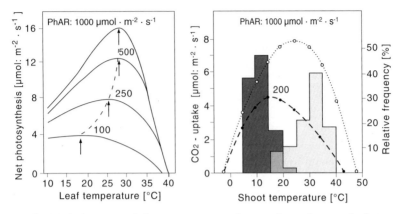

Fig. 2.56. Displacement of the temperature optimum of net photosynthesis toward higher temperatures with increasing irradiance. *Left* Model of the displacement of the temperature optimum of *Atriplex patula* (C$_3$ plant). (Hall 1979). *Right* Temperature dependence of the apparent CO$_2$ uptake of the prostrate shrub *Loiseleuria procumbens* at weak (200 μmol ·m^{-2}·s^{-1}) and strong irradiance (1000 μmol·m^{-2}·s^{-1}), referred to the area of ground covered by the dwarf shrub. *Stippled columns* Distribution of temperature frequencies in the stand at lower (*dark area*) and higher (*pale area*) irradiance. (After Grabherr and Cernusca 1977)

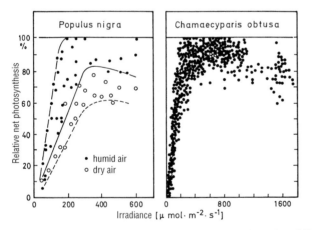

Fig. 2.57. The light response of net photosynthesis under different meteorological conditions. *Left* On cloudy days with moist air, the values for net photosynthesis in poplar leaves are higher than those attained at the same irradiance on clear days with dry air. (Polster and Neuwirth 1958). *Right* Overheating of the assimilatory shoots of *Chamaecyparis obtusa* by intense irradiation produces a steep vapour pressure gradient toward the air, leading to water stress. Therefore, measurements made in the field did not yield a light saturation curve in the form of an exponential function as in laboratory experiments, but rather an optimum curve with widely scattered individual values (Negisi 1966)

overheating and high evaporation at the time when radiation is at its strongest, the stomata close, the CO_2 concentration, C_i, inside the leaf rises (an indication of non-stomatal inhibition of photosynthesis) and the photochemical efficiency of photosystem II is lowered. At the same time, the water potential of the leaves is often reduced (see Fig. 2.50). The midday depression is the result of a combination of stresses: strong light (photoinhibition), negative water balance (loss of turgor), heat stress and in some cases probably slower translocation of assimilates away from the leaves.

Limitations to CO_2 Uptake Under Field Conditions

Under the temperature conditions prevailing at intermediate latitudes, the primary factor limiting CO_2 assimilation is *light deficiency,* due to clouds or to the low sun angle (Table 2.14). In the temperate zone, excessively *low temperatures* in late autumn, winter and spring result in a significant decrease in the carbon gains of the evergreen vegetation; deciduous woody plants are mostly not in leaf during this period and thus are less affected. Excessively *high temperatures* in the same zone, even in open, heat-exposed sites, have very little effect. In the tropics and subtropics, on the other hand, the role of heat in limiting CO_2 uptake may well play a role in selection. Worldwide, *water deficiency* is the most important environmental factor limiting assimilation. Study of diurnal fluctuations in CO_2 exchange by evergreen shrubs of maquis and bushland shows that the daily maxima of net photosynthesis are one-fifth to two-thirds lower during times of drought than during the rainy season. The magnitude of the reduction in CO_2 gain varies with the type of carboxylation (C_3, C_4, CAM) employed by the plant, its growth form and specific sensitivity, with features of the habitat (especially the amount and accessibility of water

Table 2.14. Decrease in CO_2-uptake (in %) as a result of unfavourable habitat factors during the growth period

Factor	*Pinus densiflora*[a] (young plant)	*Fagus sylvatica*[b] (forest tree)	*Prunus armeniaca*[c] (arid region)	*Hammada scoparia*[c] (C_4 desert plant)	*Carex curvula*[d] (alpine)
Light deficit at twilight and due to clouds	− 22	− 38	− 13	− 20	− 39
Temperature too low or too high	− 15	− 3	− 7	− 12	− 8
Dryness of air		− 2	− 12	− 28	
Total reduction in mean daily CO_2 uptake, in % of possible maximum	− 37	− 43	− 32	− 60	− 47

[a] Negisi (1966); [b] Schulze (1970); [c] Schulze and Hall (1982); [d] Körner (1982).

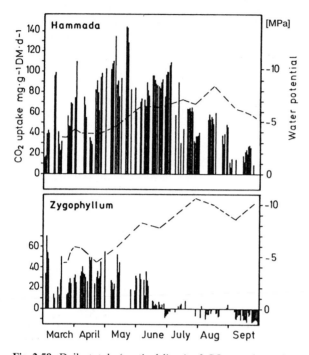

Fig. 2.58. Daily totals (*vertical lines*) of CO_2 uptake and CO_2 release by desert plants exposed to different degrees of drought stress in their habitats. *Hammada scoparia* is a C_4
leafless shrub with assimilating branches growing on loess-covered depressions, whereas
Zygophyllum dumosum is a C_3 dwarf shrub growing on south-facing slopes and stony
flats. During the dry season *Zygophyllum* sheds its leaves; the water potential of the shoots
(*dashed curve*) decreases sharply and the CO_2-exchange balance shifts to a net loss of CO_2.
Hammada is capable of withdrawing water from the soil even in dry periods, so that the
water potential remains at a moderate level and the green shoots continue to fix CO_2,
though to a lesser extent. (Lange et al. 1975). Annual courses of daily totals of CO_2 uptake
by evergreen and deciduous mediterranean tree species: Tretiach (1993)

in the deep horizons of the soil), and the duration of the drought. An example
of characteristic and deviant behaviour of desert shrubs is shown in Fig. 2.58.

2.3 The Carbon Budget of the Whole Plant

2.3.1 The Gas-Exchange Balance

The significant values for the carbon gain of a plant in its habitat are the daily
and annual totals of CO_2 uptake, rather than the transitory peak values of
photosynthesis. The net yield of carbon assimilation is the *positive balance* of
CO_2 exchange; it is not dependent on the intensity of gas exchange alone, but

also on the ratio of the mass of the photosynthetically active organs to that of the respiring tissue (*structural factor*), and on the duration of the favourable and unfavourable periods for assimilation over the course of the year (*time factor*).

The Economy of CO₂ Exchange

The carbon income of a plant consists of the gross photosynthetic yield of all its leaves, and the carbon expenditure is the CO_2 lost by the respiration of all its organs.

A look at the *CO₂ exchange balance of a single leaf* shows that the net yield of photosynthesis is the greater the higher the photosynthetic rates and the lower the respiratory activity. The ratio of the two values can be expressed as the *carbon use efficiency of a leaf (CUE_L)* at a given time [162]

$$\text{CUE}_L = \frac{\text{Ph}_g}{R} \quad (\mu\text{mol CO}_2 \text{ m}^{-2}\text{s}^{-1}/\mu\text{mol CO}_2 \text{ m}^{-2}\text{s}^{-1}) \tag{2.12}$$

Under favourable conditions, leaves of C_3 plants take up roughly three to five times as much CO_2 as they lose by dissimilatory processes during the same period of time, whereas leaves of C_4 plants, because they avoid photorespiratory losses, capture 10 to 20 times as much CO_2 as they lose. For algae and seaweeds, CO_2 exchange ratios are between 5 and 10.

The next step in arriving at a carbon balance is to consider the proportion of *green* or *photoautotrophic (assimilating)* to *heterotrophic (dissimilating)* parts (Table 2.15). The contribution of photosynthetically productive green parts to the total mass depends on the plant's growth form and alters during development, in addition to which it is influenced by environmental factors. A significant proportion of the non-green biomass in woody plants consists of dead, non-respiring sclerenchymatic elements; poplar wood, for example, contains only 8% respiring tissue [211].

Duration of the Assimilation Period

The decisive factor in carbon fixation is the time span over which a high rate of CO_2 acquisition is possible. This comprises the hours of daylight during the period of foliation, insofar as assimilation is not depressed by frost, heat or drought. From the sum total of daily carbon intake, the CO_2 released from the leaves during the night must first be subtracted. The resulting net consumption of CO_2 by photosynthesis in the course of 24 h is the daily balance, and the total of the daily balances gives the annual balance of CO_2 exchange.

The daily balance gives an idea of the production characteristics of the plant under the influence of the given environmental factors. The daily balance is the greater

1. the higher the photosynthetic rates,
2. the greater the amount of available green mass and the more favourable the orientation of the assimilating surface to incoming light,

Table 2.15. Contributions of leaves, shoots and roots (in %) to the total dry mass of different plant groups. (From measurements by numerous authors)

Plant	Green mass (photosynthetically active organs)	Purely respiratory organs	
		Aboveground shoots	Roots and belowground shoots
Evergreen trees of tropical and warm temperate forests	2 – 5	80 – 90[a]	10 – 20[a]
Palms, mature	54	38	8
Deciduous trees	1 – 2	ca. 80[a]	ca. 20[a]
Evergreen coniferous trees	4 – 10	70 – 80[a]	10 – 30[a]
Stunted woody plants at the forest limit	ca. 25	ca. 30[a]	ca. 45[a]
Young conifer plants	50 – 60	40 – 50[a]	ca. 10[a]
Ericaceous dwarf shrubs	10 – 20	ca. 20[a]	60 – 70[a]
Meadow grasses	ca. 50		ca. 50[a]
Cereals (ready to flower)	10 – 15	40 – 70	ca. 30
Steppe plants			
wet years	ca. 30		ca. 70
dry years	ca. 10		ca. 90
Desert plants			
annual herbs	80 – 90		10 – 20
shrubs and perennials	10 – 20	10 – 40	40 – 60
Plants of high mountains	10 – 20		80 – 90
Arctic tundra			
Cormophytes	15 – 20		
Cryptogams	> 95		

[a] The greater part of the mass is dead structural material.

3. the longer the daily duration of conditions favouring photosynthesis and
4. the shorter and cooler the night.

Even small changes in the relative lengths of day and night or in nocturnal temperatures are sufficient to cause considerable differences in the 24 h balance. Especially the shortage of water causes the daily totals to drop to low levels. Using simulation models the effects of numerous external and internal factors on the daily CO_2 balance can be computed, thus permitting a quantitative appraisal of their importance (Fig. 2.59). Models of this nature are of great value in developing working hypotheses. Such hypotheses approach reality more closely the more completely the adaptive capacities of the plant are taken into consideration. Simulation analyses are also a valuable aid in planning research. It must be emphasized, however, that information obtained in this way should in every case be verified by measurements in the field and/or in climate chambers, before they can be considered acceptable as a basis for practical measures or further research.

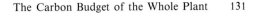

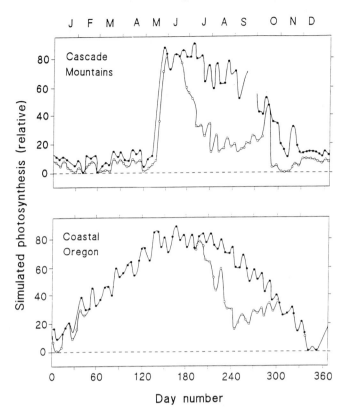

Fig. 2.59. Simulation models for annual patterns of photosynthesis of *Pseudotsuga menziesii* under (○) the temperature conditions and the consequences of drought in the Cascade Mountains and in coastal Oregon, compared with (●) potential values assuming no limitation by extremes of temperature and drought. (Emmingham and Waring 1977). A very early simulation model of the effects of different lengths of day/night on the daily CO_2 balance of a moss at various temperatures was computed from laboratory data by Stålfelt (1937)

The annual balance is the CO_2 balance for the entire growth period or for the whole year. Here, too, modelling provides the means of estimating the potential carbon gain and comparing this with the real results.

The annual balance becomes increasingly positive the longer the plants have been able to achieve large, or at least positive, daily balances. Significant yields can be obtained despite only moderate daily balances, if the total period of time favourable to assimilation is sufficiently long, which is the case in warm-temperate, subtropical and tropical humid regions. Where assimilatory activity is possible during only a relatively short period, the yields remain small, even if the plants have a high photosynthetic capacity or can make particularly efficient use of the photosynthates.

For terrestrial plants, the time available for productive activity decreases in the following sequence, according to the growth form of the plants and climatic conditions:

1. Evergreen plants of warm, humid regions, which carry on photosynthetic activity throughout the year (for example, woody plants and perennial herbs of the tropical and subtropical rainforests).
2. Evergreen plants in which there are seasonal variations in photosynthetic activity and the assimilation is low at certain times of the year, due to
   a) a cold season (for example, boreal conifers),
   b) a dry season (for example, sclerophyllous shrubs),
   c) a short photoperiod in winter (for example northern Pacific conifers).
3. Seasonally green plants (deciduous woody plants and most herbs)
   a) which utilize the fully foliated season in regions with high precipitation (for example, deciduous trees of the temperate zone),
   b) which only partially utilize the foliated season because of insufficient light (e.g. undergrowth in deciduous woodland),
   c) which only partially utilize the foliated season because of dryness (for example, trees of the steppe forest).
4. Plants which take up carbon briefly between longer unfavourable periods:
   a) vascular plants in deserts with erratic precipitation (100–200 favourable days),
   b) vascular plants of the arctic and high mountains (60–90 favourable days),
   c) mosses, lichens and aerial algae which take up carbon sporadically, after becoming wet or when humidity is high.

In the cold seas of high latitudes, the production period is limited to the polar summer, which permits photosynthetic activity for several weeks without interruption by night.

The predominance of the time factor for assimilation has long been taken into consideration for the breeding and rearing of plants. The yield of crop plants in regions with a long period favourable for growth can be increased by exploiting this period to the full. This is achieved by selection for the most precise phenological adjustment to the local weather pattern, whereas crop plants that are to be grown in regions with a shorter growth period are selected with a view to high photosynthetic capacity and economic utilization of photosynthates.

The Overall CO_2 Balance

The CO_2 balance of the whole plant involves the *total gross photosynthesis* Ph_g of the *total leaf mass* W_L (W = weight) during the total annual number of daylight hours (d), minus the total annual *respiration of leaves* (day and night) R_L, *shoots* (including flowers and fruits) R_S, and *roots* R_R [133].

$$CO_2 \text{ balance} = W_L \cdot \Sigma_d \, Ph_g - W_L \, \Sigma R_L - W_S \cdot \Sigma R_S - W_R \cdot \Sigma R_R \qquad (2.13)$$

In order to set up a complete gas-exchange balance for a plant, the CO_2 exchange of the various aboveground parts has to be measured separately, e.g. in trees: tip, interior and base of the crown, stems. The respiration of the underground parts must be measured by day and night throughout the year.

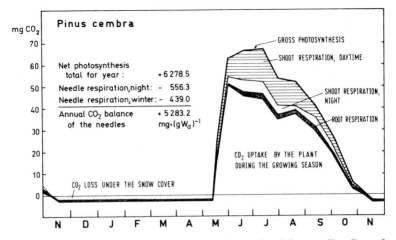

Fig. 2.60. Annual CO_2 balance of seedlings of *Pinus cembra* at the alpine tree line. Part of the CO_2 gained by photosynthesis is lost the same day by respiration of the shoots and roots. In winter the daily balance is usually negative or at best zero: the CO_2 loss during the 6 winter months is subtracted from the CO_2 acquired during the growing season, to obtain the annual balance. (From Tranquillini 1959). Estimates of overall CO_2 balances for different age classes of forest trees (*Picea, Pinus, Acer, Betula, Populus* and *Quercus*) in eastern Europe are given by Tselniker (1993)

In addition, the mass of the individual organs has to be determined. At the same time, the external factors such as distribution, leaf temperatures, the humidity of the air near the leaves, temperature fluctutations in the vicinity of stem and roots, availability of water, etc. must be recorded in order to determine their effect on the carbon balance. Figure 2.60 shows as an example the annual CO_2 balance conifer seedlings.

The annual balance can also be expressed as the *carbon use efficiency for the whole plant,* CUE_{Pl}, which provides a quantitative measure for the efficiency of organic production:

$$CUE_{Pl} = \frac{\Sigma Ph_n + \Sigma R}{\Sigma R} \; [\text{g } CO_2 \text{ g}^{-1} \text{ t}^{-1}] \qquad (2.14)$$

The calculation is based on daily balances, and the period considered, t, therefore extends over weeks up to a year, and is in any case longer than 24 h. Unlike the balance of carbon exchange of a leaf at a given moment (CUE_L), CUE_{Pl} applies to *all organs* of the plant and thus involves a time span that reveals *average* conditions and not just random situations. CUE_{Pl} is thus an average value for day and night-time, for growing and resting periods and, most important, for the CO_2 balance of the whole plant, not only for its green parts.

Balance ratios of this kind have values of roughly 2 to 4 for herbaceous crop plants (cereals, legumes and root crops). Similar values have been calculated for the trees of temperate and tropical forests (3–4), which appears surprising at first, since the heterotrophic mass of trees is usually much larger than that of herbs. Apparently the ratios of the respiring mass of the non-green parts of the plant (cortex and woody parenchyma, meristem) to the mass of the green tissues provides a well-balanced ratio between assimilation and dissimilation in the various growth forms.

The dimension of the gas-exchange balance is g or kg CO_2 per plant, per day or per year. Conversion factors can be used to express this value in terms of organic dry matter or carbon content (see p. XII), so that from the *gas exchange* of a plant, its *production of organic substance* can be calculated.

2.3.2 Dry Matter Production

Assimilated carbon that is not lost by respiration (i.e. the surplus in the CO_2 budget) increases the dry matter of a plant and can be used for growth and for accumulating reserves. There is a direct and measurable correlation between the surplus in the CO_2 budget and the increase in dry matter (Fig. 2.61). The rate of increase rises with the CO_2 gains and thus correlates with the photosynthetic capacity. C_4 plants such as sugar cane, maize and millet achieve production rates twice to three times those of C_3 crop plants such as sugarbeet, alfalfa or tobacco.

The increase in mass of a plant due to the products of assimilation (production rate, PR, or productivity), expresses the increase in dry matter per unit time (day, week) during the period of production.

$$PR = d \cdot W/dt \qquad\qquad (2.15)$$

The production rate is determined by periodic harvesting. Values for dry mass cannot be obtained simply by weighing dry plant material, because the latter contains mineral substances (an average of $3-10\%$ of the dry weight; see Table 3.1) in addition to carbon compounds. From the raw dry weight the ash weight must then be subtracted, and what remains is ash-free *organic mat-*

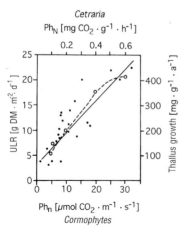

Fig. 2.61. Correlation between net photosynthesis (under optimal conditions) and dry matter increase (unit leaf rate ULR) in herbaceous C_3 plants and *Salix aquatica* (*dots, straight line*) and in the lichen *Cetraria nivalis* (*circles, dashed line*). (After Kärenlampi et al. 1975; Konings 1990). Relation between growth rate and physiological and morphological traits of tree seedlings: Walters et al. (1993). Concerning different degrees of correlation between CO_2 uptake and growth in fast- and slow-growing species see Pereira (1994)

ter. Although the production of dry matter can be determined directly by harvesting methods the procedure is destructive, it requires large amounts of material (many replicate samples), and is not feasible in every case (e.g. difficult with large trees).

The production rate of individual plants can be expressed as the relative growth rate, RGR [23], or as the unit leaf rate, ULR [24], which is synonymous with the "net assimilation rate" NAR [81]. These terms and descriptions come from **growth analysis**, in which harvesting methods are employed to investigate the production pattern of different species, especially crop plants and tree seedlings.

The *relative growth rate* expresses the increase in dry matter, DM, per unit time with respect to the initial dry weight, W, of the plant:

$$\text{PR as RGR} = \frac{dW}{dt} \cdot \frac{1}{W} \quad (\text{g org DM g}^{-1} \text{t}^{-1}) \tag{2.16}$$

The unit leaf rate (ULR) expresses the increase in dry matter for the total leaf area, A, responsible for this increase.

$$\text{PR as ULR} = \frac{dW}{dt} \cdot \frac{1}{A} \quad (\text{g org DM dm}^{-2} \text{t}^{-1}) \tag{2.17}$$

In the formula (2.17), it is assumed that the leaf area remains constant during the increase in dry matter. This is usually not the case. An exact growth analysis therefore employs dynamic formulae that take into consideration the exponential growth of the leaf area. Analytic models provide insight into the complex interplay between endogenous growth patterns and the diverse effects of environmental factors [178].

The unit leaf rate is a measure of the productivity of different plant species and varieties during their individual development under the particular environmental conditions prevailing over this period. The ULR is especially high during the phase of intensive growth and therefore production values measured at this time can only be compared with values obtained during a corresponding phase of growth. To characterize a plant group with respect to its productivity it is necessary to know the maximal values during the **main growth period** and the mean values for ULR, averaged over the period of assimilation. As is to be expected, herbaceous plants, and particularly the C_4 types and highly-bred C_3 plants, have the highest growth rates (Table 2.16).

The production rates of ecophysiologically differing *"constitutional types"* are closely related to their morphological and functional fitness to live under the conditions of the particular habitat. For instance, high productivity under favourable abiotic environmental conditions is essential for rapid growth and spreading of ruderal plants, pioneer trees and species with successful *competitive strategies* [83]. Conversely, in cold or dry habitats, or where the soil is poor in nutrients, the plants are programmed for low but reliable growth rates (e.g. evergreen dwarf shrubs, cushion or rosette plants, succulents), which

Table 2.16. Maximal and average production rates of vascular plants (ULR: g dry matter per m^2 leaf area and day). Data are from numerous authors

Plant group	Maximal ULR	Average for the growing season
C_4 plants	40 – 80	20 – 30
C_3 plants		
Rice	27	18
Temperate grains and meadow grasses	10 – 20	5 – 15
Herbaceous dicotyledons	10 – 25 (50)	5 – 10
Herbaceous legumes	14 – 18	
Floating plants		5 – 10
Tropical and subtropical woody plants	3 – 5	1 – 2
Deciduous broadleaved trees (saplings)	3 – 10	1 – 1.5
Conifers (saplings)	1 – 5	0.3 – 1
Ericaceous dwarf shrubs	1.5	0.5 – 1
CAM plants	6 – 10	2 – 5

enables them to maintain a balanced carbon, mineral and water budget (*stress strategy*).

The **production yield** Y_P is the cumulative increase in biomass and can be expressed as an integral function. It is the difference between the organic dry matter at the beginning (t_1) and the end (t_2) of the growth period.

$$Y_P = \int_{t_1}^{t_2} PR \cdot dt \tag{2.18}$$

Like the annual CO_2 balance, the production yield, too, is dependent on the assimilatory capacity of the plant as well as the length of the growth period and the effects of favourable and unfavourable environmental factors. It is thus the production yield that is the essential quantity in evaluating the inherent functional characteristics of plants, environmental effects or ecological measures, or for applied research.

Environmental factors affect dry matter production by way of their effects on CO_2 exchange and the carbon balance. Under increasing radiation (higher intensity and/or longer exposure) production is greater; like photosynthesis, dry-matter production exhibits a temperature optimum, and both water deficiency and inadequate or unbalanced provision of mineral nutrients reduce production. However, increase in biomass is not dependent on CO_2 uptake alone; the *allocation of assimilates,* which is controlled by hormones, and the specific *growth pattern* also play crucial roles. Although these processes are affected by environmental factors in the same direction as CO_2 exchange, it is not always to the same extent. For example, leaf growth is reduced earlier and to a greater extent by increasing water deficit, and particularly by nutrient deficiency, than CO_2 uptake (Fig. 2.62). Under these conditions the lower production is therefore largely due to inadequate leaf growth. The temperature patterns of the various processes involved in production and growth differ con-

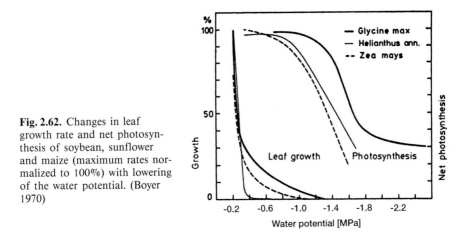

Fig. 2.62. Changes in leaf growth rate and net photosynthesis of soybean, sunflower and maize (maximum rates normalized to 100%) with lowering of the water potential. (Boyer 1970)

siderably with regard to the position of the limiting values and the width of the optimum (Fig. 2.63). Thus complete agreement between gas exchange and production rate is not always to be expected.

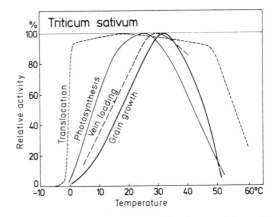

Fig. 2.63. The effect of temperature on various processes involved in the provision of photosynthates for growth, in wheat plants during grain development. The net photosynthesis of the flag leaf, the active export of photosynthates (vein loading) and the rate of carbon import by the grain proceed optimally within a narrow range of temperatures which is different in each case. Translocation in the phloem is largely independent of temperature, being interrupted only by frost or heat so great as to damage the plant. (Wardlaw 1974; Wardlaw and Passioura 1976)

2.3.3 Utilization of Photosynthates and the Rate of Growth

Plants consist largely of carbohydrates; they comprise 60% or more of the dry matter of higher plants. The carbohydrates produced in CO_2 assimilation must be distributed throughout the plant in a systematic but flexible way, to meet the varying needs of the individual organs. Thus distribution is controlled by demand (for maintenance, growth, storage), and by coordinating mechanisms, some of which involve phytohormones.

2.3.3.1 Translocation of Photosynthates

The products of CO_2 assimilation are continuously translocated from their *site of production* or *"source"* in the photosynthetically active tissues to the tissues and organs where they are utilized or stored (*"sinks"*: meristematic tissues, seeds, fruits, storage tissue). The translocation of carbohydrates within the cell and their movement from the mesophyll via transfer cells to the phloem sieve tube-companion cell complex is effected by a proton symport mechanism in some plants (e.g. Fabales, sugarbeet), or by passive transport in other species. Unloading, however, can proceed along the concentration gradient existing between "source" and "sink".

Transport of assimilates through the phloem is slow due to the high filtration resistance presented by the sieve plates. Nevertheless, because the sap is usually very concentrated, the transport capacity of the flow of assimilates is considerable. Translocation is furthered by adjustment of the conducting capacity of the phloem to the production capacity of the leaf (Fig. 2.64), and is regulated by phytohormones, the most important of which are auxin, cytokinins and abscisic acid.

Translocation follows the concentration gradient between sites where the *synthesis or mobilization of photosynthates* occurs, to those where the requirement is great (*attraction centres*). Fully developed leaves preferentially supply

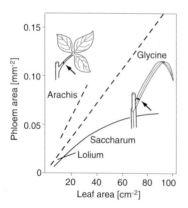

Fig. 2.64. The relationship between leaf area and cross-sectional area of the phloem in the petiole of Fabaceae (*Arachis hypogaea* and *Glycine max*) and in the base of the leaf blade of Poaceae (*Saccharum officinarum* and *Lolium perenne*). (After Lush et al., from Wardlaw 1990)

the "consumer" that exerts the greatest attraction; in cereals, for example, the lowest leaves supply the root system, and those higher up the plant supply the growth zones of the shoot and especially the flowers and ripening fruit (Fig. 2.65). In herbaceous Fabaceae the photosynthates flowing out of a *single* leaf are distributed among *several* consumers (flowers, fruit), and conversely the individual consumer draws on a combined stream from several leaves

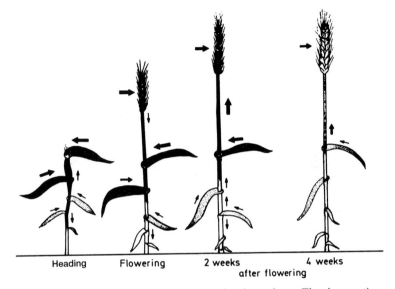

Fig. 2.65. Formation and partitioning of photosynthates in wheat plants. The *denser stippling* indicates regions of particularly productive assimilation, and the *thickness of the arrows* shows the relative rate of translocation of the products. (Stoy 1965). For carbohydrate sources and sinks in woody plants see Dickson (1991) and Kozlowski (1992)

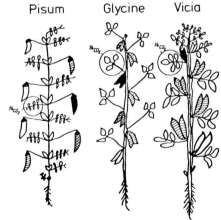

Fig. 2.66. Redistribution of photosynthates from the leaves to the ripening fruits, in pea, soybean, and broadbean plants. The black pods contain 35−60% of the assimilates formed in the leaf labelled with ^{14}C, the hatched pods 10%−35%, the cross-hatched pods 5%−10%, and the remaining pods less than 5% (Bartkov and Zvereva 1974)

(Fig. 2.66). This balances the supply of materials in the plant and ensures that the various storage tissues are equally provisioned.

For well-balanced development of the whole plant, it is important that the individual tissues and organs receive the needed amount of photosynthates at the right time. By *changing priorities,* it can be ensured that growing organs with high requirements receive adequate supplies, and resting parts of the plant are not oversupplied. Here, too, phytohormones play a role in regulating the pattern of distribution to meet the varying requirements for development and function. Detailed knowledge of the physiological mechanisms regulating the distribution of assimilates in the plant, and in particular the investment of assimilates in seeds, fruit and storage organs, is important for analysis of its *reproductive capacity.* In cereal grains the proportion of the whole-plant dry matter that is found in the harvested grain kernels (the *harvest index*) varies from about 25% (older varieties of maize, rye) to 50% (rice, barley). In Fabales the harvest index ranges from about 30% (soybean) to 60% (garden bean). By breeding to favour allocation of photosynthates to the seeds it has been possible to increase the proportion of maize kernels to the total shoot mass from 24 to 47%, while that of rice has been increased from 43 to 57% [280].

2.3.3.2 Costs and Benefits of a Leaf

Expenditure for growth and differentiation is not limited to the dry substance incorporated into tissue structures. In addition to the utilization of carbohydrates to produce supporting substance and to provide the carbon skeletons of all organic compounds in the *plant,* they have to cover the requirements for *metabolic energy* (for respiration and transport of minerals) and for all other biosyntheses. The costs for the development of a new leaf can be calculated in the form of glucose equivalents, or the carbon or energy requirements (Fig. 2.67). Long-lived and especially scleromorphic leaves are more "expensive" than short-lived soft leaves, due to their more densely concentrated structure and the elaborate composition of their organic matter, including more biochemically costly compounds like proteins, lipids, lignins and secondary metabolites. This is true not only of leaves, of course, but also holds for other parts of the plant, particularly those with a supporting function. This is why, in dry regions, where assimilation is often restricted, shoots are often lighter in structure, i.e. a supporting network instead of massive wood (e.g. the xylem of cacti).

The wide diversity of leaf structure and life span that has evolved and proved successful suggests that the advantages and disadvantages connected with a particular leaf type balance out to meet the requirements of the individual life form and living conditions. Although evergreen leaves require relatively large quantities of assimilates for their development, this, in the long run, proves to be economical for the plant as a whole; the spruce, for example, need only renew 15% of its needle mass [215] each year in order to maintain the density of its crown. If the functional life span of the leaves is short, there are

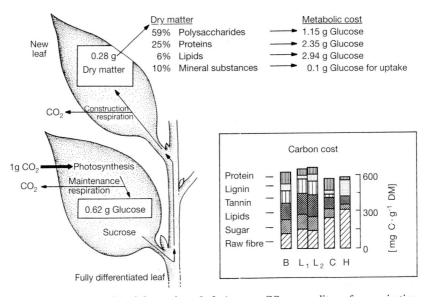

Fig. 2.67. Carbon costs of leaf formation. *Left* Average CO_2 expenditure for respiration and production of dry matter. *Above* Average chemical composition of leaf dry matter and the cost of biosyntheses (requirement in glucose units per g of the respective plant substance). (After Mooney 1972). *Below right* Carbon costs per g dry leaf tissue in different tundra plants, showing the most important classes of substances in the dry matter. B: *Betula nana* (small deciduous shrub); L_1, L_2: *Ledum decumbens*, 1-year and 2-year-old leaves (small evergreen shrub); C: *Carex bigelowii* (sedge); H: *Hylocomium splendens* (moss). (After Chapin 1989). For cost-benefit relationship and life span of leaves of tropical tree species, see Reich et al. (1991), Sobrado (1991)

the repeated costs of producing new leaves (albeit lower than for evergreen leaves), but this disadvantage is overcome by the considerably greater photosynthetic activity of newly opened leaves, both with respect to photosynthetic capacity (see Fig. 2.24) and to the greater activity of their nitrogen metabolism. By shading strawberry plants, the ageing process in their leaves can be delayed by as much as a third of their usual life span, with a resultant increase of 80% in their life-total of photosynthates [108].

Costs and benefits of a leaf, however, involve more than simply its carbon budget. The specific life spans of leaves have *coevolved with the ecosystem partners*. For their protection against herbivores and parasites, long-lived leaves require mechanical support and a wide variety of secondary metabolites (see Chap. 1.1.4.2), which means additional costs for complicated biosyntheses. Short-lived, soft but photosynthetically highly efficient leaves, if exposed to intensive pressure from herbivory, are prematurely lost to the plant and their benefits are thus considerably reduced (Fig. 2.68).

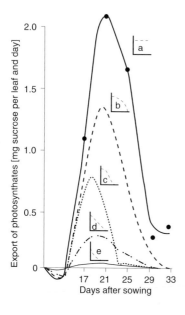

Fig. 2.68. Model of the availability of carbohydrate assimilates (sugar export from the leaves into the plant) in relation to the functional lifetime and grazing risks of leaves (patterns of leaf survivorship: small inserted diagrams). *a* No losses, functional up to senescence; *b* damage with increasing age (most common pattern in herbaceous plants); *c* sigmoid pattern for grazing-resistant older leaves; *d* equal risk of damage for leaves of all age groups (e.g. grasses); *e* greatest danger during leaf spread, risk of damage decreases with age (e.g. deciduous woody plants). (From Harper 1989)

2.3.3.3 Life Form and Utilization of Photosynthates

The level of organization and the life form of a plant determine which of a number of different *patterns of photosynthate utilization* governs its production and growth, its competitive powers and its response to habitat-specific constraints. Some examples of such different patterns are described in the following paragraphs.

Production for Population Growth: the Expansion Type

Photoautotrophic unicellular organisms are self-supporting with respect to carbon, and do not have to supply other cells. Within the cell there is a favourable ratio between compartments that **produce** and those that **consume** photosynthates: in *Chlorella,* for example, the chromatophores occupy about half of the protplasmic volume. Given this advantage it is not surprising that, with a good supply of nutrients and light, algal cells can accumulate large surpluses of photosynthates, can quickly attain their maximal size and proceed to divide. Thus an autotrophic unicellular organism uses its carbon gains to *increase the number of individuals,* i.e. for reproduction. The result of a strongly positive carbon balance is a rapid increase in population density. Since there is a direct relationship between photosynthetic rate and the number of divisions per day, the rate of growth in phytoplankton can usefully be expressed as the increase in population density or the number of divisions per unit time.

Rapid Carbon Gains: the Investment Type

The investment type of plant is characterized by high photosynthetic capacity and a high proportion of photosynthetically active tissue in its total mass (at least 50%). During the growth phase the carbon "income" of the plant is mainly used for producing the leaves which then serve to increase the carbon gains of the plant. During and after the flowering phase the distribution of photosynthates switches to favour the reproductive organs, while supplies to the other parts of the plant are reduced to little more than what is needed for maintenance purposes – the older leaves even wither and die off. Accordingly, the proportions of leaves, axial structures, roots and reproductive organs in a plant change considerably (Fig. 2.69) in the course of its life cycle.

The investment type of plant is best exemplified by *annuals*. These plants have to make full use of a short period of time during which conditions are favourable for growth, flowering and setting of fruit; they must employ their photosynthates in such a way that an abundance of tissue is formed in the shortest possible time. Annual species do this even when a fairly long period of time is available for growth. Summer grain, sunflowers, and other annual crops thus yield particularly large harvests. Under favourable environmental conditions, this way of investing the products of assimilation guarantees both luxuriant growth and prolific fruiting. On the other hand, when local conditions are less favourable, particularly if water is in short supply, or if the soil is poor in nutrients, the plant is compelled to build up an extensive root system; this is done at the cost of the shoot and leads to a smaller photosynthetic yield and lower reproductive potential.

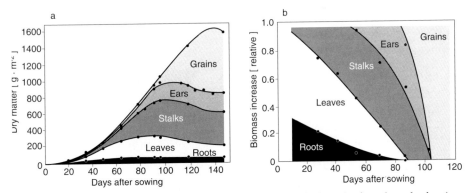

Fig. 2.69a, b. The partitioning of assimilates in an annual plant (spring-planted wheat). **a** Distribution of assimilates in the different organs during the course of development. **b** Allocation (relative increase in dry matter) to different organs. Note the changing priorities in assimilate allocation from the stage of germination, through the phase of vegetative development of the shoot, up to the reproductive phase. (After Fischer, as cited in Schulze 1982)

Saving for Safety: the Conservative Budget

Biennial and perennial herbaceous species, which budget their photosynthates in this way, achieve a lower net carbon gain and thus grow more slowly than the investment type of plant. On the other hand, they are able to survive under the unfavourable conditions of dry or cold habitats, or where the soil is poor in nutrients. Their development is at first similar to that of annuals. After their vegetative structures are formed, however, they accumulate *reserve supplies* before proceeding with flower formation. Towards the end of the first growing season the photosynthates are allocated to the shoots and above all into the underground parts of the plant, which may develop into massive storage organs. Flowers are formed only after the plant has accumulated sufficient "capital" to draw upon for that purpose.

Photosynthates stored in one year are first used in the next to extend the shoot system (Fig. 2.70). The synthesis of new substance starts up quickly in this second year and, due to the availability of stored material, is largely independent of conditions affecting photosynthesis in the spring. Once the plant is ready to flower, if the nutrient supply is satisfactory, the flowers and fruit take precedence over storage processes. Afterwards, near the end of the growing season, photosynthates move preferentially to the underground parts of the plant, which increase correspondingly in weight.

Perennial plants are at an advantage wherever the period of time favourable to production is not long enough to permit sufficient assimilation for both vegetative growth as well as the formation of flowers and fruits, and also in cases where the plants bloom so early that the necessary materials cannot be provided by the available mass of leaves. This applies, for example, to the *spring geophytes*, many of which open their flowers before the leaves have un-

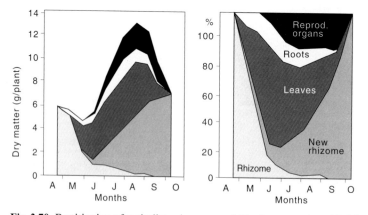

Fig. 2.70. Partitioning of assimilates in a perennial herbaceous plant (*Caltha palustris*). *Left* Distribution of assimilates within the various organs. *Right* Allocation of dry matter within the plant. (After Eber 1991). Earlier studies on *Aconitum*: Iwaki and Midorikawa (1968); on *Scilla*: Goryshina (1969); on potatoes: Sale (1974)

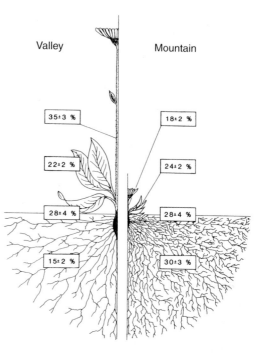

Fig. 2.71. Partitioning patterns of dry matter in perennial herbs at low altitudes (600 m a.s.l.) and in high mountains (2600–3200 m a.s.l.). The mountain plants allocate more assimilates to the belowground organs and less to flowers and fruit than do the plants in the valleys; in both situations the leaves receive similar amounts of assimilates (referred to the respective total mass). After Körner and Renhardt 1987)

folded. Plants of the *high mountains and the artic* must accomplish flowering and ripening of seeds during the short summer; their photosynthetic gain is subject to many uncertainties. Finally, these considerations also apply to *steppe plants*, which utilize the time between winter cold and summer drought to complete their life cycles. All of these plants require the presence of storage organs such as rhizomes, tubers, roots or bulbs. Moreover, such species frequently develop an extensive root system, in the case of the high-mountain species at the expense of the shoot and of flowers, although the ratio of leaf mass to total mass remains almost unchanged (Fig. 2.71).

Increasing Mass by Longevity: the Accumulation Type

A tree – the most highly differentiated and largest form among plants – manages its carbon supplies in a way suited to its long lifetime. Woody plants expend large quantities of photosynthates for the production of supporting and transporting tissues. The high investment in supporting tissues is necessitated by tree architecture; in regions where the annual assimilation period is long enough, it gives them decisive competitive advantages over the herbaceous plants, which are slowly but surely overshadowed by the ever taller woody plants.

In the first years of life, the leaf mass can make up half of the overall dry matter of a tree, but with increasing size the ratio of leaf mass to stems is altered, the leaf mass growing only slightly while trunk and branches become

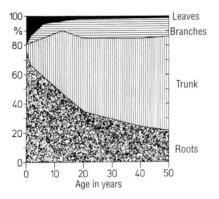

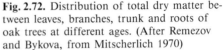

Fig. 2.72. Distribution of total dry matter between leaves, branches, trunk and roots of oak trees at different ages. (After Remezov and Bykova, from Mitscherlich 1970)

steadily thicker and heavier (Fig. 2.72). Foliage comprises only 1% – 5% of the total mass of mature trees (see Table 2.15), and these leaves must thus supply the material for maintenance and growth to parts of the tree amounting to many times their own weight. The consequence is a gradual decrease both in productivity and growth rate.

The long-term advantages of woody plants in competition with herbaceous species are counterbalanced, and sometimes outweighed, by the disadvantages of slower production and the more complicated mode of distribution of photosynthates. Because trees cannot adjust so easily to changes in their surroundings, they are often more severely affected by environmental constraints than herbaceous forms. This explains the observation that along stress gradients trees disappear sooner than the low-growing herbaceous vegetation. The polar, altitudinal and arid forest limits and treelines are not solely due to damage caused by extreme environmental conditions; even earlier there is a drop in productivity, growth and reproductive capacity as a consequence of the increasingly less favourable carbon balance.

The Seasonal Dynamics of the Allocation of Photosynthates in Woody Plants

In *deciduous* trees, the carbohydrate stores are emptied shortly before the leaves begin to unfold, the assimilates being moved to the buds and later to the new shoots (Fig. 2.73). About a third of the reserve material serves for the unfolding of the leaves which will soon contribute to the further formation of leaves and shoots in the new growth. Later, flowers and developing fruit are supplied preferentially, next in order is the cambium, and last the newly forming buds and the starch depots in roots and bark. The differentiation of flower buds is governed not only by external and endogenous signals but also by the remaining supplies. At the end of the growing period the surplus photosynthates are translocated to the woody tissues and bark of branches, trunk and roots where they are stored (Fig. 2.74). Trees in the tropics and in dry regions pass through several seasonal storage periods – four in the case of fig trees.

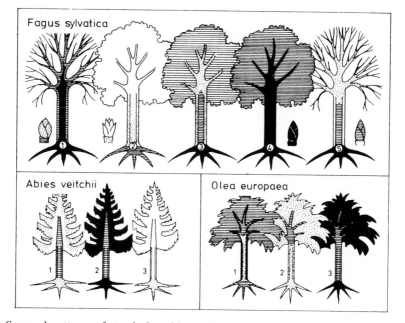

Fig. 2.73. Seasonal patterns of starch deposition and mobilization in trees. Maximal accumulation of starch is indicated by *black*, large amounts by *hatching*, and small amounts by *stippling*; in parts left *white*, starch is present in traces or not at all. *Fagus sylvatica* (central Europe): *1* Just before leaf emergence in the spring; *2* during leaf unfolding; *3* midsummer; *4* just before leaf abscission in the autumn; *5* conversion of starch to soluble carbohydrates at low temperatures during winter. (After Gäumann 1935). *Abies veitchii* (Japan): *1* During growth of new shoots in spring; *2* late summer; *3* during winter frost. (Kimura 1969). *Olea europaea* (northern Italy: *1* During shooting and flowering in spring; *2* during a dry period in midsummer; *3* in winter (frost-free) after end of the rainy season. (After Larcher and Thomaser-Thin 1988)

Evergreen woody plants of the temperate zones do not produce new shoots as soon as the winter dormant period is past, for they still have the needles or leaves of the previous year. If the weather is favourable, these continue to take up CO_2 (although in smaller amounts) during late autumn, winter and early spring. When buds do begin to open, the carbon taken up by these old organs in the spring can meet a large part of the demand, and the rest is taken from the reserves in stems and roots. Evergreen woody species, due to their extended assimilation period, usually gain dominance over deciduous species wherever a long winter, or a dry summer, restricts the growing season, i.e. in the mountains, in the northern forest belt, and in regions where aridity limits tree growth. Only in regions where the unfavourable season is extremely harsh and prolonged (subarctic, eastern Siberia, semideserts) do deciduous trees and shrubs again predominate.

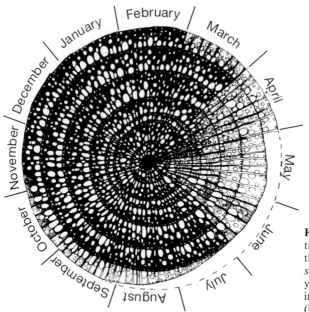

Fig. 2.74. Starch deposition and mobilization in the root wood of *Acer saccharum* throughout the year. Tissues containing starch are *black*. (From Wargo 1979)

2.4 The Carbon Budget of Plant Communities

2.4.1 The Productivity of Stands of Plants

The quantity of dry matter formed by the vegetation covering a given area is called the *primary production* (PP) of a plant community. Primary production is greater the higher the production rates of the species of which the stand is composed, the more completely the light passing through an extensive system of *assimilating surfaces* is absorbed and the longer the time in which the plants can maintain a positive gas-exchange balance (duration of the *assimilation period*).

The *production* of a *plant stand* is expressed as organic dry matter per unit time, and refers to the *area of ground covered by the vegetation* (not to the leaf surface as in the case of single plants). The **production rate of a plant stand,** or the *crop growth rate* (CGR) is derived from the unit leaf rate of the individual plant (ULR) and the leaf surface per ground area (LAI; leaf area index) of the plant stand.

$$CGR = ULR \cdot LAI \text{ (g org DM m}^{-2} \text{ ground area t}^{-1}) \tag{2.19}$$

It should not be forgotten that both parameters change continuously during the course of development.

Stands of C_4 grasses in the tropics and subtropics achieve maximal values of $50-60$ g DM m^{-2} d^{-1} during the main growing season, whereas with the lower level of irradiation in the temperate zone, maximal values are about $20-30$ g DM m^{-2} d^{-1}. Stands of C_3 crop plants produce at most $15-30$ g DM m^{-2} d^{-1}, depending on the species, though the average for the entire growing season is only one-third to half of this [157]. Table 2.17 shows maximum yields of dry matter for various crops and plantations.

As the **leaf area index** increases, a greater photosynthetically active surface area becomes available, and it would therefore be expected that the production rate would be greater the higher the LAI. This is indeed so for low LAI values, but when the plants become more closely spaced and the foliage overlaps in many places, then the light in the most shaded places is no longer sufficient to maintain a positive CO_2 balance at all times. The production of the entire stand, the CGR, is therefore reduced (Fig. 2.75).

Thus there is an *optimal* leaf area index for production, which, as a rule, is attained when the radiation is most evenly absorbed in its passage through the leaf canopy. In herbaceous stands with horizontal leaves this is reached at a LAI of 4 to 6, and in grasses at values of 8 to 10. Employing remote measurement of the NIR/red ratio ($0.7-1.1$ μm/$0.6-0.7$ μm), it is possible to determine the LAI of many square kilometres of vegetation.

The leaf area index and the density of the plants (that is, the degree of plant cover) are not only important factors affecting production: each is itself affected by production. Table 2.18 lists the most common and the extreme LAI values for a variety of plant communities. In very dense forests an increase in the LAI above 15 m^2 m^{-2} is limited by lack of light. In open plant communities on poor stony soil, or where the ground is too dry, too cold or too salty, the LAI drops to minimal values, but in this case the drop is chiefly due to the decreasing degree of plant cover (Fig. 2.76).

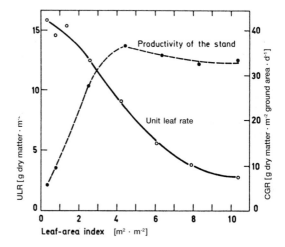

Fig. 2.75. Relationship between unit leaf rate (ULR) of single maize plants and the production rate of stands of maize (crop growth rate, CGR) as a function of leaf-area index. (After Williams et al., as cited in Baeumer 1991)

Table 2.17. Dry matter production and harvest index of stands of crop plants. The annual production yield is shown in kg total dry matter per m² of ground, the harvest index expresses the economically relevant contribution (S = seeds and grains, F = fruits, M = aboveground mass, R = storage organs, W = wood) to the total annual increase in dry matter. (Based on compilations by Lieth 1962; Loomis and Gerakis 1975; Cooper 1977; Osmond et al. 1982; Schulze 1982; Jarvis and Leverenz 1983; Nobel 1988; data from Sirén and Sivertsson 1976; Wolverton and McDonald 1979; Eagles and Wilson 1982; Nobel et al. 1992a)

Plant species	Maximal yields ($kg\ m^{-2}\ a^{-1}$)	Harvest index[a]
C_4 grasses		
Sugar cane	6 – 8	0.85 (M)
Maize (subtropics and tropics)	3 – 4	
Maize (temperate zone)	2 – 4	0.4 – 0.5 (S)
Tropical millet	4 – 5	0.4 (S); 0.88 (M)
Tropical fodder grasses	3 – 8	0.85 (M)
C_3 grasses		
Rice	2 – 5	0.4 – 0.55 (S)
Wheat	1 – 3	0.25 – 0.45 (S)
Barley	ca. 1.5	0.32 – 0.52 (S)
Meadow grasses	2 – 3	0.7 – 0.8 (M)
Swamp grasses	5 – 10	
Leguminosae		
Lucerne	3	
Soybeans	1 – 3	0.3 – 0.35 (S)
Root crops		
Manihot esculenta	3 – 4	0.7 (R)
Sugar beet	2 – 3	0.45 – 0.67 (R)
Potatoes	2	0.82 – 0.86 (R)
Ipomoea batatas	2	
Helianthus tuberosus	2	0.75 (R)
Oil palm	2 – 3	
Forests plants		
Cryptomeria japonica	5.3	0.65 (W)
Pinus radiata	4.6	0.66 (W)
Pseudotsuga menziesii	2.8	0.71 (W)
Pinus nigra	2.5	0.46 (W)
Picea abies	2.2	0.61 (W)
Fagus sylvatica (60-yr)	1.3	0.70 (W)
Intensive short rotation forestry		
Willows	5.0	0.6 (W)
Poplar hybrids	3.5 – 4	0.5 (W)
Eucalyptus grandis	4.1	
Hevea brasiliensis	2.5 – 3.5	
CAM plants		
Pineapple plantations	2 – 3	
Agaves	1 – 2.5 (4[b])	
Cacti	0.8 – 1.7 (4.5[b])	0.2 – 0.35 (F)

Table 2.17 (continued)

Plant species	Maximal yields (kg m^{-2} a^{-1})	Harvest index[a]
Aquatic plants		
Eichhornia crassipes	15 – 20	
Waste water algae	3.5 – 9	
Seaweeds	3 – 5.5	

[a] Harvest index = (economic yield/production yield) × 100.
[b] With fertilizations and daily irrigation.

A relationship similar to that between LAI and stand production exists between the latter and the amount of chlorophyll per m^2 of ground (see Table 2.18). Another measure used to characterize the degree of overlapping of photosynthetically active layers in bodies of water is the chlorophyll content of the plankton living in the column of water below 1 m^2 of water surface. The use of spectral analysis to measure chlorophyll density per unit water surface has made it possible, knowing the NIR/red ratio, to calculate the production potential of large-scale areas of the oceans.

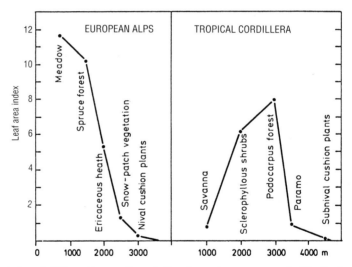

Fig. 2.76. Leaf area index of different types of vegetation through a sequence of altitudinal belts in the central Alps and in the coastal Cordillera of Venezuela. Where closed communities give way to open plant communities the leaf area index decreases sharply. (After Vareschi 1951, 1953)

Table 2.18. Annual net primary production, leaf area index and chlorophyll content of biomes. (Whittaker and Likens 1975; for details see Ajay et al. 1979; Schulze 1982)

Type of vegetation	Area (10^6 km^2)	Net primary production			Leaf area index		Chlorophyll concentration Mean (g m^-2)
		Range (kg m^-2 a^-1)	Mean (kg m^-2 a^-1)	Area total (10^12 kg)	Range (m^2 m^-2)	Most frequent value (m^2 m^-2)	
Continents	**149.0**		**0.78**	**117.5**		**4.3**	**1.5**
Tropical rainforests	17.0	1 – 3.5	2.2	37.4	6 – 16	8	3.0
Deciduous woodland (semiarid)	7.5	1.6 – 2.5	1.6	12.0		5	2.5
Deciduous forests (temperate)	7.0	0.4 – 2.5	1.2	8.4	3 – 12	5	2.0
Evergreen temperate forests	5.0	1 – 2.5	1.3	6.5	5 – 14	12	3.5
Boreal forests	12.0	0.2 – 1.5	0.8	9.6	7 – 15	12	3.0
Dry scrub and sclerophylls	8.5	0.3 – 1.5	0.7	6.0	4 – 12	4	1.6
Savannas	15.0	0.2 – 2	0.9	13.5	1 – 5	4	1.5
Meadows and steppes	9.0	0.2 – 1.5	0.6	5.4		3.6	1.3
Tundra	8.0	0.01 – 0.4	0.14	1.1	0.5 – 2.5	2	0.5
Shrub deserts	18.0	0.01 – 0.3	0.9	1.6		1	0.5
Dry and cold deserts	24.0	0 – 0.01	0.003	0.07		0.05	0.02
Agricultural crops	14.0	0.1 – 4	0.65	9.1	4 – 12	4	1.5
Swamps, marshes	2.0	1 – 6	0.3	6.0		7	3.0
Inland waters	2.0	0.1 – 1.5	0.4	0.8			0.2
Oceans	**361.0**		**0.155**	**55.0**			**0.05**
Open ocean	332.0	0.002 – 0.4	0.125	41.5			0.03
Upwelling zones	0.4	0.4 – 1	0.5	0.2			0.3
Coastal zones	26.6	0.2 – 0.6	0.36	9.6			0.2
Reefs and tidal zones	0.6	0.5 – 4	2.5	1.6			2.0
Brackish water	1.4	0.2 – 4	1.5	2.1			1.0
Global total (earth)	**510**		**0.336**	**172.5**			**0.48**

2.4.2 The Net Primary Production of the Vegetation of the Earth

The *yield of the dry matter production of a plant community* (Y_{PP}) is called **net primary production** and is expressed as kg dry matter per m^{-2} of ground area. Estimations of the net primary production on a regional and global scale are still difficult and imprecise, and the estimates made by different authors diverge considerably. Nevertheless, more recently good progress has been made in obtaining data by harvesting techniques and remote spectral analysis (Fig. 2.77), and in the use of this data for computer calculations.

High **net primary production** is limited to those regions of the continents and oceans that offer the plants a favourable combination of water, warmth and nutrients. This occurs on *land* in the humid tropics, and in water, in the zone between the latitudes 40° and 60°N and S (Fig. 2.78a). The most abundant production, however, is found in belts where land and water meet – in shallow water near the coasts and on coral reefs, in swamps and swamp forests in warm climates (Table 2.18 and Fig. 2.79). In warm, humid regions, production is limited in places by shortage of mineral substances, and in very dense stands also by deficiency of light. The greater part of the earth's surface area, both land and water, permits only moderate production. On 41% of the land, the crucial factor limiting yield is water shortage (Fig. 2.80), and on 8% unfavourable temperatures are responsible for lower production (short growth peri-

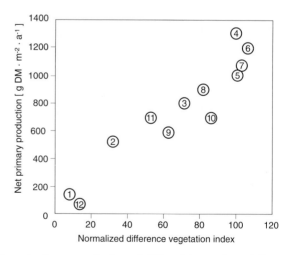

Fig. 2.77. Relationship between the net primary production of different types of vegetation and the Normalized Difference Vegetation Index based on spectral reflection measurements via satellites (Advanced Very High Resolution Radiometer). *1* Tundra; *2* tundra-taiga ecotone; *3* boreal coniferous belt; *4* humid temperate coniferous forests; *5* transition from coniferous to deciduous broadleaved forests; *6* deciduous forests; *7* oak-pine mixed forests; *8* pine forests; *9* grassland; *10* agricultural land; *11* bushland and similar woodland formations; *12* deserts. (Goward et al. as cited in Field 1991)

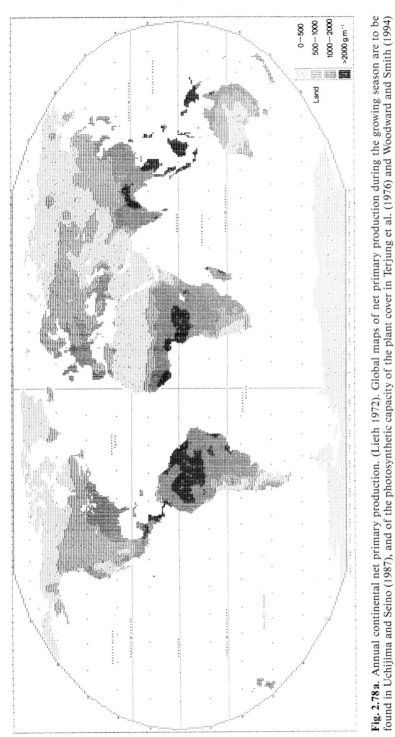

Fig. 2.78a. Annual continental net primary production during the growing season are to be found in Uchijima and Seino (1987), and of the photosynthetic capacity of the plant cover in Terjung et al. (1976) and Woodward and Smith (1994) production. (Lieth 1972). Global maps of net primary

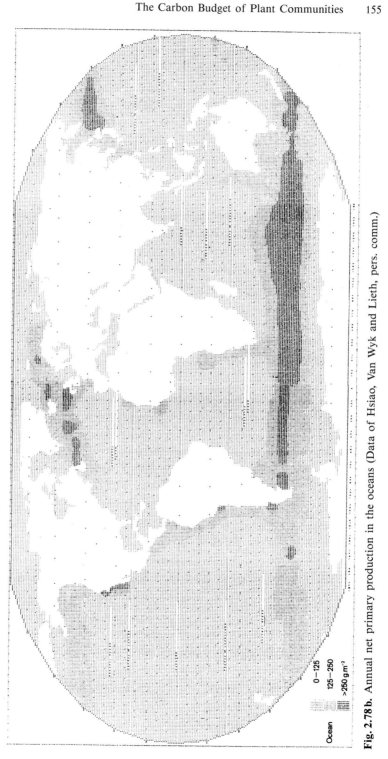

Fig. 2.78b. Annual net primary production in the oceans (Data of Hsiao, Van Wyk and Lieth, pers. comm.)

Ocean

0–125
125–250
>250 g.m⁻²

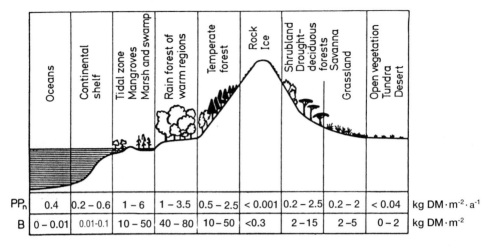

PP$_n$	0.4	0.2 – 0.6	1 – 6	1 – 3.5	0.5 – 2.5	< 0.001	0.2 – 2.5	0.2 – 2	< 0.04	kg DM·m^{-2}·a^{-1}
B	0 – 0.01	0.01-0.1	10 – 50	40 – 80	10 – 50	<0.3	2 – 15	2–5	0 – 2	kg DM·m^{-2}

Fig. 2.79. Annual net primary production (PP_n) and phytomass (B) in different biomes over the earth. (Based in part on E. P. Odum 1971, with values from calculations by Bazilevitch and Rodin 1971; Whittaker and Likens 1975; Ajtay et al. 1979)

od due to cold; cool summers). In tropical *oceans* it is deficiency of nutrients, and in seas near the poles inadequate light, that limits productivity (Fig. 2.78 b).

Plant breeding and cultivation in combination with pest control measures can achieve yields far in excess of those of the primary production of naturally occurring plant plant communities of a particular region (see Table 2.17). The highest known dry matter production has been achieved by algae in culture tanks (equivalent to about 10 kg m^{-2} a^{-1} [115]). For land plants, too, (including trees) the annual production of biomass can be increased to twice the normal value by the employment of special measures, such as continuous irrigation and fertilization at levels adjusted to the requirements of each growth

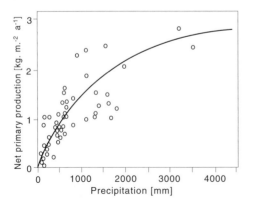

Fig. 2.80. Relationship between annual net primary production and average total annual precipitation. (Lieth 1975 a)

stage. Nevertheless, the average worldwide agricultural yields still remain far below those that could achieved, chiefly due to extensive instead of intensive use of the land, incomplete utilization of the natural production period, and the use of inferior seed varieties.

High **yields of biomass** open up the possibility of employing plants with an especially high productive capacity as sustainable sources of raw material, and for the production of bioenergy. In many cases, however, not all of the produced mass can be used. The utilizable portion of the total dry matter (*harvest index*) is 30–60% in the case of seeds, 85% for the green mass, 50–70% for wood (see Table 2.17). In every attempt to maximize the productivity of crop plants it should be remembered that intensive cultivation invariably requires considerable investment in technical equipment and involves high energy costs. If the efficiency of the investment necessary to achieve a substantial increase in production is calculated (in terms of energy input per unit energy of utilizable biomass produced), the efficiency of primitive methods of farming is seen to be 50-fold, whereas it is only 2- to 5-fold with modern agrotechnical methods. Thus although biomass production per unit area is far higher than in former times, the expenditure involved has risen by a still higher factor.

2.4.3 The Carbon Balance of Plant Communities

2.4.3.1 Gross Production and the Respiration of a Plant Community

The **gross primary production** (PP_g) of a plant community cannot be measured directly; therefore a rough estimate is computed from the net primary production (PP_n as the Y_{PP}), and the respiration of the plant community (ΣR).

$$PP_g = PP_n + \Sigma R \qquad\qquad (2.20)$$

The **"operating cost" of respiration** is expressed either as a percentage or as a decimal. Herbaceous plant communities consume 20–50% of the gross carbon uptake for respiration, whereas forests and dwarf shrub heaths, with a relatively large proportion of photosynthetically unproductive mass, use 40–60% of their gross carbon gain for respiration in temperate zones, and roughly 70% in the warm and humid tropics (Table 2.19).

2.4.3.2 The Fate of the Net Yield of Primary Production

The net primary production is used for building up organic matter, part of which is lost as litter (L) over the course of the year or is grazed by consumers (G). These losses include the shedding of leaves, flowers, fruits and dead branches, the decay of dead roots, consumption by animals and parasites, and the release of photosynthates to symbionts (e.g. mycorrhizal fungi) and via root excretions. The remaining net yield goes to increase the existing phytomass

Table 2.19. Productivity and loss of organic dry matter in forests (annual balance); all data refer to kg m^{-2} ground area. Detailed balances for coniferous forests of the temperate zone are given in Vogt (1991)

Stand	Beech forest 60-year-old Denmark[a]		Tropical rainforest Thailand[b]	
LAI	5.6		11.4	
Annual increase in stand biomass	ΔB	in % PPg	ΔB	in % PPg
Foliage	0		0.003	
Stems	0.53		0.29	
Roots	0.16		0.02	
Total	0.69	35	0.313	2
Annual litter L				
Foliage	0.27		1.2	
Stems	0.1		1.33	
Roots	0.02		0.02	
Total	0.39	20	2.55	20
$PP_n = \Delta B + L$	1.08	55	2.86	22
Annual consumption by respiration				
Foliage	0.46		6.01	
Stems	0.35		3.29	
Roots	0.07		0.59	
Total	0.88	45	9.89	78
$PP_g = PP_n + \Sigma R$	1.96	100	12.75	100
$CUE_{PP} = \dfrac{PP_n + \Sigma R}{\Sigma R}$	2.23		1.29	

[a] Mar-Möller et al. (1954).
[b] Kira et al. (1964); Yoda (1967).

per unit area of ground (increase in biomass $= +\Delta B$). The **ecological production equation** [22] describes the distribution of the carbon yield as follows:

$$PP_n = \Delta B + L + G \tag{2.21}$$

All the quantities in this production equation, i.e. change in biomass, litter, and losses to consumers, can be determined directly; their sum is the measure normally used to express net primary productivity (for an example of such a calculation see Table 2.19). Measurements of all components are not simple to make in natural stands of plants; thus, production data for forests and other perennial, multi-layered plant communities should be considered as guidelines only, unless they are confirmed by other procedures applied at the same time.

The fraction of the annual net primary production represented by the losses L and G is a critical factor in the carbon balance of the vegetation of an ecosystem (Fig. 2.81). The *annual litter production* [1, 175] is highest in tropical

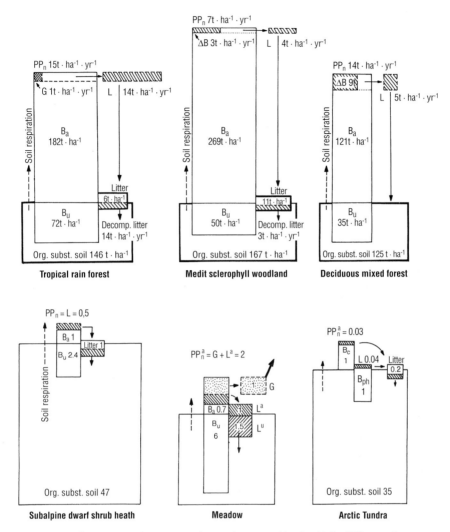

Fig. 2.81. Standing mass and turnover of organic matter (dry basis) in different plant communities: Tropical primary rainforest in Puerto Rico (Odum and Pigeon 1970), evergreen broadleaved woodland in Southern France (Rapp 1971), mixed deciduous broadleaved forest in Belgium (Duvigneaud and Denaeyer-De Smet 1970), subalpine dwarf shrub heath in the Alps (Larcher 1977, Schmidt 1977) meadow (Festucetum) in Moravia (Rychnovská 1979), arctic tundra in Canada (Bliss 1975). *Areas in heavy outline.* Pools of organic dry matter in $kg \, m^{-2}$; *hatched areas* annual turnover of dry matter in $kg \, m^{-2} \, a^{-1}$; B_a aboveground biomass; B_u underground biomass; B_c biomass of cryptogams; B_{ph} biomass of phanerogams (tundra); PP_n net primary production; PP_n^a aboveground net primary production; ΔB annual increase in biomass; L annual loss of dry matter as detritus; L^a aboveground losses; L^u underground losses; G annual loss of dry matter by grazing

forests, where it averages $2\,kg\,m^{-2}a^{-1}$ (although in the rainforests of southeast Asia it can be as much as $2.5\,kg\,m^{-2}a^{-1}$) and in savannas $(1.5\,kg\,m^{-2}a^{-1})$. In the majority of closed plant formations such as forests and meadows of the temperate zone, reeds, mangroves, macchia and dwarf shrub heaths $0.6-1.2\,kg$ of litter fall per m^{-2} each year. In deciduous woods of regions with occasional dry periods the litter amounts to $0.3-0.6\,kg\,m^{-2}a^{-1}$, and in steppes, semi-desert regions and tundra $0.1-0.5\,kg\,m^{-2}a^{-1}$ falls. The amounts of biomass lost to consumers differ enormously, depending on the type of vegetation; in any case losses are far smaller in forests than in grass communities.

Whether the phytomass of a plant community remains unaltered ($\Delta B = 0$), increases ($+\Delta B$), or temporarily decreases ($-\Delta B$) depends on the yield of the PP_n and the magnitude of the material losses suffered. Which of these possibilites is realized in a particular ecosystem is determined mainly by its species composition, dynamics (age of the stand and successional stage), and the severity of natural and anthropogenic constraints within this ecosystem. The richest and most persistent accumulation of biomass occurs in forests; more then three quarters of the carbon taken up by all terrestrial plants is stored in the wood produced by the earth's forests.

2.4.3.3 The Production Dynamics of Various Plant Communities
Forests, Woodland and Dwarf Shrub Heaths

In early successional stages or in reforestation, *even-aged* stands of woody plants go through a *growth phase*. As long as the plants are young, the mass of stems and roots to be fed by the foliage is relatively small and net primary production is accordingly large. This means that there is a considerable surplus of organic matter, and the total phytomass of the stand increases rapidly year by year. As the stand grows older this productive *growth phase* gradually gives way to the *phase of maturity* [170]; in this phase ΔB is at first still positive but later fluctuates around zero. This reduced rate of increase is brought about by the decline in net production as development of the stand proceeds. The larger the trees grow, the smaller is the ratio of green to non-green tissues. As a result, the photosynthetic yield suffices only for renewing the leaves and for the greatly enlarged mass of shoot and root systems. In deciduous forests net wood increase comes to a halt when the leaf mass makes up less than 1% of the total mass.

In self-seeding communities of woody plants, such as virgin forest, bush formations in dry regions, or dwarf shrub vegetation, individuals of all age classes are present simultaneously. Development and subsequent decline of the stand take place in smaller, mosaic-like patches instead of throughout the entire stand [189, 196]. An overall steady state is reached for the whole community: production and losses are balanced, so that averaged over many years ΔB is zero (*protective stage* [170]). However, a stagnating production of biomass is not necessarily an indication of a "protective" state of equilibrium. When growth of the stand is limited by recurrent losses of biomass as a result of herbivory, or freezing of plant parts not protected by a covering of snow in winter (see subalpine dwarf shrub heath in Fig. 2.81), the biomass remains fairly constant for many years – but in this case as a result of external constraints.

Grassland and Other Herbaceous Communities

In the course of the production period there is a *rapid* increase in phytomass, but at the same time parts of the shoots and roots die off or are removed by consumers. In herbaceous stands ΔB fluctuates around zero. In dry regions with variable precipitation ΔB fluctuates in successive years between large positive and negative values, but when averaged over many years it is also approximately zero.

In herbaceous plant communities left in their *natural state,* the leaves turn yellow and dry up at the end of the growing season, and parts of both the shoot and root systems are withdrawn. In the steppes, this loss accounts for more than half of the phytomass formed during the year, and in desert plant communities consisting primarily of ephemeral species [137], it may be as much as 60 – 100%. In grasslands *that are regularly mown or grazed,* the biomass is continually removed during the growing season, so that G may exceed L (see Fig. 2.81). The great difference between losses due to grazing and those due to shedding of foliage consists mainly in the fact that when plants are grazed they are still at a stage in which they are capable of full photosynthetic activity; excessive grazing (G more than half of PP_n) therefore endangers the existence of the stand. On the other hand, once the growing season is over, the entire biomass aboveground can die off without danger to the vegetation.

Plankton Populations

In water, the first consequence of net primary production is an increase in the numbers of (usually short-lived) algae in the euphotic zone. This supply of matter in the community serves to feed a horde of consumers; G is high, *consumption* averaging two thirds of the net primary production. Loss as litter may be less, but is hard to estimate. In the case of phytoplankton, *detritus* consists of those cells that sink below the compensation depth, either drawn by gravity or carried by water currents. The *velocity of sinking* depends upon the size, shape and specific weight of the organisms, as well as upon water movements and temperature. In cool waters at 6 °C most algae sink at an average rate of 3 m per day; at 20 °C they sink twice as fast. A characteristic of aquatic ecosystems is the sequential appearance of a productive growth phase and a protective equilibrium phase (comparable to the successions in terrestrial communities), as well as a high proportion of loss due to grazing (as in mown or grazed grassland). Thus in an aquatic ecosystem $G \geq L$ and the sum of the two is greater than ΔB.

2.5 Energy Conversion by the Plant Cover

2.5.1 Energy Conversion by Photosynthesis

Photosynthesis is the most efficient process so far known for the conversion of solar energy to chemical energy. The **efficiency of photosynthesis** in the utilization of solar energy, known as the **radiation use efficiency**, RUE_{Ph}, indicates the percent of the absorbed radiant energy fixed in the form of chemical bonds.

$$RUE_{Ph} = \frac{\text{stored chemical energy}}{\text{absorbed radiant energy}} \; 100(\%) \qquad (2.22)$$

The radiation use efficiency of photosynthesis can be calculated from the *quantum requirement* for the photochemical process. The number of quanta of energy, i.e. photons, required to liberate one molecule of O_2 is 8; the production of one molecule of glucose requires 48 photons. From the ratio of 2874 kJ (the energy content of 1 mol of glucose) to 8642 kJ (contained in 48 mol photons in the wavelength range 400–700 nm) a theoretical value of 33.2% is obtained for the maximal possible efficiency of photosynthesis. The value obtained experimentally, using isolated chloroplasts or cell suspensions under optimal conditions and CO_2 saturation, amounts to only roughly 30%, however, since there is an unavoidable loss of energy in the photochemical process, primarily in the form of heat. Under similar optimal conditions in the laboratory the efficiency of gross photosynthesis was found to be approximately 24% in leaves of C_4 plants, and about 14% in the leaves of C_3 plants. In a *natural environment* where, because of the low ambient CO_2 content, secondary processes are limiting for photosynthesis, the radiation use efficiency of *gross photosynthesis* is estimated at between 8 and 10%. After respiratory

Table 2.20. Energy losses (in % total global radiation) due to carbon assimilation in plants. (Beadle and Long 1985)

Energy losses due to:	Relative loss
Solar energy lying outside photosynthetically active wavelengths	50
Reflection and transmission	5 – 10
Absorption by photosynthetically inactive tissues and structures (cell walls, non-photosynthetically active pigments)	2.5
Absorption of radiation in photosystems I and II (heat, fluorescence)	8.7
Electron transport and secondary processes in carbon assimilation	19 – 22
Photorespiration	2.5 – 3

Dark respiration
  C_3 plants
  C_4 plants
	
	3.7 – 4.3
	4.9 – 5.8

losses have been subtracted, the remaining RUE of net photosynthesis is one third to one half that of gross photosynthesis (Table 2.20).

2.5.2 The Energy Content of Plant Matter

The energy content of the phytomass produced can be determined by calorimetry of representative samples. The ash-free dry matter contains between 15 and 35 kJ g^{-1}, although more usually the values are between 19 and 20 kJ g^{-1} organic dry matter. The greater the carbon content of a substance, the higher is its energy content; the most important plant substances with a very high carbon content are lipids, lignin and a variety of products of secondary metabolism, such as isoprenoids and their derivatives (Table 2.21). It follows that above-average quantities of energy are present wherever such substances occur in a plant, for example in scleromorphic and resinous leaves, in lignified organs and in seeds rich in lipids.

The principal factors determining the *energy content of dry matter* (Fig. 2.82) are the level of organization, life form and the genetically determined chemical characteristics of a plant. Woody plants are richer in energy than herbaceous species, and as a rule aquatic plants have less energy than terrestrial herbs. Unicellular organisms are richer in energy than foliose thallophytes, the phytomass of conifers usually contains more energy than that of angiosperm trees; monocots are poorer in energy than dicotyledonous woody plants. It seems that *evolutive development proceeds in the direction of economical investment of energy* so that the energy density of the dry mass of more derived forms is lower than that of more primitive types.

The influence of environmental factors on energy content is mainly of an indirect nature. Latitudinal gradients in the energy contents of plant biomass are not so much due to the direct effect of climate on the energy content of the individual plant, but are rather a reflection of the zonal and regional alterations in type of vegetation. If the average values for the energy contents of

Table 2.21. Energy content of plant substances. (Paine 1971; Lieth 1975b)

Plant substance	Energy content (kJ g^{-1})
Oxalic acid	2.9
Glycine	8.7
Malic acid	10.0
Pyruvic acid	13.2
Glucose	15.5
Polyglucans	17.6
Proteins	23.0
Lignin	26.4
Lipids	38.9
Terpenes	46.9

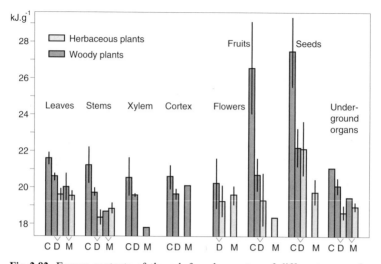

Fig. 2.82. Energy contents of the ash-free dry matter of different parts of coniferous (*C*), dicotyledonous (*D*) and monocotyledonous (*M*) plants. *Vertical lines* on the bars show the 95% confidence limits. (Pipp and Larcher 1987). For lists of data for seeds and fruit of different families, see Jordano (1992)

leaves of broadleaved woody plants are plotted against *latitude*, conspicuously high values are seen to occur between 30° and 32°N and between 43° and 45°N; they are due to the especially high proportion of sclerophyllous plants in the species spectrum. Between 50° and 75°N the raw energy content (caloric values apply to dry matter including ash content) is higher than that of the phytomass of other regions, but this is due to the low ash content of arctic plants; if referred to organic dry matter, the caloric values for the dry matter of plants of the high arctic are lower than those of species from 50°N. The increase in energy content with altitude observed in certain cushion- and rosette plants in *mountainous regions* is probably due to greater storage of fat.

2.5.3 Energy Efficiency of the Primary Production of Plant Communities

The *radiation use efficiency of primary production,* RUE_{PP}, is computed from the energy yield of gross primary production and the radiation absorbed per unit area of ground in the same time. The *efficiency* of the use of energy is usually expressed as the energy content of the annual net production as a percent of the total PhAR.

With respect to *net primary production*, C_4 plants growing under agricultural conditions achieve at their peak of productivity *maximal efficiencies* of 3% (maize) to 6% (*Panicum maximum*); corresponding values for C_3 crop plants range from 1.5 – 2% (mostly Fabales) to 2 – 4% for grasses, cereals and root crops [12, 148]. Under the varying temporal and spatial conditions affecting assimilation, the efficiency of energy utilization by plants, when *averaged over the production period,* is low. Even under the very favourable conditions prevailing in rainforests or wetlands efficiencies are less than 2%, and for most forests and grass communities less than 1%. Relatively unproductive plant communities such as steppes, semi-deserts and tundra achieve efficiencies between 0.2 and 0.5%, and the same is true for a large part of the oceans. In the regions exposed to the highest irradiance (near the Tropics of Cancer and Capricorn) there is little precipitation and productivity is therefore low. The sparse vegetation cannot make full use of the incoming radiation and efficiencies attain values of only 0.05% or less (Fig. 2.83).

The energy stored in the earth's phytomass at the present time (roughly 30×10^{21} J) is approximately of the same order of magnitude as the known reserves of coal, natural gas and mineral oil (approximately 25×10^{21} J [87]). Each year, about ten times more energy is fixed by photosynthesis than the present worldwide consumption, so that exploitation of the phytomass might appear to be the answer to meeting our energy requirements. The utilization of **phytomass as primary energy** can, however, be considered only in special situations and on a local, restricted scale – and even then only after careful analysis of the costs and benefits. Steppes, karst regions and deserts bear witness to the negative consequences of centuries of extensive consumption of phytomass. Increasingly, land that is not required for food production is being

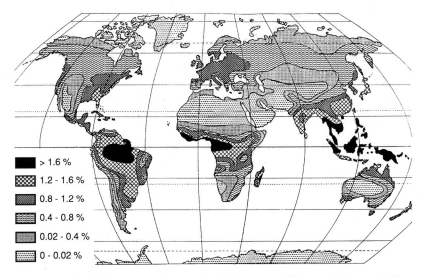

Fig. 2.83. Global distribution map of annual radiation use efficiency of the vegetation. Energy efficiency is expressed as the energy content of net primary production as a percent of the annual global photosynthetically active radiation. (Based on models of Uchijima and Seino 1987)

3 The Utilization of Mineral Elements

Plants require a large number of elements which are either derived from minerals or are mineralized during the biological breakdown of organic matter. The mineral nutrients are taken up in the form of ions and incorporated into the plant structure or stored in the cell sap. The inorganic (i.e. mineral) components remain as ash after combustion of dry plant material in the laboratory.

3.1 The Soil as a Nutrient Source for Plants

3.1.1 Mineral Nutrients in the Soil

Mineral nutrients occur in the soil in both dissolved and bound form. Only a small fraction (less than 0.2%) of the nutrient supply is dissolved in the soil water. Most of the remainder, almost 98%, is bound in organic detritus, humus and relatively insoluble inorganic compounds or incorporated in minerals. These constitute a nutrient reserve which becomes available very slowly as a result of weathering and mineralization of humus. The remaining 2% is adsorbed on soil colloids. The soil solution, the soil colloids, and the reserves of mineral substances in the soil are in a state of *dynamic equilibrium,* which ensures continued replenishment of supplies of nutrient elements.

3.1.2 Adsorption and Exchange of Ions in the Soil

Colloidal clay particles and humic substances (see Table 1.2), because of their surface electrical charges, attract ions and molecular dipoles and bind them reversibly. Both clay minerals and humic colloids have a negative net charge, so that they attract and absorb primarily cations. There are also some positively charged sites where anions can accumulate. How tightly a cation is held depends on its charge and degree of hydration. As a rule, ions with the higher valence are attracted more strongly, for example Ca^{2+} more strongly than K^+, and among ions with the same valence those with little hydration are retained more firmly than more strongly hydrated ions. The tendency for adsorption decreases in the order Al^{3+}, Ca^{2+}, Mg^{2+}, NH_4^+, K^+ and Na^+ for cations, and for anions it decreases from PO_4^{3-} through SO_4^{3-} and NO_3^- to Cl^-.

Heavy metals can also be adsorbed, although usually present only in trace amounts.

The swarm of ions around particles of clay and humus acts as an intermediary between the solid soil phase and the soil solution. If ions are added to or withdrawn from the soil solution, *exchange* takes place between solid and liquid phases. Adsorptive binding of nutrient ions offers a number of advantages: nutrients liberated by weathering and the decomposition of humus are captured and protected from leaching; the concentration of the soil solution is kept low and relatively constant, so that the plant roots and soil organisms are not exposed to extreme osmotic conditions; when needed by the plant, however, the adsorbed nutrients are readily available.

3.2 The Uptake of Mineral Nutrients

Aquatic plants take up nutrients over their entire surfaces, whereas *terrestrial species* acquire their mineral substances via a root system specialized for this function. Small amounts of minerals can also enter through the surface of the shoot. For example, in warm, humid regions leaves take up the nitrogen excreted by nitrogen-fixing bacteria and cyanobacteria adhering to their surface (*phylloplane*). In the same way, dry as well as wet air pollutants settling on the shoot surface, sprayed chemicals such as fertilizers (*leaf fertilization* with trace elements, especially Fe for the treatment of chlorosis due to nonavailability in the soil) and pesticides, also enter plants via the epidermis.

3.2.1 The Uptake of Mineral Nutrients from the Soil

A root takes mineral nutrients from the soil in the following ways (Fig. 3.1) by:

1. **Absorption of nutrient ions from the soil solution:** these ions are available directly, but their concentrations in the soil solution are very low: the most abundant is NO_3^-, in concentrations as high as $5-10$ mM, followed by SO_4^{2-}, Mg^{2+} and Ca^{2+} in concentrations up to $2-5$ mM, K^+ up to $1-2$ mM, and PO_4^{3-} up to $4\,\mu M$.
2. **Exchange absorption of adsorbed nutrient ions:** by releasing H^+ and HCO_3^- as dissociation products of respiratory CO_2, the root promotes ion exchange at the surface of the clay and humic particles, obtaining in return the nutrient ions. Excretion of H^+ and acids depends upon the intensity of respiration and thus upon the availability to the roots of oxygen and carbohydrates, and upon the temperature. The efflux of H^+ from dicotyledon roots can usually increase if the plant is deficient in mineral substances (Fig. 3.2). The roots are able to adjust the pH of the soil solution according to the form in which nitrogen is available. An example of this is seen in maize roots; if exclusively supplied with NO_3-N the pH in the rhizosphere

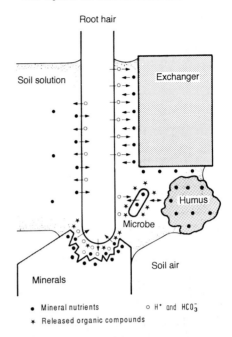

Fig. 3.1. Mobilization of mineral nutrients in the soil and the uptake of mineral elements by the root. (After Finck 1969, modified)

- • Mineral nutrients   ○ H^+ and HCO_3^-
- * Released organic compounds

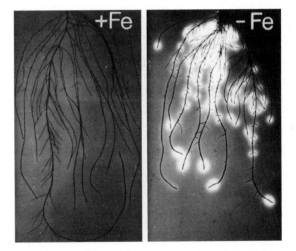

Fig. 3.2. Increased efflux of H^+ from cucumber roots in response to ion deficiency (*right*); on the *left*, the case in which supplies of iron are good (FeEDTA). Acidification can be demonstrated by addition of the pH indicator bromocresol purple to the substrate; the *pale areas* surrounding the roots have a pH of 4.5 or less. (From Römheld and Kramer 1983)

increases to roughly 7.5, but if NH_4-N predominates the pH drops to as much as pH 4.

3. **Mobilization of chemically bound nutrients** via excreted H^+ ions, by increasing the reductive capacity of the roots, and by releasing into the soil low-molecular weight organic compounds capable of forming soluble complexes, known as chelates, with nutrients that would otherwise not be able to enter the roots, especially iron and trace elements (Fig. 3.3). Particularly

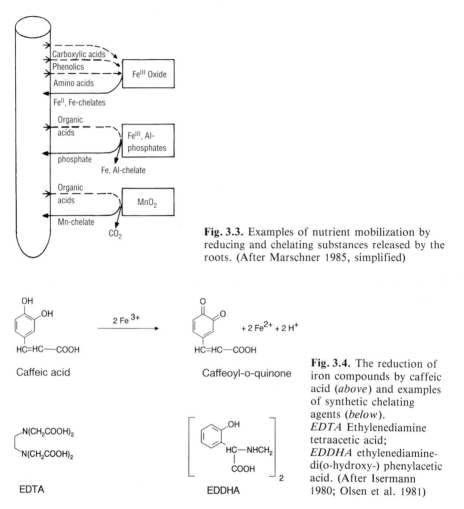

Fig. 3.3. Examples of nutrient mobilization by reducing and chelating substances released by the roots. (After Marschner 1985, simplified)

Fig. 3.4. The reduction of iron compounds by caffeic acid (*above*) and examples of synthetic chelating agents (*below*). *EDTA* Ethylenediamine tetraacetic acid; *EDDHA* ethylenediamine-di(o-hydroxy-) phenylacetic acid. (After Isermann 1980; Olsen et al. 1981)

important, especially for monocotyledons, is the formation of metal chelates, which are complexes of metals with organic acids (malic acid, citric acid, amino acids) and phenols (caffeic acid, Fig. 3.4). Chelation protects mineral nutrients from being bound again, and also facilitates their entry into the roots.

3.2.2 Ion Uptake into the Cell

Because of their hydration, ions would be hindered in entering the living cell if it were not for the special properties of most membrane systems that assist in ion transport. Usually such transport depends on large protein molecules or complexes of such molecules, which can either move within the membrane (*carrier systems*) or occupy fixed positions in it (ion channels; *tunnel proteins*). Because the ions are electrically charged, passage through the membrane is possible only if two ions of opposite charge are transported at the same time (co-transport or *symport*), if an ion within the cell is exchanged for another with the same charge (counter transport or *antiport*), or if there is already an electrical potential difference across the membrane (Fig. 3.5). The binding sites of the carrier systems in many cases are specific for certain ions or for groups of closely related ions.

Ion transport occurs not only down a concentration gradient (passive transport), but also frequently plants accumulate ions by transporting them against a concentration gradient (active transport). Such transport requires a supply of energy, thus depending on the energy-providing processes of respiration and photosynthesis. In certain cases there is a measurable increase in respiration during ion uptake (salt respiration). ATP-hydrolyzing proteins in the biomembranes (*membrane ATP-ases*) function as proton pumps, establishing electrochemical gradients across the membranes. The membrane potential built up by the flow of electrons provides energy for the transport of the various ions.

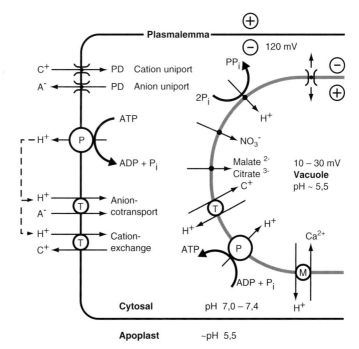

Fig. 3.5. Model of ion transport across plasmalemma and tonoplast.][= ion channels; *P* proton-transporting membrane-ATPases ("proton pump"); *C* carrier system; *M* calmodulin-dependent transport; *PD* electric potential differences; A^- anion; C^+ cation; PP_i pyrophosphate; P_i inorganic phosphate. (After Pitman and Lüttge 1983; Kaiser et al. 1988; Martinoia 1992)

Membrane ATP-ases are active over the entire plasmalemma and the tonoplast, and at every site within the cell where transport of ions between compartments takes place.

The following characteristics of the uptake of mineral nutrients result from interplay between passive and active transport.

The ability to concentrate ions: plant cells are able to take up ions against a concentration gradient and to accumulate them, particularly in the vacuoles, at concentrations much higher than those in the surrounding solution. This is especially important for aquatic plants, which must take up their nutrient elements from extremely dilute solutions.

Preference: plant cells can take up preferentially certain ions that are required in larger amounts. Thus cations have preference over anions in the uptake process, and among the cations some are accumulated in higher concentrations than others. When necessary, electrical neutrality can be maintained by ion exchange (H^+, HCO_3^-).

Limits to selectivity: plant cells cannot entirely exclude salts that are not required or are even injurious; biomembranes are not highly permeable to ions, but are never entirely impermeable. Consequently, when there is a pronounced difference in concentration on the two sides of a membrane, any ion can leak through. Especially when the outside concentration is high, as it is in saline soils, the cells may be flooded with ions (e.g. Na^+ and Cl^-) in unfavourable quantities.

3.2.3 Supplying the Root with Ions

The *rate at which nutrients are supplied* to a plant depends on the concentration of available minerals in the rooted volume of soil, on the ion-specific rates of diffusion, and on mass flow. Nitrate, as a rule, reaches the root surface rapidly, whereas phosphate and potassium ions, with lower diffusion coefficients, move more slowly.

The efficiency with which nutrients are taken up by the roots and the preference for certain ions, are genetic characteristics of a plant species. In addition to the specific affinities of the ion transport- and binding processes, the rate of nutrient uptake depends on the capacity of the roots to adjust to the quantities of nutrients available. This is achieved by a combination of chemotropic growth along the concentration gradient in the soil, a greater density of root hairs, and, in dicotyledons, an enormous increase in the surface area of the plasma membranes of the rhizodermis cells (*transfer cells*). A 20-fold increase in the surface available for the entry of ions can be achieved in this way (Fig. 3.6).

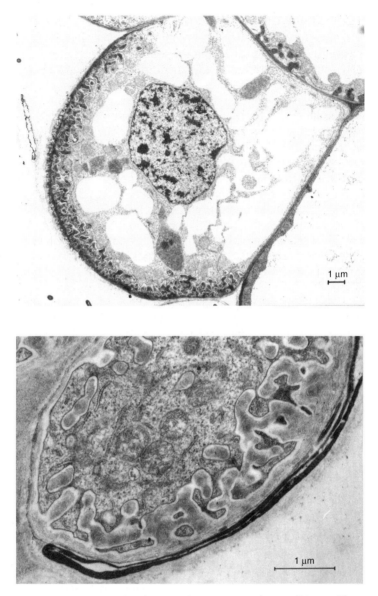

Fig. 3.6. Transfer cells in the rhizodermis of *Helianthus annuus* under conditions of iron deficiency. *Above* general view of a rhizodermal cell showing the wall labyrinth; *below* portion of the membrane labyrinth at higher magnification. (Kramer et al. 1980)

3.2.4 The Transport of Ions in Roots

The ions of nutrient salts from the soil solution first move, with the inflowing water, into the interconnected system of cell walls and intercellular spaces in the parenchyma of the root cortex (*apoplastic transport*); there they are adsorbed, owing to the charges on the surfaces of the cell walls and at the outer membrane of the protoplasts. This is a purely passive process, following the concentration and charge gradients between the soil solution and the interior of the root. The adsorption capacity of the root apoplast is called the "apparent free space". Ion uptake into the cytoplasm itself occurs chiefly in the root cortex. Mineral ions entering the vacuoles remain there until they are actively returned to the cytoplasm (Fig. 3.7).

The intercellular transport of ions bypasses the vacuoles, and proceeds along a continuous chain of living protoplasts that are in direct contact with one another via the plasmodesmata. This continuum of living protoplasts is called the *symplast* [163]. In contrast to apoplastic transport along the cell walls, which is interrupted by hydrophobic or impermeable elements embedded in the walls of the root endodermis (i.e. *Casparian strips*), the symplast route leads all the way to the central stele. In addition, ions flow passively along the concentration gradient into the water-filled vessels and tracheids, or they are actively excreted into the vessels by parenchyma cells.

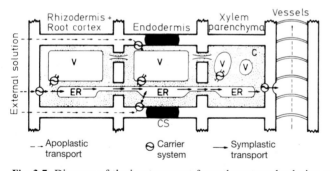

Fig. 3.7. Diagram of the ion transport from the external solution to the long-distance conducting system (*vessels*) in the central cylinder of the root. Ions, together with water, are transported in the cell walls and in the water-filled intercellular spaces from the rhizodermis (root epidermis) to the endodermis (Casparian strip, *CS*). After their active uptake (⊖) into the living protoplasts (cytoplasm, *C*) the ions are transported through the endoplasmatic reticulum (*ER*) and the plasmodesmata. Vacuoles (*V*) are spaces for the excretion and storage of substances. The pathways for transport in the opposite direction, which is mainly passive, are not shown. (After Lüttge 1973; Läuchli 1976)

3.2.5 Long-Distance Transport of Minerals in the Plant

In the xylem, ions are rapidly distributed by the **transpiration stream** from the root to the shoot. The rate-limiting stage in the nutrient translocation chain is the uptake, conduction and release of ions by the symplast in the roots; the transpiration stream is usually capable of carrying much greater quantities of mineral nutrients. Thus even if the velocity of the transpiration stream in the xylem is low, it is sufficient to move the nutrients absorbed by the roots up to the shoot. The other long-distance transport system, the phloem, plays an equally important role in the distribution of nutrients. Phloem and xylem systems are linked at many sites, particularly in the roots and in the nodes of the stems.

In the shoot, the nutrients diffuse out of the vessels and are actively taken up by the bundle parenchyma. Here, too, *transfer cells* promote the transport of ions from the vascular system to the parenchymal cells. Local transport also takes place via the symplast pathway; again, part of the nutrient ions are retained in the vacuoles.

The main function of phloem transport is the *retranslocation* of mineral substances already incorporated into the plant. These substances differ considerably in the ease with which they can be redistributed (Fig. 3.8, see also Table 3.4 and 3.5); nutrients, like N, P and S, that are bound in organic compounds, are easily translocated, as are the alkali ions, especially K^+. These more mobile elements are most highly concentrated in young leaves, and are gradually moved elsewhere with ageing of the leaf. Less easily shifted are the heavy metals and the ions of alkaline earths, especially calcium which thus accumulates steadily in the leaves at the end of the xylem translocation route. This leads to a rise in the Ca:K ratio in the leaves with increasing age. Redistri-

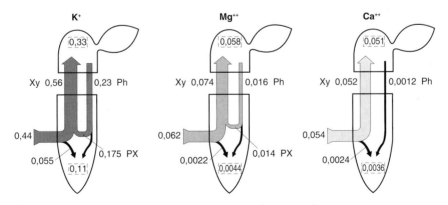

Fig. 3.8. Transport and retranslocation of K^+, Mg^{2+} and Ca^{2+} between the root and shoot of *Lupinus albus*. *Boxed numbers* Ion accumulation during growth; *other numbers* flux rates (μmol ions g^{-1} fresh weight h^{-1}); *Xy* transport via xylem; *Ph* via phloem; *PX* changeover from phloem to xylem in the root. (After Jeschke et al. 1985)

bution of minerals is a usual occurrence over the course of the year; in herbaceous species it takes place chiefly from ageing leaves into growing shoot tips and reproductive organs, and in woody plants it is directed towards the buds in spring, into storage tissue in summer and autumn.

3.3 Utilization and Deposition of Minerals in the Plant

3.3.1 The Ash Content of Dry Matter and the Composition of Plant Ash

Every chemical element that occurs in the lithosphere can be found in plant ash. A survey of the *ash content* of various plant groups is given in Table 3.1; the average *composition of plant ash* is shown in Table 3.2. The elements N, K, Ca, and in some plants Si, are present in quite large quantities $(10-50 \text{ g kg}^{-1}$ of the dry matter). Mg, P and S are found in amounts between a few to 10 g kg^{-1}, and the trace element content lies between 0.2 g kg^{-1} (Fe) and a few mg kg^{-1} DM.

The proportions of the various bioelements are highly characteristic for certain *plant species* and families, as well as for specific *organs* and *developmental stages*. Many herbaceous plants contain more K than N, whereas in nitrophytes the opposite is true. Most plants contain slightly more P than S, although in *Brassica* there is much more S than P. The ratio Ca:K is particularly characteristic; in Caryophyllaceae, Primulaceae and Solanaceae potassium predominates, and in Crassulaceae and Brassicaceae there is more calcium. Halophytes (e.g. Chenopodiaceae, Brassicaceae and Apiaceae) accumulate large quantities of Na^+, which is an element normally found at the bottom of the list, just ahead of the trace elements. Grasses, sedges, palms and horsetails take up Si in such amounts that it accounts for three quarters of the total ash; in diatoms, the skeletons of which are composed of silicates, Si can comprise more than 90% of the ash.

Within a given plant the leaves and cortical tissues contain the most ash and the woody organs the least. The elements incorporated preferentially in the foliage are N, P, Ca, Mg, S, and, in the case of grasses and palms, Si as well. Flowers and fruits store mainly K, P and S; the bark of tree trunks contains relatively large amounts of Ca and Mn, and the wood of some (especially tropical) species stores Si and Al.

From the ash content and its composition — apart from the characteristic properties of the plant species — some information can be derived about the *mineral supply and composition of the soil at the site of growth*. Plants growing on soils particularly low in nutrients and above all on acid soils, are low

Table 3.1. Average ash content (in %) of the dry matter in various groups of plants and microorganisms. (From measurements by various authors; for details see Pipp and Larcher 1987)

Thallophytes	
Bacteria	8 – 10
Fungi	7 – 8
Planktonic algae (without skeletal material)	ca. 5
Diatoms	up to 50
Seaweeds	10 – 20
Mosses	2 – 4
Herbaceous Cormophytes	
Meadow- and forest herbs	6 – 10
Grasses and sedges	7 – 10
Geophytes	
aboveground parts	5 – 10
rhizomes	ca. 6
Succulents	
leaf succulents	10 – 20
stem succulents	10 – 15
Halophytes	10 – 20
Tundra herbs	ca. 5
Swamp plants	5 – 15
Aquatic plants	5 – 12
Epiphytes	ca. 3
Woody plants	
Broadleaved trees and shrubs	
deciduous leaves	4 – 6
evergreen leaves	5 – 9
cortex and outer bark	5 – 8
wood	0.5 – 3
Coniferous trees	
needles	ca. 4
cortex and bark	3 – 4
wood	0.5 – 0.7
Lianas (leaves)	ca. 8

in ash (1 – 3% of the dry matter), as are epiphytes. Conversely, plants growing on saline soils commonly have a high ash content (up to 55% of the dry matter) and the ash contains above-average amounts of Na, Mg, Cl and S.

The different *geochemical plant groups* can be characterized by radial diagrams (Fig. 3.9). Since plants are able to absorb certain elements preferentially, but cannot prevent the uptake of any one of them, the composition of their ash reflects the geochemical nature of the soil on which they grow (Table 3.3). High concentrations of nitrogen are found particularly in plants growing on the nitrogen-rich soils of flood-plain forests, meadows and pastures, and on ruderal sites; calcium accumulates in plants on calcareous soils and in the vegetation of dry subtropical regions, and there are elevated amounts of Al, Fe and Mn in plants on acidic soils, of Si in tropical rainforests and savannas, and of Cl^- and SO_4^{2-} in the halophytes typical of saline soils. Plants

Table 3.2. Average content of mineral elements (in g kg^{-1} dry matter) in the soil and in the phytomass of land plants, together with the average mineral nutrient requirements. (Epstein 1972, 1994; Bowen 1979; data for various plant groups are given by Altman and Dittmer 1972; Baumeister and Ernst 1978; Lieth and Markert 1988

Element	Soil mean	Plants' range	Requirements
Si	330	0.2 – 10	
Al	70	0.04 – 0.5	
Fe	40	0.002 – 0.7	ca. 0.1
Ca	15	0.4 – 15	3 – 15
K	14	1 – 70	5 – 20
Mg	5	0.7 – 9	1 – 3
Na	5	0.02 – 1.5	
N	2	12 – 75	15 – 25
Mn	1	0.003 – 1	0.03 – 0.05
P	0.8	0.1 – 10	1.5 – 3
S	0.7	0.6 – 9	2 – 3
Sr	0.25	0.003 – 0.4	
F	0.2	up to 0.02	
Rb	0.15	up to 0.05	
Cl	<0.1	0.2 – 10	>0.1
Zn	0.09	0.001 – 0.4	0.01 – 0.05
Ni	0.05	up to 0.005	
Cu	0.03	0.004 – 0.02	0.005 – 0.01
Pb	0.03	up to 0.02	
B	0.02	0.008 – 0.2	0.01 – 0.04
Co	0.008	up to 0.005	
Mo	0.003	up to 0.001	<0.0002

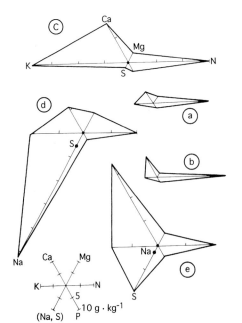

Fig. 3.9a–e. Radial diagrams showing the characteristic proportions of the principal mineral nutrients in leaves **a** of a tropical rainforest **b** an temperate oak forest, **c** of herbaceous plants of moist temperate regions, **d** of halophytes, and **e** plants growing on dry gypsum soils. Na and S are represented on the same axis. The data refer to g kg^{-1} dry matter. (After Boukhris and Loissaint 1975; Klinge 1976)

Table 3.3. The mineral contents of different types of vegetation. (After Rodin and Bazilevich 1967)

Type	Ash characteristics	Mineral content	Rate of litter breakdown	Vegetation
Nitro-boreal	N > (K, Mn)	Slight	Slow	Tundra
	N > Ca	Slight	Slow	Boreal coniferous forests
	N > Ca (Si, Mg)	Moderate	Slow	Boreal birch forests
Nitro-arid	N > Ca (Na, Cl)	Moderate	Very rapid	Shrub deserts
Nitro-subtropical	N > Ca (Si, Al, Fe)	Moderate	Rapid	Deciduous forests
Calco-temperate	Ca > N	Moderate	Delayed	Oak-beech forests
Calco-subtropical	Ca > Si (Al, Fe)	Moderate	Very rapid	Subtropical desert vegetation
Silico-semiarid	Si > N	Moderate	Very rapid	Steppes
Silico-arid	Si > N (Na, Cl)	Moderate	Very rapid	Desert annuals, semishrubs
Silico-tropical	Si > N (Fe, Al)	Moderate	Very rapid	Savannas
	Si > N (Al, Fe, Mn, S)	Moderate	Very rapid	Equatorial rainforests
Haline	Cl > Na	High	Very rapid	Halophytic vegetation

growing near ore deposits are characterized by elevated concentrations of heavy metals. *Analysis of ash* to measure habitat-dependent mineral accumulation can help in detecting the presence of nutrient deficiencies and incorrect fertilization of crop plants; furthermore, knowledge of the mineral content of wild plants allows them to be used as indicators of nutrient availability and the presence of ore deposits.

3.3.2 Nutrient Requirements and the Incorporation of Mineral Substances

The individual *bioelements* are incorporated into the plant tissues, becoming components or activators of enzymes, or regulators of the degree of hydration of the protoplasm and thus of the activity of the enzymes. A special position is occupied by calcium: in cooperation with calmodulin, a regulator protein, it functions as a secondary messenger in reactions involved in metabolism, development and responses to stimuli [198].

A survey of the specific ways in which elements are incorporated and operate, and of the sites in the plant where they are concentrated, is given in Tables 3.4 and 3.5. The *principal nutrient elements,* N, P, S, K, Ca and Mg, all of which are required in considerable amounts, and the *trace elements,* Fe, Mn, Zn, Cu, Mo, B and Cl are vitally essential to the plant and cannot be replaced by any other element. Certain elements are required only by some groups of plants, for example Na by Chenopodiaceae, Co by Fabales with symbionts, Al by ferns, Si by diatoms and Se by certain planktonic algae.

During the growing season, much of the total uptake and **incorporation of minerals** has been completed before the rapid increase in mass begins. The most important nutrient elements must be made available at an early stage, and it is clear that an inadequate supply of minerals restricts the production of organic matter from the very start. When the rate of uptake of minerals during the phase of elongation can no longer keep pace with the assimilation of carbon, the ratio of organic dry matter to mineral incorporation begins to increase. The mineral content of the dry matter now appears to be lower ("*dilution effect*"; see Fig. 3.11). If, however, the mineral content is calculated for the individual leaf then there is only a drop in concentration if minerals are transported away from the leaves. When the vegetative part of the plant is completely formed, carbon assimilation and mineral incorporation attain a state of equilibrium.

Young woody plants, during their main growth period, behave in a similar way to herbaceous species. In short-rotation forestry, the best utilization of mineral nutrients is obtained by the intermittent application of fertilizers in quantities exactly matched to the exponential growth of the plants [102]. In adult trees the minerals required for the new spring growth are drawn from reserves of mineral substances stored in the previous year. Thus an application of fertilizer after the foliage has matured only becomes fully effective for production, as a rule, in the following year.

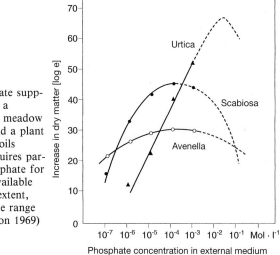

Fig. 3.10. The effect of phosphate supply on dry-matter production by a ruderal plant (*Urtica dioica*), a meadow plant (*Scabiosa columbaria*) and a plant adapted to nutrient-poor acid soils (*Avenella flexuosa*). *Urtica* requires particularly large amounts of phosphate for growth; *Avenella* utilizes the available phosphate only to a moderate extent, but fairly uniformly over a wide range of concentrations. (After Rorison 1969)

According to the **Liebig "Law of the Minimum"**, it is the element present in the least relative amount that sets the limit to growth and yield [141]. However, yield is not determined by this nutrient *alone*. For a plant to achieve well-balanced metabolism, high production and unimpeded development, not only must all major and trace elements be taken up in sufficient quantities, but they must also be taken up in *balanced proportions*.

Plant species differ greatly in their mineral requirements. Although the needs of crop plants have been studied in considerable detail, few experimental studies have been devoted to the specific nutrient requirements of wild plants. This is surprising, because comparative analyses might help to elucidate the causes underlying characteristic floristic distribution patterns. An example of the different phosphate requirements of nutritionally contrasting species is shown in Fig. 3.10.

3.3.3 Mineral Nutrient Status

According to the extent of mineral incorporation into a plant, three basic nutritional states can be distinguished: deficiency, adequate supply and unfavourable excess (Fig. 3.11).

Plants suffering from *mineral deficiency* are stunted, and their development proceeds abnormally. If, during the main growth period, the uptake of minerals cannot keep pace with the production of organic matter, the concentration of mineral substances decreases. Since it is the *concentration* of nutrients in the tissues (and not the *quantity*) that is important for metabolism, symptoms of deficiency often develop with excessively rapid growth. Scanty supplies of

Table 3.4. Occurrence, uptake, distribution, incorporation and function of macronutrients. (Compiled from measurements made by numerous authors, after Finck 1969)

Bio-element	Bound form in soil	Accessible form in soil	Taken up as	Incorporation in plant	Function in plant	Sites of accumulation	Transport-ability
N	Organically bound, nitrate, ammonium	Supplied by microbial decomposition; NH_4^+ adsorbed on clay minerals and humus; NO_3^- in solution	NO_3^-, NH_4^+	Free as NO_3^- ion (vacuoles), in organic compounds, in protein, nucleic acids, secondary plant substances	Essential component of protoplasm and enzymes	Young shoots, leaves, buds, seeds, storage organs	Good, primarily in organically bound form
P	Organically bound, phosphates of Ca, Fe, Al	As PO_4^{3-}, HPO_4^{2-}, rel. insoluble, adsorbed and in chelated complexes, Microbial release slight	HPO_4^{2-} / $H_2PO_4^-$	Free as ion, in esteric compounds, nucleotides, phosphatides, phytin	Basal metabolism and synthesis (phosphorylation)	More in reproductive organs than in vegetative (pollen granules)	Good, in organically bound form
S	Organically bound, sulphur-containing minerals, sulphates of Ca, Mg and Na	SO_4^{2-} readily soluble, little adsorbed	SO_4^{2-} from soil (SO_2 from air)	Free as ion, bound as SH- or SS-group and as ester, in protein, coenzymes, secondary plant substances	Component of protoplasm and enzymes	Leaves, seeds	Good in organic form, poor as ion
K	Feldspar, mica, clay minerals	Adsorbed > >dissolved	K^+	Dissolved as ion (primarily in cell sap) and adsorbed	Regulation of hydration (synergists: NH_4^+, Na^+; antagonist: Ca^{2+}), electro-	Meristem, young tissue, bark parenchyma, sites of intense metabolism	Good

Element	Compounds	Behavior in soil	Chemical form	Form in plant	Physiological function	Deposition	Utilization
Mg	Carbonate (dolomite), silicate (augite, hornblende, olivine), sulfate, chloride	dissolved > adsorbed; deficient in acid soils, in excess in serpentine soils	Mg^{2+}	As ion dissolved and adsorbed, bound in complexes, organically bound in chlorophyll and pectates, component of enzymes and ribosomes	Regulation of hydration (antagonist to Ca^{2+}), basal metabolism (photosynthesis, phosphate transfer) synergists: Mn^{2+}, Zn^{2+}, chemical effects (membrane potential, osmoregulation), enzyme activation	Leaves	Good in part
Ca	Carbonate, gypsum, phosphate, silicate, (feldspar, augite)	Adsorbed > dissolved; deficient in very acid soils	Ca^{2+}	As ion, as salt dissolved, crystallized and incrusted; as chelate; organically bound in pectates	Regulation of hydration (antagonists: K^+, Mg^{2+}); enzyme activator (amylase, ATPase); regulator of growth in length; signal substance (via calmodulin)	Leaves, tree bark	Very poor
Fe	Sulphides, oxides, phosphates, silicates (augite, hornblende, biotite)	Adsorbed > mobilized; fixed in chalk soils	Fe^{2+}, Fe(III)-chelate	In metal-organic compounds; component of enzymes (heme, cytochrome, ferredoxin, catalase, peroxidase, nitrate reductase)	Basal metabolism (redox reactions), nitrogen metabolism, chlorophyll synthesis	Leaves	Poor

Table 3.5. Occurrence, uptake, distribution, incorporation and function of trace elements

Bio-element	Bound form in soil	Accessible form in soil	Taken up as	Incorporation in plant	Function in the plant	Site of accumulation	Transport-ability
Mn	Amorphous oxide (MnO_2), carbonates, in silicates	Adsorbed > dissolved; better available in acid soils; accumulates under reducing conditions	Mn^{2+} Mn^- chelate	In metal-organic compounds and complexes; component of enzymes (pyruvate carboxylase)	Basal metabolism (photosynthesis, phosphate transfer), stabilizes chloroplast structure, nucleic-acid synthesis, synergists: Mg, Zn	Leaves	Poor in part
Zn	Phosphates, carbonates, sulphides, oxides, in silicates	Adsorbed > soluble; mobilization acid > basic	Zn^{2+} Zn^- chelates	Component of enzymes (carbonic anhydrase, alcohol dehydrogenase)	Chlorophyll formation, enzyme activator, basal metabolism, protein breakdown, biosynthesis of growth regulators (IAA)	Roots, shoots	Poor
Cu	Sulphides, sulphates, carbonates	Adsorbed, mobilization acid > basic strong fixation of humus	Cu^{2+} and Cu^-	Bound as complexes (plastocyanin), component of enzymes (cytochrome oxidase, phenol oxidase)	Basal metabolism, nitrogen metabolism; sec. metabolism	Woody shoots	Poor

Mo	Molybdates, in silicates	Adsorbed, mobilization basic > acid	MoO_4^{2-}	In metal-organic compounds; component of enzymes (nitrate reductase, nitrogenase)	Nitrogen fixation, phosphorus metabolism, iron absorption and translocation		Poor
B	Tourmaline, borates	Adsorbed > soluble, availability acid > basic	HBO_3^{2-} $H_2BO_3^-$	Bound to carbohydrates as complexes; esteric binding	Carbohydrate transport and metabolism; phenol metabolism, activation of growth regulators (growth of pollen tubes)	Leaves, tips of shoots	Poor
Cl	Salt, silicates	Soluble > adsorbed	Cl^-	Free as ion, mostly stored in cell sap	Chemico-colloidal effect (strongly increases hydration); enzyme activation (photosynthesis)	Leaves	Good

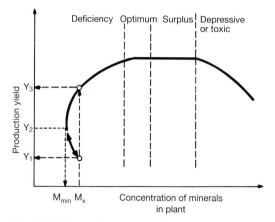

Fig. 3.11. Schematic representation of the relationship between concentration of minerals in the plant and the production of dry matter. If the plant remains small (Y_1), even a modest uptake of nutrients is sufficient for M_x to be reached ("accumulation effect"). If adequate mineral nutrients are taken up and their incorporation during the period of rapid growth is well balanced, there is a considerable increase in plant substance ($Y_1 \rightarrow Y_3$), without the need for appreciable alterations in the mineral concentrations in the tissues (M_x). However, if, during a rapid spurt of growth, the uptake of mineral substances is slower than the increase in dry matter ($Y_1 \rightarrow Y_2$), the concentration of mineral substances drops temporarily from M_x to M_{min} ("dilution effect"; Steenbjerg 1951; Steenbjerg and Jakobsen 1963). The plant develops best with an optimal concentration of mineral substances. If nutrient elements are taken up beyond this concentration they bring no additional benefit ("luxury" surplus nutrition), and excessive quantities of many mineral substances have a *depressive* or *toxic* effect. The transition from favourable to harmful concentrations is gradual in the case of the principal nutrient elements, whereas for trace elements it occurs within a narrow range of concentrations. (After Drosdoff; Prevot and Ollagnier, as cited in Smith 1962; Bates 1971)

minerals need not necessarily result in a greatly reduced mineral concentration in the tissues, however. If growth is much restricted by other factors (genotypic dwarfs, water shortage, cold), the same mineral concentration is reached with the lesser production of organic matter as would be attained with a greater production with correspondingly better supplies of mineral nutrients. Dwarf growth as a *deficiency-stress strategy* [83] is thus a means of concentrating minerals in the tissues in habitats with poor supplies. If the deficiency involves individual elements, or if the plant species requires extraordinarily large amounts of certain elements, *specific* deficiency symptoms can appear. These are best known for cultivated plants and forest trees (Table 3.6).

With an adequate mineral supply, the actual amounts of nutrients available can vary over wide ranges without noticeable effects on yield. Once the plant's requirements have been met, any excess seems to offer no further advantage for growth (luxury nutrition), although the possibility cannot be excluded that other ecophysiologically important properties conferring competitive advantages, such as resistance to parasites or extreme climatic situations, may in fact be promoted.

Table 3.6. Symptoms of deficiency in crop plants and forest trees. (After Wallace 1951; Bergmann 1983; Mengel 1984; Marschner 1986; Hartmann et al. 1988; Walker and Gessel 1990; Walker 1991)

Deficient element	Herbs and broadleaved woody plants	Conifers
N	Stunting (dwarfism), scleromorphism; shoot/root ratio shifted towards roots; premature yellowing of old leaves	Chlorophyll deficiency, discoloration, less growth (shorter needles and new shoots), premature loss of needles and browning of assimilation shoots
P	Disturbance of reproductive processes (delayed flowering), spindly appearance, dark green or bronze-violet discoloration of leaves and stalks	Reddening of needles and young shoots, necrosis without previous chlorosis
S	Similar to N-deficiency, intercostal chlorosis of young leaves	Chlorosis of young needles and shoots
K	Disturbed water balance (tip drying), curling of edges of older leaves	Tips of needles dry out, needles drop prematurely
Ca	Disturbances in growth by division (small cells), tip drying, leaf deformation, impaired root growth	Drying of buds; young shoots and root tips die off; chlorosis of the tips of fir trees, followed by browning of needles
Mg	Stunted growth, intercostal chlorosis of older leaves	Chlorosis, mainly of older needles and scales, in firs also discoloration of tips of needles (yellow to brown), lower branches become bare
Fe	Straw-yellow intercostal chloroses, in extreme cases young leaves turn white (veins green), apical bud formation suppressed	Young needles yellow to white, older needles green
Mn	Inhibition of growth, chloroses and necroses on young leaves	Young needles chlorotic, tip drying of shoots and tree
Zn	Stunted growth, white-green discoloration of older leaves, disturbances in fructification	Young needles first chlorotic, then necrotic
Cu	Tip drying, leaf curl, spotty chloroses of young leaves	Chlorosis of young needles, dieback of shoots and tree
B	Impaired growth (meristem necroses), diminished branching of roots, phloem necroses, disturbances in fructification	Terminal buds dry out, diminished growth in length, lateral branches dense and bent (nest-like), branch roots die

In the range of *excessive concentrations,* inorganic nutrients can be deleterious or even toxic, particularly if only one is present in excess. Over-fertilization with nitrogen results in too rapid growth of the shoot, inadequate supporting tissue, poorly developed root system, delayed reproductive development, insufficient resistance to climatic stress and greater susceptibility to parasitic fungi and insect pests. Excessive concentrations of alkali- and alkaline earth ions may cause imbalances or other depressive effects, and heavy metals may also be toxic (see Chaps 6.2.5 and 6.3.2.2).

3.4 The Elimination of Minerals

The ascending sap carries minerals into the shoot were they gradually accumulate. Most of the deposited mineral substances are eliminated in the course of shedding of various parts of the plant, especially leaves, flower debris and ripe fruits. The regular loss and replacement of leaves and bark thus provide a necessary process of elimination in perennial plants.

Minerals are also removed as components of various materials eliminated by the plant. Three **processes of direct elimination** can be distinguished (Fig. 3.12).

Recretion is the elimination of minerals in the form in which they were taken up. Considerable quantities of mineral elements leave the plant by this process, which takes place over the entire surface of the plant, where the salts can be washed away by rain. K^+, Na^+, Mg^{2+} and Mn^{2+} are easily leached out. Recretion is enhanced by acid rain and mist. Many species growing in saline habitats have glands specialized for the elimination of salt, others eliminate minerals substances through hydathodes situated at the tips or along the mar-

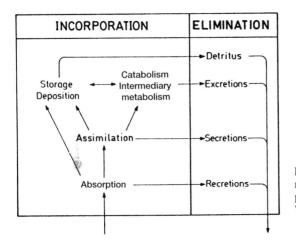

Fig. 3.12. Diagram of the turnover of inorganic matter in plants. (Based in part on Frey-Wyssling 1949)

gins of leaves. This also explains the presence of efflorescences of $Ca(CO_3)_2$ on the dentate leaf margin of certain saxifrage species.

Mineral elements, mainly nitrogen and sulphur, are also released from plants in the form of organic compounds. When these compounds are products of assimilation, as are the amino acids in root exudates, the process is known as *secretion*; when they are products of secondary metabolism or end products of dissimilatory metabolism it is *excretion*. The ecological significance of secretions and excretions lies not so much in regulating the composition of the plant as in biotic interaction (see Chap. 1.1.4.2).

3.5 Nitrogen Metabolism

Among the macronutrients, nitrogen is especially important. In terms of quantity in the phytomass, it is fourth among the bioelements, after C, O and H. The shoots of herbaceous plants contain on an average $2-4\%$ N, the leaves of deciduous trees $1.5-3\%$, evergreen needles and sclerophyll foliage $1-2\%$, herbaceous shoots, and roots $0.5-1\%$ N; planktonic algae contain roughly $5-8\%$ N, and proteins $15-19\%$.

A close relationship exists between nitrogen supply and increase in biomass; this can be expressed as the nitrogen use efficiency of production [NUE_P; see formula (2.11)]. The energy and the molecular structures required for the incorporation of nitrogen are derived from carbon metabolism which, because it depends on photosynthesis, is itself dependent on nitrogen-containing compounds such as chlorophyll. Thus increase in the mass of a plant is chiefly limited by the supply of nitrogen. When insufficient nitrogen is available, greater amounts of carbohydrates are converted into storable forms (starch and fat) or utilized in secondary metabolism (e.g. increased lignin synthesis). If the nitrogen deficit is severe, plants are stunted, the individual cells are small and their walls thickened (*nitrogen-deficiency sclerosis* or *peinomorphosis*), and, as a rule, reproductive events and senescence set in before their normal time.

3.5.1 Nitrogen Uptake by the Plant

Green plants utilize inorganically bound nitrogen; that is, they are autotrophic with respect to nitrogen as well as carbon. Nitrogen is taken up from the soil as nitrate or ammonium ions. Most plants can meet their nitrogen requirements with either NO_3^- or NH_4^+, as long as the pH in the rooting zone is suitable. At a low pH, ammonium uptake is impaired less than the uptake of nitrate and of other cations. Like all ion absorption, that of nitrogen requires energy and is thus dependent on respiration, which explains why plants growing on cold, poorly aerated soils often suffer from nitrogen deficiency.

3.5.2 Nitrogen Assimilation

The nitrogen taken up is incorporated into carbon compounds in amino groups, forming amino acids and amides. Amino acids are the basic compounds from which proteins, nucleic acids and the nitrogen compounds of secondary metabolism are synthesized. All of these compounds are important starting materials for building up cell substance. Nitrogen (just as sulphur and phosphorus) is an *organogenic bioelement,* i.e. not only is it incorporated, it is also assimilated (Fig. 3.13).

The first step in nitrogen assimilation is the reduction of nitrate to nitrite, catalyzed by *nitrate reductase* (NR). This is an enzyme complex (with the cofactors FAD, cytochrome b and Mo-pteridine) present in the cytosol. When needed, NR can be synthesized rapidly; it is inactivated by end-product inhibition (e.g. by NH_3). Because of its very high turnover rate, NR is able to play a decisive role in regulating **nitrate reduction**. The activity of nitrate reductase (NRA), as the key enzyme in nitrogen assimilation, is a measure of *the habitat-dependent nitrate utilization* of a plant species (Table 3.7). In herbaceous species its greatest activity is usually in the leaves; in woody plants in the roots

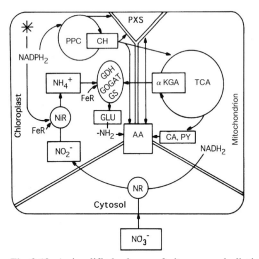

Fig. 3.13. A simplified scheme of nitrogen assimilation in a chlorenchyma cell. The energy-providing systems are mitochondrial respiration (reduction equivalents: $NADH_2$) and photosynthesis ($NADPH_2$). Organic carbon skeletons are provided by the tricarboxylic acid cycle [*TCA,* most important metabolites: a-ketoglutaric acid (*aKGA*), pyruvate (*PY*) and various carboxylic acids (*CA*)], and by the pentose phosphate cycle (*PPC,* Calvin cycle). The carbohydrate pool (*CH*) supplies the processes involved in mitochondrial respiration and the glycolate metabolism in the peroxisomes (*PXS*), as well as providing the carbon skeletons for transaminations. Nitrate reduction is catalyzed by nitrate reductase (*NR*), nitrite reduction by nitrite reductase (*NiR*), with the participation of ferredoxin (*FeR*). Glutamate dehydrogenase (*GDH*), glutamate synthase (*GOGAT*) and glutamate synthetase (*GS*) are involved in ammonium assimilation. The glutamate (*GLU*) formed acts as an amino-group donor ($-NH_2$), and is the starting point for the synthesis of the various amino acids (*AA*). For differentiated representations of nitrogen assimilation in root and mesophyll cells, see Stulen (1986), and in Fabaceae, see Streit and Feller (1982)

or in twigs and leaves, depending or species and habitat. The maximal activity of NR occurs during the juvenile stage and in growing organs; it is promoted by increasing supplies of nitrate (substrate induction), is stimulated by cytokinin, and is regulated in some way by the daily changeover from light to dark (its peak of activity is reached about halfway through the light part of the day).

Nitrite reductase is involved in reducing nitrite to NH_4. The energy and reducing power for assimilative nitrite reduction is provided by respiration ($NADH_2$) and, in chloroplast-containing cells, by photosynthesis ($NADPH_2$).

The *crucial process in assimilation* is the *reductive amination of α-keto acids*; in higher plants the first of these is α-ketoglutaric acid, an intermediate product in the respiratory citric acid cycle. From the glutamic acid and glutamine that are formed, NH_2 groups are transferred to other α-keto acids produced in glycolysis and the citric acid cycle (transamination). The other amino acids are derived from the primary amino acids; their carbon chains are obtained from intermediate products of carbohydrate metabolism, including the Calvin cycle and the oxidative pentose phosphate cycle. Amino acids are also formed during photosynthesis (glycine, serine, and alanine in C_3 plants, aspartic acid in C_4 plants) and during photorespiration (glycine, serine).

Protein metabolism proceeds at an extremely high rate that is also organ-specific and age-dependent. Organs and tissues that are growing and storing materials typically synthesize protein at an especially high rate, whereas in ageing leaves and parts of flowers it is protein breakdown that predominates. Environmental factors affecting protein metabolism are chiefly temperature and various stresses (e.g. drought and salinity).

The *temperature optimum* for protein synthesis, as a rule, is limited to a narrow band, because all the metabolic events that precede or participate in

Table 3.7. Average activity of nitrate reductase (NRA in $\mu mol\ NO_2^-\ g^{-1}$ dry matter h^{-1}) in plants in different habitats. (Gebauer et al. 1988; Stadler and Gebauer 1992; Downs et al. 1993)

Plants and habitats	Leaves	Underground organs	Reproductive organs
Ruderal plants	14	1	4.5
Nitrogen-rich habitats	7 – 13	0.6 – 1	1.6 – 4
Herbs in riparian vegetation	9	0.6	1.6
Meadows and pastures	3 – 4	0.3 – 0.5	0.5 – 0.7
Herbs of alpine pastures	1	0.3	0.9
Grassland on poor soils			
chalk soil	0.8	0.1	0.2
silicate soil	0.2	0.1	0.3
Ericaceous dwarf shrubs	0.06 – 0.1	0.04 – 0.07	0.8
Tree seedlings			
Pines	0.4	1.6	
Acer rubrum	1.5	0.7	
Trees (10 – 15-year)			
Fraxinus excelsior	1.4	0.1	

Triticum aestivum

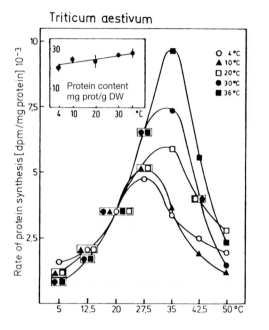

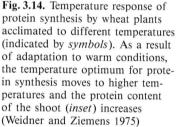

Fig. 3.14. Temperature response of protein synthesis by wheat plants acclimated to different temperatures (indicated by *symbols*). As a result of adaptation to warm conditions, the temperature optimum for protein synthesis moves to higher temperatures and the protein content of the shoot (*inset*) increases (Weidner and Ziemens 1975)

the production of protein (active nitrogen uptake and translocation, basic metabolic processes that provide metabolites, amino-acid synthesis, transcription and translation) are themselves temperature-dependent, with different temperature coefficients, despite which their turnover rates must be mutually compatible. As a result, protein synthesis is characterized by the capacity for highly flexible and prompt adaptation (Fig. 3.14). This is a central prerequisite for temperature adaptation of the plant at all levels: molecular, functional and morphological.

In a plant under stress, protein synthesis is inhibited and protein breakdown accelerated, so that there is a considerable increase in free amino acids and amides. A distinguishing characteristic of disturbances in protein metabolism is alteration of the proportions of amino acids, especially a drastic increase in proline concentration (see Chap. 6.1.3).

3.5.3 Nitrogen Partitioning in the Plant

Some of the inorganic nitrogen compounds taken up are assimilated in the roots, while the others are carried by the sap flow into the shoot and there incorporated into organic compounds. In addition, nitrate is stored in the cell sap of root and shoot. The amino acids may be used where they are formed, for the biosynthesis of macromolecules (protein, nucleic acids), or may be transferred to other tissues and organs. The most important *translocatable forms* are the amino acids glutamic and aspartic acid, as well as their amides glutamine and asparagine; in certain families and species citrulline and allantoic acid are important. When carbohydrate is abundant or nitrogen limited, amino acids predominate; when there is a good supply of nitrogen, amides are the predominant form.

Translocatable organic nitrogen compounds are produced on a larger scale during the transition from the vegetative to the reproductive phase (Fig. 3.15), and during mobilization of storage protein for the new shoots of perennial plants. In connection with protein breakdown in ageing or stress-damaged parts of a plant, soluble organic nitrogen compounds shift into younger tissues and perennial organs. In leaves, growing parts of shoots and ripening fruits, the organic transport forms of nitrogen provide amino groups for the synthesis of amino acids and for transamination, and serve as building blocks in protein synthesis and cell growth.

Vascular plants exhibit various characteristic **patterns of nitrogen partitioning** (Fig. 3.16), depending on the site at which nitrate reduction is most active, the nature of the chief nitrogen compounds transported and stored, and the intensity and direction of protein metabolism at different times.

Annual herbs, and especially the *nitrophytes* (with their preference for nitrogen-rich habitats, e.g. *Urtica, Lamium album, Anthriscus cerefolium, Rumex alpinus*), many *ruderal plants* (e.g. species of Chenopodiaceae and Asteraceae) and various crop plants (e.g. spinach, cucumber, cotton plants)

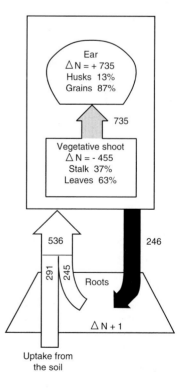

Fig. 3.15. Simplified diagram of the translocation and utilization of nitrogen compounds in a wheat plant one week after commencement of flowering. *Unshaded arrow* Transport of nitrogen compounds via the xylem (directly from the soil *via* the root); *black arrow* phloem transport from the shoot into the root; *stippled arrow* translocation from the vegetative parts into the ear via xylem and phloem. ΔN shows the increase (+) or decrease (−) in the nitrogen content of the different organs. The % values indicate relative partitioning of nitrogen. (Nicolas et al. 1985)

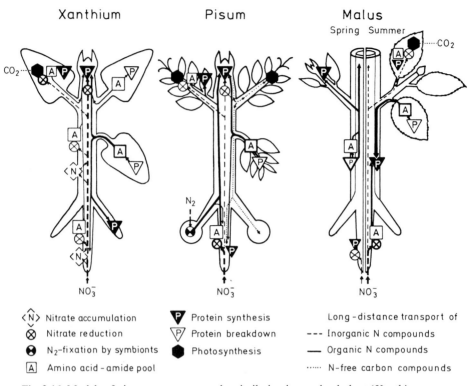

Fig. 3.16. Models of nitrogen transport and assimilation in a ruderal plant (*Xanthium pennsylvanicum*), in a plant with N_2 fixation in the root nodules (e.g. *Pisum sativum* or *Lupinus albus*), and in a temperate-zone deciduous tree (e.g. *Malus domestica* or *Fagus sylvatica*). For the herbaceous plants, the situation during the main growth phase is shown; during flowering and ripening of fruit the reproductive organs receive most of the mobile nitrogen compounds. For the apple tree (*Malus domestica*) two phases are shown: on the *left* the movement of amino acids; amides and soluble proteins from storage tissues in trunk and branches into the tips of the new shoots, and on the *right*, the replenishment of the storage tissues in summer and autumn. The *thickness of the arrows* indicates the relative intensity of nitrate reduction and transport processes. Long-distance transport of photosynthates, which supplies the nitrogen-assimilating tissues with carbon skeletons, is shown only for *Pisum*. (Simplified diagram based on data of Thomas 1927; Gäumann 1935; Wallace and Pate 1967; Pate 1976; Beevers 1976)

accumulate large amounts of nitrate in the cell sap of the roots and older parts of the shoot. At the time when nitrogen is most actively taken up, just before flowering, nitrate in the stems may amount to 20–40%, and in the roots to 15–20% of the total nitrogen content of the respective organ [200]. Most of the nitrogen taken in by the roots is translocated in *inorganic* form, as nitrate, in the xylem to the leaves and tips of the shoots, where nitrate reductase activity is greatest in this type of plant. The growing parts of the plant also receive, via the phloem, amino acids and amides that are either produced in excess amounts by leaves with an especially high photosynthetic capacity, or are liberated by protein breakdown in the senescing leaves below. The phloem is also the pathway by which the root is supplied with organic nitrogen compounds when excess amounts of these are produced in the shoot.

In Fabaceae with root nodules containing nitrogen-fixing symbiotic bacteria, inorganic nitrogen is assimilated primarily in the root, and is conducted to the rest of the plant in the *organic* nitrogen transport form (amides in *Pisum, Vicia, Lupinus,* ureides in tropical genera such as *Glycine* and *Vigna*). The main products of the nodules are amines, which are carried into the shoot through the xylem. The amount of nitrogen compounds made available to plants by symbiotic nitrogen-fixing organisms can be quite considerable: 10–50% of the nitrogen incorporated into the leaves of *Acacia* and *Prosopis* species in arid regions comes from the root nodules [217, 224].

In most *trees,* amino acids are synthesized mainly in the roots, but to some extent also in the shoot. In midsummer and autumn the amino acids and amides are gradually shifted to the trunk and branches, where they accumulate chiefly in the bark. Prior to abscission, the products of protein breakdown migrate out of the leaves and are stored in the axial tissues of the tree. Thus, considerable reserves of amino acids and proteins are available for the growth of new leaves and shoot in the following spring.

3.5.4 Nitrogen Fixation by Microorganisms

There are organisms that can make use of the exceedingly inert atmospheric nitrogen ("diazotrophic organisms"). These N_2-fixing organisms represent the most significant performance level of N autrophy. All of them are prokaryotic – bacteria, cyanobacteria and actinomycetes – some living free in the soil and others as symbionts.

Among the **free-living microorganisms**, nitrogen fixation was first demonstrated for the soil bacteria *Clostridium pasteurianum* and *Azotobacter chroococcum* and *A. agilis.* But there are many other species of bacteria that incorporate molecular nitrogen, among them the photoautotrophic bacteria and certain H_2-oxidizing bacteria living in water, as well as some cyanobacteria of genera that form heterocysts, for example, *Synechococcus, Pleurocapsa, Oscillatoria, Nostoc, Anabaena, Calothrix* and *Mastigocladus* [21]. These microorganisms are self-sufficient with respect to organogenic elements, being autotrophic for carbon as well as for nitrogen. Nitrogen-fixing cyanobacteria are found in water and are among the first to colonize raw soils, particularly in mountains and the arctic, in thermal springs and in other extreme habitats. As the third partner, heterocyst-forming cyanobacteria account for 3–6% of the algal components of symbiosis in lichens. *Nostoc* is the one most commonly encountered in the algal layer of the lichen thallus as well as in cephalodia. *Nostoc* and *Anabaena* also can have trophic interactions with mosses (*Anthoceros, Blasia*), cycads and *Gunnera.*

The performance of free-living nitrogen-fixing organisms is especially good in warm, permanently moist habitats such as rice paddies, where cyanobacteria fix $50-70$ kg ha^{-1} a^{-1} [77]. In the temperate zone, green-manuring of fields can be expected to add only a few kg N_2 ha^{-1} a^{-1}, whereas in the tropics and subtropics the fixation may be up to 100 kg N_2 ha^{-1} a^{-1}. Nitrogen-fixing microorganisms also live as epiphytes on the leaves of tropical trees, where they can fix as much as 5 kg N_2 ha^{-1} a^{-1} [31]. In raw humus soils of cold regions the rate of fixation is low; in the arctic and subarctic, depending on location, it amounts to $0.5-3$ kg N_2 ha^{-1} a^{-1} [33, 80].

Symbiotic carbon-heterotrophic nitrogen-fixing organisms solve the problem of carbohydrate supply by living in the cells of autotrophic plants, or by taking up root exudates in the rhizoplane. This enables them to fix much greater quantities of atmospheric nitrogen than free-living microorganisms (Table 3.8).

The most important symbiotic N_2-fixing organisms are bacteria of the genera *Rhizobium* and *Bradyrhizobium*, comprising only a few species with many physiologically distinct races inhabiting nodules of the roots of legumes. The two genera differ in their affinity for certain host plants: *Rhizobium* forms root nodules on *Trifolium, Lotus, Melilotus, Vicia, Pisum,* and *Phaseolus,* as well as on *Glycine soja*, the wild form of the soybean plant; *Bradyrhizobium* nodulates on lupines, *Glycine max, Vigna* spp and other tropical Fabales. In *Sesbania rostrata,* a tropical woody plant belonging to the Fabaceae family, the symbiotic bacteria induce small nodules on the shoot cortex instead of on the roots. Another group of nitrogen-fixing symbionts is the Actinomycetes [21, 77], primarily those of the genus *Frankia,* which form nodules on *Alnus, Myrica, Hippophae, Elaeagnus, Casuarina, Ceanothus,* and certain other woody plants. All of these trees and shrubs are pioneer plants living on nitrogen-deficient soils; in symbiosis with Actinomycetes their annual fixation of nitrogen is in the order of $50-150$ kg ha^{-1} a^{-1}.

A looser type of symbiosis (*association*) is maintained between nitrogen-fixing bacteria colonizing the root network of various plants and the mycorrhizae of trees. Nitrifying bacteria, in association with tropical grasses, fix $5-30$ kg N_2 ha^{-1} a^{-1}; *Anabaena azollae* in association with the aquatic fern *Azolla* fixes as much as $60-120$ kg N_2 ha^{-1} a^{-1} [236]. In East Asia

Table 3.8. Symbiotic N_2 fixation (kg N ha^{-1}a^{-1}) in Fabaceae. (Werner 1992)

Plants	Minimum/maximum	Mean
Lens	$50-150$	80
Trifolium	$45-670$	250
Pisum	$50-500$	150
Medicago	$90-340$	250
Lupinus	$140-200$	150
Vicia	$100-300$	200
Glycine	$60-300$	100
Arachis	$50-150$	100
Sesbania	$600-800$	700
N_2-fixing trees	$80-500$	150

green-manuring with *Azolla*, which has been practiced since early times, provides a very important source of nitrogen in tropical rice paddies.

Nodulation of Roots and the Process of Nitrogen Fixation

Rhizobia are obligatory aerobes that live as saprophytes in the soil of the regions where their host plants are found. Infection of the young plants and the formation of root nodules is induced by a *chain of signals* controlled by the bacterial genes [207]. When a suitable host has been recognized on account of its specific lectins, and contact between host plant and symbiont has been established, the nearest root hair curls around the bacterium. Local dissolution of the cell wall of the hair allows the infection thread to enter. This penetrates into the cortical parenchyma of the root where, by means of signal substances, the bacterium induces the beginning of meristematic cell division. The resulting tissue proliferation leads to formation of root nodules. Development of the nodules is promoted by high concentrations of IAA, cytokinins and gibberellins. Synthesis of nodule-specific proteins, glutamine synthetase, uricase, PEP carboxylase and leghaemoglobin is initiated by the endogeneous regulatory genes of the host cell. At this stage of infection the multiplying bacteria are provided with nutrients entirely by the host plant. Later, the rod-like bacteria change into larger bacteroids and are enclosed in peribacteroid membranes of the host cells. Functional units or "symbiosomes" [201] similar to cell compartments now become recognizable within the cytoplasm (Fig. 3.17). By derepression of the nitrogen-fixing *nif* genes of the immobilized bacteroids,

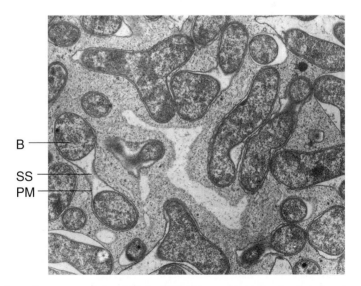

Fig. 3.17. Section through a root nodule of Vicia faba with symbiosomes, i.e. functional units consisting of *B* bacteroid, *SS* symbiosomal space and symbiosomal membrane, and *PM* peribacteroid membrane. Electron micrograph by D. Werner; 20 000×

synthesis of the nitrogenase system is initiated, and N_2 fixation can begin. When flowering of the host plant is over the nodules senesce, their leghaemoglobin is broken down, and finally the bacteroids are transformed into bacteria capable of infecting other plants.

Fixation of atmospheric nitrogen begins with the reductive splitting of the N_2 molecule. This strongly endergonic reaction is catalyzed by the *nitrogenase system*, a complex of two proteins, one containing iron, and the other containing molybdenum and iron. The energy and the electrons necessary for reduction are supplied by respiration (Fig. 3.18). The host plant supplies the bacteria with carbon compounds, provides the enzymes necessary for assimilation of ammonium, protects the bacteroids from excessive concentrations of oxygen by the presence of leghaemoglobin and layers of suberin, and ensures them a constantly moist environment.

During the vegetative stage in annual Fabales (pea, lupin soybean, cowpea), and up to the onset of flowering, one- to two thirds of the carbohydrates formed in the shoot are translocated to the roots and their nodules; of this amount about one quarter to one third is respired, roughly one fifth used for nodule growth, and the remaining 40–50% is returned to the shoot with the fixed nitrogen [62, 268]. For every gram of nitrogen, in the form of amino acids and amides, 4 g of carbon in the form of carbohydrates are required at the time

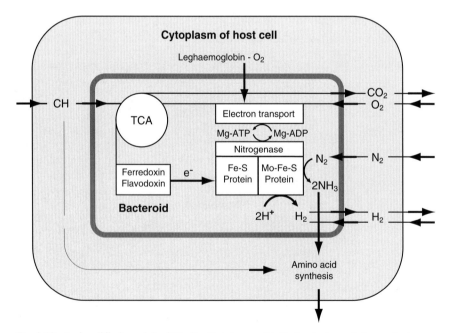

Fig. 3.18. A simplified model of N_2-fixation by symbiotic bacteria and the exchange processes involved. *TCA* Respiratory tricarboxylic acid cycle. (After Evans and Barber, as cited in Marschner 1986; Werner 1992b)

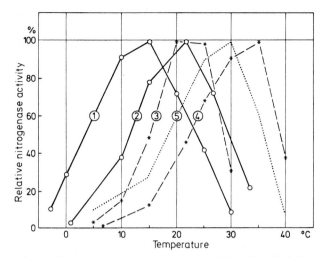

Fig. 3.19. Temperature dependence of nitrogenase activity in root nodules with N_2-fixing symbionts and in cyanobacteria. *1: Astragalus alpinus*, Norway; *2: Medicago sativa*, greenhouse culture at 16 °C; *3: Alnus glutinosa*, England; *4: Casuarina equisetifolia*, Malaysia; *5: Anabaena cylindrica*, Cambridge collection. (After Granhall and Lid-Torsvik 1975; Waughman 1977)

when the host plant and nodule bacteria show the greatest synthetic activity [176].

The amount of N_2 fixed by the symbionts depends very much on the supply of photosynthates. When photosynthetic yield of the host plant is low, bacterial nitrogen fixation falls off, whereas an increase in photosynthetic production (for example by exposure to elevated CO_2 concentrations) can triple the yield of organically bound nitrogen. The temperature of the surrounding soil and its water content play an important role in nitrogen fixation as well. The temperature-activity relationship of the nitrogenase is adjusted to the temperatures prevailing in the area of distribution of its host plant (Fig. 3.19). On account of the cold sensitivity of the nitrogenase system, however, N_2 fixation comes to a halt at low, but still above-zero temperatures.

3.6 Habitat-Related Aspects of Mineral Metabolism

The *chemical milieu in the rhizosphere* determines the supply of mineral substances to which the metabolism of the plant must be adjusted. Mineral composition of the soil, pH, and availability of ions are principally dependent on the nature of the underlying rock. The problem of the wide variety of habitat patterns encountered by plants within very small distances is solved by their exploitation of propitious soil horizons, avoiding unsuitable areas by means of positive or negative chemotrophic root growth, by metabolic flexibility, and by the development of chemoecotypes, races and vicarious species, the distribution of which is disjunctive. Not only does the functional analysis of habitat-related peculiarities in mineral metabolism constitute a productive field of comparative ecophysiological research, it also forms an important basis of causal analytical geobotany.

Soils that exhibit marked peculiarities with respect to quantity and combination of mineral substances are inhabited by a characteristic *spectrum* of specialist species with typical metabolic traits. Examples are provided by species with a high tolerance to acidity, or those with a high resistance to excessive accumulation of minerals, such as salt plants (halophytes), plants living on soils rich in heavy metals (metallophytes), as well as species that manage to survive on soils poor in mineral nutrients, or where the mineral substances present are unsuitable (e.g. serpentine soils). A few examples selected from the multitude of possible variants will be discussed in the following section.

3.6.1 Plants Growing on Acidic and Basic Substrates

The **hydrogen-ion concentration** in the surroundings of a plant is an important factor influencing its nutrition and distribution. The reaction of the soil depends on the hydrogen-ion concentration in the soil solution (*actual acidity*) and the adsorption of H^+ on exchange substances (*potential acidity*). Most soils in humid regions are weakly acid (pH $5-6.5$) to neutral (pH $6.5-7$), raised bog soils are strongly acid (pH <4). A low pH can also occur locally in the presence of acid-forming minerals such as pyrite. In dry regions the soils are basic due to the accumulation of alkali- and alkaline-earth ions and carbonates; neutral salt soils are weakly alkaline (pH $8-9$), and the pH of sodic soils is 10 or more.

Acidification of the soil can arise in a variety of ways; it may be due to leaching of exchangeable basic cations, to the presence of organic acids released by plant roots and by microorganisms, to humic and fulvic acids travelling downwards from overlying layers of raw humus, or due to the dissociation of carbonic acid accumulating in the soil as a product of respiration and fermentation processes. Additionally, acid enters the soil in precipitations, and with acid-forming air pollutants (principally SO_2). In most cases, the soil can buffer itself to a certain pH range, depending on the nature of the rock from which it originates and the degree to which the adsorption complexes are saturated with cations. However, the pH of the soil does change over the course of the year, mainly in connection with the distribution of precipitations, and it may differ within very small distances, particularly from one soil horizon to the next. Therefore characterization of any habitat must include measurements of the soil pH throughout the year and through the entire soil profile.

The majority of vascular plants, mosses and lichens tolerate a *soil reaction* ranging from weakly acid to weakly alkaline; in other words, they are *amphitolerant* over the pH range $3.5-8.5$. There are, nevertheless, some excep-

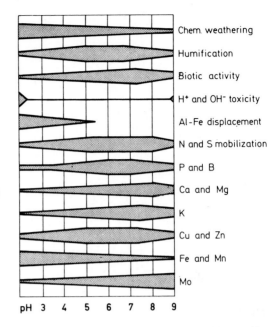

Chem. weathering

Humification

Biotic activity

H^+ and OH^- toxicity

Al-Fe displacement

N and S mobilization

P and B

Ca and Mg

K

Cu and Zn

Fe and Mn

Mo

pH 3   4   5   6   7   8   9

Fig. 3.20. Influence of soil pH on soil formation, mobilization and availability of mineral nutrients, and the conditions of life in the soil. The *width of the bands* indicates the intensity of the process or the availability of the nutrients. (After Truog, from Schroeder 1969)

tional species that are characterized by narrower limits of tolerance; one of these, *Avenella flexuosa*, which exhibits optimal development between pH 4 and 5, serves as an indicator for acid soil. Other acidophilic species are certain Ericaceae (e.g. *Calluna* and many *Rhododendron* spp.) and species growing on sites containing extremely little nutrient matter (plants of the raised bogs; lichens on acid bark).

Among *microorganisms,* many bacteria show a preference for a neutral to weakly alkaline milieu, whereas many fungi favour surroundings that are neutral to weakly acid. Some species, however, exhibit a wide range of tolerance, from pH 2 to 10 (for example, fungi of the species *Aspergillus, Penicillium* and *Fusarium*), and others are specialized for growth on extremely acid or alkaline substrates. Sulphur bacteria of the genera *Thiobacillus* and *Sulfolobus* and fungi of the genera *Acontium, Cephalosporium* and *Trichosporon* live in acid thermal springs at pH values between 1 and 3. Nitrifying bacteria and certain ammonifying bacteria are basitolerant up to pH 11. Cyanobacteria and algae, too, occur not only in strongly acid waters (*Cyanidium caldarium,* optimum at pH 2 – 3; *Dunaliella acidiphila* at pH 1 – 2), but also in those that are strongly alkaline (*Dunaliella salina* at pH 11; *Plectonema nostocorum* up to pH 13) [193].

The hydrogen-ion concentration of the surrounding milieu has a direct effect on the metabolism of *plant cells,* because the optimal pH range for many biochemical processes is a very narrow one (usually about pH 6 – 7). To keep the pH of the intracellular milieu within a suitable range, the differences in pH between the external and internal milieu of the cell must be overcome by continuous proton transport, with a corresponding expenditure of energy. Under

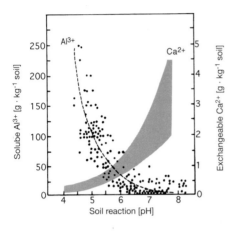

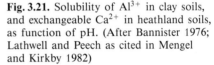

Fig. 3.21. Solubility of Al^{3+} in clay soils, and exchangeable Ca^{2+} in heathland soils, as function of pH. (After Bannister 1976; Lathwell and Peech as cited in Mengel and Kirkby 1982)

certain circumstances pH gradients of as much as four orders of magnitude may build up, i.e. $1:10^4$ [78].

The reaction of the soil also affects plants by its influence on the *availability of nutrients* (weathering, humification, nutrient mobilization, ion exchange). The most important of these relationships are shown in Fig. 3.20. In very acid soils, too many Al, Fe and Mn ions are liberated, and Ca^{2+}, Mg^{2+}, K^+, PO_4^{3-} and MoO^{2-} are depleted or occur in a form difficult for plants to take up (Fig. 3.21). Frequently, too, the NH_4^+/NO_3^- ratio is higher in acid soils. In more alkaline soils, phosphate and Fe, Mn, and other trace elements (particularly Zn) are bound in relatively insoluble compounds, so that the plants receive too little of these nutrients. Borates are known to exert toxic effects in alkaline soils.

3.6.2 Calcicole and Calcifuge Plants

Some species of plants are found exclusively on calcareous soils, while others occur only on siliceous and sandy, calcium-deficient soils. According to their *substrate preference* such plants are called *calcicole* and *calcifuge*, respectively. It can be seen from Table 3.9 that closely related species may differ in their areas of distribution (vicarious species). Furthermore, different ecotypes may occur within a species. The attempt to discover the underlying causes of this striking degree of habitat dependence in the distribution of plants was one of the earliest tasks undertaken by analytical plant ecologists.

Calcareous soils differ from calcium-depleted soils in the following ways: they are usually more permeable to water and therefore warmer and drier than silicaceous soils, but their primary distinction is that they contain much greater amounts of Ca^{2+} and HCO_3^-. This means calcareous soils are buffered towards a higher pH than other soils, and show a weakly alkaline reaction. Nitrogen is more rapidly mineralized in calcareous soils; P, Fe, Mn and most heavy metals are less available than in acid soils. Chalk soils are by definition also

Table 3.9. Examples of habitat preference of vicarious plant species in the European Alps. (Landolt 1971; Ellenberg 1986)

Genus	Calcicole species	Calcifuge species
Achillea	*atrata*	*moschata*
Carex	*firma*	*curvula*
Doronicum	*grandiflorum*	*clusii*
Gentiana	*clusii*	*kochiana*
Hutchinsia	*alpina*	*brevicaulis*
Pedicularis	*rostratocapitata*	*kerneri*
Primula	*auricula*	*hirsuta*
Pulsatilla	*alpina*	*sulphurea*
Ranunculus	*alpestris*	*glacialis*
Rhododendron	*hirsutum*	*ferrugineum*
Saxifraga	*moschata*	*exarata*
Soldanella	*alpina*	*pusilla*

carbonate soils. There are also soils rich in HCO_3^- (e.g. on dolomite rock, which contains $MgCO_3^-$), on which certain calcicoles do not occur (e.g. *Helianthemum nummularium, Plantago media*), and where calcifuge plants are even encountered (e.g. *Anthyllis montana, Saxifraga longifolia* in the Pyrenees).

Siliceous soils and other substrates poor in bases are acid, and, if high in clay content, they are dense, moist and therefore cooler than more porous soils. Iron and manganese are plentiful and accessible, and aluminium compounds enter readily into solution. It is reasonable to assume that all of these edaphic factors affect the inorganic metabolism and the vigour of plants, and that failure to satisfy the interspecifically differing substrate requirements must result in lower yields, greater susceptibility to climatic and other constraints, and reduced ability to compete.

The **response of plants** to unsuitable substrates is extremely varied. *Calcicolous plants* are obviously able to take up phosphorus and trace elements from calcareous soils. Many plants (e.g. of the Fabaceae, Brassicaceae, Solanaceae, and Cucurbitaceae families, in some cases only certain varieties of a species) are particularly efficient in taking up iron from the soil. This is achieved by an increased efflux of protons in the root region; iron is thereby reduced to the bivalent transport form, and its uptake is enhanced by the formation of larger numbers of transfer cells (Fig. 3.22, see also Fig. 3.2). If *calcifuge* plants are transferred to a chalky soil they develop signs of phosphorus and iron deficiency ("lime chlorosis"; Table 3.10). This applies especially to crop plants and ornamental species, among which *Citrus* species and many Ericaceae (*Rhododendron* species) are particularly susceptible. Lime chlorosis is a highly complex metabolic disturbance involving nitrate, phosphate and acid metabolism. Strict calcifuges are also hypersensitive to HCO_3^- and Ca^{2+}. Peat mosses and calcifuge grasses such as *Avenella flexuosa* produce large quantities of malate in their roots if the concentration of

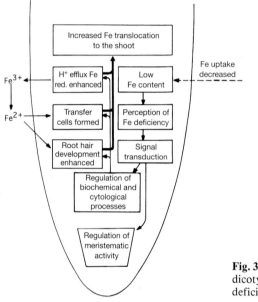

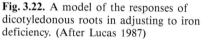

Fig. 3.22. A model of the responses of dicotyledonous roots in adjusting to iron deficiency. (After Lucas 1987)

HCO_3^- is too high; this has the effect of inhibiting growth and may result in root damage.

On a low-calcium soil, calcicolous plants would be damaged by the iron, magnesium, and especially by the alumium ions which are liberated preferentially in acid soils. Excessive uptake of Al^{3+} in acid soils is especially toxic to calcicole species (Table 3.10). Calcifugous plants, in contrast, form innocuous complexes with heavy metal ions, and are not damaged by large amounts of Al^{3+}.

According to the **type of calcium metabolism**, a distinction can be made between two "physiotypes" [117]: *calciotrophic* species accumulate water-soluble

Table 3.10. Examples of plant species with susceptibility to lime chlorosis and to aluminium toxicity (Grime and Hodgson 1969)

Calcifuge plants susceptible to lime chlorosis on chalky soils	Susceptible to Al toxicity on siliceous soils (calcicole species)
Avenella flexuosa	*Hordeum vulgare*
Holcus mollis	*Agrostis stolonifera*
Paspalum dilatum	*Festuca pratensis*
Lathyrus linifolius	*Beta vulgaris*
Galium saxatile	*Medicago sativa*
Eucalyptus dalrympliana	*Asperula cynanchica*
Eucalyptus gomphocephala	*Scabiosa columbaria*
Eucalyptus gunnii	*Lactuca sativa*

calcium bound to malate and citrate; the cell sap therefore contains more Ca^{2+} than K^+. Many Brassicaceae, Fabaceae, Geraniaceae, Euphorbiaceae and, as far as is known, all Crassulaceae are calcitrophic. In *calciophobic* species high concentrations of Ca are avoided by its precipitation in the form of calcium oxalate and/or by binding to pectin (Fig. 3.23). Cactaceae, Polygonaceae, Chenopodiaceae, Lamiaceae and most Caryophyllaceae are oxalate types; oxalic acid production is stimulated by calcium uptake, and if their acid metabolism can provide oxalate in sufficient quantities, members of these families can grow on chalky soils. Yet another mechanism for Ca regulation is recretion; this is carried on actively via the hydathodes in some species of Saxifragaceae and Plumbaginaceae.

In many cases, there is a close connection between Ca physiotype and distribution of a species (Fig. 3.24), although there is no general correlation. That both calcicolous and calcifugous species are known to occur in calciotrophic families and genera is a good illustration of the fact that ecological reality is

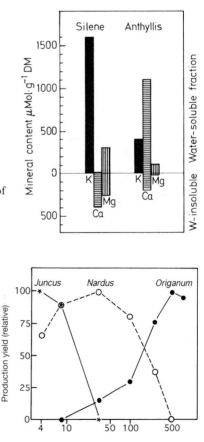

Fig. 3.23. Contents of potassium, calcium and magnesium in the leaves of *Silene vulgaris* and *Anthyllis vulneraria* from calcareous habitats. The calciophobic species *Silene vulgaris* binds the Ca taken up to produce water-insoluble oxalate (oxalate type); such plants contain large amounts of K. In contrast, the calciotrophic species *Anthyllis vulneraria* shows a high level of soluble Ca, with a correspondingly low level of K. (Horak and Kinzel 1971). The distinction of these two physiotypes is due to Iljin (1940)

Fig. 3.24. Growth response of an acidophilic calcifuge plant, *Juncus squarrosus*, the calcifuge plant *Nardus stricta*, and the calcicole *Origanum vulgare*, to nutrient cultures containing different amounts of Ca. Under hydroponic conditions the responses of these species are much the same as under the ecological conditions in their natural habitat. (After Jefferies and Willis, from Kinzel 1982)

seldom the result of an *isolated* physiological response, and emphasizes the necessity for considering all of the many and varied relationships between plants and their environment. It is not unlikely that genetic differences in the ability to take up essential bioelements, such as iron, play a greater role than calcium metabolism in determining the distribution of species and ecotypes on chalky, basic soils.

3.6.3 Plants Growing in Oligotrophic Habitats

In oligotrophic habitats the predominant environmental factor determining plant growth and the growth form spectrum of the vegetation is *general nutrient deficiency*. Especially serious is the lack of available organogenic elements in the soil, chiefly nitrogen and phosphorus. The deficiencies may be the result of a primary lack of mineral substances, as in quartz sand, or to degradation and leaching of the soils, but the main reason is slow and incomplete breakdown of organic litter. On poor, often also acid and sandy soils, heaths and sclerophyll shrub formations (macchia, chapparal, mattoral, Fynbos, Campos Cerrados) develop. The cold soils of the arctic tundra and high mountains are poor in nitrogen and phosphorus on account of the slow mineralization of litter, the acidity of the raw humus layers, and the frequent waterlogging (Fig. 3.25). Mineral concentrations in bogs and in peat soils are extremely low due to the high content of humic substances, very low pH and stagnation of the groundwater. In such places *"ombrotrophy"* (Table 3.11), which is input of mineral substances carried by rain or dust, plays an appreciable role.

The majority of plants living in **habitats poor in minerals** exhibit a special types of organization and growth form: i.e. lichens in cold regions (some with N_2-fixing phycobionts, and epiphytes); mosses in the tundra and in bogs; graminaceous plants, mainly tussock-forming grasses and sedges; dwarf perennial dicot herbs, especially rosette plants; dwarf shrubs and bushes, often with small, evergreen leaves; sclerophyll plants, their leaves with a high *scleromorphy index* (% raw fibre/% raw protein [147]). Chemotaxonomically, Ericaceae and related families appear to be of the physiotype adapted to nitrogen-deficient soils; their nitrate reductase activity is extremely weak, many representatives have leathery, lignaceous leaves, cellular metabolism tends strongly towards lipid accumulation, and the uptake of mineral substances is assisted by specific rhizofungal symbioses (ericoid and arbutoid mycorrhiza).

In areas extremely low in mineral substances, highly specialized plant forms occur, such as vascular epiphytes and a variety of carnivorous plants.

Epiphytes live on an organic substrate, but for their supply of minerals they are dependent on substances deposited from the atmosphere. The collection of moisture and detritus is assisted by modifications of the leaves to form niches or urns (heterophylly in e.g. *Platycerium* and *Dischidia*), or (for example in *Asplenium* species and Bromeliaceae) funnel-forming rosettes and cisterns. Absorption scales on the leaves of Bromeliaceae, for example, facilitate the en-

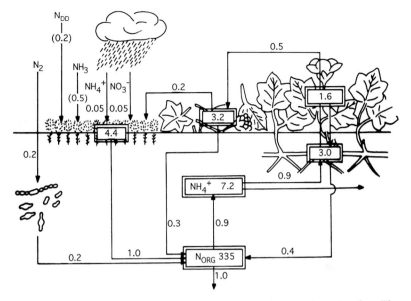

Fig. 3.25. The nitrogen budget of an ombrotrophic palsa mire in northern Sweden. The vegetation consists of vascular plants (in this case dominated by *Rubus chamaemorus*), lichens and mosses. The nitrogen input from the atmosphere is made up of NO_3^- and NH_4^+ from rain $(0.1 \text{ g } N \text{ m}^{-2} \text{ a}^{-1})$; NH_3 $(0.5 \text{ g } N \text{ m}^{-2} \text{ a}^{-1})$; dry deposition $(N_{DD}, 0.2 \text{ g } N \text{ m}^{-2} \text{ a}^{-1})$; and N_2 fixation by diazotrophic microorganisms $(0.2 \text{ g } N \text{ m}^{-2} \text{ a}^{-1})$. Additions from litter amount to $1.6 \text{ g } N \text{ m}^{-2} \text{ a}^{-1}$. Pools in the standing phytomass are expressed as $\text{g } N \text{ m}^{-2}$ and are shown in *double boxes*. (From Rosswall et al. 1975). For mineral and carbon cycling in a subarctic mire see Sonesson (1980)

try of ions and water. The different morphological forms are functional adaptations for securing nutrients. The bacteria, fungi and animals in the litter collecting in and around epiphytic ferns and flowering plants break down the organic substrate, releasing nutrients that can be taken up by the plant.

Table 3.11. Import of mineral elements $(\text{g } m^{-2} \text{ a}^{-1})$ by rain. (Golley et al. 1975; Kallio and Veum 1975; Likens et al. 1977

Element	Tropical rainforest (Central America)	Deciduous broad-leaved forest (North America)	Northern tundra (Scandinavia)
N		2.07	0.07 – 0.1
S		1.88	0.5 – 0.6
P	0.10	0.0004	
Ca	2.93	0.22	0.25 – 0.54
Mg	0.49	0.06	0.05 – 0.15
K	0.95	0.09	0.08 – 0.12
Na	3.07	0.16	0.1 – 0.4
Fe	0.30		

Carnivorous plants secrete digestive enzymes (proteases, peptidases) from glandular cells with a dense endoplasmic reticulum and lysosome-like vesicles, and they take up amino acids and ions via the same glands. In addition to this, the partially digested remains of trapped animals are broken down and mineralized by bacteria. Carnivory is a means by which plants growing in dystrophic habitats receive supplementary nitrogen and phosphorus. The *Pinguicula* species of high latitudes, for example, obtain 20–60% of their nitrogen and 35–80% of their phosphorus by protein digestion [111]. The direct utilization of organic nutrient sources makes possible not only the rapid replacement of leaves, but also the accumulation of important organogenic nutrients for new growth and seed production [216].

Plants that are permanently confined to nutrient-deprived habitats employ **efficiency strategies** [59] to achieve the necessary metabolic activity required for maintaining their competitive ability despite shortage of nutrients. This can be done by increasing the efficiency with which mineral substances are taken up (*absorption efficiency*; e.g. by increased root growth or formation of transfer cells) and by improving the availability of nutrients in the immediate vicinity of the roots (*mobilization efficiency*; e.g. excretion of acids and chelating substances by the roots). On soils with a general deficiency of mineral substances, and on acid soils, especially efficient utilization of nutrients can be achieved by their redistribution within the plant for the production of new organs (*retranslocation efficiency*); this ensures their conservation for longer periods. For example, during seed formation in annual species, macronutrients, especially phosphorus, are transferred in disproportionately large quantities to the reproductive organs at the expense of the vegetative organs. In perennial species bioelements are temporarily stored in the surviving organs, and can be retranslocated, ensuring the individual plant its basic requirements for survival in a habitat poor in mineral nutrients.

In cold regions, perennial species are at an advantage compared with annuals. The *longevity* of evergreen leaves has a similar effect; only part of the foliage has to be replaced each year, and the high losses of mineral substances occasioned by leaf abscission are avoided. A genotypic, morphogenetic or hormone-regulated *dwarf growth form* facilitates the maintenance of adequate concentrations of mineral substances in the tissues, and makes possible better

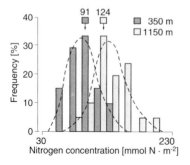

Fig. 3.26. Frequency distribution of leaf nitrogen content in herbaceous plant species from lowland (350 m a.s.l.) and higher altitudes (1150 m a.s.l.) in northern Sweden. *Arrows* Arithmetic means for each habitat. The plants growing under less favourable climatic conditions at the higher altitude have smaller leaves and a statistically greater nitrogen content (referred to leaf area and to dry matter). (From Körner 1989)

utilization of the available supplies of nitrogen and phosphorus for dry matter production. In such ways, herbaceous plants on cold and therefore nitrogen-poor montane soils attain higher concentrations of nitrogen than comparable lowland species (Fig. 3.26). The most successful species in habitats where mineral substances are in short supply, such as high mountains, tundra or heaths, are those with the *nutrient-conserving strategy*. Under abiotically restrictive environmental conditions, the most successful strategy is to avoid high production, and by limiting growth to obtain an harmonious decrease in all metabolic processes.

3.7 Mineral Cycling in Plant Communities

3.7.1 The Mineral Balance of a Plant Community

The mineral balance and carbon budget in the vegetation are closely interdependent. The uptake of mineral nutrients controls the increase in plant mass, and carbon assimilation makes available the material in which the elements are incorporated.

Using the production equation, the yearly mineral uptake by vegetation can be determined if the content and composition of the ash in the plant material are known. A certain fraction of the *total quantity of minerals absorbed* (M_{abs}) by the vegetation per area of ground in the course of a year is lost during the same year due to *recretion* (M_r) and is washed away from the shoots by rain or snow. The remainder, M_i, is the *quantity of minerals incorporated* in the phytomass.

$$M_{abs} = M_i + M_r \ (kg \ ha^{-1} \ a^{-1}) \ . \tag{3.1}$$

The minerals fixed in the plant material are apportioned, as in the production Eq. (2.21), to the *increase in biomass* ($+\Delta B$) occurring during the year and to the annual *losses* of dry matter as *detritus* (M_L) and from *removal by consumers* (M_G).

$$M_i = \Delta M_B + M_L + M_G \ (kg \ ha^{-1} \ a^{-1}) \ . \tag{3.2}$$

The terms in this equation represent the content of minerals or the ash content of the dry matter. The various mineral elements in a plant are non-uniformly distributed, accumulated, fixed and eliminated; moreover, mineral concentration depends on the state of development of the vegetation. For these reasons the **mineral balance** of a stand of plants cannot be found simply by multiplying the terms of the production equation by the average ash content of the plant mass. As in deriving carbon balance, the computation must be based on the separate contributions of the individual organs at the different sampling times. Table 3.12 shows the mineral partitioning of a deciduous forest and an evergreen broadleaved forest.

Table 3.12. Mineral contents and turnover in forests. All values are $g\ m^{-2}\ a^{-1}$

Stand	Deciduous oak-beech-hornbeam mixed forest with undergrowth, Belgium, 30–75 years old[a]					Evergreen oak forest (*Quercus ilex*) southern France, ca. 150 years old[b]				
	N	P	K	Ca	Total[c]	N	P	K	Ca	Total[c]
Mineral content of phytomass above ground	40.6	3.2	24.5	86.8	163.2	76.3	22.4	62.6	385.3	550.5
Amount fixed in new growth annually (ΔM_B)	3.0	0.22	1.6	7.4	12.78	1.32	0.26	0.89	4.27	6.83
Mineral content of the annual loss (M_L)	6.1	0.41	3.6	12.0	22.8	3.28	0.28	1.62	6.39	12.03
Mineral incorporation $M_i = \Delta M_B + M_L$	9.1	0.63	5.2	19.4	35.57	4.6	0.54	2.51	10.66	18.86
Leaching (M_r)	0.09	0.06	1.7	0.71	3.18	0.05	0.08	2.57	1.94	4.87
Annual absorption of minerals $M_{abs} = M_i + M_r$	9.19	0.69	6.9	20.11	38.75	4.65	0.54	5.08	12.6	23.73
Turnover coefficient $k_M = \dfrac{M_L + M_r}{M_{abs}}$	0.68	0.68	0.77	0.63	0.67	0.72	0.67	0.83	0.66	0.71

a Duvigneaud et al. (1969).
b Rapp (1969, 1971).
c Total amount of mineral in the dry matter; this value is greater than the sum of N, P, K and Ca since it includes the other elements in the ash.

The proportion of the total amount of mineral nutrients taken up that is incorporated in the annual increase in wood, and the fraction lost by leaching and the shedding of plant parts can be expressed as the **recycling coefficient** k_M [257]:

$$k_M = \frac{M_L + M_r}{M_{abs}} . \tag{3.3}$$

In a spruce forest with sparse undergrowth, 80% of the K taken up, 70−75% of the N, P, Ca and roughly 60% of the Mg were turned over annually [258]. Values for the nutrient turnover of a beech forest with a dense herbaceous layer were roughly 90% for N, P, K and about 80% for Ca and Mg. For a tropical rainforest, turnover factors for Ca and Mg were found to be 95−100% [171] (one third of this by leaching from the crown), for K 85% (half by leaching) and for Fe 50% [79].

3.7.2 The Cycling of Minerals

In returning the greater part of its mineral uptake to the soil every year with the fallen leaves, the plant cover makes an important contribution to the **circulation of minerals**. Some are withdrawn from deep layers of the soil by the roots, translocated into the shoot, then returned in leaf fall to the soil surface. In particular, deep-rooted deciduous trees, because of their extensive root systems, raise nutrient minerals that have sunk quite deep into the ground; once the litter from the trees has been broken down, the mineral substances are available to plants of the herbaceous stratum with shallower roots (Fig. 3.27).

The crucial component in the turnover of mineral substances between plant communities and the soil is the **recycling mechanism**; if the litter (of which there is more as the productivity of the plant cover increases) were removed, the store of nutrients represented by the mineral substances incorporated in it would be lost. Recognition of this fact is the basis of well-planned fertilization; the nutrient depletion brought about by harvesting of crops or removal of litter must be compensated by the provision of fertilizers in amounts of the appropriate order of magnitude. Calculation of the turnover of mineral nutrients provide objective figures for this purpose.

A useful measure of the recycling rate in a given ecosystem is the *net rate of mineralization* of the organic nitrogen compounds in the litter (Table 3.13). The net rate of mineralization is the excess of mineral nitrogen remaining after subtraction of the losses of gaseous compounds of nitrogen arising in denitrification and the requirements of the microorganisms.

Mineral nutrients enter the functional system "plants-microorganisms-soil" more or less continuously, while at the same time others leave it by one means or another. *Weathering* is a continuous process, adding mineral nutrients to the soil and making them available to plants in quantities that are hard to estimate, but that undoubtedly are significant. Mineral substances are moved into the

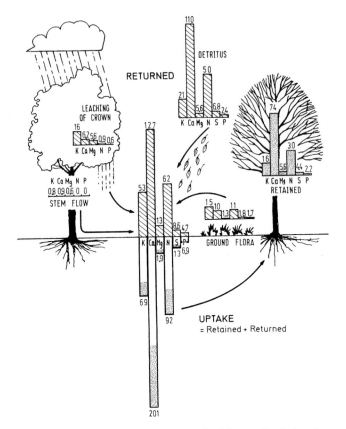

Fig. 3.27. The cycling of minerals in a mixed forest of oak, hornbeam and beech in Belgium. All quantities are given in units of kg ha^{-1} a^{-1}. The yearly uptake of minerals by the stand was calculated by summation of the quantities retained in the phytomass (*stippled*) and those returned to the soil (*hatched*) by way of washing and leaching of the canopy, stem flow, and litter from the trees and ground flora (cf. the carbon turnover of the same stand in Fig. 2.81). (After Duvigneaud and Denaeyer-De Smet 1970). For examples of mineral budgets of temperate coniferous forests, see Johnson et al. (1982); of deciduous and evergreen forests in Japan, Tsutsumi (1987); of tropical rainforest, Edwards (1982); of savannas, Egunjobi (1974); of temperate grasslands, Rychnovská (1993a); of Atlantic heaths, Gimingham (1972); of alpine ericaceous heaths; Larcher (1977); of arctic tundra, Bliss et al. (1981), Chapin and Shaver (1985); of wetlands, Dykyjova and Kvĕt (1978)

root zone by groundwater, water seeping down slopes, and by water rising by capillarity. Moreover, *precipitation* carries inorganic materials contained in the atmosphere as gases, dust, fog or aerosols, onto the plants and into the soil. In the opposite direction, *export of mineral substances* results from removal of phytomass (consumers, harvesting, removal of litter), by erosion, wind transport of organic litter and humus, as well as by the leaching or seepage of soluble substances.

Table 3.13. Rates of nitrogen mineralization in the upper layers of soil and the potential release of mineral nitrogen from litter during the growing season (g N m^{-2}). (From summaries of data in Rodin and Bazilevich 1967; Ellenberg 1977; and data given by Lossaint and Rapp 1971; Janiesch 1978; Rehder and Schäfer 1978; Nadelhoffer et al. 1992)

Vegetation	Net rate of mineralization	Release of mineral nitrogen (potential)
Tropics		
Rainforest	10 – 80	26
Dry deciduous forests	10 – 20	22.5
Gallery forests	7	
Savannas	0.3 – 0.5	
Agricultural land (shifting cultivation)	7 – 10	
Temperate zone		
Mediterranean sclerophyllous woodland		2 – 2.5
Evergreen coniferous forests	3 – 12	5 – 18
Deciduous broadleaved forests	(2) 10 – 20 (25)	
Fraxinus-Aceretum	15 – 38	
Riparian woods	(2.5) 10 – 20 (50)	
Atlantic ericaceous heaths	(0.5) 1 – 3 (5)	
Meadows (manured)	14 – 26	
Meadows		13 – 23
Wet meadows, sedges	(0) 1 (4)	
Dry grassland, steppes	1 – 3	2 – 6.5
Ruderal localities	4 – 30	
Dune vegetation	1 – 3	
Halophyte vegetation	0.2 – 1	1 – 1.5
Cold regions		
Boreal and mountain forests	1 – 5	1.5 – 2.5 (8)
"Krummholz" scrubwood	1 – 2	
Alder vegetation	15	
Alpine dwarf shrub heaths	0.1 – 1	
Alpine grasslands	(0.5) 2 – 10	
Arctic tundra	0.03 – 0.5	2 – 5
Heather moors and bogs	0 – 0.5	

4 Water Relations

Life evolved in water, and water remains the essential medium in which biochemical processes take place. Protoplasm displays signs of life only when provided with water – if it dries out, it does not necessarily die, but it must at least enter an inactive (anabiotic) state, in which vital processes are suspended.

Plants are composed mainly of water. On an average, protoplasm contains 85–90% water, and even the lipid- and protein-rich cell organelles such as chloroplasts and mitochondria contain 50% water. The water content of fleshy fruits is particularly high (85–95% of the fresh weight), as is that of soft leaves (80–90%) and roots (70–95%). Freshly cut wood contains about 50% water. The parts of plants having the least water are ripe seeds (usually 10–15%); some seeds with large stores of fat contain only 5–7% water [123].

4.1 Poikilohydric and Homoiohydric Plants

In terrestrial plants, the aerial parts of which continually lose water by evaporation, the establishment of suitable *water relations* is the first requirement for survival. Depending on their ability to compensate for short-term fluctuations in water supply and rate of evaporation, terrestrial plants may be classified as poikilohydric or homoiohydric [264].

Poikilohydric plants match their water content to the humidity of their surroundings. Prokaryotes, fungi, and some algae and lichens have small cells that lack central vacuoles; when they dry out, these cells shrink very uniformly, without disturbance of the protoplasmic fine structure, so the cells remain viable. As the water content decreases, the vital functions, e.g. photosynthesis and respiration, are gradually suppressed. When sufficient water has been imbibed again, such plants resume normal metabolic activity.

The minimal water potential necessary for activity is species-specific and determines the range of distribution of the various species. Thus, depending on the species, most soil bacteria and fungi, in order for metabolism and cell division to proceed, require water potentials somewhere in the range of -5 (or higher) to -30 MPa (corresponding to about 80–95% RH; see Table 4.1); but moulds can grow at relative humidities of about 75–85%, and species of *Xeromyces* can grow well at RH as low as 60% [183]. Halophilic bacteria are still active at around -40 MPa. Many *lichens* remain capable of photosynthesis as long as the water potential in the thallus does not drop much below

Table 4.1. Values of relative humidity (in the air above a solution) and the osmotic pressure (MPa) of the solution, when the two phases are enclosed and allowed to equilibrate at 20 °C. (Recomputed from Walter 1931)

% RH	MPa	% RH	MPa
100	0	93.0	9.8
99.5	0.67	92.0	11.2
99.0	1.35	91.0	12.6
98.5	2.03	90.0	14.1
98.0	2.72	80.0	30.1
97.5	3.41	70.0	48.1
97.0	4.10	60.0	68.7
96.0	5.50	50.0	93.3
95.0	6.91	0	∞
94.0	8.32		

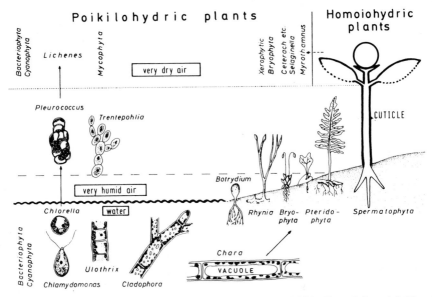

Fig. 4.1. Evolution of the water relations of plants to terrestrial life. *From left to right* Transition from aquatic lower algae with non-vacuolated cells to primarily poikilohyric aerial algae; development of vacuole in aquatic green algae and Characeae; transition from vacuolated thallophytes to homoiohydric vascular plants (hygrophytic mosses are still restricted to habitats with high air humidity, and in dry habitats become secondarily poikilohydric; there are also secondarily poikilohydric forms among the pteridophytes and angiosperms, but not among gymnosperms). Most vascular plants, because they are equipped with a cuticle that limits transpiration and because their cells are considerably vacuolated, are homoiohydric. (Walter 1967)

−3 MPa [203]. Poikilohydric forms are found not only among thallophytes, but also among *mosses* of dry habitats, certain *vascular cryptogams* (especially *Selaginella* species and various ferns), and in a few angiosperms (see Chap. 6.2.4.3). Pollen grains and embryos in seeds are poikilohydric stages of homoiohydric plants.

Homoiohydric plants are descended from green algae with vacuolated cells; a large central vacuole is a common characteristic of all homoiohydric plants. Because the water content is stabilized within limits by the water stored in the vacuole, the protoplasm is less affected by fluctuating external conditions. However, the presence of a large vacuole also means the loss of the cell's ability to tolerate dehydration. This is why the predecessors of homoiohydric land plants are found pressed close to wet soil or living in permanently moist habitats (Fig. 4.1). Only with the evolution of a protective cuticle to slow down evaporation, and of stomata to regulate transpiration, were plants able to control their water economy adequately; with these adaptations, and an extensive system of roots, the protoplasm can be maintained in an active state despite sudden changes in humidity. Thus plants became sufficiently productive to form a closed cover over large areas, and ultimately to produce the enormous phytomass now covering the continents.

4.2 Water Relations of the Plant Cell

4.2.1 The Water in the Cell

The water in plant cells occurs in several forms; it is a chemically bound *constituent* of protoplasm; as *water of hydration* it is associated with ions, dissolved organic substances and macromolecules, filling the gaps between the fine structures of the protoplasm and of the cell wall; as a reserve, it is *stored* in vesicles and vacuoles; finally, as *interstitial water* it serves as a transport medium in the spaces between cells and in the conducting elements of the xylem and phloem systems (*vascular water*).

Water of hydration. In accordance with their *dipole character,* water molecules aggregate at polar surfaces in clusters (*structured water;* Fig. 4.2). Strongly charged ions of about the same size was water molecules bind water more firmly, the greater their charge and the smaller the radius of the ion. The water molecules are bound tightly at the surfaces of the ions by electrostatic forces. A similar situation occurs at the surfaces of protein molecules and polysaccharides. Water molecules become associated with polar groups (hydroxyl, carboxyl and amino groups) and form several layers of structured water; the water molecules are more readily displaceable the further they are from the polar group. Hydration water accounts for only 5−10% of the total cell water, but this amount is absolutely essential for life; a slight decrease is sufficient to cause severe alterations in protoplasmic structure.

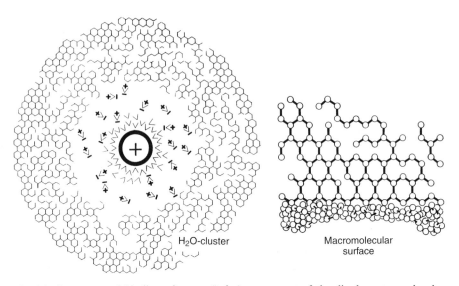

H₂O-cluster   Macromolecular
                                    surface

Fig. 4.2. Structure and binding of water. *Left* Arrangement of the dipole water molecules around a positively charged ion; the more distant water molecules, no longer under the influence of the ion, organize via hydrogen bonds into short-lived microcrystals, continuously forming and re-forming. Between these two layers of structuralized water is a "chaotic" zone in which the water molecules are pulled in both directions. *Right* Water molecules structured into a lattice formation by the apolar groups of macromolecules, e.g. biomembranes. (Karow and Webb 1965)

Most of the water of hydration is bound by *capillary* forces in the protoplasm and cell wall. The cell walls of plants hold water with a tension of 1.5 – 15 MPa, depending on the density of the fibrils. The forces holding water to the surface of structural elements in a matrix (cell wall, plasma colloids) can be expressed in terms of the *matric pressure* τ.

Stored water. The most easily translocated water is that in those cell compartments specialized as reservoirs for solutions. More than half the water present in leaves is stored in this way; but even this water is not completely mobile since it is osmotically bound to dissolved substances such as sugars, organic acids, secondary plant substances and ions. The *osmotic pressure* π of a solution, according to Van't Hoff, is given approximately by

$$\pi^* = n \cdot R \cdot T = 2 \cdot 27 \cdot n \, \frac{T}{273} \, [\text{MPa}] \tag{4.1}$$

The osmotic pressure of a solution increases with absolute temperature T and with the number of dissolved particles, n. An ideal solution of 1 osmol kg⁻¹ at 25 °C has an osmotic pressure of 2.48 MPa (equivalent to 98.2% relative humidity). Macromolecular substances can be present in considerable amounts, in terms of weight, without raising the osmotic pressure appreciably. Through the polymerization of small molecules to macromolecules – for example the conversion of sugar to starches and the reversal of this procedure by hydrolysis – the cell can rapidly alter its osmotic pressure and the net influx of water can be regulated. In protoplasts with central vacuoles there is a close connection between the osmotically bound water in the cell sap and water availability in the protoplasm. The osmotic pressure of the cell sap can thus be regarded as an indicator of the state of hydration within the cell, and hence of the protoplasm.

4.2.2 The Water Potential of Plant Cells

It is the *thermodynamic state of the water* rather than its total quantity that influences the biochemical activity of protoplasm. The thermodynamic state of the water in a cell can be compared with the chemical potential of pure water and the difference expressed in terms of potential energy. The relative *water potential* Ψ is the work necessary to raise the bound water to the potential level of pure water [227]. The dimensions of water potential are energy per unit mass ($J\,kg^{-1}$), or per unit volume, and are convertible into pressure via the relationship $1\,J\,g^{-1} = 1\,MPa$.

The water in solutions is osmotically bound, and only becomes available if energy is added. The *osmotic water potential* Ψ_π is lower than that of pure water and is therefore always negative. Water bound to colloids and hydrophilic surfaces also has a negative potential, the *matric potential* Ψ_τ. If water is placed under pressure, its free energy increases, and the pressure potential Ψ_P is positive with respect to non-pressurized water. The availability of water is expressed as the water potential of the aqueous system (of the cell, cell compartment, or external solution) with respect to the potential of pure water. This means that the availability of water is less, the more negative the potential of the system under consideration.

4.2.3 The Dynamic Equilibrium of Cellular Water

Between sites differing in water potential there is a *potential difference* $\Delta\Psi$. This situation is analogous to that in an electrical system, in which different points are at different voltages. Just as there is a tendency for current to flow from a point of higher voltage to one of lower voltage, in living cells there is a tendency for water to move from regions of higher to those of lower water potential.

Every potential difference offers the possibility of a reduction of this difference. In the cell this occurs by the *transport of water* or other substances; water transport always proceeds in the direction of a lower water potential. As long as there is no obstacle to diffusion, a thermodynamic steady state is rapidly attained within the cell, and between the cells and their surroundings. A high water vapour deficit in the air, or a hypertonic surrounding medium (e.g. sea water or saline soil water) can cause water to leave the cell and thus lower its water potential. Conversely, water flows from the surroundings into a cell with a more negative water potential.

The laws governing cellular water transport were already recognized in the last century. The fundamental processes are shown in Fig. 4.3. The water potential for the whole cell Ψ_{cell} at a given state of hydration is given by the difference between the *osmotic water potential* Ψ_π and the *pressure potential* Ψ_P.

$$\Psi_{cell} = (-)\Psi_\pi + (+)\Psi_P \tag{4.2}$$

The osmotic potential Ψ_π is invariably negative, whereas the pressure potential Ψ_P can be positive, zero, or in exceptional cases even negative. A negative water potential indicates that the cell as a whole is under tension. But since forces of hydration are also active inside the cell, formula (4.2) must be expanded to include the *matric potential* $(-)\,\Psi_\tau$ of the protoplasm and cell wall. Usually, however, the matric potential is so small as to be negligible (except in poikilohydric cells).

When **water-saturated**, the protoplast has attained its greatest volume and exerts the greatest pressure on the cell wall. Due to the internal pressure (*turgor pressure*) the cell wall is maximally distended. Then the resulting wall pressure compensates for the osmotic effect of the cell sap so that net water uptake into the cell is stopped. At this point of water saturation $\Psi_{cell} = 0$ and $\Psi_\pi = \Psi_P$. The *water-storing capacity of the cell* (C_{cell}, hydraulic capacitance [40]) is large, if a slight alteration in the water potential suffices to produce a large increase in cell volume (V_{cell}):

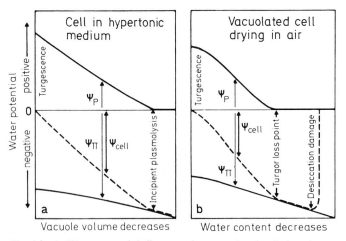

Fig. 4.3 a, b. Water-potential diagrams for vacuolated cells in a hypertonic medium (**a**) and for vacuolated cells of a leaf drying in air (**b**). As water loss proceeds, the pressure potential, Ψ_P falls from positive values to zero, and the osmotic potential Ψ_π becomes more negative. (Schematic representation after Höfler 1920; Barrs 1968; Kyriakopoulos and Larcher 1975; Pospišilová 1975)

$$C_{cell} = \frac{\Delta V_{cell}}{\Delta \Psi_{cell}} \; [m^3 \, MPa^{-1}] \tag{4.3}$$

This property is dependent on the elasticity of the cell wall, which is defined by the volumetric *modulus of elasticity*, ε [40]:

$$\varepsilon = \frac{\Delta P}{\Delta V/V} \; [MPa] \tag{4.4}$$

where ΔP indicates the change in turgor pressure connected with a relative change in cell volume (ΔV/V). A low bulk elastic modulus (up to 1–5 MPa, as for example in the soft leaves of herbaceous species) indicates that the cell walls readily distend in response to changes in cell volume, and therefore the turgor pressure rises slowly; cells of this kind can store large quantities of water. On the other hand, for leaves of trees and shrubs, which are usually more rigid than those of herbs, the maximal values measured for the bulk elastic modulus at high tissue turgor are 10–20 MPa for deciduous leaves, and 30–50 MPa for evergreen leaves, indicating lower ability to store water [199].

Loss of water leads to reduction of the vacuolar volume and a rise in cell sap concentration. Increasingly less pressure is exerted on the protoplast by the cell wall, until the cell volume has diminished to a threshold value, beyond which the cell wall can shrink no further (*zero turgor point*). If the cell is in an aqueous medium the protoplast begins to pull away from the cell wall; this stage is called *incipient plasmolysis* (Fig. 4.3a). When the cell is in this condition $\Psi_P = 0$, so that $\Psi_{cell} = \Psi_\pi$. In land plants exposed to the air, cells lose water by evaporation from the surfaces of the cell walls. As the cell wall dries out, water flows out of the protoplast and the cell loses its turgor; but plasmolysis does not occur, since the cell walls are impermeable to air. The walls follow the protoplasts; they are drawn inward and may wrinkle and fold up in the process (*cytorrhysis*). If desiccation is so extreme that the biomembranes are damaged, the osmotic system breaks down and the ability to absorb water is lost (Fig. 4.3b).

When cells are not isolated, but form a **tissue**, their wall pressure may be increased or decreased by that of their neighbours; thus tissue tension is a factor to be taken into consideration. If the adjacent cells *reinforce* the cell wall pressure the cells become turgid at a lower water content (with a smaller vacuolar volume). This is important in tissues with delicate cell walls, for under such conditions they can maintain turgor without excessive filling with water. If, on the other hand, the cell walls form a rigid surface, they will not readily follow the shrinking protoplasts. Under these conditions, a negative tissue potential of as much as −1 to −2 MPa may develop (e.g. palm leaves, and *Zygophyllum dumosum* [172]. This means that in the event of poor water supply, translocation of water may occur from cells with elastic walls (e.g. water-storing tissues in succulent leaves and shoots) to adjacent cells with less elastic walls (e.g. palisade parenchyma cells), given the necessary osmotic gradient (Fig. 4.4).

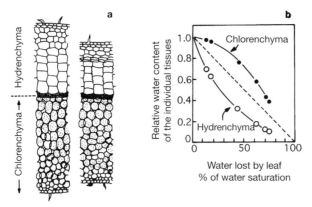

Fig. 4.4 a, b. Water translocation from a water-storing tissue (hydrenchyma) to the assimilating tissue (chlorenchyma) in the succulent leaves of *Peperomia* during desiccation. **a** Cross-section of a leaf of *P. trichocarpa* in the water-saturated state (*left*) and after 3 days of drying (*right*). The hydrenchyma shrinks more than the chlorenchyma. **b** Unequal decrease in water content of the hydrenchyma and the chlorenchyma of *P. obtusifolia,* relative to the water loss [expressed as water saturation deficit; see formula (4.14)] of the leaf during the course of desiccation. When the leaf has lost 50% of its water the hydrenchyma contains only 25%, but the chlorenchyma still has 75% of its saturation water content. (After Haberlandt 1924; Schmidt and Kaiser 1987)

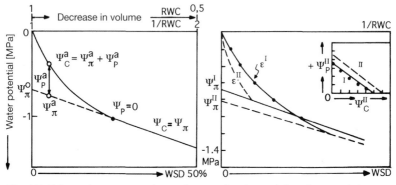

Fig. 4.5. Schematic pressure-volume diagram for determining characteristic parameters of water relations in plants. *Left* Depression of tissue water potential with progressive decrease in cell volume [a quantitative measure of which is given by the relative water content RWC or the water saturation deficit WSD; see formulae (4.13) and (4.14)]. Ψ_c^a = actual cellular water potential; Ψ_P^a = actual turgor potential; Ψ_π^a actual osmotic potential. Beyond the zero turgor point ($\Psi_P = 0$) the measured water potential corresponds to the osmotic potential. If the straight part of the curve is extended to the origin of the abscissa, the point at which it intersects the ordinate gives the osmotic potential at turgidity, Ψ_π^0. *Right* Variant types of cell water relations. Tissues with rigid cell walls (high elasticity modulus ε^{II}) reach the zero turgor point sooner than those with more elastic walls (lower ε^I). Shifts in the pressure-volume diagram also provide an indication of active depression of the osmotic potential (e.g. due to enzymatic degradation of starch to sugar or as a result of ion transport). If the osmotic potential falls from Ψ_π^I to Ψ_π^{II}, the turgor range increases (difference $\Psi_c - \Psi_\pi$). The extent of the active osmotic processes can be assessed by comparing Ψ_P with Ψ_c (inset). (Adapted from data and graphics of Richter 1978; Schulte and Hinckley 1985; Richter and Kikuta 1989)

The characteristic relationship between the **water potential of plant organs** and the state of hydration can be illustrated by pressure-volume diagrams (P/V diagrams [256]) in which, in accordance with the Boyle-Mariotte Law ($P \cdot V$ = constant), the decrease in cellular pressure with progressive loss of water is related to the decrease in volume. The water potential Ψ_{cell} is substituted for pressure, and the relative water content [RWC, see formula (4.13)] is substituted for volume (Fig. 4.5). From the P/V diagram the osmotic water potential at water saturation can be read off (by extrapolaton of the linear portion to RWC = 1), as well as the water potential at zero turgor pressure (at the point where the curve becomes linear). The *elastic modulus* ε can be derived from the shape and slope of the curve between full turgidity and the turgor zero point: in tissues with rigid cell walls ε is high, which means that a higher turgor pressure is necessary to distend the cell walls during water inflow, but on the other hand Ψ_P sinks rapidly at the slightest loss of water. The P/V diagram also allows the recognition of osmotic adjustments; an active increase in the osmotically effective constituents of the cell sap causes a shift in the zero turgor point towards lower RWC values.

4.3 Water Relations of the Whole Plant

The shoot of terrestrial plants steadily loses water to the air surrounding it; this water must be replaced from the soil. Transpiration, water uptake and conduction of water from the roots to the transpiring surfaces are inseparably linked processes in water balance. The vapour-pressure deficit of the air is the driving force for evaporation, and the water in the soil is the crucial quantity in water supply. The water balance is maintained by a continuous flow of water, and is thus in a state of *dynamic equilibrium*.

4.3.1 Water Uptake

Plants can absorb water over their entire surfaces, but the greater part of the water supply comes from the soil. In higher plants water uptake occurs by way of the roots, which are the specialized organs of absorption. Lower plants are rootless and thus dependent upon the direct uptake of water via organs above ground.

4.3.1.1 Direct Water Uptake by Thalli and Shoots

Thallophytes draw water by capillary action from damp substrates and from the rain, dew and fog wetting their surfaces, and swell in the process. When saturated with water, most mosses contain three to seven times, and lichens two to three times [240] as much water as in the dry state, but fruiting bodies of fungi, gelatinous lichens and peat moss contain up to 15 times as much water as when dry. The water is usually taken up rapidly, so that the thalli of some species are already soaked within a few minutes and maximally swollen in half an hour. Bacteria, lower fungi, fungal hyphae, and some algae and lichens can also take up water from *humid air*; but never as much as when their surfaces have been wetted, and as a rule several days are required to bring their moisture content up to that of the surroundings.

Vascular plants are protected from evaporative water loss by a cuticle. To the same degree that the *cuticle* and cutinized layers slow down the loss of water by the shoot, they interfere with the entry of water when the surface is wetted. Therefore direct water intake by such shoots occurs mainly, if at all, through specialized parts of the plant such as *hydathodes* (water-permeable openings in the epidermis) and non-cutinized attachment points of wettable hairs. In epiphytic Bromeliaceae, a significant part of the water supply enters through *scales* specialized for imbibition. The leaves of floating plants take up most of their water and mineral substances via modified stomata.

4.3.1.2 Water Uptake from the Soil

The Water in the Soil
Water infiltrates the soil following precipitation and gradually seeps down to the water table. In highly permeable soils the rate of percolation is several metres per year, in loamy soils $1-2$ m per year, and in very dense soils only a few decimetres per year. A portion of the infiltrating water, the *capillary water*, is held back and stored in the pore spaces of the soil. How much water is retained as capillary water in the upper layers of the soil, and how much penetrates through them as gravitational water depends on the nature of the soil and the distribution of pore sizes within it (Fig. 4.6). Pores up to about 10 µm in diameter hold water by capillary action, whereas coarser pores (over 60 µm) let it seep down rapidly.

The water-storage capacity of the soil: the water content at saturation of the soils in their natural locations, after the gravitational water has percolated through, is called the *field capacity* and is expressed in g H_2O per 100 g of soil (% of dry weight) or per 100 ml of soil. Fine-grained soils, as well as soils with a high colloid content and those rich in organic substance, store more water than coarse-grained soils. The field capacity thus increases according to the following sequence: sand, loam, clay, peat. After long periods of rain, and immediately after the snow has melted in spring, some of the gravitational water remains longer in the upper layers of the soil and is thus available for the plant roots (*available gravitational water*). For plants, a soil with a high water-retaining capacity is of great advantage for surviving periods of drought.

How water is retained by the soil: the water that remains in the soil after the passage of gravitational water is held in the pore space by capillary action; it is also attached to the soil colloids by surface forces and (especially in saline soils) it can be osmotically bound to ions. Thus soil water, like the water within a plant, is not entirely free. In most soils the contribution made to the total

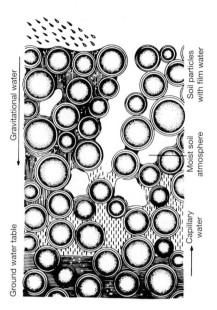

Gravitational water

Ground water table

Soil particles with film water

Moist soil atmosphere

Capillary water

Fig. 4.6. The status of water in saturated (*left half*) and in aerated soils (*right half*). (After Lerch 1991)

water potential by osmotic binding to ions is negligibly small, as is also Ψ_P, the hydrostatic pressure of the water in the pore space. The crucial component of the soil's water potential is the *matric (or "capillary") water potential Ψ_τ*, the energy with which the capillary water is held by surface forces. The matric potential can become particularly large in soils with fine pores. The capillary component of the matric potential is described by the formula [72]

$$\Psi_{cap} = -\frac{4\,\sigma}{d} \approx -\frac{290}{d} \ [J\ kg^{-1}] \tag{4.5}$$

where σ is the *surface tension* of the water and d is the *pore diameter* (in μm). The force with which the water is held increases greatly during drying-out, as the large-diameter pores are emptied and capillary water remains only in the finer (less than 0.2 μm) pores. In sandy soils with a coarse granular structure the transition is particularly sharp, whereas in loamy and clay soils, in which the pore sizes range from average to the smallest, the water potential declines less abruptly (Fig. 4.7).

Water Uptake by Roots

A plant can withdraw water from the soil only as long as the water potential of its fine roots is more negative than that of the soil solution in their immediate surroundings. The rate of water uptake is the greater, the larger the absorbing surface of the root system and the more readily the roots can withdraw water from the soil. This is described by the formula (4.6)

$$W_{abs} = A \cdot \frac{\Psi_{soil} - \Psi_{root}}{\Sigma r} \tag{4.6}$$

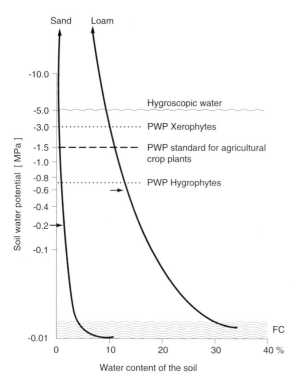

Fig. 4.7. Diagram showing the relationship between the water potential of soil and its water content (for a sandy and a loam soil). Conventional threshold value for field capacity (FC) is −0.015 MPa, for permanent wilting (PWP) −1.5 MPa; at values more negative than −5 MPa and there is only hygroscopically bound water. The average values shown for the different plant types depend on the type of soil (soil texture, pore size) and the vegetation; they may also be lower due to adjustment of the plants to water deficiency. The *arrow* is referred to in the text. (After Kramer 1949; Laatsch 1954; Slavíková 1965)

That is, the amount of water the roots can absorb per unit time (W_{abs}) is proportional to the *exchange area* (A) in the region penetrated by the roots (active root area) and the *water potential difference* between root and soil; it is inversely proportional to the *transfer resistances* (Σr) to water movement within the soil and in the passage from soil into plant. In the course of the growth period the older (proximal) parts of the root system suberize, which leads to alterations in the diffusion properties of the roots. But, above all, the active root area increases due to continuous growth at the root tips, and therefore the changing capacity of the root system to take up water must always be taken into consideration.

Roots usually develop **negative water potentials** of a few tenths MPa, which is nevertheless quite sufficient to withdraw the greater part of the capillary water from most soils. This effect can be seen in Fig. 4.7: with a drop in potential to only −0.2 MPa the roots can withdraw more than two thirds of the water storable in a sandy soil; a loamy soil, which holds water more tightly because of the fineness of its pores, can give up half of its capillary water to roots with a water potential of only −0.6 MPa. To a limited extent, some plants can obtain more water from the soil by actively lowering their root potential. *Hygrophytes* (i.e. species restricted to permanently moist sites) can at most lower the water potential of their roots to −1 MPa, crop plants in humid regions to −1 to −2 MPa, *mesophytes* to −4 MPa [228], and plants growing in dry regions (*xerophytes*) to, at most, −6 MPa. For forest trees −2 to −4 MPa is considered to be the limit [269].

As a result of extraction of soil water from the immediate surroundings of the roots, water is drawn from moister places in the soil. This **movement of water** takes place by capillarity, is very slow and can occur only over short distances (a few mm to cm). As the water in the pore system becomes increasingly exhausted the *transfer resistance* rises considerably, particularly in clay soils. In large-pored sandy soils the columns of water held up by capillarity break under even slight tension, so that the supply of water is interrupted.

Water can also move in the soil in the form of *water vapour*. During the nocturnal cooling of the uppermost soil layers, water vapour diffuses from the warmer underlying layers to the surface and condenses in the root region (*thermocondensation*). In dry regions, the water made available to the upper layers of the soil in this way, together with dew, is in some places sufficient for dry farming. A further means by which water can be moved upwards under dry conditions is by *internal conduction in the root system*. During the night, water drawn from the deep, moist soil horizons is conveyed not only to the shoots but also to the roots nearer the surface. If the water potential in the uppermost soil layers is lower than that of these roots the latter lose water to the surrounding soil (Fig. 4.8). This *nocturnal hydraulic lift* in deep-rooting plants helps other plants with more superficial root systems to survive periods of drought.

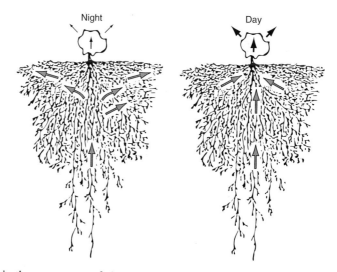

Fig. 4.8. Hydraulic lift in the root system of *Artemisia tridentata*. Water adsorbed by deep roots in moist soil moves through the roots, is released in the upper soil layer at night, and is stored there until it is resorbed by roots the following day. (From Richards and Caldwell 1987, Caldwell 1988; for different life forms, see Dawson 1993)

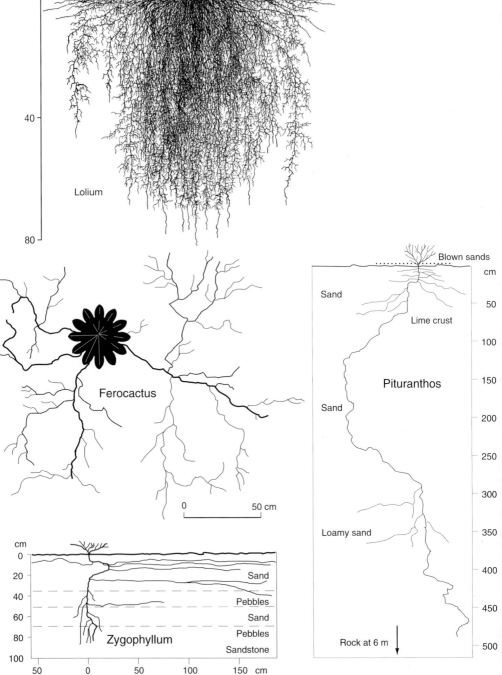

cm

40

80

Lolium

Ferocactus

0   50 cm

Blown sands

cm

Sand

50

Lime crust

100

150

Pituranthos

200

250

Sand

300

Loamy sand

350

400

cm
0

20   Sand

40   Pebbles

60   Sand

80   Pebbles

Zygophyllum   Sandstone

100

50   0   50   100   150   cm

450

Rock at 6 m

500

   The **root system of a plant** develops in accordance with its species-specific morphological pattern and to the extent that local conditions permit (soil structure and depth), as a shallow, deep, or multi-layered system (Fig. 4.9). The form taken by the root system of a particular species is closely connected with its functional type: for example palms and many, but not all, grasses have *intensive root systems*, in which the active root surface is enlarged by an extremely dense growth of fibrous roots; *extensive root systems* penetrate large volumes of soil in search of moisture, either by means of widely spreading roots (many plants growing in open habitats, cacti), or by roots stimulated to grow toward the water table (e.g. species growing in the desert or on rocks).

   Roots respond constantly by growth toward water sources. As the soil gradually becomes drier, some parts of a root system may die and dry up, while in other places the root is at the same time continuing to grow for many metres and is branching prolifically. Continuous *root growth* in an important prerequisite if the plant is to make full use of the water in the soil. Growth and spread of roots are impeded in shallow, impacted and wet soils, and delayed in cold soils. The chief reason for poor water supply to the plants on such soils is inadequate root growth.

   It becomes obvious that the availability of water to the plant cannot be judged solely on the basis of soil parameters. Although conventional threshold values for water availability in the soil (for example the *"permanent wilting point"*) continue to provide useful guidelines for agricultural purposes, for investigations in the field of ecophysiology and geobotany, the water status of the plant should be recorded. For this purpose the leaf water potential under steady-state conditions (e.g. *predawn water potential*), or when transpiration is artificially suppressed, is especially informative, since it corresponds approximately to that of the rhizosphere.

4.3.2 The Plant in the Water-Potential Gradient Between Soil and Atmosphere (Soil-Plant-Atmosphere Continuum)

The plant bridges the steep water-potential gradient between soil and air. Because the shoot is exposed to the vapour-pressure deficit of the air (i.e. to a low water potential) a flow of water through the plant is set in motion. The steepest *water-potential gradient* is that between the shoot surface and dry air. Here also is the largest transfer resistance in homoiohydric plants (Fig. 4.10); this is due to the high energy requirement for the evaporation of water $(2.45 \text{ kJ g}^{-1} \text{ H}_2\text{O}$ at $20 \,^\circ\text{C})$ and to the epidermal resistance to diffusion.

Fig. 4.9. Examples of intensive (*Lolium multiflorum*) and extensive root systems: vertical extension of the tap root of *Pituranthos tortuosus* in a Wadi in Egypt; horizontal spread of the root system of *Zygophyllum album* in a salty depression in the Algerian desert, and of *Ferocactus wislizenii* in Arizona. (After Cannon, as cited in Kutschera 1960; Walter 1960; Kausch 1959, 1968).

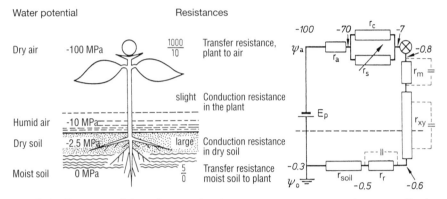

Fig. 4.10. Water-potential gradients and resistances to water transport between soil, plant and atmosphere. *Left* Order-of-magnitude estimates of potential gradients between soil and atmosphere, and of the transfer resistances in the system. (Kausch 1955). *Right* Circuit diagram: E_P potential evaporation (source of tension); Ψ_0 water potential of the liquid phase; Ψ_a water potential of the atmosphere; *italicized numbers* water potential in MPa; r_{soil} hydraulic resistance in the soil; r_r transport resistance in the secondary roots and root cortex; r_{xy} conduction resistance in the xylem of roots, shoot, leaf petioles and veins; r_m transport resistance in the mesophyll; r_c cuticular resistance; r_s stomatal resistance (variable); r_a boundary layer resistance at the surface of the shoot; *capacitor symbol* storage capacity in the apoplast and symplast of the root, in the wood and cortex and in the leaves; $\otimes$ transition from liquid to vapour phase. (After Cowan 1965; Boyer 1974; Kreeb 1974a; Schulze 1986; Nobel 1991 a)

The *movement of water* is governed by rules analogous to those for the flow of electricity, as described by Ohm's law. The soil-plant-atmosphere system can therefore be represented schematically by analog circuit diagrams. The potential gradient in the *soil-plant-atmosphere continuum* is the driving force for water transport through the plant. The water potential Ψ_Z at a particular location in the plant is lower (i.e. more negative), the lower the water potential in the soil, the greater the effect of gravity (Ψ_g), the greater the hydraulic resistances (r_i), between the soil and the point of reference (z) in the shoot, and the more water flowing through the plant (sum of the partial fluxes Σj_i) [197].

$$\Psi_z = \Psi_{soil} + \Psi_g + \Sigma_{soil}^z j_i \cdot r_i \tag{4.7}$$

From this formula it can be deduced that the plant would be expected to exhibit a steep gradient in water potential only when large quantities of water are flowing through it, i.e. when conditions promote intensive transpiration (Figs. 4.11 and 4.12).

The Path of Water in the Plant

Within the plant water movement takes place along the water potential gradient by diffusion from cell to cell (*short-distance transport*) and by conduction through the xylem (*long-distance transport*).

Diffusion through the tissues proceeds along the hydrostatic gradient from cell to cell, and through the cell walls (apoplastic translocation). The path tak-

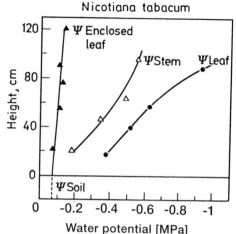

Fig. 4.11. Vertical gradients of the water potentials in leaf (Ψ_{Leaf}) and stem (Ψ_{Stem}) of tobacco plants. When the rate of transpiration is high there is a steep hydrodynamic gradient along the shoot. If transpiration is restricted by enclosing the leaves in polyethylene bags, a shallow hydrostatic steady-state gradient develops between the soil (Ψ_{Soil}) and the enclosed leaves ($\Psi_{[leaf]}$). (From Begg and Turner 1970)

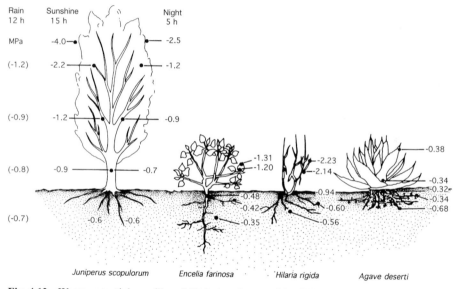

Fig. 4.12. Water-potential profiles (MPa) in plants with different growth forms. For *Juniperus scopulorum*, a tree-like juniper of the arid regions of western N. America, measurements were made in September on a sunny and a rainy day, and at night following a sunny day. For *Encelia farinosa*, a C$_3$ subshrub, for *Hilaria rigida*, a C$_4$ tussock grass, and *Agave deserti*, a leaf succulent of the semidesert in California, the water potentials were measured at the time of maximal transpiration. (After Wiebe et al. 1970; Nobel and Jordan 1983)

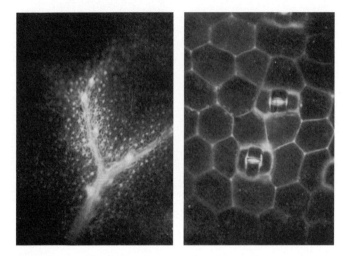

Fig. 4.13. Extrafascicular water translocation. *Left* Movement of water from the bundle sheath into the lower epidermis of a *Soleirolia soleirolii* leaf; *right* water transport in the anticlinal walls of the lower epidermis and guard cell walls of *Tradescantia albiflora*. The apoplastic translocation of water in the cell walls is demonstrated by means of a fluorescent dye, berberin sulphate. (From Strugger 1938)

en by water, its speed of transfer, as well as possible obstacles to its progress, can be shown up by employing suitable indicators (Fig. 4.13).

In the root the water passes to the endodermis via the parenchyma of the cortex (which acts as a kind of reservoir, buffering short-term fluctuations in the supply of water from the soil). In the endodermis the apoplastic transport is blocked by hydrophobic barriers in the radial walls (Casparian strips) or by woody cell walls, thus ensuring that all of the inflowing water is channeled to the special passage sites in the endodermis.

In the central stele of the root the water enters the vascular system, where the long-distance transport (conduction) through the plant takes place. The conducting vessels are specialized for the rapid movement and distribution of water, most of it moving by mass flow through the lumina of the vascular elements (*vascular conduction*). Water transport can also be shown to take place in cell walls in the long-distance system, but the amounts transported in this way are insignificant. In the leaves the bundles divide into fine branches, and through the terminal tracheids the water passes to the vascular parenchyma, whence it is distributed to the mesophyll cells by diffusion.

The Rate of Sap Flow

The amount of water moved through the vascular system in unit time is dependent on the specific properties of the xylem, such as the conducting area (cross-sectional area of the vessels) and flow resistances, on the physiological state of the plant (e.g. degree of stomatal opening) and on environmental conditions. The *water flow* (J^{H20}) is greater, the larger the conducting area (A) and the higher the flow velocity (v), as described by the formula

$$J^{H20} = v \cdot A \ [\text{kg s}^{-1})$$ (4.8)

The **conducting area** in a shoot axis or a petiole is the sum of the cross-sectional areas of all the xylem elements. In the trunks and thicker branches of conifers and of broadleaved trees with diffuse-porous wood and wood with diameters of conducting vessels below 100 μm (microporous), water is conducted through a thick mantle of sapwood and none flows through the heartwood; a similar situation is encountered in the macroporous wood (with large pore diameters) of tropical trees. In the cycloporous (ring-porous) wood of ash, robinia, elm, sweet chestnut and oak of the temperate zone, only a few of the annual rings are fully functional (Fig. 4.14), and most of the water is transported in the outermost annual ring.

In considering the water relations of the whole plant the *relative* xylem conducting area is more informative than the total cross-sectional area. The **relative conducting area** ("Huber value" [99, 281] is the ratio of the xylem conducting area (A_{xyl}) and the leaf surface area (leaf area A_l) or leaf mass (e.g. fresh weight of the leaves FW_l) of the transpiring parts of the plant that are supplied with water by this conducting tissue. The relative conducting area (xylem transverse area/distal leaf area A_{xyl}/FW_l) is particularly large in plants that lose large quantities of water by transpiration [63]: the average relative conducting area of desert plants is $2-3 \ \text{mm}^2 \ \text{g}^{-1}$, and in extreme cases

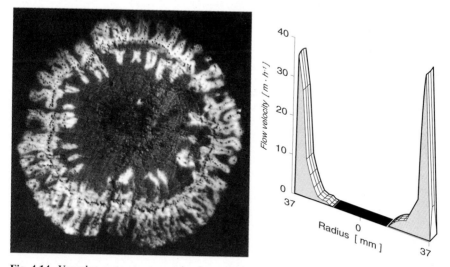

Fig. 4.14. Vascular water transport in the cycloporous wood of *Quercus robur. Left* The water-conducting regions of the hydrosystem of a twig are indicated by fluorescence (marker substance: berberin sulphate). (Braun 1970). *Right* Velocity profile of the ascending transpiration flow in an oak stem (diameter 8 cm). (Čermák et al. 1992). For visualization of ascent of sap in woody species by a thermal imaging technique see Anfodillo et al. (1993)

even $5-7$ mm^2 g^{-1}. For plants of the steppes, herbs growing in sunny locations, ericaceous dwarf shrubs and mediterranean woody species the corresponding values range between 0.5 and 2 mm^2 g^{-1}; the values for most trees of the temperate zone are approximately 0.5 mm^2 g^{-1}, and those of sciophytes are around 0.2 mm^2 g^{-1}. Especially small conducting areas of only $0.02-0.1$ mm^2 g^{-1} are found in aquatic plants, and also in succulents. Furthermore, the relative conducting area is matched to the size of the plant, and it also modifies during growth to meet local moisture conditions.

The **specific hydraulic conductivity** depends on the diameter of the conducting element and its type of perforation. According to the Hagen-Poiseuille Law, conductance is proportional to the fourth power of the radius of a tube [35]. Conducting elements with a narrow lumen and high filtration resistance (such as conifer tracheids, and vessels with simple or scalariform perforations) lower the water conductivity. Thus the specific hydraulic conductivity of conifer wood is only half that of evergreen broadleaved trees, and the latter in turn is half that of deciduous trees (Table 4.2). Roots, with their large-diameter vessels, conduct water particularly well, as do lianas. The specific conductivity can also be referred to the leaves to be supplied (leaf-specific conductivity [281]). In trees, the leaf specific conductivity is highest in the trunk, and is much reduced in the branches from the point of attachment onwards

Table 4.2. Hydraulic conductivity of the xylem of various types of plants (cm^2 s^{-1} MPa^{-1}). (From data of Berger 1931; Huber 1956; Zimmermann and Brown 1974; Raven 1977; Ogino et al. 1986; Lösch 1990). Conductivities of secondary xylem of ancient (Carboniferous) woody plants were comparable to those in recent tree species: Cichan (1986)

Conifer wood	$5-10$
Ericaceous wood	$2-10$
Evergreen broadleaved trees with microporous wood	(2) $3-15$
Deciduous broadleaved trees with microporous wood	$18-50$
Macroporous and cycloporous angiosperm woody species	$100-350$
Liana wood	$300-500$
Root wood of deciduous broadleaved trees	$200-1500$
Fibrous roots	$1-2$
Vascular bundles of herbaceous plants	$30-60$ (250)

Table 4.3. Maximum velocity of sap flow (m h^{-1}) of various types of plants. (Data taken from Berger 1931; Rouschal 1938; Huber 1956; Zimmermann and Brown 1974; Hinckley et al. 1978; Rychnovská et al. 1980; Yoshikawa et al. 1986)

Conifers	$1-2$
Sclerophyllous woody plants	$1.5-3$
Broadleaved trees of humid tropical forests	(9) $18-34$
Deciduous broadleaved microporous woody plants	$1-4$ (6)
Cycloporous broadleaved woody plants	(6) $20-45$ (60)
Herbaceous plants	$10-60$
Lianas	150

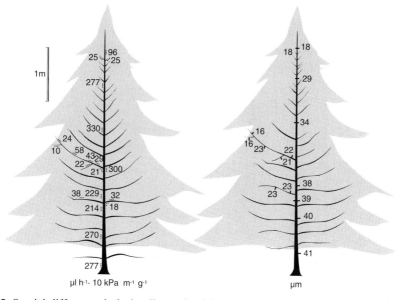

Fig. 4.15. Spatial differences in hydraulic conductivity in *Abies balsamea* trees. *Left* Leaf-specific conductivity of the xylem ($\mu l\ h^{-1}$ water flow at $10\,kPa\,m^{-1}\,g^{-1}$ DM of needles). *Right* Differences in the diameters of the tracheid lumina (μm) in *Abies balsamea*. Vertical scale shows 1 m. (After Ewers and Zimmermann 1984). For hydraulic conductivity profiles for *Betula*, see Zimmermann (1983), for *Toxicodendron* vine and shrub, see Gartner (1991)

(Fig. 4.15). Just as valves are employed to reduce flow in non-living systems, the flow of water through the axial system of the tree is adjusted by means of anatomical differentiation of the cells and in the branching.

The maximal attainable **transpiration flow velocity** varies according to plant type and the anatomy of the conducting system (Table 4.3 and Fig. 4.16). Flow velocity differs in various parts of a plant (branches, trunk and root), as well as over the axial cross section (maximal flow in wide vessels, slower mass flow in the rest of the xylem). As long as water uptake is unimpeded, the velocity of flow in the xylem increases with increasing intensity of transpiration. In larger trees, movement of water begins in the morning at the top of the crown and the tips of the branches, pulling up the column of water that extends from the roots to the base of the trunk. Then the sap begins to flow rapidly, in accordance with the transpiration rates in the crown. In the evening flow becomes slower, but until late at night there can be a slow influx of water into the trunks, so that reserves are replenished.

According to formula (4.9) the *actual* sap flow velocity (v) is linearly proportional to the transpiration intensity per unit area (Tr, as flux J^{H_2O} through the leaf) and inversely proportional to the relative conducting area (A_{xyl}/A_l):

$$v = \frac{Tr}{A_{xyl}/A_l}\ [m\ s^{-1}] \tag{4.9}$$

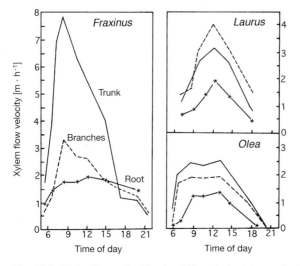

Fig. 4.16. Water flow velocities in different plant types. In the deciduous *Fraxinus ornus*, with cycloporous wood in trunk and branches, the water flow velocity in the trunk is higher than in the branches (*broken line*) and the diffuse porous root (*). This sequence of flow velocities was designated "oak type" (trunk > branches) by Huber (1956). The evergreen *Olea europaea* is of the same type, but xylem flow through its diffuse porous wood and the narrow lumina of the vessels is considerably slower than in cycloporous trees. In *Laurus nobilis* the flow velocity is lower in the trunk than in the branches ("birch type" according to Huber's system). All measurements were made on sunny July days in Istria. (After Rouschal 1938)

The transpiration flow adjusts remarkably quickly to transpiration intensity. This means that it reflects even brief fluctuations in evaporation and thus recordings of sap flow velocity made in the trunk of a tree give information about the transpiration of the whole crown (Fig. 4.17).

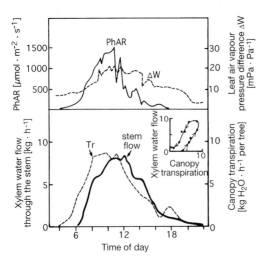

Fig. 4.17. Water flow through a 16 m-high spruce (in Moravia) and canopy transpiration (*Tr*) throughout a summer day. Stem flow (xylem water flow): water volume per unit time, or the product of the flow velocity and cross-sectional area of the conducting vessels. *Upper figure* Irradiance (PhAR) and difference in water vapour pressure (ΔW) between the needles and the atmosphere; *inset* the relation between average xylem flow and canopy transpiration shows that water conductance lags behind the loss of water by evaporation. (After Schulze et al. 1985; for *Eucalyptus grandis*, see Dye and Olbrich 1993)

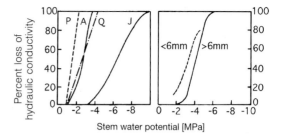

Fig. 4.18. Loss of hydraulic conductivity in various species due to cavitation resulting from low xylem water potential: *P*: *Populus deltoides*; *A*: *Abies balsamea*; *Q*: *Quercus rubra* and *Juniperus virginiana*. The diagram on the *right* shows the greater susceptibility of the thinner branches (<6 mm) of *Acer saccharum*. (After Tyree and Sperry 1989; Tyree and Ewers 1991). The critical tension (T_{crit}) for bubble growth can be calculated from the formula T_{crit} [MPa] = $2t/r$, where t is the surface tension of water (7.28×10^{-8} MPa m^{-1}) and r the bubble radius [m]. (Sperry and Sullivan 1992)

Under conditions of increasing strain in the water transport system (such as impeded supplies in connection with drought or ground frost), or due to mechanical strain (buffering and bending of the shoots in strong wind), freezing and thawing of the water in the vessels, or as a result of xylem injuries, the cohesion in the water columns may suddenly break down, the negative pressure is abolished and the columns are said to cavitate [47, 152]. Air enters the vessels (*embolism*) and causes local interruptions to the flow in the xylem (Fig. 4.18). Until they are dispersed by refilling of the vessels and the surrounding tissues, embolisms obstruct the water flow. Woody plants with a high specific xylem conductivity are more susceptible to embolisms than those with narrow conducting elements. In macroporous and cycloporous tree species of N. America the hydraulic conductivity drops in late winter by, on an average, 55% of the summer values, in diffuse-porous species by 17% and in conifers by only 0–8% [267]. Embolisms also occur in herbaceous species.

The **movement of water in the wood of trees** can also be maintained by osmotic forces if the potential gradient between plant and atmosphere is very low. This situation occurs in deciduous trees in spring when ascent of sap begins prior to unfolding of the leaves, and also in trees of the rain- and cloud forests where the air is saturated with moisture. In such cases, mobilization of starch stored in the rays and the wood parenchyma may release soluble carbohydrates into the conducting vessels and creates osmotic gradients along which water then can flow (osmotic water shifting).

4.3.3 Water Loss from Plants

Plants lose water in vapour form by evaporation (*transpiration*) and occasionally also, in small quantities, in liquid form (*guttation*). The quantitative contribution of guttation to water balance is negligible, so that in the following discussion any mention of water loss refers to transpiration.

4.3.3.1 Evaporation from Moist Surfaces

An exposed water surface loses more water vapour per unit time and area, the stepper the *vapour-pressure gradient* between this surface and the air. A vapour-pressure gradient arises when the water-vapour content of the air ($mol\ m^{-3}$ or $g\ m^{-3}$) at the evaporating surface is greater than at some distance from this surface. This is always the case when the evaporating surface is adequately supplied with water and is warmer than the air. Strong irradiation *warms the surface* and thus leads to a steeper vapour-pressure gradient and to more rapid evaporation (Fig. 4.19).

Evaporation under conditions of unlimited water supply and unimpeded removal of the water vapor is called the *potential evaporation* (E_p. In the arid subtropics potential evaporation reaches $10-15\ mm\ d^{-1}$ ($= 10-15\ kg\ H_2O\ m^2\ d^{-1}$), during the dry season in a mediterranean climate evaporation amounts to $5-6\ mm\ d^{-1}$, and in the equatorial zone $3-4\ mm\ d^{-1}$. In the temperate zone, potential evaporation on clear summer days can reach as much as $4\ mm\ d^{-1}$, whereas the average value over the growing season is around $2\ mm\ d^{-1}$ and in winter only $0.1-0.2\ mm\ d^{-1}$.

The *actual evaporation* from moist surfaces (soil) is usually less than the potential evaporation, because water is almost never replenished as rapidly as it is lost.

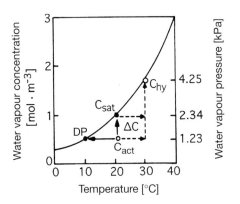

Fig. 4.19. Water vapour concentration in the air at different temperatures, and the vapour-pressure gradients ($\Delta C = C_{sat} - C_{act}$) between saturated and unsaturated air at the same temperature and between an overheated moist surface (e.g. hydrated thallophytes or leaf mesophyll under high irradiance) and the surrounding air. Example: water-saturated air (C_{sat}) at 20°C contains 0.96 mol $H_2O\ m^{-3}$, or 17.3 g $H_2O\ m^{-3}$. If the actual water vapour concentration (C_{act}) at this temperature is 0.52 mol m^{-3}, then the relative air humidity is 53.7%. If the temperature drops to 10°C, this water vapour concentration is sufficient for saturation (dew point, *DP*). If, by absorption of radiation, plants are warmed up by as much as 10 °C relative to the ambient air, the concentration gradient between the transpiring surface (C_{hy}) and the unsaturated air (C_{act}) increases. Plants that would not transpire in saturated air under isothermic conditions lose water in proportion to $C_{hy}^{30} - C_{sat}^{20}$ when their temperature rises above that of the air.

4.3.3.2 Transpiration as a Physical Process

Transpiration in plants proceeds according to the laws governing the evaporation of water from moist surfaces. Water evaporates from the entire outer surface of a plant, and from all interior surfaces that come into contact with air. In the case of thallophytes it is the outer surfaces of the thallus that are involved, whereas in vascular plants external transpiration takes place via the cutinized epidermis (*cuticular transpiration*) and suberized surfaces (*peridermal transpiration*). Within the organs of the plant, water evaporates from the surfaces of cells bordering on the intercellular air spaces. Here, the water is first converted from the liquid phase to the vapour phase, after which the water vapour escapes through the stomata (*stomatal transpiration*). From the surface of the plant the vapour diffuses into the boundary air layer and thence into the open air (Fig. 4.20).

Transpiration is affected by *external factors* to the extent to which they alter the *steepness of the vapour pressure gradient* between the plant surface and the surrounding air. Thus the intensity of transpiration rises with increasing *dryness of the air* **and** with rising *temperature* (Figs. 4.21 and 4.22). Also warming of the leaf surface by strong irradiance leads to a steeper vapour-pressure gradient, so that transpiration can also occur despite high air humidity, even if

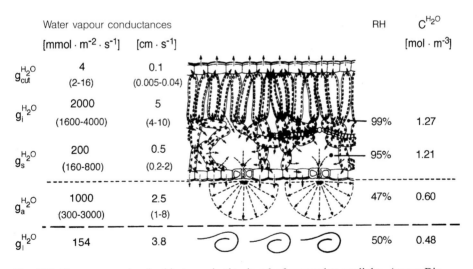

Water vapour conductances			RH	c^{H_2O}
$[mmol \cdot m^{-2} \cdot s^{-1}]$	$[cm \cdot s^{-1}]$			$[mol \cdot m^{-3}]$
$g_{cut}^{H_2O}$ — 4 (2-16)	0.1 (0.005-0.04)			
$g_i^{H_2O}$ — 2000 (1600-4000)	5 (4-10)		99%	1.27
$g_s^{H_2O}$ — 200 (160-800)	0.5 (0.2-2)		95%	1.21
$g_a^{H_2O}$ — 1000 (300-3000)	2.5 (1-8)		47%	0.60
$g_l^{H_2O}$ — 154	3.8		50%	0.48

Fig. 4.20. The processes involved in transpiration in a leaf exposed to sunlight. *Arrows* Distribution of water in the tissues, evaporation in the intercellular spaces and via the cuticle, exit of water vapour via the stomata, shells of water vapour in the boundary layer. *Right* Temperature (T) relative air humidity (*RH*) and water vapour concentration (C^{H_2O}) in the intracellular space, in the substomatal chamber, in the boundary layer and in turbulent air. *Left* Typical average values and ranges for diffusive conductance of water vapour (at an air temperature of 20 °C and leaf temperature of 25 °C): $g_{cut}^{H_2O}$ cuticular; $g_i^{H_2O}$ intercellular; $g_s^{H_2O}$ stomatal; $g_a^{H_2O}$ boundary layer; $g_l^{H_2O}$ total leaf conductance. (After Nultsch 1991; Nobel 1991 a)

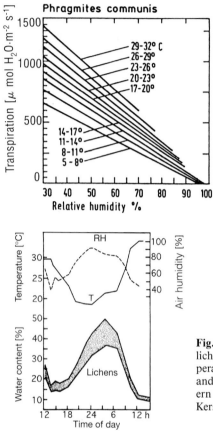

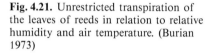

Fig. **4.21.** Unrestricted transpiration of the leaves of reeds in relation to relative humidity and air temperature. (Burian 1973)

Fig. **4.22.** Changes in the water content of a lichen community in relation to the thallus temperature and relative air humidity during the day and at night. *Shaded band* Values for five northern species of *Cladonia*. (After Heatwole, from Kershaw 1985)

it reaches saturation. This is important for the transport of water and minerals in the plant in humid regions.

Wind removes the vapour-saturated layer of air from the epidermis and replaces it with fresh, unsaturated air. In calm air, the boundary layer resistance (see Fig. 2.15) for large leaves, such as those of banana, is three times that of small, incised, or needle-like leaves. Such differences are levelled out as wind strength increases: above a wind velocity of about 2 m s^{-1} the boundary layer resistance is less than 0.5 s cm^{-1}, and is then negligible as compared with the stomatal resistance. Inside closed plant stands, dense tree crowns, tussocks of grass, and cushion-plants the force of the wind is greatly reduced, boundary layers become thicker and transpiration is therefore less intense.

Maximal Transpiration

Maximal transpiration is defined as the unimpeded intensity of evaporation from plants under the *regularly occurring conditions of evaporation* in their natural habitat. Average maximal values for terrestrial vascular plants with fully open stomata are listed in Table 4.4. In exposed seaweeds, lichens and moss-

Table 4.4. Maximal total transpiration from the leaves of morphologically and ecologically different plant types and from the shoot surface of leafless succulents under the evaporative conditions prevailing in the habitat. These are typical values, drawn from the data in original works of many authors. Transpiration intensity is given in μmol $H_2O\ m^{-2}s^{-1}$

Plant type	Transpiration with stomata open
Humid tropics	
Rainforest trees	up to 1800
Cloud forest trees	400 (2000 – 3000)
Lianas	up to 2000
Semiarid tropics	
Palms	1200 – 1800 (2800)
Dry woodland	800 – 1400 (2000)
Mangroves	600 – 1800
Shrubs and semishrubs of subtropical deserts	2800 – 7000 (10000)
Mediterranean sclerophylls	(600) 1500 – 3000 (4000)
Deciduous forest trees of the temperate zone	
Light-adapted species	(1500) 2500 – 3700
Shade-adapted species	(780) 1200 – 2200
Evergreen conifers	1400 – 1700
Dwarf shrubs	
Tundra	150 – 450
Alpine heath	1800 – 3000
Herbaceous dicotyledons	
Tall forbs	9000 – 11000 (16000)
Heliophytes	5200 – 7500
Sciophytes	1500 – 3000
Mountain plants	(1500) 3000 – 6000
Grasses, sedges and rushes	
of the tundra	200 – 350
in meadows	3000 – 4500
in reeds	5000 – 10000
on dry sites	(1800) 4500 (9300)
on coastal dunes	2000 – 4000
Desert plants	1000 – 5000 (8000)
Halophytes	1200 – 2500 (4500)
Succulents	
Leaf succulents	800 – 1800
Cacti	600 – 1800
Floating plants	5000 – 12000

es in a fully saturated condition, evaporation is governed mainly by morphological characteristics such as surface/volume ratio, surface structure, branching type, and cushion shape.

The rate of maximal transpiration appears to be connected with the specific morphology and life-form of a plant (such as herbaceous dicots versus grasses;

evergreen vs. deciduous woody plants; succulents) and with the specific ecological habitat typical of the species (for example, hydrophytes, helophytes, sciophytes, heliophytes and xerophytes). The highest transpiration intensities have been recorded for tall herbs of the meadows on river banks in northeastern Asia, and for floating and swamp plants. Herbaceous plants in sunny habitats also transpire large quantities of water, whereas shade plants and deciduous trees transpire at only half the intensity, and conifers and sclerophyll species even less. Under identical conditions of evaporation most succulents would be at the bottom of the list, but in their natural habitats, where irradiance is very high, they are exposed to far greater evaporative demand than plants in closed stands or in regions where the air humidity is higher. The strikingly low transpiration maxima of trees of the rain- and cloud forests as compared with temperate forest trees, is due to the humid environments, which allow only low evaporation.

Transpiration as a Diffusion Process

Water vapour escapes from plant tissues by diffusion and by mass flow (due to pressure differences between the ambient and the intercellular air). If transpiration is considered as a process of diffusion, the transpiration rate (Tr, as *flux* J^{H_2O} is given by the product of the *vapour-pressur difference* ΔC^{H_2O}, expressed as H_2O per volume (e.g. $g\,m^{-3}$ or $mol\,m^{-3}$) and the *conductance* of water vapour, g^{H_2O}, expressed in terms of velocity values ($cm\,s^{-1}$) or molar flux values ($mmol\,m^{-2}\,s^{-1}$): thus

$$J^{H_2O} = g^{H_2O}\,\Delta C^{H_2O}\ (mmol\ H_2O\ m^{-2}\,s^{-1})\ .\qquad(4.10)$$

Transpiration rates are expressed as water loss per unit surface area (in the case of leaves usually referred to the total leaf surface, upper *and* lower sides) and per unit time.

The conductance of water vapour g^{H_2O} is the inverse of the diffusion resistances ("transpiration resistances" [221]). For leaves the diffusion resistance for water vapour is the sum of the boundary layer resistance and the epidermal leaf resistance (see Fig. 4.20). The epidermal *leaf resistance* (r_l) consists of the parallel resistances, the stomatal (r_s) and the cuticular (r_{cut}) resistances for water vapour. Thus

$$\frac{1}{r_l} = \frac{1}{r_s} + \frac{1}{r_{cut}}\qquad(4.11)$$

or

$$g_l^{H_2O} = g_s^{H_2O} + g_{cut}^{H_2O}\ (mmol\,m^{-2}\,s^{-1}\quad or\quad cm\,s^{-1})\qquad(4.12)$$

The leaf conductance for water vapour, $g_l^{H_2O}$, increases linearly with the degree of opening of the stomata. The *maximal stomatal conductance* is determined by anatomical features such as the size, structure (Fig. 4.23), arrangement and density of the stomata of the plants concerned (species, variety, provenance, even individuals), and is thus a specific parameter for maximal

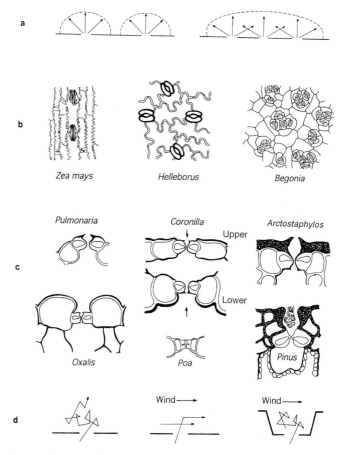

Fig. 4.23 a–d. Different types of stomatal apparatus and the diffusion of water vapour. **a** *Top left* Single, widely separated stomata each build up a hemispherical humid shell from the surface of which water is lost to the drier air. *Top right* Above clustered stomata the humid shells overlap and diffusion of the water molecules into the open air is hampered. **b** Distribution patterns of stomata on the leaf surface: *Zea mays*, a graminaceous type with serial arrangement of stomata; *Helleborus foetidus*, with regular distribution of stomata, which is the most frequent type among dicotyledons; and *Begonia semperflorens* with clusters of stomata. **c** Median cross section through the stomata of different types of plants: *Pulmonaria officinalis*, a sciophyte with raised guard cells (often found also in hygrophytes and floating plants); *Oxalis acetosella*, a shade plant with good adaptability to bright light, with the guard cells situated in depressions in the succulent epidermis; *Coronilla varia*, a heliophilic meadow mesophyte with different forms of stomata on upper and lower leaf surfaces; *Poa annua*, a characteristic stomatal apparatus for grasses and sedges, with parallel guard cells, a narrow central slit, and wide, thin-walled accessory cells; *Arctostaphylos uva ursi*, with a thick-walled stomatal apparatus peculiar to scleromorphic leaves; *Pinus merkusii*, the coniferous type with deep-set guard cells and wax plugs in the stomatal entrance. **d** Diffusion routes of water molecules after leaving via the stomatal slit: in calm air (*left*) the water molecule bounces against air molecules but gradually leaves; in wind (*centre*) the water molecule is removed quickly, but (*right*) if the stomata are sunken, the water molecules are shielded from the wind and are not carried away so quickly. (From Haberlandt 1924; Esau 1953; Oehlkers 1956; Pisek et al. 1970; Braune et al. 1987; Mauseth 1988)

transipiration. Crop plants and heliophytes have a particularly high diffusion conductance ($300-500$ mmol m^{-2} s^{-1}), the maximal values for grasses lie between $250-400$ mmol m^{-2} s^{-1}, for broadleaved trees and shrubs between $160-250$ mmol m^{-2} s^{-1}, and the lowest conductances are those of succulent leaves (about 100 mmol m^{-2} s^{-1} [122]. Very low values in conifer needles are partially due to the spongy wax plug in the stomatal cavity, which reduces water vapour diffusion to a third of that otherwise possible. Flowers, as a rule, have few or non-functional stomata, or sometimes none at all (like tropical orchids [93] and certain rose varieties [85]). They lose water more slowly than deciduous leaves.

Cuticular transpiration can be regarded as diffusion through a hydrophobic medium, since the water molecles must pass through the cutinized layers of the outer wall of the epidermis and through the epicuticular wax lamellae. The cuticular conductance for water vapour, $g_{cut}^{H_2O}$, is very small. Reference values for herbaceous crop plants are $10-20$ mmol m^{-2} s^{-1}, for deciduous leaves of woody plants are $6-10$ mmol m^{-2} s^{-1}, for evergreen leaves and conifer needles $3-5$ mmol m^{-2} s^{-1}, and for desert succulents 0.5 mmol m^{-2} s^{-1} [122]. The conditions under which the individual plant grows greatly affect the development and efficacy of the cuticular resistance, e.g. the leaves of plants grown in dry air and soil have thicker layers of cuticle and a thicker wax coating than those of plants of the same species that have grown in humid habitats. Dehydration of the outer epidermis causes the hydrophobic layers to move close together, thus doubling the cuticular diffusion resistance. There is a drop in the conductivity of the cuticular layers at low temperatures, but an increases at high temperatures, both effects being the result of phase transition and alterations in density. Acid precipitation damages leaves and needles by destroying the waxy layers of their protective cuticle.

The intensity of **peridermal transpiration** through the surfaces of *suberized* shoots and roots depends on the structure of the periderm, the density and permeability of the lenticels, and the presence or absence of cracks in the bark. This is why the trunks and branches of poplar, oak, maple and pine give off more water than those of spruce, beech and birch trees, which have smoother and denser bark (Fig. 4.24). In trees of the temperate zone peridermal transpi-

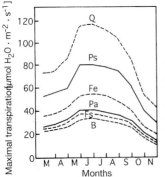

Fig. 4.24. Peridermal transpiration of forest trees of the temperate zone with different types of cortex and bark. *Q*: *Quercus rober*; *Ps*: Pinus sylvestris; *Fe*: *Fraxinus excelsior*; *Pa*: *Picea abies*; *Fs*: *Fagus sylvatica*; *B*: *Betula pendula*. Transpiration rates apply to the temperature conditions in the habitat and completely dry air (maximal evaporative capacity). (After Geurten 1950)

ration in summer is in the order of $0.015-0.15$ mmol H_2O m^2 s^{-1}, or about 1% of the potential evaporation. So far no comparative studies of peridermal transpiration in smooth-barked trees of the wet tropics or of thick-barked woody plants of dry regions have been carried out.

4.3.3.3 Physiological Control of Transpiration

Transpiration is strictly dependent on the physical conditions affecting evaporation only as long as the degree of stomatal opening does not change, that is, as long as the stomata remain open to a *fixed* degree or firmly closed. Only under these conditions is the amount of water lost proportional to the evaporative power of the air. The ability of a plant to **regulate stomatal opening** enables it to modulate the rate of transpiration to the requirements of its water balance. Quantitative changes in stomatal opening can be determined *porometrically*. The involvement of physiological regulatory mechanisms can also be assumed (Fig. 4.25) when the transpiration rates no longer keep step with evaporation. A temporary reduction in the degree of stomatal opening is elicited by a decrease in light intensity, dry air (particularly in connection with wind), water deficit, extremes of temperature, and toxic gases.

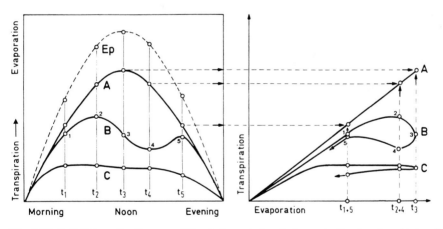

Fig. 4.25. *Left* Schematic diagram of the daily variation in transpiration by *A* a plant well supplied with water (unrestricted transpiration), *B* a plant with water supply limited during the middle of the day and *C* a plant under continual regulation of stomatal opening, as compared with the potential evaporation (E_p). *Right* The transpiration/evaporation ratio corresponding to the figure on the left. For *each point* representing evaporation, the corresponding (simultaneous) value for transpiration is noted. When transpiration is unrestricted, the *line joining these points* is straight, i.e. transpiration is linearly proportional to evaporation. Restrictions on the rate of transpiration are reflected in a departure from proportionality, the rate of transpiration falls from time t_2 to time t_3, although the rate of evaporation is rising during that interval, so that the diagram forms a loop. Any *departure from a diagonal* in the Tr/E_p diagram indicates stomatal regulation of transpiration. The scale for potential evaporation is reduced by a factor of 2. (Original)

The *response threshold*, the *rapidity*, and the *effectiveness* of stomatal regulation vary among species and with the degree of adaptation to the habitat. Sciophytes narrow their stomatal openings very rapidly in response to even slight water deficits, whereas herbs of sunny habitats restrict their stomatal transpiration only under much drier conditions, and even then closing is slow. The way stomata respond may differ from one plant species to another and even from one individual to the next within the same species. Even the leaves of a single plant vary quite considerably in this respect, depending on their age (stomatal response during unfolding differs from that in senescence), the conditions under which they developed (abnormal response if air humidity was continuously high), and their positions on the shoot.

The specific **efficiency of the process of closure** can be expressed as the *modulation amplitude of transpiration,* which is the ratio of maximal transpiration to cuticular transpiration. In plants of shady and moist habitats cuticular transpiration accounts, on an average, for roughly one-third of total transpiration. In sclerophylls, evergreen conifers and desert shrubs, water losses can be reduced to $3-10\%$ of maximal transpiration, and in succulent species to $1-2\%$ This means that in the plants with the best protection against transpiration the cuticular transpiration is reduced to $0.1-0.05\%$ of the potential evaporation from a moist surface [181]. The values for arriving at these figures are obtained from measurements made under the evaporation conditions in the habitat concerned. In order to arrive at characteristic values for a plant species or variety, or for a state of adaptation, independent of the actual weather conditions, the efficiency of stomatal closure is better expressed as the *ratio* of cuticular conductance to total leaf conductance, $g_{cut}^{H_2O}/g_l^{H_2O}$. The values thus obtained range from 0.04 to 0.1 for herbaceous plants, $0.03-0.06$ for deciduous woody plants, and $0.01-0.02$ for evergreen trees and shrubs; values for succulents are less than 0.005.

4.3.4 The Water Balance of a Plant

4.3.4.1 Water Balance: a Dynamic Equilibrium

The basic processes involved in the water balance of a plant are water uptake water conduction and water loss. Only if the rates of these processes are suitably adjusted − at least in the long term − is it possible for a satisfactory water balance to be maintained. How well the intake and output of water are matched can be seen from a comparison of the quantities of water taken up by a plant and those lost by evaporation during a given interval of time. The difference between absorption and transpiration, i.e. the **water balance**, indicates the direction and size of any deviation from equilibrium. The balance becomes *negative* as soon as the uptake of water is insufficient to meet the requirements of transpiration. If the stomata narrow as a result of this deficit, so that transpiration is decreased while uptake is unchanged, the balance is restored after a transient overshoot to *positive* values. Thus the water balance of

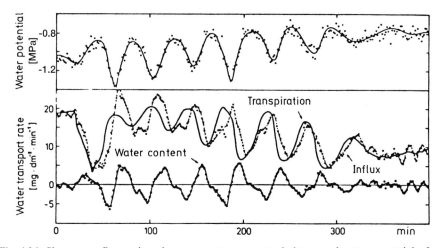

Fig. 4.26. Short-term fluctuations in water turnover, water balance and water potential of cotton leaves. During the phase of rapid transpiration the *water content* of the leaf falls and its water potential becomes more negative. The amount of water passing through the petiole (*Influx*) follows a curve 180° out of phase with that for water potential. The short-term fluctuations in transpiration are brought about by oscillation of the stomatal apertures. (Lang et al. 1969)

a plant oscillates continually between positive and negative deviations, but it is instructive to distinguish between short-term oscillations and long-term disturbances of this equilibrium. *Short-term fluctuations* reflect the interplay of the various water-regulating mechanisms, particularly the changes in stomatal aperture (Fig. 4.26). More marked departures from equilibrium occur over the *course of the day*, particularly at the changeover between day and night (Fig. 4.27). During the day the water balance almost always becomes increasingly negative; it is restored in the evening or during the night (if there are sufficient water reserves in the soil). When a negative water balance begins to develop in the leaves, as an immediate short-term regulatory measure water is transferred to them from well-supplied tissues, such as the cortical and phloem parenchyma. This is the explanation for the measureable fluctuations in thick-

Fig. 4.27. Schematic diagram of the gradual lowering of the water potential of the leaves, roots and soil during one week of drought. The greatest daily fluctuation occurs in the leaves, which are exposed to transpiration stress throughout the day. The water balance is no longer fully restored during the night (*shaded blocks*), so that the dawn water potential becomes less from day to day. (After Slatyer, from Kozlowski et al. 1991)

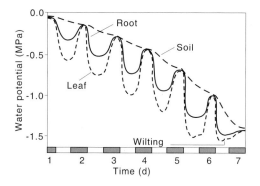

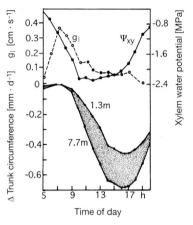
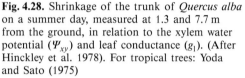

Fig. 4.28. Shrinkage of the trunk of *Quercus alba* on a summer day, measured at 1.3 and 7.7 m from the ground, in relation to the xylem water potential (Ψ_{xy}) and leaf conductance (g_l). (After Hinckley et al. 1978). For tropical trees: Yoda and Sato (1975)

ness of tree trunks over the course of the day (Fig. 4.28). Another useful water reservoir is provided by juicy fruits (e.g. *Citrus* fruit), which shrink and swell according to the state of the plant's water balance. During dry periods the water content often is not entirely restored overnight, so that the deficit accumulates from day to day, until the next rainfall (*seasonal* fluctuations in water balance; see Fig. 4.37).

4.3.4.2 Maintenance of Positive Water Balance: Regulatory Mechanisms

The strategy adopted by plants of *humid* regions to meet a deteriorating water balance is to reduce the *degree and duration of stomatal opening*. This is illustrated in Fig. 4.29: at first, transpiration is reduced only during the hottest part of the day; as the water deficit increases the stomata also remain shut in the

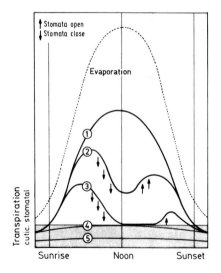

Fig. 4.29. Diagram of daily changes in transpiration with decreasing soil moisture (curves *1–5*). The *arrows* indicate the stomatal movements elicited by changes in the water balance. The *stippled area* shows the range in which transpiration is exclusively cuticular. *1* Unrestricted transpiration; *2* limitation of transpiration during the middle of the day as the stomata begin to close; *3* full closure of the stomata at midday; *4* complete cessation of stomatal transpiration by persistent closure of the stomata (only cuticular transpiration continues); *5* considerably reduced cuticular transpiration as a result of membrane shrinkage (Stocker 1956a)

afternoon, opening only in the morning. Finally, transpiration takes place exclusively via the cuticular route. Such measures safeguard the plant from injurious deficits in its water balance. Figures 4.30 and 4.39 show examples of the effects of dry periods extending over a number of days.

Plants of *arid regions* usually possess roots extending far enough into the ground to reach the ground water, or they have water-storing tissues, and therefore a drastic reduction of transpiration (and of CO_2 uptake at the same time) is not immediately necessary. This is an expression of their habitat adaptation. Nevertheless, when for weeks on end no rain has fallen and the water reserves in the soil are declining, the plants reduce transpiration more and more, opening the stomata less widely and for shorter periods of time during the day (Fig. 4.31).

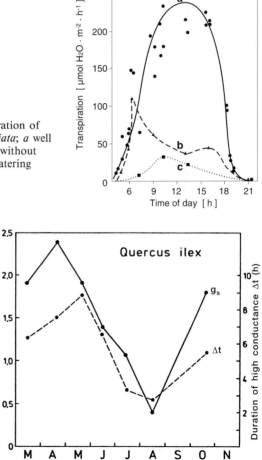

Fig. 4.30. Diurnal course of transpiration of two-year-old seedlings of *Pinus radiata*; *a* well supplied with water; *b* after 9 days without watering; *c* after 12 days without watering (Kaufmann 1977)

Fig. 4.31. Maximal stomatal conductance (g_s) and duration (Δt) of the daily period of stomatal opening of *Quercus ilex* in southern France over the course of the year. During the dry season from July to September the stomata open less widely and for a shorter time. (After Lossaint and Rapp 1978)

Closure of the stomata in connection with water shortage is brought about by a combination of several processes. Shortage of water means that the water flow inside the leaves is insufficient to maintain full turgor in their epidermis. The tendency of the stomata to shut down under these circumstances often is enhanced by an increasing water vapour deficit of the air (Fig. 4.32). In addition, the reduction in turgor and increase in cell sap concentration cause inhibition of photosynthesis, and this indirectly elicits closure of the stomata via the CO_2 control circuit. An important role is also played by stomata-regulating hormonal and electrophysiological signals [68] based on action potentials and transmitted from the roots via the phloem. As soon as the soil begins to dry out, the phytohormone abscisic acid is quickly dispatched from the

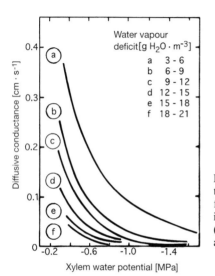

Fig. 4.32. Decrease in diffusive conductance of the needles of Douglas fir (seedlings) as a function of predawn xylem water potential and increasing atmospheric water-vapour deficit (a–f). (After Hällgren, as cited in Lassoie et al. 1985)

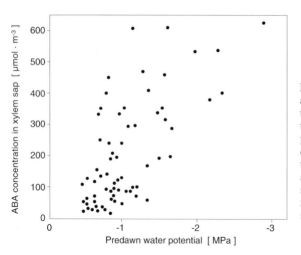

Fig. 4.33. Increase in abscisic acid (ABA) concentration in the xylem water of almond trees as a function of water potential during progressive drying of the soil. (Wartinger et al. 1990; Hartung and Davies 1991). Model for stomatal control by ABA in xylem sap and leaf water status: Tardieu et al. 1993)

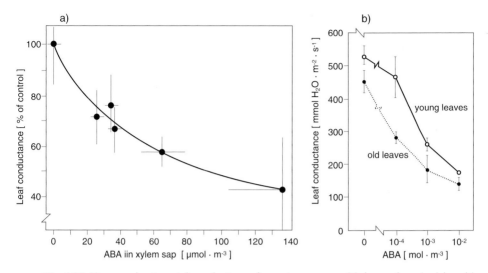

Fig. 4.34. Decrease in stomatal conductance for water vapour with increasing abscisic acid concentration in the xylem water. *Left* Relative leaf conductance as a function of the increasing ABA in the xylem sap of maize plants during progressive drying of the soil. *Right* Effect of experimental application of ABA via the transpiration stream on the leaf conductance of young and older leaves of *Lupinus lutens*. (Zhang and Davies 1990; Henson and Turner 1991)

roots via the xylem flow to the leaves, where it elicits stomatal closure (Figs 4.33 and 4.34).

4.3.4.3 Indicators of the State of Water Balance

Water balance can be *computed* from quantitative determinations of water uptake and transpiration. However, the necessary measurements are still difficult to obtain under field conditions and are also inexact; this applies to a particular degree in the case of water uptake by the roots. Therefore it is customary to make an *indirect estimate* of the water balance through its effect upon the water content or water potential of the plant. A negative balance always manifests itself in a decrease in turgidity and water potential of the plant tissues.

Change in water content as an indicator of water balance: water deficit can be demonstrated by means of repeated measurements of the water content of leaves or other parts of the shoot. The actual water content (W_{act}) at a given time must be given with respect to a standard measure, such as the water content of the leaves under conditions of saturation (W_{sat}). The water content at any particular time of observation can be expressed as the *relative water content* (RWC) [169], a percentage of the water content at saturation:

$$RWC = \frac{W_{act}}{W_{sat}} \cdot 100 \ (\%) \tag{4.13}$$

A measure of water *deficiency* is the *water saturation deficit* (WSD) [237]. A water saturation deficit indicates how much water a tissue lacks as compared with complete saturation.

$$\text{WSD} = \frac{W_{sat} - W_{act}}{W_{sat}} \cdot 100 \ (\%) \tag{4.14}$$

Changes in water potential: fluctuations in water content affect the concentration of the cell sap and the turgor of the cells. The *osmotic potential*, $\Psi\pi$, rises when the water balance is negative. However, $\Psi\pi$ is not only altered by changes in water content, but by osmoregulating processes as well (accumulation of sugar, proline and ions in the vacuoles). As an indicator for the state of the water balance the actual value of $\Psi\pi$ is compared with its optimum value (that is, when transpiration and uptake are balanced), and the osmotic maximum under conditions of extreme water shortage (Fig. 4.35).

A more sensitive measure of changes in water balance than the osmotic potential is the leaf water potential, Ψ_l (Fig. 4.36). The immediate result of water deficit is a loss of turgor accompanied by a distinct decrease of water potential

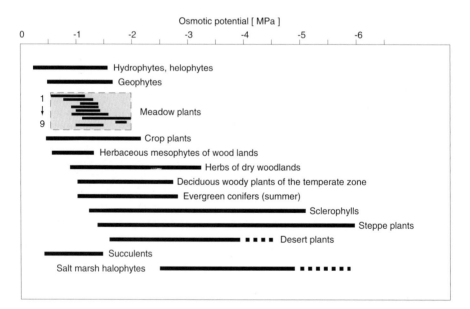

Fig. 4.35. Ranges for the osmotic potential of leaves of ecologically different types of plants. The subranges (enclosed in *stippled box*) for meadow plants illustrate how to interpret the osmotic range shown for each plant group: it is derived from the difference between the lowest and the highest osmotic potentials found among all individual species studied in the particular group. *1: Taraxacum officinale; 3: Galium mollugo* and *Campanula rotundifolia; 4: Achillea millefolium; 5: Tragopogon pratensis; 6: Poa pratensis; 7: Melandrium album; 8: Cynodon dactylon* and *Lolium perenne; 9: Arrhenatherum elatius.* (After Walter 1960, supplemented by data from Sveshnikova 1979 and Nobel 1988)

Fig. 4.36. Diurnal changes in total water potential and osmotic potential in the leaves of desert shrubs (*Hammada scoparia*) at the end of the dry period, under natural drought conditions and irrigated. (After Kappen et al. 1976). For variations in the water potential of evergreen and deciduous trees in the semihumid tropics over the course of the year: Sobrado (1986); for responses of evergreen *Quercus* species across a gradient from mesic to xeric sites: Rambal (1992)

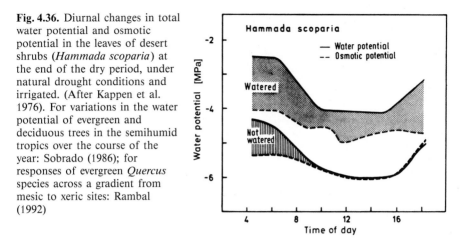

in the tissues: especially in the range of slight water deficits this changes more rapidly than the osmotic potential.

4.3.5 Water Relations in Different Plant Types

4.3.5.1 Hydrostable and Hydrolabile Water Balance

According to their habitat and the functional traits of plants, it is possible to distinguish two contrasting variants [234] of water balance:

Hydrostable plants can, to a considerable extent, maintain a favorable water content, the water balance changing very little throughout the day. Hydrostable plants include aquatic species, succulents, sciophytes, certain grasses and trees of humid regions (Table 4.5). Their stomata respond with great sensitivity to lack of water, and their root systems as a rule are extensive and efficient. An additional factor contributing to stabiliziation of the water balance is the presence of water reserves in storage organs, roots, and the wood and bark of the stems or trunks; these function as capacitive elements in the soil-plant-atmosphere system (capacitors in the analogue circuit scheme, see Fig. 4.10), providing the means for internal adjustment of the water balance. The diurnal and seasonal fluctuations in osmotic potential and leaf water potential remain within narrow limits.

Hydrolabile plants can afford to risk quite large losses of water and the consequent rise in cell sap concentration. Many herbs of sunny habitats, steppe grasses, as well as woody plants, particularly pioneer species, are hydrolabile, and all poikilohydric plants are extremely so. Such plants can tolerate strong fluctuations in water potential (they are *euryhydric*), as well as temporary wilting. Their recovery from such adverse situations is rapid on account of their usually favorable root/shoot ratios and efficient water transport systems.

Table 4.5. Minimal water potential values (MPa) of assimilative organs of plants of ecologically different groups. (Richter 1976; Doley 1981; Nobel 1988)

Plant group	Ψ_{min}
Aquatic plants	-1.2
Herbaceous mesophytes	-1.5 to -2.5
Graminoid mesophytes	-2 to -3 (-4.5)[a]
Trees of tropical rainforests	-1.5 to -4
Woody plants of the temperate zone	
Deciduous trees and shrubs	-1.5 to -2.5
Conifers	-1.5 to -2.2 (-6)
Plants of periodically dry regions	
Sclerophylls	-3.5 to -5
Arid bush	-3.5 to -8.5
Mediterranean low shrubs	-4 to -8
Xerophytes	
Desert shrubs	-5 to -8 (-16)
Succulents	-0.8 to -2
Mangroves	-5 to -6
Halophytes	-3 to -6 (-9)

[a] Extreme values in brackets.

4.3.5.2 Functional Types of Water Economy

In the earliest period of physiological field studies, comparative analysis of the phenomena connected with the different types of water economy constituted the central theme. In order that this pioneer research be not forgotten, some of the figures in this chapter are devoted to results of these early eco-physiological studies.

By the middle of the present century, much knowledge had already accumulated concerning the specific characteristics of most morphologically and ecologically distinct plant groups from all climatic zones and every type of vegetation. It soon became apparent that the variety of structural and functional types exceeds the variety of habitats or, in other words, that among the species inhabiting a particular area many different ways of adapting to the prevailing habitat conditions can be encountered. At the same time, genetically fixed characteristics of the species, genera and families are retained. O. Stocker recognized that this required a critical evaluation of the significance of every characteristic considered to be of ecological relevance for the existence of a plant in a particular habitat. Among this wealth of behavioral patterns it is nevertheless possible to distinguish *ecophysiologically different types* [121, 238] *and life strategies*, on the basis of characteristic responses to the prevailing environmental conditions in their habitats. Some examples will be discussed in the following sections: namely, trees, sclerophyll shrubs, dwarf shrubs and cushion plants, herbaceous dicotyledons, grasses, hemiparasites and epiphytic vascular plants. In addition, studies have also been made of the water relations of succulents, shrubs with chlorophyllous stems, lianas and palms.

Trees

Tall trees, with their extensive evaporating surfaces and with the long distances water must travel from roots to leaves, present a classic example of a type of organization that cannot allow appreciable water deficits to develop; any disturbance in the water balance must be dealt with from its first appearance. On sunny days, around midday, in many trees water uptake does not keep pace with the rate of water loss, so that the guard cells – which in most trees respond even to very small water deficits – temporarily restrict transpiration. Later in the day, when the water balance has been restored, the stomata open again and the rate of transpiration increases (Fig. 4.37). The midday depression of transpiration does not occur simultaneously throughout all regions of the crown; the leaves of the top of the crown frequently respond later. Not all trees are as hydrostable as the conifers and understorey shade trees growing below the deciduous canopies of the temperate zone. There are also trees, for example *Fraxinus* and *Robinia,* in which a strongly negative water balance may develop.

Sclerophyll Trees and Shrubs

In regions that experience summer drought and equinoctial rainfall or winter rains (in the mediterranean region, the African Cape, California, Chile, Southwest Australia), as well as in the semiarid marginal regions of the tropics, a type of vegetation (Matorral: Macchia, Garrigue, Chaparral, Fynbos) is encountered in which the dominant species are shrubs and low trees with small evergreen, usually leathery ("scleromorph"), leaves (Matorral, Macchie, Chaparral, Fynbos). These evergreen woody species are adjusted to a moderate but continuous assimilative activity under climatic conditions that vary greatly from season to season. Sclerophyll shrubs of the mediterranean type carry out

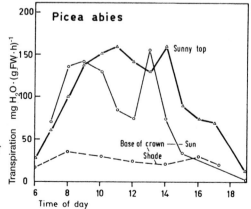

Fig. 4.37. Daily fluctuations in the transpiration of spruce shoots on a sunny August day. As water supply becomes inadequate, the shoots in the shade at the base of the crown are the first to reduce their water loss, then the twigs in the sun at the lower margin of the crown, and finally the shoots in the sunny top of the crown (Pisek and Tranquillini 1951)

most of their growth during the period following the autumn rains into the early summer when the soil contains adequate moisture; after this, until the end of summer, their metabolic activity is limited by shortage of water.

Among the sclerophyll shrubs, representatives of hydrostable as well as hydrolabile functional types are encountered. Evergreen *Quercus* species, *Laurus* and *Arbutus* are of the former type, buffering their water status by sensitive adjustments of transpiration; *Olea, Phillyrea, Myrtus* and *Ceratonia,* on the other hand, are more tolerant of fluctuations in water content and are slower to limit their water consumption. As a rule, the stomata of their shade leaves remain open longer than those of the peripheral leaves exposed to stronger irradiance.

The extent to which the water content and water potential of a plant fluctuate does not depend exclusively on the reactivity of the stomata; the nature of the root system plays an even greater role. In a favourable locality the osmotic potential in most sclerophyll shrubs decreases by roughly 2 – 3 MPa by the end

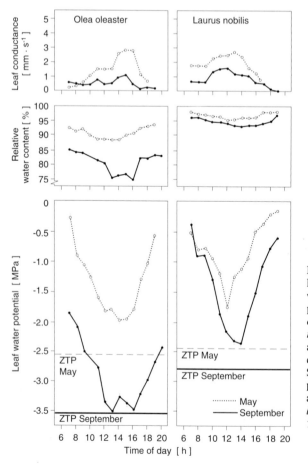

Fig. 4.38. Daily courses of leaf conductance, relative water content and water potential of leaves of *Olea oleaster* and *Laurus nobilis* before (May) and at the end of summer drought (September) in Sicily. *ZTP* Zero turgor point in May (*broken line*) and in September (*solid line*). LoGullo and Salleo 1988)

of the summer drought. On dry sites, however, water deficits are much more severe (Fig. 4.38); the plants (particularly euryhydric species) respond to this situation by increasing their resistance to desiccation (see Fig. 6.68).

An outstanding feature is the high velocity of flow achieved by the transpiration stream as soon and as long as the roots take up sufficient water. The wood of all sclerophyll shrubs and trees is diffuse-porous and its vessels very narrow, so that even at extremely low xylem potentials there is little danger of cavitation.

Dwarf Shrubs and Cushion Plants of Temperate and Cold Regions

Dwarf shrubs of heaths and tundra and the cushion plants of alpine and arctic regions grow on open, wind-exposed sites. Characteristic for such plants is that they are densely branched, and bear small, closely arranged leaves, often rolled up, or in the form of scales or needles. Their most striking characteristic, however, is a prostrate growth form. The distribution of this type of plant is governed not so much by hydrophysiological features, but rather by the heat balance, the depth of snow cover in winter, and the availability of nitrogen. Nevertheless, this combination of genetically determined growth form and the plant's development in a wind-dominated climate, must necessarily exert a significant influence on the water economy.

Dwarf shrubs like *Loiseleuria procumbens* and low-lying willows (e.g. *Salix serpillifolia*) create an equable phytoclimate by means of their carpetlike growth. Although the stomata of such species are highly sensitive to dry air, the protection afforded by this growth form, and the humid air in the immediate surroundings, enable the stomata to remain open to a considerable degree without overtaxation of the water balance by excessive transpiration.

A similar effect, which can be described as a high aerodynamic diffusion resistance, also occurs in *cushion plants*. In addition, these plants are well provided with water by their deep tap roots and by the dense growth of roots put out by the shoots into the detritus and dust which collects inside the cushion and also serves as a moisture reservoir. The water status of hemispherical cushion forms is in any case safeguarded by the favourable ratio of transpiring surface to moisture-conserving interior.

Herbaceous Mesophytes and Hygrophytes

Among herbaceous plants all intermediate gradations between hydrostability and extreme hydrolability can be found. This diversity of types is particularly evident in sunny habitats with a tendency to dry soil (Fig. 4.39).

When the water supply is plentiful, **mesophytes** transpire without restriction throughout the day at a rate determined by the evaporative power of the air. Wasteful transpiration of this kind can be afforded only by plant species that can either draw upon abundant water reserves through an extensive root system, and rapidly convey this water to the leaves, or by species fitted to tolerate a high degree of desiccation. An important side-effect of such uninterrupted transpiration is heat regulation by the cooling associated with evaporation. Thus plants of this functional type are mainly found on open and hot places,

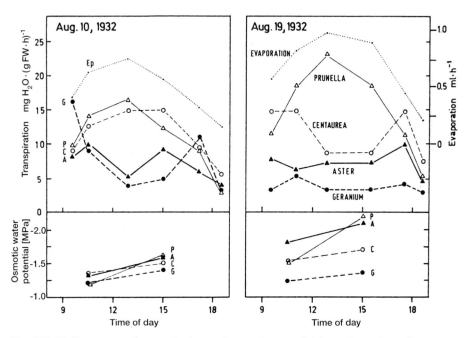

Fig. 4.39. Daily courses of transpiration and osmotic potential in various plants from a xerothermic habitat at the beginning of a dry period (*left*) and after 9 days without rain (*right*). *Prunella grandiflora* is a shallow-rooted herb that barely reduces its rate of transpiration and thus is subject to considerable fluctuations in its water balance; *Centaurea scabiosa* has moderately deep roots and reduces stomatal transpiration only when it becomes difficult to obtain adequate water; *Aster amellus* also has moderately deep roots but responds to slight changes in water balance and restricts transpiration at an early stage; *Geranium sanguinum* is a very shallow-rooted plant which at the slightest shift in its water balance drastically narrows its stomatal apertures, thus largely avoiding a drop in osmotic water potential even after several days of dryness. (Müller-Stoll 1935)

such as slopes exposed to strong irradiance. There are other functional types in which stomatal narrowing occurs at the beginning of water deficit around midday or during the latter half of the day. The stomata of these plants reopen in the late afternoon if the water balance is restored by a more favourable ratio of water uptake to water loss.

Tall forbs are unique with respect to their water content and water turnover. The dense growth of tall rhizomatous forbs (up to several metres), like *Heracleum lanatum, Filipendula camtschatica*) in flood areas and river terraces of islands and peninsulas in the region of the Sea of Okhotsk transpire on sunny days in summer an average of 2 litres of water per plant. Species with either very large leaves (*Petasites amplus*) or very extensive growth (*Reynoutria sachalinensis*) transpire as much as 6 litres per plant [14]. Although the leaves have an unusually high water content (90–94% of the fresh weight) and the soil does not dry out to any great degree, and in spite of good stomatal regulation, water deficits of about 20% may still develop, causing the leaves to wilt temporarily (Fig. 4.40). The conducting system is perhaps not always able to cope with the enormous throughflow of water.

Hygrophytes, species occurring in the undergrowth of forests and in shady habitats with moist air, respond to the ever-changing pattern of passing sun flecks with strongly fluctuating stomatal movements. Within a short space of time transpiration may vary by as much as three- to fivefold [55]. Hygrophytes are unable to tolerate prolonged exposure to the sun because even after closure of the stomata they continue to lose considerable amounts of water by cuticular transpiration. Their root systems are usually poorly developed and the conducting system inadequate for supplying the leaves under these conditions. It is not surprising, therefore, that their xylem vessels tend to cavitate under severe evaporative stress (especially *Impatiens* spp.).

Spring geophytes, in their water relations, occupy a place somewhere between the sciophytes and the heliophytes, although tending rather towards the hydrostable type. Geophytes of the steppes, which complete their aboveground development at a time when there is sufficient moisture in the soil, are also hydrostable, with low transpiration and only very slight fluctuations in the water content of the leaves. Before large water deficits can develop the plants shed their leaves.

Grasses and Sedges

In the course of their evolution, graminoid species have successfully colonized the entire span of possible habitats, ranging from aquatic environments to steppes and deserts. Correspondingly, they exhibit considerable functional diversity in their reactions to water availability. Some species (although there may be deviant varieties within the species) continuously consume large amounts of water (above all, swamp and riparian plants); in others transpira-

Fig. 4.40. Transpiration of tall forbs of the flood areas and river terraces on Sakhalin. *Top of diagram* Precipitation (mm) and air temperature (°C) above the stand. *Fc: Filipendula camtschatica f. typica*; *Rs: Reynoutria (Polygonum) sachalinensis*; *Au: Angelica ursina.* *Insets* Daily course of transpiration in June with highest transpiration at slight water saturation deficit (up to 5% in *F.c.*, up to 6% in *A.u.*) and in July at higher water saturation deficit (up to 23% in *F.c.*, up to 15% in *A.u.*). (After Morozow and Belaya 1988)

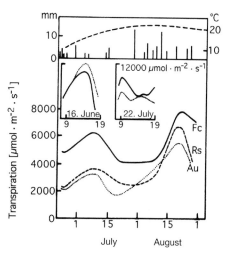

tion is always depressed (xerophytic grasses with permanently curled leaves); in some species the water balance is maintained by gradual and timely closure of the stomata (Fig. 4.41), whereas others are hydrolabile and euryhydric (especially C$_4$ grasses). Many grasses reduce the transpiring surface by folding and rolling their leaves in response to loss of turgor (Fig. 4.42). In *Stipa tenacissima*, transpiration can be reduced in this way to 40% of that measured in normally open leaves [205]. As the leaves age, which in savanna grasses is at the beginning of the dry season, the stomata become rigid and the plants

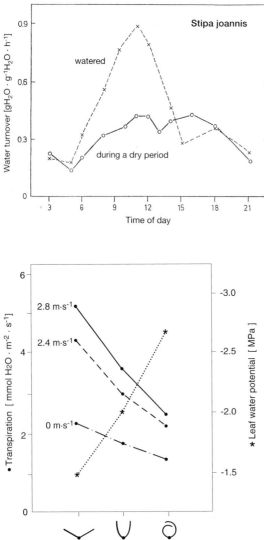

Fig. 4.41. Daily course of transpiration of the steppe grass *Stipa joannis* during a dry period in July, as compared with the transpiration of watered plants. Here, transpiration is given in terms of water turnover (water transpired per water content of the leaves). (After Rychnovská 1965)

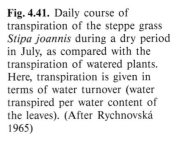

Fig. 4.42. The effect of leaf-rolling in *Oryza sativa* on transpiration in still air and when exposed to wind velocities of 2.4 m s^{-1} and 2.8 m s^{-1}. (*...*) = leaf water potential. (After O'Toole et al. 1979; Hsiao et al. 1984)

therefore lose control over their water balance; although the soil is dry, the plants continue to transpire until the leaves dry out.

Desert Plants

Deserts harbour a remarkable diversity of plant forms: small shrubs with narrow, scale-like and/or water-storing leaves, shrubs and canes with assimilatory and often succulent cortical tissue, prostrate cushions and subshrubs, geophytes, tussock grasses, succulents and ephemeral therophytes of the rain flora (see Fig. 6.64).

Some *ephemeral* species do not check their evaporation at all and transpire copiously until they dry out. The *perennial* species employ two strategies in succession: in spring, as long as the soil is moist, they are hydrolabile, whereas in summer they enter a hydrostable phase, in which the danger of desiccation is avoided by the greatest possible limitation of transpiration. During and after the rainy season in winter and spring the plants obtain water through a root system extending many metres downwards or spreading horizontally over considerably distances to the still moist soil horizons, or to water pockets. Dry air presents no problem to such plants as long as the soil provides sufficient water. At this time of year transpiration is only slightly lowered, at least never by drastic narrowing of the stomatal apertures (Fig. 4.43). Maintenance of a high stomatal conductance is necessary for carbon assimilation, and water deficits (not more than 20–25% at this time [241]) and a transient drop in water potential do little harm.

During the dry summer period, safety takes precedence over production; the stomata remain closed for most of the time, and the structures and mechanisms that contribute towards limiting transpiration (wax layers, thickened, cutinized epidermal structures, sunken stomata, rolling of leaves) exert their full effects (Table 4.6). In this way the danger of excessive desiccation can be

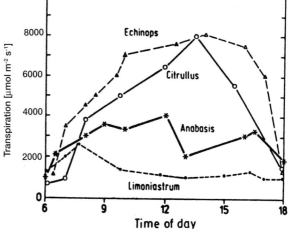

Fig. 4.43. Transpiration of desert plants in the southern Algerian Sahara at the beginning of the dry season. *Citrullus colocynthis* is a soft-leaved geophyte with a deep root; *Echinops spinosus* is a sclerophyllous dwarf shrub; *Anabasis aretioides* is a deep-rooted cushion plant of compact habit; *Limoniastrum feei* is a succulent rosette plant capable of growing in extremely dry habitats. (After Stocker 1974b)

Table 4.6. Total daily transpiration of desert plants with an adequate water supply and during prolonged drought, under the evaporation conditions prevailing in the habitat. (Stocker 1970, 1974; Caldwell et al. 1977; Nobel 1977)

Plants	Total daily transpiration[a]		Reduction in transpiration (%) as percent of the value in the rainy season
	Rainy season	Dry season	
North African desert plants			
Nitraria retusa	210	165	*78*
Zilla spinosa	240	150	*62*
Zygophyllum coccineum	165	80	*48*
Pennisetum dichotomum	165	65	*39*
Haloxylon persicum	280	100	*36*
Hammada scoparia	(4)	(1.5)	*38*
Anabasis articulata	(3.1)	(1.0)	*32*
Retama retam	270	80	*29*
Artemisia herba-alba	(6)	(1.6)	*27*
Noea mucronata	(5.5)	(1.0)	*18*
Halophyte desert in Utah			
Atriplex confertifolia (C_4)	155	30	*19*
Ceratoides lanata (C_3)	154	2.2	*1.4*
Cactus desert in California			
Ferocactus acanthodes	17	0.35	*2*

[a] Without brackets: mol H_2O $m^{-2} d^{-1}$. In brackets: g H_2O g^{-1} fresh weight d^{-1}.

avoided, despite the extreme evaporative conditions from heat and dryness of the air.

Hemiparasites
Hemiparasites, which obtain their water supplies via their host plants, are frequently encountered in regions with dry periods and in xerothermic habitats. As with all parasitic forms, they must be well adjusted to their hosts in order for both to survive. Loranthaceae on trees sink haustoria into the water-conducting system of the host, and follow (at least as far as timing is concerned) the specific patterns of the water relations of the tree. Because the principal advantage for a mistletoe consists in obtaining mineral substances via the transpiration stream, it must necessarily be more hydrolabile and extravagant than its host, requiring a copious flow of water to ensure its receiving adequate supplies of minerals. This is achieved by a greater evaporative loss of water: the stomata of the hemiparasite are opened wider than those of the host and shut down later if the water balance becomes negative (Fig. 4.44). Hemiparasites such as *Rhinanthus* and *Striga* [222], that tap the roots of herbs and grasses, represent a much greater danger to the water balance of their hosts. They transpire far more water than the host plants because their stomata remain open even when the leaves wilt.

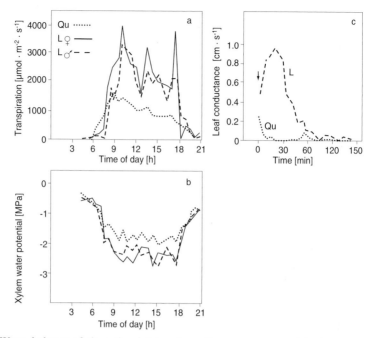

Fig. 4.44a–c. Water balance of the oak mistletoe *Loranthus europaeus* and its host tree *Quercus robur.* **a** Transpiration and **b** xylem water potential of female (L♀) and male mistletoe (L♂) and of oak (*Qu*) on a sunny July day in eastern Austria, under moderately dry soil conditions. At low water potentials the mistletoe transpires more intensely than the host tree. **c** Stomatal narrowing in leaves of mistletoe and oak following experimental reduction of the water flow in the supporting oak branch. (After Glatzel 1983)

Epiphytic Vascular Plants

Epiphytes in their natural habitat are dependent for their water supply upon sporadic rainfall, in between which they are exposed to gradual desiccation. The poikilohydric forms among them, such as lichens, mosses, Hymenophyllaceae and certain ferns dry out very soon when water is scarce, but quickly take it up as soon as there is mist or rain. Many epiphytes (especially the homoiohydric vascular species) have developed special means of taking up and storing precipitation or runoff water, examples of which are the nest-like form of ferns, the funnel-shaped leaf bases with specialized scales for taking up water in bromeliads, the velamentous roots of Araceae and of orchids, and water-storing tissue in axial organs, petiolar bulbs and in leaves. More than half of the water intercepted in submontane rain forests (see Sect. 4.4) is caught by the epiphytes, despite the fact that they make up only 7% of the total phytomass [184]. Their high water-storing capacity delays the decrease in water content and water potential in metabolically important tissues such as the mesophyll (see Fig. 4.4). In Orchidaceae, Bromeliaceae and other families with epiphytic species employing CAM, the diurnal stomatal dynamics provide an additional

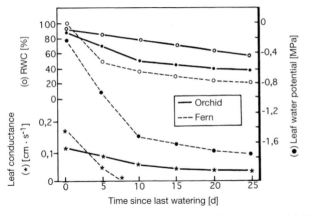

Fig. 4.45. Relative water content (*RWC*), leaf water potential (daily minima) and leaf conductance for water vapour (daily maxima) in epiphytic vascular plants subjected to a period of 25 days without water. The fern *Pyrrosia angustata* (*broken line*) loses half its saturation water content within a week, after which its stomata remain closed until water is again available. The orchid *Eria velutina* (*solid line*) dries out much more slowly due to its water storage tissues and CAM behaviour (stomata open mainly at night); it retained the ability to open its stomata intermittently throughout the entire period of observation. (After Sinclair 1983)

means of keeping evaporation at a low level, so that such plants are able to endure longer periods without precipitation (Fig. 4.45).

4.4 Water Economy in Plant Communities

4.4.1 The Water Balance of Stands of Plants

4.4.1.1 The Water Balance Equation

The state of water balance in a stand of plants, and in the soil penetrated by its roots, can be expressed by the *water balance equation*, a formula similar in structure to those for the carbon balance (2.21) and mineral balance (3.2) of ecosystems. All the quantities involved are referred to unit ground area and are given as precipitation equivalents in mm H_2O (i.e. litre per m^2).

Under the simplifying assumption that the only input to the plant cover is precipitation, and that there has been no lateral inflow, the water intake (the mean *total precipitation*, Pr) when averaged over years and decades, is accounted for by evaporation from plants and soil (loss by *evapotranspiration*, L_E) and by surface *runoff* and *percolation* through the soil (L_{RP}); see Fig. 4.46. Over shorter periods, however, the *water stores in the ecosystem* increase ($+\Delta W$) or decrease ($-\Delta W$), since sometimes more rain water falls than is lost by evapotranspiration, runoff and percolation, or because, at times, not

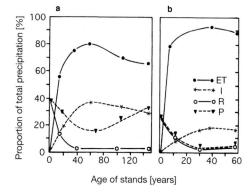

Fig. 4.46 a, b. Hydrological budgets for **a** spruce forest in the Taiga zone and **b** oak stands in the eastern European forest steppe. *ET* Evapotranspiration; *I* interception; *R* runoff; *P* percolation. (After Molchanov 1971)

enough falls to meet the requirements of the plants. The water balance equation is thus

$$Pr = \Delta W + L_E + L_{RP} \ (mm) \tag{4.15}$$

In the hydrology literature, ΔW is considered to include only the water reserves in the soil, that is, the amount of capillary water and the available gravitational water. The amount of *water contained in the soil* is greatest in the temperate zone after snow melt or spring rains. During the summer the water content usually decreases steadily, despite occasional replenishment by precipitation, until a minimum is reached late in summer. Then, especially in dry regions, the water reserves are filled up during autumn rains, a process requiring weeks until even the deeper layers of the soil are thoroughly wet.

For ecological purposes, ΔW must also include the water stored in the phytomass and in the layer of litter. More than three quarters of the green plant mass in herbaceous plant communities, and half of the fresh weight of the phytomass of forests is water; the water content of the layer of vegetation fluctuates during the course of the day and year, and is at its maximum when the plants are in leaf. Figure 4.47 shows all components of the water balance of a deciduous forest of the temperate zone, both for the time when the trees are in leaf and when they are bare.

4.4.1.2 Available Precipitation

The precipitation available to the plants for maintenance of their water balance is the amount reaching and penetrating the soil. In dense stands of plants not all of the precipitation falling on an area (*total precipitation*) reaches the ground. Instead, the water reaching the ground is only the amount that falls through gaps in the plant canopy (*canopy throughfall*), drips off the leaves, and runs down the stems (*stemflow*). This is the amount of precipitation available to the plants, and is termed the *available or net precipitation*. The resulting local unevenness in the distribution of precipitation is particularly pronounced in woodland. The greater amounts received under gaps in the

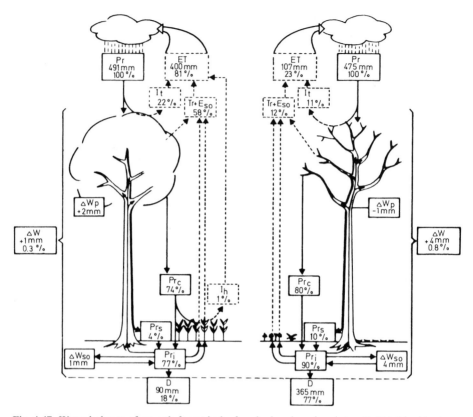

Fig. 4.47. Water balance of an oak forest in leaf and when bare in winter. *Pr* Total incident precipitation; *Pr_c* canopy throughfall; *Pr_s* stemflow; *Pr_i* infiltration (water soaking into the soil); *D* drainage water; *ET* evapotranspiration; *Tr* transpiration of the stand; *E_so* evaporation from the soil; *I_t* interception by the tree canopy; *I_h* interception by the herbaceous layer; *Δ W* total water content of the stand; *Δ W_p* water content of the phytomass; *Δ W_so* water content of the soil. On the average over the year 966 mm precipitation falls on this forest; 52.5% of this returns to the atmosphere by evaporation of intercepted water, transpiration and evaporation from the soil, 47% drains away, and 0.5% is retained in the increased biomass. (After Schnock 1971, simplified; for tropical rainforests, see Brünig 1987)

crowns of trees and under the outer parts of the crown have a considerable influence on the spreading of the roots of trees and of the understorey (Fig. 4.48). The amount of stemflow is greater, the steeper the angle of the branches and the smoother the bark; at the bases of beech trunks more than 1.5 times as much water infiltrates the soil as in the open [153].

Only a negligibly small fraction of the water wetting the trees is taken up directly through the leaves and bark. By far the greater part of the intercepted water (*crown interception*) is lost by evaporation, so that for practical purposes all of the water retained by the vegetation can be treated as a loss (*loss by interception*). A different situation can arise in regions where mist occurs frequent-

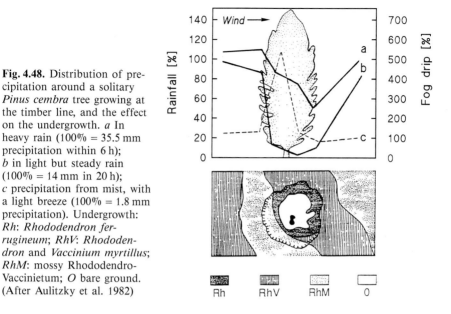

Fig. 4.48. Distribution of precipitation around a solitary *Pinus cembra* tree growing at the timber line, and the effect on the undergrowth. *a* In heavy rain (100% = 35.5 mm precipitation within 6 h); *b* in light but steady rain (100% = 14 mm in 20 h); *c* precipitation from mist, with a light breeze (100% = 1.8 mm precipitation). Undergrowth: *Rh*: *Rhododendron ferrugineum*; *RhV*: *Rhododendron* and *Vaccinium myrtillus*; *RhM*: mossy Rhododendro-Vaccinietum; *O* bare ground. (After Aulitzky et al. 1982)

ly (montane belts, and coastal areas near cold ocean currents). As a result of the moisture "combed" by the vegetation from low clouds and passing billows of mist, the available precipitation can be even greater than the precipitation in the open field, that is, there is a *gain by interception*.

The size of the *loss by interception* depends upon the composition and density of the plant cover and upon the meteorological conditions prevailing during and after precipitation. Dense crowns of trees with small, easily wettable leaves or needles retain more precipitation than open crowns with large, smooth leaves; and of course the degree to which leafing-out has progressed is also very important. On an average, the loss from interception [48, 52, 174] in coniferous forests amounts to 20–35% of the total precipitation, or as much as 50% in very dense stands; in deciduous forests of the temperate zone values are 15–30%, in subtropical shrub formations 5–15%, in palm groves 10–15% and in tropical forests 15–70%. Forest undergrowth intercepts an average of 10% (5–20%) of the precipitation, dwarf-shrub heaths as much 50%, grassland only 3–5%. Interception on cultivated fields and barren land is less than 10%. The changeability of weather conditions means that the fraction intercepted varies over a wide range, according to the amount and type of precipitation (rain, dew or snow), the prevailing temperature and the wind. In general, more precipitation is intercepted, the finer the drops and the smaller the total amount of water (Fig. 4.49). A certain amount of water is required to wet the plants thoroughly, and only after this has happened does the water begin to drip from leaves and twigs. In broadleaved forests rain only penetrates the canopy and reaches the understorey when about 1 mm (when the trees are in leaf) or about 0.5 mm (when the trees are bare) has fallen; in coniferous forests

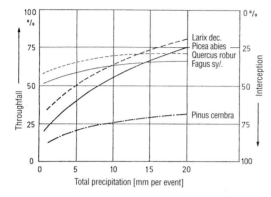

Fig. 4.49. Relative amount of throughfall (*left ordinate*) and interception (*right ordinate*) by the canopies of trees under different amounts of precipitation. The crown of *Pinus cembra* is very dense and even under heavy rainfall intercepts a large proportion of the precipitated water. (After Ovington 1954; Aulitzky 1968)

2 mm are necessary before penetration occurs (*interception storage capacity*). Heath vegetation and grassland hold back roughly 1 – 2 mm, a peat moss cover about 15 mm, before precipitation reaches the ground.

4.4.1.3 Evapotranspiration from a Stand

Water consumption by stands of plants is approximately proportional to the green mass (Fig. 4.50), although the rate of transpiration of individual leaves decreases with increasing stand density because the microclimate within the stand tends to restrict evaporation (chiefly due to shielding from radiation and wind, and to the high humidity in the stand). Therefore in very dense stands the transpiration curve departs from proportionality (Fig. 4.51).

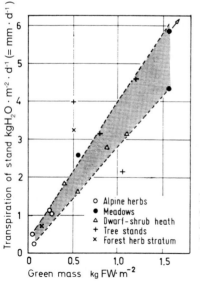

Fig. 4.50. Average daily water consumption of different plant stands as a function of the green mass (fresh weight of the non-woody aboveground parts). (After Pisek and Cartellieri 1941; Polster 1967)

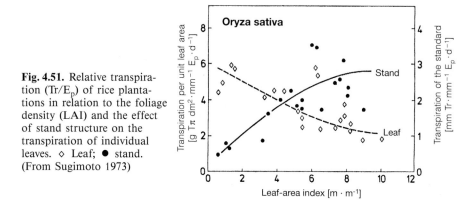

Fig. 4.51. Relative transpiration (Tr/E_p) of rice plantations in relation to the foliage density (LAI) and the effect of stand structure on the transpiration of individual leaves. ◇ Leaf; ● stand. (From Sugimoto 1973)

Restriction of transpiration on account of structural properties of the stand is expressed as the *aerodynamic exchange resistance,* R_{ae}. The *canopy resistance,* R_c, is given by the sum of all transpiration resistances in the stand, i.e. the aerodynamic resistance, R_{ae}, plus the sum of the diffusion resistances of the leaves R_l [156]:

$$R_c = R_l + R_{ae} \ (cm \ s^{-1}) \tag{4.16}$$

The *stand (or canopy) transpiration* can be described by a formula analogous to (4.10), substituting for ΔC the concentration gradient for water vapour between the plants of the stand and the open air above it. All parameters are referred to the unit ground area of the stand. Representative data for various plant stands are given in Fig. 4.52.

Similar relationships hold for the transpiration of individual *tree crowns.* The total transpiration of a tree, when not modified by stomatal regulation, correlates roughly with the total leaf surface or leaf mass (Table 4.7), but it is affected by mutual shading of individual parts of the crown and by the leaf specific conductance of the xylem.

Fig. 4.52. Simulation model of the dependence of canopy transpiration on canopy conductance for water vapour ($1/R_c$) under exchange conditions ($1/R_{ae}$) typical for the particular vegetation. Assumptions for the calculation: available energy 400 W m^{-2}, water vapour deficit 1 kPa, air temperature 15 °C. *F* Deciduous forests; *C* field crops; *H* heathland and short grass. (After Jarvis 1981). Further data for coniferous forests and grassland in Kelliher et al. 1993)

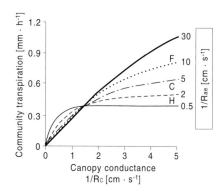

Table 4.7. Water transpired by trees of different sizes, under natural conditions

Trees	Height (m)	Trunk diameter at base (m)	Leaf area (m²)	Leaf mass (kg DM)	Water transpired		
					Maximal per hour kg H_2O h^{-1}	Daily total kg H_2O d^{-1}	kg H_2O during growth period
Tropical rainforest[a]							
Upper storey	ca. 20	0.5			20–100	400–1000	
		0.1–0.2			10–12		
Understorey	ca. 5–10	0.002			0.5	2–3	
Evergreen broadleaved trees of warm temperate regions							
Laurus azorica[b]	15	0.16	40	5.5	6–7	up to 50	
Eucalyptus spp.[c]	20–23	0.06–0.09				50–100 (200)	
Deciduous trees of temperate zone[d]							
Loose-crown trees	ca. 12		60–70	4.5–5.5		130–140	>5000
Trees in dense forests	ca. 12		30–55	2.5–3.5		30–70	<4000
Young trees	ca. 3		3–5			3–4 (9)	130–350
Salix fragilis (polycorm)[e]	ca. 10	0.4	190	13	76	463	
Conifers[d]	ca. 15			4–10		30	2500–3000

[a] Kline et al. (1970), Jordan and Kline (1977), Ogino et al. (1986).
[b] Jiménez M.S., Morales D., Kučera J., Čermák J; pers. comm.
[c] Doley and Grieve (1966).
[d] Ladefoged (1963), Braun (1977), Künstle and Mitscherlich (1977).
[e] Čermák et al. (1984).

In calculating average values for stand transpiration, whether for the **growing season or for the whole year,** the temporal variability in evaporation stress and water availability have to be taken into account. The very detailed simulation models now available for this purpose include all basic hydrological parameters as well as leaf area indices, the latter obtained by remote sensing. The *daily* and *annual patterns of evapotranspiration* can be calculated for homogeneous plant stands of limited extent, using data from the energy balance or by summation of lysimeter findings. Figure 4.53 gives an example of the changes in evapotranspiration over a year for a stand for reeds. This plant community is always *adequately supplied with water,* but the amounts transpired vary considerably from day to day, depending on the conditions for evaporation. Such fluctuations become much larger when stomatal regulation is involved as a result of *water deficiency,* which is to be expected in land plants. In this case, it is necessary to determine not only the average daily transpiration, but also the minimal water turnover when the water supply is restricted. These limiting values are particularly important data when the annual water consumption of the vegetation in regions with dry seasons is to be determined.

In forests, fluctuations in transpiration over the year are also affected by the development of the foliar mass, the degree of differentiation and ageing of the leaves, and by growth processes in shoot and root systems. Transpiration from the crowns of deciduous trees increases rapidly in spring when the leaves unfold, and reaches a maximum towards the end of the main growth period (Fig. 4.54). The situation is similar for evergreen trees of the temperate zone, for the new shoots transpire more strongly than the leaves or needles from the

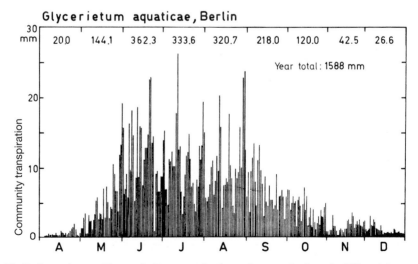

Fig. 4.53. Daily and monthly totals for transpiration of a stand of reeds (*Glycerietum aquaticae*) in central Europe during the growing season. The daily water consumption fluctuates according to the weather conditions. (From Kiendl 1953)

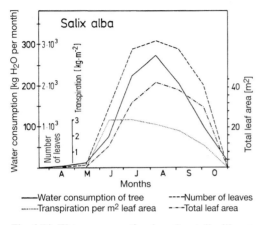

Fig. 4.54. Water consumption by a 3 m-tall willow tree (*Salix alba*) well supplied with water throughout the growing season, and the seasonal variability of transpiration per unit leaf area, per total leaf area and per number of leaves. The water consumption of the whole tree is greatest at the height of summer, which is when leaf area and number of leaves attain their maximum values. The transpiration per unit leaf area is at its highest in early summer, when the leaves are expanding. (After Braun 1974). Transpiration estimates for *Picea abies* during a growing season: Cienciala et al. (1992)

previous year. In autumn and winter transpiration falls to a minimum. This lower rate is due primarily to the lower evaporation power of the atmosphere, but also to the fact that evergreen woody plants do not open their stomata continually or fully during winter dormancy, and especially not in freezing weather.

Table 4.8 summarizes the data for **water consumption of stands** in various climatic regions. Under similar climatic conditions, forests (because of their greater mass) transpire appreciably more than grasslands, and these in turn transpire more than heath. The greatest water turnover is always found in stands of plants with access to groundwater; they sometimes transpire more water than is brought in by precipitation (see Table 4.9). In dry regions, as water becomes less plentiful the stands are more open, but the distance between individual plants is only greater aboveground. In the soil, the root systems extend over wide areas and thus the distance between the trees is determined not by meeting of the crowns but rather by the radius of the root systems of the individual trees. For regions with dry seasons there is little point in giving annual data; instead, it is more useful to know the average value for the rainy period and the minimal value for the dry period.

Quantitative data concerning the average and minimal water consumption of stands of plants are valuable as a basis for decision-making in forestry and landscape management, as well as for irrigation projects, if they include the expected precipitation. For example, it has been calculated that open stands of trees can exist only when they receive at least 110 mm precipitation in a year (10–12 mm per month during the growth period). The water-balance equation

Table 4.8. Total annual and daily transpiration of stands of plants. (From data of numerous authors)

Type of vegetation	Community transpiration	
	mm per year	mm per day
Stands of woody plants		
Tropical tree plantations	2000 – 3000	
Tropical rainforests	1500 – 2000	
Deciduous forests of the temperate zone	500 – 800	4 – 5
Evergreen conifer forests	300 – 600	2.5 – 4.8
Sclerophyllous woodland	400 – 500	
Forested steppe	200 – 400	
Ericaceous heath	100 – 200	2 – 5
Grassland and herbaceous vegetation		
Sedge and reeds	1300 – 1600	6 – 12 (20)
Tall forbs	800 – 1500	
Wet meadows	1100	8 – 15
Grain fields	400 – 500	
Prairies and savannas		4 – 6
Meadows and pastures	300 – 400	3 – 6
Steppes	ca. 200	0.5 – 2.5
Open vegetation		
Halophyte communities		2 – 5
Mountain screes	10 – 20	0.3 – 0.4
Lichen tundra	80 – 100	
Arid deserts		0.01 – 0.4

(4.15) indicates the point at which afforestation becomes uneconomical. Conversely, the enormous transpiration of trees that tap the groundwater (e.g. poplars, eucalyptus) can be employed to lower the water table and to increase air humidity.

4.4.1.4 Surface Runoff and Percolation

Not all of the water reaching the ground is available for evapotranspiration. Some of it runs off the *surface* of the ground, and another fraction *percolates* into deeper layers and joins the groundwater, which in humid regions is a subterranean form of drainage, appearing at the surface here and there as springs, maintaining connections with the rivers, and eventually draining to sea level. The surface runoff is relatively easy to measure, particularly when circumscribed catchment areas of a river are studied. Subterranean drainage, however, must be estimated indirectly.

The amount of *water drained away* depends primarily on the slope of the terrain and the type and density of vegetation. In Table 4.9 precipitation, evapotranspiration and drainage are compared for various stands of plants,

Table 4.9. Water balance of extensive stands of plants. (Duvigneaud 1967; Stanhill 1970; Mitscherlich 1971; Grin 1972; Doley 1981; Brünig 1987)

Type of vegetation	Region	Precipitation mm per year	Evapotranspiration L_E in % of precipitation	Drainage L_{RP} (surface and groundwater) in % of precipitation
Forest regions				
Tropical primary rainforest	N. Australia	3900	38	62
Tropical rainforest	Africa, SE Asia	2000 – 3600	50 – 70	30 – 50
Tropical deciduous forests	SE Asia	2500	70	30
Bamboo thicket	Kenya	2500	43	57
Savanna with scattered trees	Congo basin	1250	82	18
Deciduous forests (lowland)	Central Europe	600	67	33
	NE Asia	700	72	28
Coniferous forests (lowland)	Central Europe	730	60	40
	NE Europe	800	65	35
Mountain forests	Southern Andes	2000	25	75
	Alps	1640	52	48
	Central Europe	1000	43	57
	N. America	1300	38	62
Grassland				
Savannas	Tropics	700 – 1800	77 – 85	15 – 23
Reeds	Central Europe	800	>150	–
Pastures	Central Europe	700	62	38
Alpine pastures	*Annual:*	1000 – 1700	10 – 20	80 – 90
	Growing season:	500 – 600	25 – 40	60 – 75
Steppes	E. Europe	500	95	5
Wasteland				
Semideserts	Subtropics	200	95	5
Arid deserts	Subtropics	50	>100	0
Tundra	N. America	180	55	45
Dry mountain grassland	N. Argentine	370	70 – 80	20 – 30

each covered by uniform vegetation. In regions with low amounts of precipitation, the water is soaked up by the soil and little drains off, but where precipitation is heavy it is important for water retention that the rainfall rapidly penetrate the litter layers and the soil itself. Such conditions are best provided by loose forest soils covered by a thick layer of litter. This is yet another reason why deforestation leads to higher peak values for runoff after rainfall, which is detrimental to the regional water balance. Water penetrates much more slowly into soils of grasslands with thick mats of roots, and into pasture soils compacted by the weight of the grazing animals. Frozen soil, too, slows the percolation of water from melting snow and ice, so that the water accumulates in depressions or runs off slopes.

Where there are *steep slopes*, more than half the precipitation flows off over the surface, and where the precipitation is heavy and vegetation sparse, as much as two thirds to three quarters may be lost in this way. The high gravitational energy on the steep areas in mountainous regions, and increased runoff combined with decreased water-storage capacity of the soil and vegetation, greatly increase the danger of floods, erosion and landslides.

4.4.1.5 Other Water Supplies to the Plant Cover

Precipitation is not the only source of water for the plant cover. From its surroundings a stand may receive additional water, brought in by the groundwater flow, by streams and by irrigation. In dry regions a permanent cover of vegetation can become established only by drawing upon the groundwater or runoff from higher levels, and even in humid climates trees absorb a great deal of groundwater. By exploiting deep-lying stores of water, plants accelerate the circulation of water in the biosphere by directly pumping back into the atmosphere water which otherwise would have to flow to the sea and return via the much longer route of global water circulation.

5 Environmental Influences on Growth and Development

The essence of plant life is lifelong growth. Throughout the entire life of a plant apical meristems in the buds and root tips remain active, for the production of new shoots, leaves and flowers, and for uninterrupted extension of the root system. Secondary meristems ensure the continuous expansion of the conducting system, and are responsible for keeping intact the insulating outer layers of the bark. Not every part of a plant grows, nor does growth take place all the time, but the plant retains its ability to develop as long as it is alive. Even if the shoot tip is destroyed, the high regenerative capacity of a plant enables it to put out new shoots from dormant buds, from remaining regions of meristem tissue, and by dedifferentiation of specialized cells.

Development is the term used to describe the changes in structure and function of the plant and its parts in the course of genesis, growth, maturation and decline within the individual (ontogeny) and in the succession of generations (phylogeny). Plant development involves cell multiplication (growth by division), increase in volume (elongation growth) and the differentiation of organs and tissues. *Growth* is the permanent increase in substance and volume of living parts. In the course of time this irreversible process in certain species leads to the production of giant forms, such as mammoth trees (*Sequoiadendron*) or huge seaweeds (*Macrocystis*); the production and maintenance of the immense reserves of phytomass in the biosphere is the result of the continuous growth of the plants of the earth.

Developmental physiology has achieved striking success in elucidating the mechanisms underlying endogenous and induced processes of growth and morphogenesis. This has provided an important foundation for ecophysiological studies. However, still too little is known about the actual course taken by processes involved in development of plants in natural surroundings, due to the complexity of the stimuli and the manifold variations in the reactions of plants. Unlike a physiologist working in the laboratory, who necessarily plans and performs experiments under controlled conditions, the ecophysiologist is confronted with the problem of recording and measuring the overall response of the plant, and the changeability of events in the habitat. In view of the highly sensitive and sophisticated methods of developmental physiology, the relevant field research necessary for a strict analysis of underlying causes is fraught with difficulties. This explains why attempts in this direction have, up to this time, scarcely progressed beyond the analysis of quantitative observations. At this level, however, much has been learned about the effects of external factors on growth and development. This applies particularly to crop and forest plants, due to the practical importance of the results for improving plant cultivation and yields. Experimental studies on wild plants in the field have been principally concerned with germination, elongation growth and morphogenesis. The influence of radiation, temperature, gravity and chemical gradients on development and on the underlying mechanisms are dealt with in detail in textbooks of general plant physiology and developmental physiology, so that in the following sections only a few processes, including some that have so far received little attention, will be discussed with respect to their ecophysiological significance.

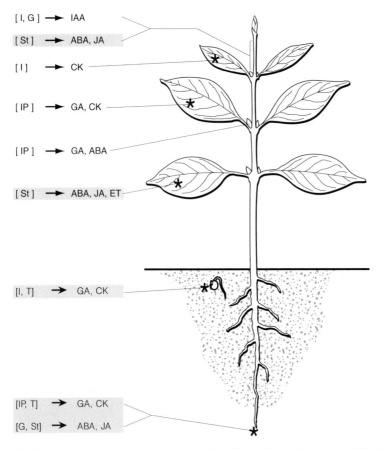

Fig. 5.1. Environmental factors triggering the effects of phytohormones. *I* Quality, intensity, direction and duration of radiation; *IP* photoperiod (short day/long day); *T* temperature; *G* position with relation to gravitiy; *St* stress (cold, heat, drought, flooding). Phytohormones: *IAA* auxin; *CK* cytokinin; *GA* gibberellin; *ABA* abscisic acid; *JA* jasmonic acid; *ET* ethylene. ★ = site of synthesis. (Modified after Matthysse and Scott 1984; Parthier 1991)

5.1 Regulation of Growth and Development

5.1.1 The Role of Phytohormones

Growth and development of plants are regulated by both endogenous and external factors. Endogenous factors are active not only at the molecular and cellular levels, affecting metabolic processes via transcription and translation, they also play a coordinating role within the whole organism through the action of phytohormones. The ecological significance of phytohormones lies in their role as transducer substances; following the perception of an environmental stimulus, every part of the plant is informed of the situation either by the synthesis of, or changes in concentration of, one or more phytohormones. Which phytohormones are called into action depends on the state of development and activity of the plant and the nature of the external stimulus, the part of the plant receiving the stimulus, and the timing of its impact (Fig. 5.1). The reactions elicited, either synergistic or antogonistic, can vary greatly depending on the organ concerned and predisposition of the plant. Together with external factors, phytohormones start the processes of growth and differentiation, and synchronize the plant's development with the seasonal changes in the environment. Additionally, they regulate the intensity and direction of growth, metabolic activity, and the transport, storage and mobilization of nutrient materials.

5.1.2 The Effect of External Factors

External factors such as intensity, duration and spectral distribution of radiation, temperature, gravity, and the forces imposed by wind, water currents and snow cover, as well as a wide variety of chemical influences, affect the growth and development of plants in different ways: by *induction,* by initiating or terminating developmental processes and thus effecting temporal regulation; *quantitatively,* by affecting the speed and extent of growth; *formatively,* by influencing morphogenesis and tropisms. The different modes of action are interwoven with one another, and the end result involves the interplay of a large number of processes.

5.1.2.1 The Effect of Light on Developmental Processes

Light affects development in a number of ways: by *photostimulation* of biosynthesis (such as the formation of chlorophyll from protochlorophyllid, enzyme syntheses, anthocyanin synthesis), by determining the direction of growth (*phototropism* and *solar tracking*), by acting as a "timer" in *photonasty* and as a trigger initiating the different stages of development in the course of a plant's life (*photoinduction*). Radiation affects the differentiation and

hence the structure of the plant at the subcellular (e.g. chloroplast differentiation), cellular and organ levels (*photomorphogenesis,* see Fig. 1.34 and Table 2.9); in conjunction with endogenous rhythms it is also the most important
synchronizer for sequences of development and the periodicity of growth
(*photoperiodism*).

A photocybernetic effect is exerted by blue to ultraviolet radiation, and by
red light up to near infrared (Table 5.1). In this case, the photoreceptors are the
pigments phytochrome, cryptochrome, and a UV-absorbing pigment. *Phyto-
chromes* are photoconvertible chromoproteins, in which the chromophore
group is an open-chain tetrapyrrol closely related to phycobilins. They occur
in two interconvertible forms. The red-absorbing form, Phy_r or Phy_{660}, is converted by absorption in the spectral region $620-680$ nm to the biochemically
active far-red absorbing form. Phy_{fr} or Phy_{730}; in dark red light
$(700-800$ nm) the unstable Phy_{730} reverts to Phy_{660}. The ratio of the two
forms of phytochrome depends on the proportion of red to far red in the radiation. *Cryptochrome,* which is a flavone, acts as a blue light receptor in fungal
hyphae and moss protonemata.

Photoreceptors are incorporated in the peripheral biomembranes. In vascular plants, the receptor organs may be buds, flower parts, fruit coats, and
seeds, but above all the leaves. Through the orientation of the receptor molecules the plant is able to recognize the direction of incoming light, and bring
about **photomodulations** accordingly, i.e. reversible reactions to changes in the
light environment. Photonastic movements are caused by changes in cell turgor
in the petioles. By means of such *movements* leaves can adjust their position
with respect to incoming radiation so as to capture the amount of radiation

Table 5.1. Effect of radiation on developmental processes in plants. (After Salisbury 1985;
Kronenberg and Kendrick 1986)

Process	Mode of action[a]	Spectral range[b]	Type of period[c]	Delay of response[d]
Seed germination and bud break	I	R/FR, B	P	h – d
Stem elongation	Q, F	R/FR	P	min
Stem orientation	Q, F	B		min
Leaf orientation	Q	R/FR	C	min
Flowering process	I	R/FR	C	h-weeks
Development and filling of storage organs	I	R/FR	P	
Dormancy	I	R/FR	P	
Enzyme syntheses	I	R/FR		h
Enzyme activation	I	R/FR		min
Membrane potentials	I	R/FR		s

[a] I = Inductive; Q = quantitative; F = formative.
[b] B = Blue light; R/FR = red to far-red ratio.
[c] P = Photoperiodism; C = circadian rhythm.
[d] Interval between commencement of irradiation and reaction.

necessary to obtain good photosynthetic yields, while at the same time avoiding excessive irradiance and overheating ("solar tracking"). Flowers also exhibit heliotropic movements. In the Arctic this is important for obtaining the warmth necessary for the flowers to open (and thus for pollination) and for subsequent seed growth.

Photomorphogenesis is brought about by the activation of enzymes and regulation of gene activity. The extracellular signals are transduced by mediators in such a way that they can be understood by the intracellular structures. An important second messenger for the interpretation of red light signals is the Ca^{2+}-binding protein *calmodulin.* Changes in light bring about an increase in the intracellular Ca^{2+}; this causes activation of calmodulin, which in turn activates certain enzymes.

Ecologically important **regulations of developmental processes** in which the phytochrome transduction system is involved are photoinduction of germination and the photoperiodic control of differentiation, both of them important turning points in the life cycle.

There are many plant species, particularly those of open habitats and forest clearings, with seeds that germinate only when exposed to light in which red wavelengths predominate (*light-promoted germination*). In the open, the ratio of red/far-red (660/730 nm) in natural light is $1:2-1:3$, whereas below a closed canopy the amount of far-red can be $2-10$ times that of red [50]; thus seeds requiring more red cannot germinate until the quality of the light changes (photodormancy), either by leaf fall or by thinning of the upper layers of vegetation. Also seeds that have been exposed to far red before becoming buried in the soil require exposure to red light in order to germinate. In such cases the postponement of germination serves to regulate the next generation − it has a population-ecological effect. Many individual variations to this kind of light reaction have been recognized. Some seeds even require less light for germination after they have been in shade for a longer time. Temperature also influences the light requirements for germination; for example, seeds of certain photodormant species are rendered more sensitive by cold, so that they need less light for the induction of germination.

The phytochrome system responds to the day/night cycle. For the plant, *seasonal photoperiodism* is an astronomical, weather-independent signal coordinating the endogenous circadian rhythm with a variety of processes involved in development, from germination to bud-break, expansion of leaves, branching, development of storage organs and, above all, the induction of flowering (Table 5.2). All photoperiodic phenomena serve the purpose of alerting and preparing the plant for the approaching and unavoidable alterations in external conditions. In regions with alternating rainy and dry seasons some plants produce two types of leaves, differing quite distinctly in structure and function − they exhibit *seasonal dimorphism.* The larger and softer winter leaves of *Phlomis fruticosa* [167] and *Sarcopoterium spinosum* in mediterranean regions, and of *Prosopis glandulosa* on river banks in dry regions of N. America, have greater photosynthetic capacities than the small summer leaves, which are much more economical in their use of water (see Fig. 2.26).

Table 5.2. Plant responses to photoperiodism. (After Salisbury 1982)

Developmental process	Plants (examples)	Promoted by
Induction during seed development	*Chenopodium album*	SD
Seed germination	*Betula pubescens*	LD
	Nemophila menziesii	SD
Stem elongation	Most vascular plants,	LD
	Alstroemeria cv. Regina	SD
Leaf expansion	*Glycine max*	LD
Leaf succulence	*Kalanchoe blossfeldiana*	SD
Tillering in grasses	*Hordeum vulgare*	SD
	Oryza sativa	LD
Branching	*Oenothera biennis*	SD
Development of storage organs	*Solanum tuberosum*	SD
	Allium cepa	LD
Flowering	*Rudbeckia hirta*	LD
	Cosmos sulphureus	SD
Favours female flowers	*Cannabis sativa*	SD
	Spinacia oleracea	LD
Clonal reproduction	*Bryophyllum* species	LD
Leaf abscission	Broadleaved trees of intermediate and higher latitudes	SD
Bud break	Broadleaved trees of intermediate and higher latitudes	LD

SD = short days. LD = long days.

Precision of timing is of particular ecological significance for successful reproduction. The many and often complicated forms of *photoperiodic induction of flowering* and initiation of sporangia and gametangia of cryptogams are the means by which the fine adjustment of the reproductive phase is effected. The time of flowering must be synchronized with climatic conditions that are favourable for pollination, so as to coincide with the activity of the pollinators. There must be sufficient time for the seeds to ripen, and the appropriate agents for fruit dispersal must be available. There are more than 20 qualitatively and quantitatively differing patterns of flower development among the long-day and short-day plants, in which the species, varieties and ecotypes differ with respect to their critical day lengths [206]. Such diversity promotes not only differentiation within a population, it also favours the selective coevolution of plants and their pollinators.

5.1.2.2 The Effects of Temperature on Development

Sufficient but not excessive heat is a basic prerequisite for life. Each vital process is adjusted to a certain temperature range, but optimal growth can only be achieved if the diverse processes involved in metabolism and development are harmoniously attuned to one another. Thus temperature has an indirect influence on growth and the course of development — due to its *quantitative* effect on the supply of energy from basal metabolism and on biosyntheses — and a direct effect via regulatory processes such as *thermoinduction, thermoperiodism,* and *thermomorphism.*

The **effect of temperature on germination** is of great importance for population ecology. For spores and seeds to be able to germinate, their "cardinal temperatures" must correspond to external conditions that ensure sufficiently rapid development of the young plants. The temperature range for the onset of germination is broad in species that are widely distributed and in those adapted to large temperature fluctuations in their habitat (Table 5.3).

After the minimum threshold temperature has been reached, the rate of germination increases exponentially with rising temperature. An ecological connection frequently exists between the *speed of germination* and climatic conditions. In species germinating in summer (usually those of northern origin), as

Table 5.3. Minimal, optimal, and maximal temperatures (in °C) for the germination of seeds and spores. These data are representative, but in particular cases may be changed considerably by various factors, both external (light, soil moisture, thermoperiod) and internal (stage of maturation, age, and readiness of seeds to germinate). (After data from a number of authors)

Plant group	Minimum	Optimum	Maximum
Fungal spores			
Plant pathogens	0 – 5	15 – 30	30 – 40
Most soil fungi	ca. 5	ca. 25	ca. 35
Thermophilic soil fungi	ca. 25	45 – 55	ca. 60
Grasses			
Meadow grasses	3 – 4	ca. 20	ca. 30
Temperate-zone cereals	(0) 2 – 5	20 – 25 (30)	30 – 37
Rice	10 – 12	30 – 37	(35) 40 – 42
C_4 grasses of tropics and subtropics	(8) 10 – 20	32 – 40	(40) 45 – 50
Herbaceous dicotyledons			
Plants of tundra and high mountains	(3) 5 – 10	ca. 20	
Meadow herbs	(1) 2 – 5	15 – 20	35 – 45
Cultivated plants in the temperate zone	1 – 3 (6)	15 – 25 (30)	30 – 40
Cultivated plants in the tropics and subtropics	10 – 20	30 – 40	45 – 50
Desert plants			
summer-germinating	10	20 – 30 (35)	
winter-germinating	0	10 – 20	ca. 30
cacti	10 – 20	20 – 30	30 – 40
Temperate-zone trees			
Conifers	4 – 10[a]	15 – 25	35 – 40
Broad-leaved trees	below 10[a]	20 – 30	

[a] After cold stratification.

opposed to winter-germinating species (from regions with mild winters), germination proceeds extremely slowly at low temperatures; only after the seedbed has warmed up to more than 10 °C does the process accelerate, but then soon makes up for lost time (Fig. 5.2). Thus synchronization is achieved with the season most favourable for development of the young plants, thus improving their chances of survival and continued growth. In some plants there are complicated *thermoregulatory mechanisms which prevent germination* at unfavourable times. The seeds of many Rosaceae, Primulaceae, Iridaceae and some forest trees (*Betula, Fagus, Tilia, Fraxinus, Picea, Pinus, Thuja*) germinate more readily if they have been exposed in an imbibed state to low temperatures or mild frost for a period of several weeks to some months (cold stratification at 0 to +8 °C). In some other species, germination is started off by high temperatures; for example, exposure of the dry seeds of rice and oil palm to 40 °C rapidly leads to breaking of dormancy.

For **shoot and root growth**, active *cell division and expansion* of the above-ground organs of temperate-zone plants begins before the temperature reaches 10 °C, whereas in tropical plants growth does not begin below 12 – 15 °C [17]. Plants in the Arctic, mountain plants, and spring-blooming plants show signs of growth at 0 °C [116]. However, the most *vigorous* growth by cell division requires considerable warmth. The temperature optimum for cell division (at which the duration of the cell cycle is shortest), is about 30 °C for most herbaceous crop plants, and is thus near the temperature maximum for growth [67]. The temperature at which shoot elongation proceeds most rapidly lies between 30 and 40 °C for plants of the tropics and subtropics [242], and between 15 and 30 °C for other plants. *Cell differentiation,* on the other hand, can proceed at low temperatures, albeit very slowly. Thus the differentiation of bud meristems and the initiation of flower primordia is suspended in winter only during especially cold periods.

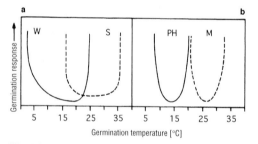

Fig. 5.2. a, b. Temperature dependence of the time required for germination of 50% of the seeds of plant species with different distribution areas and season of germination. **a** Caryophyllaceae of northern origin germinate in summer (*S*; e.g. *Lychnis flos cuculi*), their germination rate is greater at higher temperatures; species of southern origin germinate in winter (*W*; e.g. *Silene secundiflora* from southern Spain), and their germination rate is greater at lower temperatures. **b** The seeds of the subshrub vegetation in the eastern mediterranean region ("phrygana" *PJ*; e.g. *Cistus, Sarcopoterium, Phlomis*) germinate at cooler temperatures, whereas the seeds of sclerophyll woody plants of the macchia (*M*; e.g. *Myrtus, Nerium, Ceratonia*) germinate at warmer temperatures. (After Thompson, as cited in Bannister 1976; Mitrakos 1981)

Root elongation is usually possible over a wide range of temperatures. In woody plants of the temperate zone the minimal limiting temperature for root growth is rather low, at between 2 and 5 °C. It is thus not surprising that roots begin to grow before the buds sprout and that they continue growing until late in the autumn (see Fig. 5.16). Plants of warmer regions require higher temperatures: *Citrus* roots grow only above 10 °C. A major factor preventing tropical and subtropical species from advancing into cooler regions is probably the insufficient warmth for root growth.

Flower formation is induced within certain temperature thresholds, and still other temperatures are effective in bringing about the development and unfolding of the flowers. Winter grains and biennial rosette plants, as well as the buds of certain woody plants (e.g. peach, but also olive trees and *Citrus unshiu*) require a period of low temperatures in order to flower normally in the following year (chilling requirement, vernalization). Flower induction only occurs in these species after their apical meristems have been exposed to temperatures between −3 and +13 °C, ideally between +3 and +5 °C, for some weeks [86, 165]. If the cooling period is too short, comes at the wrong time, or is interrupted by warming above 15 °C, the effect does not appear. In bulbous plants from the steppes of the Near East, and in their cultivated forms, such as garden tulips and hyacinths, the leaf and flower primordia are initiated at temperatures above 20 °C, whereas low soil temperatures (around 10 °C), like those prevailing in winter in the regions where these plants naturally grow, are more favourable for final differentiation of the shoot apex in the bulb (Fig. 5.3). Higher temperatures, again, are required when the inflorescence axis begins to grow out of the bulb. It is generally true that the course of development of a plant is critically affected not so much by one particular optimal temperature range, but by a *sequence* of optimal temperatures.

Fruits and seeds usually require more heat to ripen than is necessary for growth of the vegetative parts of the plant. In habitats with both shorter and cooler growing seasons, a plant can maintain itself better if it has the alternative of *asexual reproduction,* provided by the formation of runners, stem

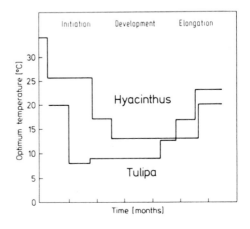

Fig. 5.3. Optimal temperatures for initiation, development and elongation of the flowering shoots of tulips and hyacinths. (After Hartsema et al. 1930; Luyten et al. 1932)

bulblets, or by fragmentation of aboveground and underground parts of the shoot.

A *thermoperiod* (temperature alternation between day and night) almost always favours growth and development; such effects show a distinct adaptation to the amplitude of the diurnal temperature alternation in the habitat [274]. Plants growing in continental regions, where the temperature fluctuates over a wide range during 24 h, develop best when the night is about 10–15 K cooler than the day; for cacti and other desert plants and amplitude of 20 K is favourable. For most temperate-zone plants the optimal amplitude of the diurnal thermal cycle is 5–10 K. Tropical plants, in accordance with the stable temperature regime in equatorial regions, are adjusted to low-amplitude oscillations (about 3 K).

5.2 Stages in the Life of a Plant

The life of any organism begins with a reproductive process. This is followed by vegetative development, including growth and the formation of organs, fol-

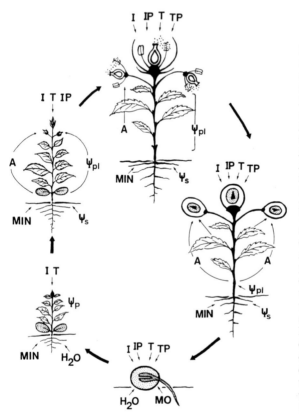

Fig. 5.4. The life cycle of annual plants and the influence of environmental factors on their development. *I* Quality, intensity and duration of irradiance; *IP* photoperiod; *T* temperature; *TP* thermoperiod; *MIN* mineral substances; *MO* action of microorganisms in breaking dormancy; Ψ_{pl} water potential of the plant; Ψ_p turgor potential; Ψ_s water potential of the soil; *A* production and allocation of assimilates. At each phase of its life, a plant is exposed to changing environmental conditions, to which it reacts in different ways. For example, the intensity and quality of available light varies with the microclimate, and alters throughout the different stages of development, from the seed up to the fruit-bearing mature plant. (After Evenari 1984, modified)

lowed in turn by the reproductive processes leading to the next generation. The life cycle is then complete. All these phases of development proceed according to a genetically set norm, coordinated by hormones and induced and modified by environmental factors (Fig. 5.4). Each phase takes up a certain portion of the plant's lifetime (Table 5.4, Fig. 5.5), and has its own particular external form and functional characteristics, regulated by the interplay of differential gene activity with the effects of the immediate environment. At each stage the plant has special requirements as to resources and environmental conditions, and responds differently to external influences. Of course it would be wrong to consider each of the successive stages of development on its own; the events of a previous phase have a preconditioning effect on the succeeding phases. Thus, the nutritional state of the mother plant affects the amount of nutrients

Table 5.4. Life span and transition to flowering in different plants. (Altman and Dittmer 1973; Van Valen 1975; Harper 1977; Kramer and Kozlowski 1979; Wareing and Phillips 1981; Tomlinson 1990; Brunstein and Yamaguchi 1992; Lyr et al. 1992)

Plants	Earliest flowering (years)	Life span (years)
Annual herbs	Weeks	Up to 1
Perennial herbs	(1) 2 – 10	10 – 40
Dwarf shrubs	5 – 10	50 and longer
Shrubs	5 – 20	50 – 100
Tree species of first stage of succession		
Alnus	ca. 10	80 – 150
Populus, Salix	5 – 15	80 – 150
Betula	5 – 20	100 – 120
Robinia	10 – 20	100 – 200
Fraxinus	10 – 40	100 – 250
Deciduous forest trees		
Ulmus	15 – 30	200 – 400
Acer	15 – 30	150 – 500
Fagus	30 – 50 (70)	300 – 900
Tilia	15 – 25	700 – 1200
Quercus	20 – 40 (75)	500 – 1400
Conifers		
Juniperus	10 – 20	300 – 2000
Cupressaceae	10 – 20	300 – 2000
Sequoiadendron giganteum	15 – 50	2000 – 4000
Taxus	≥ 10	up to 2000
Larix	10 – 15	200 – 400
Picea abies	20 – 40	200 – 500
Pseudotsuga menziesii	15 – 20	500 – 1500
Pinus sp. (subtropical)	5 – 8	100 – 300
Pinus sp. (temperate)	10 – 20 (40)	300 – 500
Pinus aristata (Rocky Mts.)	?	2000 – 4000
Pinus longaeva (Great Basin)	?	>4000
Palms	Up to 50 – 80	50 – 100

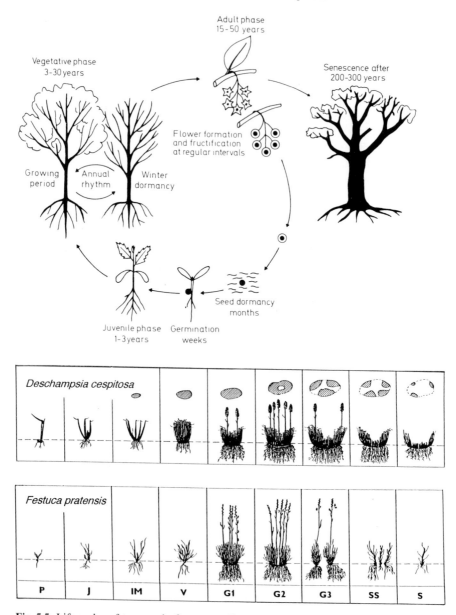

Fig. 5.5. Life cycles of trees and of grasses. *Above* Deciduous forest trees of the temperate zone. The lengths of time shown are average values, from which there may be considerable individual deviations, depending on species and locality. (After Hess 1991). *Below* Development of perennial grasses. *P* Seedlings; *J* juvenile plants (prior to tillering); *IM* immature plants (tillering); *V* fully grown plants, but not ready to flower; *G1–G3* young to mature generative plants (development stages of sexual reproduction); *SS* subsenescent plants; *S* senescent plants (declining). Perennial grasses may persist in the generative phase for 30 years or more. (After Uranov and Serebryakova, as cited in Rychnovská 1993b). Detailed explanations in Rabotnov (1978)

available for translocation to the seeds; the temperature and light conditions before, during and shortly after germination influence the shape and size of the plant and the process of flowering. Again, nutrient and water supplies during the vegetative phase affect the abundance of flowers and the vitality of the succeeding generation.

5.2.1 The Embryonic Phase: Imprinting by the Mother Plant

The embryonic phase, the period between fertilization and ripeness of the seed, is marked by intensive cell division, differentiation of the primordial organs and storage of carbohydrates, lipids, proteins and mineral substances in the embryonic tissue and in the endosperm. An important role in the regulation of these processes, and particularly the transfer of material in the growing seeds, is played by the phytohormones indoleacetic acid (IAA), gibberellic acid (GA), cytokinin (CK) and abscisic acid (ABA); the amounts and the activities of the various hormones involved at any time vary according to the state of development.

The phase of embryonic development is also one of *prenatal imprinting* by the mother plant. Even the condition of the plant at the time of flower formation, particularly in the case of *megasporogenesis,* determines to some degree the fate of the progeny. Weak, senescent, and environmentally stressed plants produce ovules that are either underdeveloped, or are unable to develop into normal seeds. The cones of immission-stressed spruce trees, for example, may contain abnormally low numbers of embryo sacs. Additional disturbances may occur in *microsporogenesis,* so that too little or even sterile pollen may result; this happens in cold-sensitive varieties of rice, in which low summer temperatures arrest pollen development at the tetrad stage (cold-summer sterility [245]), and in millet (*Sorghum*) if it is too hot and dry. Important factors for *fertilization* as such, but also for the subsequent stages of embryogenesis, are efficiency of pollination and growth of the pollen tube. That unfavourable weather conditions endanger pollination has long been realized in horticulture, agriculture and forestry. A specific example has been observed in *Tilia cordata* [180] growing near its northern limit of distribution, where unduly low summer temperatures delay or even disrupt growth of the pollen tubes. Studies on the effect of environmental factors on the embryogenesis of wild plants could yield valuable information about reproductive potentialities and problems especially when the climate is changing.

After fertilization has taken place, the genetic information in the zygote determines the *development of the embryo*, which is nevertheless still influenced by the mother plant. Surrounding the embryo is the triploid endosperm (in the genome of which only one third comes from the pollen), the integuments and the pericarp — which are maternal tissues. During embryogenesis the pericarp determines the red/far-red ratio of the phytochrome system in the seed. The mother plant is also largely responsible for providing both mechanical and chemical inhibitors of germination. Above all, depending on environmental conditions and degree of adaptation to the dominant external factors, the mother plant provides the growing seed with building materials and reserves for subsequent germination.

The size of the *seed*, the state of differentiation of the embryo and the food reserves play decisive roles in determining the seed's capacity to germinate, and also the vigour of germination. Here, even slight influence play a role, such as the time at which the seed is formed, or its position on the inflorescence; thus different sizes of seeds are often found in fruit clusters of herbaceous plants, and also within different regions of the crown of a tree.

The embryo *itself* is also able to pass on information to the mother plant via its own hormones, and thus cause the maternal metabolism to adjust to meet the requirements of the ripening fruits. This hormonal coordination also exerts a selective effect among the ripening seeds and fruits, because discarding of weak fruits ensures better provisions for the remaining seeds. The seeds become independent at the time of ripening, when the vascular bundle connecting the fruit to the mother plant is broken. By this time the information necessary for regulating seed dormancy and its termination is already stored.

5.2.2 Germination and Establishment: to Be or not to Be

The *phase of germination* commences with imbibition of water and the activation of metabolism in the embryonic tissue. The first step is the production of energy via glycolysis; reduction equivalents and the metabolites for syntheses are provided by the pentose phosphate cycle. Phytohormones give the signal for de novo synthesis of enzymes (e.g. by GA in the aleurone layer of barley caryopses), leading to the mobilization of reserve substances in the endosperm. This is followed by the synthesis of hormones promoting cell division and elongation (CK, IAA), by reorganization of the ultrastructure of the protoplasm, intensification of mitochondrial respiration and protein synthesis and, lastly, by the growth processes that result in the appearance of the radicle. This event signals, by definition, the beginning of germination.

The process of germination is complete when nutrition no longer depends upon reserve materials, but is autotrophic. By this time the root has secured a hold in the soil, the cotyledons or, in the case of hypogeal germination, the primary leaves have unfolded, and the seedling has attained independence. This is the primary condition for the plant to become *established*.

The *duration of germination* is the time elapsing between hydration of the seed and the appearance of the radicle; the *germination rate* is the percent increase in germinating seeds per unit time. There are certain species of plants in which the seed population begins and completes its germination very uniformly. This is particularly so with fast-germinating species, including many herbs and grasses, tree species such as willow, poplar and other pioneer woody species; this strategy permits the rapid exploitation of conditions favourable for germination.

In contrast, germination is slowed in many species, for varying reasons. There are certain plant species, such as orchids and palms, whose seeds contain undeveloped embryos. Commonly, in other species including some in the genera *Anemone, Caltha, Ficaria, Heracleum, Gentiana, Fraxinus,* embryonic development is still incomplete in the freshly dispersed seeds and must be continued [230]. In many seeds, germination is prevented by the presence of hard coats or by inhibitory substances, and often, too, by external factors such as the far-red influence, all of which impose a state of *dormancy.* The seeds of these species germinate very irregularly, which means that over a long period only small portions of the seed reserves are used up at one time. In this way, the plants emerge at different times, so that part of the progeny avoids any unfavourable weather and severe attacks by pests.

For the *breaking of dormancy* a variety of influences may be effective. For species with light-promoted germination this can be the exposure of covered or buried seeds to light by the clearing of forests, or, in steppes, by the activities of burrowing animals such as voles, squirrels or moles. The seeds of many trees in regions with a cold winter season, and also those of mountain plants, require stratification; for the seeds of desert plants to germinate, leaching of soluble inhibiting substances or salts from fruit- and seed coats by heavy rains is necessary. Hard-coated seeds often need the decomposing action of microorganisms; others are only able to germinate after inhibitory substances have been broken down in the digestive tracts of animals (after endozoochoric distribution). In some cases, seeds germinate spontaneously at periodic intervals, possibly via a signal from a molecular biological clock. Under natural conditions, so many factors exert an influence on the onset and progress of

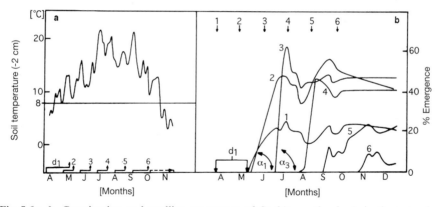

Fig. 5.6a, b. Germination and seedling emergence of *Scabiosa columbaria* in the natural habitat. The seeds were sown at monthly intervals from spring to late autumn on a xerothermic south slope in England. **a** Soil temperature at a depth of 2 cm; *numbered arrows d 1−6* start on the abscissa at time of sowing and their tips indicate the beginning of germination. **b** Germination rate (α), and seedling emergence and subsequent survival from May to December. The temperature minimum for germination is 8 °C. In early summer, the germination rate is greater ($\alpha_3 > \alpha_1$) and the seedling emergence is at its best, due to the higher soil temperatures and water content. (Rorison and Sutton 1976)

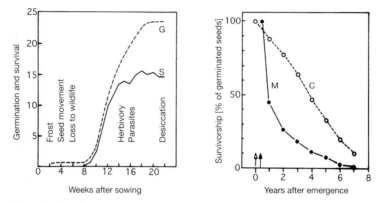

Fig. 5.7. Germination, establishment, hazards and survival of young plants of tree species. *Left* Germination (*G*) and survival of the seedlings (*S*) of *Pseudotsuga menziesii* in N. America, after sowing in February. *Right* Survival of young plants of the evergreen Fagaceae *Machilus thunbergii* (*M*) and *Castanonopsis cuspidata* (*C*) in a broadleaved forest in southern Japan. (After Lawrence and Rediske, from Kozlowski 1971; Tagawa 1979)

germination that seeds of one and the same species may germinate at very different times depending on the external conditions (Fig. 5.6).

The process of emergence and the *seedling stage* represent a particularly sensitive period. During this phase the seedling requires plentiful nutrients to cover the increased amounts of energy and metabolites needed for biosyntheses, as well as sufficient water to maintain turgor during rapid elongation growth and cell-wall differentiation. Seedlings are often not only particularly sensitive to drought, extreme temperatures and biotic stress factors, they are also more susceptible to other dangers. The greatest climatic extremes, and in dry regions the highest salt concentrations, are encountered in the contact zone between the soil surface and the air layer near the ground. Accordingly, the loss of progeny during this phase of life is at its highest (Fig. 5.7). The seedling stage is thus the decisive life phase for the survival of the individual plant and for the spread of a population, because a species can only permanently occupy those habitats in which the most sensitive stages of life can be survived.

5.2.3 The Vegetative Phase: Period of Greatest Growth

Juvenile plants, and others in the developing period before the reproductive phase, grow rapidly both in length and girth. As they increase in size they gradually take on their typical form and achieve a well-balanced *shoot/root ratio*. As long as there are no drastic differences between conditions in the rhizo-, sphere and the atmospheric environment, a logarithmic-linear correlation between shoot- and root mass is maintained ("allometric growth"). The dynamic balance existing between shoot and root is a morphogenetic regulatory system ensuring the supply of mineral substances and a favourable water economy; it is elicited by hormonal signals from the root.

During the *main phase of growth* plants are at the peak of their metabolic activity (photosynthesis, respiration, uptake of mineral substances). In view of the competition for space in plant communities, early and fast growth of the shoot, roots and parts responsible for clonal reproduction (runners, offsets, suckers) are of decisive importance for the future of the individual. It is during the vegetative phase of growth that the characteristics of phenotypical plasticity and, above all, modificative adaptations to conditions in the habitat manifest themselves.

Important factors affecting the differentiation of organs with limited growth, such as leaves, flowers and fruits, are primarily the regulation of mitotic activity in the bud primordia, the time interval between the initiation of successive leaf primordia (plastochrons [58]), the progress of cell enlargement, and the speed of differentiation under the influence of external factors, either direct (e.g. light, which promotes the extensibility of the cell walls) or indirect. The *number of cells* in the leaves depends on the frequency of cell division and the size of the leaf primorida. After increasing to $10-50$ times the size of the cells initiated in the primordia, leaf extension is complete. In leaves that have not attained a good size during the growth period, because of low temperatures, inadequate nutrients or continued repression (due to grazing, wind abrasion, bonsai treatment) the cells are not necessarily smaller than those of normal-sized leaves of the same species (Fig. 5.8). The specific maximal size of the cells is genetically controlled; it seems that a higher DNA content per cell nucleus, such as is commonly encountered in plants of wet and cool regions, leads to the production of larger cells [100].

The process of *cell extension* involves the interaction of a number of factors, including turgor pressure, phytochrome and blue light receptors, hormonally regulated (IAA, CK, GA) extensibility and cell wall "loosening" factors such as H^+ ions (Fig. 5.9). The relative increase in cell volume $[(dV/dt)\cdot 1/V]$ is proportional to the extensibility modulus of the cell wall (m), and the dif-

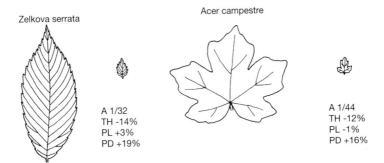

Fig. 5.8. A comparison of leaves of bonsai plants and normal leaves of the same species. *A* Leaf area (bonsai leaf compared with normal leaf); *TH* leaf thickness (% of the normal leaf); *PL* length of palisade cells; *PD* diameter of palisade cells. (Körner et al. 1989)

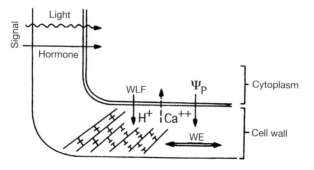

Fig. 5.9. Diagram of processes involved in elongation growth. Signals such as light or phytohormones stimulate the cytoplasm to produce and release a wall-loosening factor (*WLF*); its function is to loosen bonding between the cellulose strands in the the cell walls. The cell walls are then stretched (wall extension, *WE*) by turgor (Ψ_p). (Cleland 1986)

ference between the prevailing turgor potential (Ψ_P) and the threshold pressure (Ψ_P^{lim}) required to stretch the wall irreversibly. For sunflowers, soya, rice and maize characteristic values are $0.4-0.7$ MPa [248]. In a simplified form the relationship can be expressed [144] as

$$\frac{dV}{dt}\cdot\frac{1}{V} = m\,(\Psi_P - \Psi_P^{lim})\;(s^{-1}) \tag{5.1}$$

Shortage of water means that cell extension cannot occur since turgor pressure is inadequate, or, if already begun, the process of cell enlargement slows down and its prematurely terminated, perhaps as a result of elevated ABA levels. Thus with drought the cells are unable to attain their potential maximal size (see Fig. 6.65). Photosynthetic yield and the allocation of assimilates (see Chap. 2.3.3) are important factors also, because growth depends on an adequate supply of carbohydrates for maintenance and structural metabolism.

5.2.4 The Reproductive Phase: Flowering and Fruiting

Transition from the vegetative to the adult phase, or maturity, is marked by the capacity of the plant to flower, which results from changes in the state of the apical meristems in the buds. A prerequisite for this ability to initiate and form flowers is *floral induction*. In self-inducing plants this occurs by itself, either when a genetically determined age for flowering is reached (endogenous timing), or after the formation of a certain number of leaf primordia, or when the vegetative parts of the plant have reached a certain size ("size effect" e.g. storage organs) and when the carbon/protein balance of the plant is favourable. In many plants, however, the initiation of flowers requires induction by external signals such as light, temperature or an incipient water shortage. Phytohormones and nucleotides participate in the activation and derepression of the genes responsible for development of the flower primorida. Following

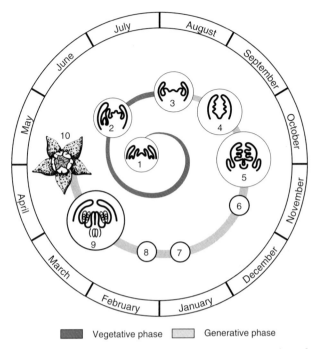

Vegetative phase ▨   Generative phase ▨

Fig. 5.10. Development of the flower buds of fruit trees in central Europe. *1* Formation of primordia of leaf- and axillary shoots; *2* formation of the vegetative buds; *3* beginning of flower differentiation by broadening of the apical meristem; *4* invagination of the floral cup and differentiation of the stamen primordia; *5* appearance of primordia of gynoecium and further development of the stamens; *6* onset of winter dormancy; *7* deep winter dormancy; *8* release of dormancy and readiness for bud development; *9* differentiation of flower organs almost complete, tetrad stage of anthers; *10* buds ready to open. (After data and diagrams in Zeller 1958)

floral induction, flower formation may progress either rapidly or with longer interruptions, such as during the winter or in dry periods (Fig. 5.10). When the flower buds have fully opened, the process of flowering is complete.

Environmental factors, in conjunction with endogenous regulating mechanisms, influence the frequency of flowering, fruit set, and the ripening of seeds, mostly by effects on the nutrient status. The energy and building materials required for flowering and for fruit formation ("reproductive effort" [11]) are provided by concurrent photosynthetic activity and incorporation of mineral substances, as well as by mobilization of reserve materials and the recycling of breakdown products from senescing leaves. Abundance in flowering and fruiting is therefore in competition with vegetative growth and with the laying up of reserves for renewed bud formation in case of biomass losses caused by animals or other external forces. In perennial plants the latter leads, in turn, to a reduction in reproductive capacity. It is therefore not surprising that the coldest and the driest regions of the earth are settled almost exclusively by cryptogams, which have a minimal investment in their reproductive organs.

Annual plants draw the carbohydrate compounds for their reproductive requirements mainly from their current dry matter production (in annual grains up to 65% from the photosynthesis of green glumes and of the uppermost leaf); and some 50–90% of the necessary nitrogen-containing compounds and phosphorus are drawn from the vegetative organs [34]. Because, in many annuals, vegetative growth is already coming to an end when seed development begins, the quantity and quality of the seeds are influenced by the environmental factors prevailing shortly before and during the reproductive phase. When nutrients and/or water are in poor supply, depending on the plant type, flower primordia may not be initiated at all, or vegetative development may be given priority and the reproductive organs under-supplied, with the result that the unripe fruits drop off ("reproductive-vegetative switch", Fig. 5.11).

In *biennial herbs,* which in the first year form a rosette and underground storage organs to ensure a rapid start for development in the second year, the major part of the energy and building material for flowering and fruit formation comes from the plant's reserves. The size of the overwintering rosettes and storage organs correlates well with the number of flowers and fruits. *Perennial herbs* behave in the same way. In habitats where the growing season is limited, a compromise is reached between assimilation, reproductive performance and provision for the progeny; although fewer flowers are formed, a higher percentage of the reduced assimilation yield is invested in reproductive structures (Fig. 5.12). The saving connected with the reduction in length of the inflorescence axis alone (see Chap. 2.70) means that sufficient reserves are available for the overwintering organs, for the new shoots, for the necessary regenerative processes and for clonal reproduction. In many perennial plants of polar regions and high mountains, flower formation proceeds very slowly, taking as long as 1 or even 2 years before the opening of flowers (see Fig. 5.17).

In *woody plants,* flower formation, frequency of flowering, quantity of fruit and ripening of the seeds are regulated by a combination of nutritional factors, allocation of assimilates, and endogenous control mechanisms. Thus, a large abundance of fruits competes with growth of supporting tissue, and if the photosynthetic yield is poor, only vegetative buds and no flower buds are formed for the following year.

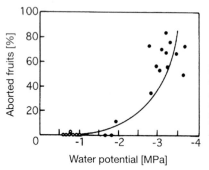

Fig. 5.11. Loss of unripe siliques (in percentage of set fruit) of *Capsella bursa-pastoris* as a function of leaf water potential (predawn values) following a dry period. (Pyke 1989). For defoliation effects on herbaceous perennials and woody plants see Obeso (1993)

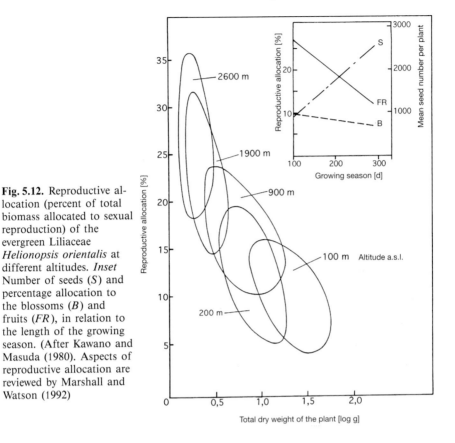

Fig. 5.12. Reproductive allocation (percent of total biomass allocated to sexual reproduction) of the evergreen Liliaceae *Helionopsis orientalis* at different altitudes. *Inset* Number of seeds (*S*) and percentage allocation to the blossoms (*B*) and fruits (*FR*), in relation to the length of the growing season. (After Kawano and Masuda (1980). Aspects of reproductive allocation are reviewed by Marshall and Watson (1992)

The *expenditure of assimilates for reproduction* ("reproductive effort") can be quite considerable; in pines it amounts to 5–15% of the total dry matter production, in mediterranean evergreen oaks 12% [139], in beech 20% and more, in the palm *Corypha elata* 16% [250], in apple trees 35% and in *Citrus* species up to 50% [138]. In dioecious plants, the amounts of photosynthates required for the reproductive processes of the two forms differ greatly. In female individuals of the evergreen shrub *Simmondsia chinensis*, for the maximum fruit set (initiation and development of fruit) 30–40% of the plant's assimilates are diverted to flowers and fruits, whereas male individuals require only 10–15% of the assimilates for flowering. Under desert conditions this difference has a marked effect on shoot growth, so that female and male individuals can be distinguished by size as a result of *sexual dimorphism* [262].

If well supplied with assimilates, tropical and subtropical woody plants such as coffee, cocoa, *Artocarpus*, *Carica papaya,* coconut palms and also lemon trees bear fruit throughout the year. In regions with a seasonal climate there are tree species that come into full flower and bear fruit every year (*Populus, Salix, Alnus, Carpinus, Tilia, Acer* and many more). Other trees of the temperate zone are only able to produce large quantities of fruits at intervals of several years ("alternately bearing"); for deciduous trees this usually occurs every 2−3 (5) years, for conifers after 2−6 (10) years. These intervals between full fructification become much longer, the nearer to the polar limits of distribution, and in mountains. Under chronic stress few seeds are produced, and the germination of these is poor.

5.2.5 The Phase of Senescence: Orderly Withdrawal

The phase of ageing, or senescence, is characterized by slowing down of metabolic activity, decreasing apical and cambial growth, smaller leaves, fewer flowers and seeds, the latter with reduced powers of germination. The organism is also more sensitive to abiotic stress, and its susceptibility to parasitic attack is greater. Annual and monocarpic plants, which flower only once, begin to age as soon as flowering is ended; in species that survive for several years, with alternating periods of vegetative growth and flowering, the parts of the plant that are regularly replaced die off at the end of the growth period. Only when the apical meristems begin to degenerate, after the completion of a large number of cell cycles, does the entire plant enter a general phase of ageing. Trees may spend as much as one third of their life span in this phase, so that primaeval forests contain a high percentage of very old trees, many of them with a large number of dead branches. The limits of growth for trees are set by diminishing shoot renewal and cambial cell division. The resulting unfavourable ratio of productive leaf mass to total mass means that the carbon balance gradually deteriorates, increases in wood are less, and thus the long-distance water transport in the tall trees is inadequate.

Ageing and decay accompany life: every growing plant has certain parts in which all of the tissues become fully differentiated and then age. Even at the seedling stage and in the juvenile phase there are short-lived cells and tissues (such as root hairs, epidermal layers) that soon collapse, and meristematic cells that divide and are soon transformed into dead conducting and supporting elements. Cotyledons and primary leaves, too, change colour and are shed at an early date. But also the foliage, flowers and fruits undergo a fast process of ageing, which ends with abscission from the plant. Limitation of the functional life of metabolically active parts of a plant by means of programmed ageing is an economy measure ensuring timely transition to a period of dormancy in regions where the growing season is limited.

The *ageing of leaves* can follow different patterns (Fig. 5.13): in some species the leaves turn yellow in succession, in the order in which they opened; in others, all of the leaves of one period of leaf formation become senescent at the same time. Leaf fall in continuously growing plants (many herbs, certain tropical phanerophytes) is usually of the former type, whereas geophytes lose their leaves simultaneously. Among woody species there are examples of every degree of transition between the sucessive and the simultaneous patterns of leaf senescence.

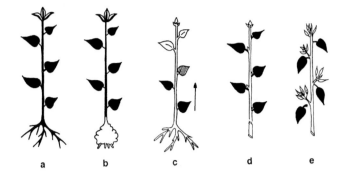

Fig. 5.13a−e. Different types of senescence. **a** Simultaneous whole-plant senescence (e.g. rain-ephemeral desert annuals). **b** Programmed die-back of the aboveground shoot (e.g. geophytes). **c** Sequential leaf senescence (e.g. in many annual herbs and grasses, and some woody plants of the humid tropics). **d** Simultaneous leaf senescence and leaf fall *before* bud break (deciduous trees in regions with marked seasonal patterns of temperature or humidity). **e** Leaf fall associated with bud break or at the beginning of leaf expansion (e.g. leaf-exchange pattern of some evergreen woody plants and green-overwintering herbs). (Adapted from Wareing and Phillips 1981; Longman and Jeník 1987)

Regulation of senescence is effected via a genetic programme (e.g. during differentiation of sclerenchyma cells; there are also genes for senescence). In monocarpic plants the signal for ageing to begin comes from the ripening seeds; removal of the flowers and unripe fruits therefore prolongs the life of the plant. In perennial species the signal for ageing is often provided by external factors such as short day, the occurrence of certain threshold temperatures, or by stress situations (Table 5.5). Again, coordination at the level of the whole plant, is effected by phytohormones, chiefly abscisic acid and jasmonic acid (which also initiates the events leading up to leaf abscission), and ethylene (which accelerates senescence). A delaying effect on senescence is exerted above all by a high concentration of cytokinin in the leaves, but in some plants also by auxins and gibberellins.

Ageing cells show characteristic alterations; early in the process, stroma proteins (RuBP carboxylase) and later chlorophyll as well, are broken down in the chloroplasts, the thylakoid structures disappear, and large plastoglobules occur in the gerontoplasts. The cytoplasm, and particularly the endomembrane system, shrinks; as a result of physical and chemical changes in the phospholipids, the biomembranes become permeable to ions, soluble carbohydrates and amino acids; the activities of hydrolases, peroxidases, polyphenoloxidases and proteases increase; as a result of the accumulation of catalytic enzymes the vacuole assumes the character of a lysosome. Even before the breakdown of compartmentalization a climacteric rise in respiration in leaves and fruits may occur.

These events lead to a *disproportionality in protein metabolism.* Because protein breakdown outweighs its synthesis, soluble amino compounds accumulate and are diverted into what can be termed centres of attraction (seeds,

Table 5.5. Influence of environmental factors on leaf abscission in trees. (Addicott 1968)

Factor	Promotion	Retardation
Radiation		
Deficiency or surplus of PhAR		×
Longer photoperiod		×
Shorter photoperiod	×	
Temperature		
Moderate	×	
Light frost	×	
Heat or severe frost		×
Water		
Drought	×	
Flooding	×	
Minerals		
Nitrogen fertilization		×
Deficiency of minerals	×	
Excess Zn, Fe, Cl	×	
High soil salinity	×	×
Gases		
Ethylene	×	
Noxious gases	×	

younger parts of the shoot). Up to 60% of the protein of the leaf can be withdrawn for reuse, and valuable bioelements like nitrogen, phosphorus and sulphur are recovered. This stepwise breakdown of the leaf proteins to the advantage of the remaining parts of the plant, prior to completion of the life cycle and before external conditions become unfavourable (drought, winter), is of great significance for the material balance of the plant. The process of ageing and leaf abscission, which annually account for the disappearance from the earth of 1.2×10^9 t chlorophyll [92] can also be seen from this aspect; conservation of the pigments is not important, but rather the nitrogen in the chlorophyll-binding proteins. If these proteins were broken down first, photodynamic chlorophyllids would be released, which in turn would result in photooxidation and cellular death before the valuable nitrogen compounds could be withdrawn. This is why it is necessary for the pigment-protein complexes to be dissociated successively by enzyme action (Fig. 5.14). The prophyrin rings are opened and the harmless breakdown products (e.g. linear tetrapyrrols, lipofuscin-like compounds) are actively transported into the vacuole. These water-soluble catabolites are colourless, which is why the green colour of the leaves disappears. Some breakdown products are blue- or green-fluorescent, which enables the progress of chloroplast senescence to be traced fluorometrically (Fig. 5.15).

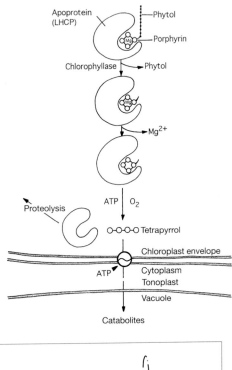

Fig. 5.14. Successive steps in the breakdown of the pigment-protein complex in ageing chloroplasts of yellowing leaves, and the translocation of the chlorophyll breakdown products into the vacuole. (Matile et al. 1989; Matile 1991)

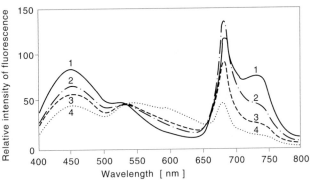

Fig. 5.15. Spectral changes in the fluorescence of senescing beech leaves in connection with chlorophyll breakdown. *1* Normal green leaves (ratio of chlorophylls/xanthophylls + carotenes = 4.6); *2* pale green (chl/x+c = 3.5); *3* yellowish green (chl/x+c = 1.9); *4* yellow leaves (chl/x+c = 0.4). The changes are due both to decreasing amounts of chlorophyll and to an increase in fluorescent catabolites. (Lang and Lichtenthaler 1991)

5.3 The Seasonality of Growth and Development

In the course of their development, plants adjust to the seasonal periodicity of insolation, day length, temperature and precipitation (its onset and amounts). The transition from one phase to another are the critical steps for temporal adjustment of the life cycle to suitable and unsuitable periods for growth. Quite often, there is a delay at the end of a developmental phase, or before the beginning of another, such as seed dormancy, or because of the need for floral induction. Only plants with a short life cycle and those that grow under continuously favourable external conditions can thrive without predetermined interruptions.

5.3.1 Different Patterns of Life History

5.3.1.1 Plants with Continuous Growth

In regions with a pronounced *seasonal climate* (summer-winter, wet-dry seasons) plants that grow continuously are necessarily short-lived: summer annuals in the temperate zone, winter annuals in winter-rain regions, and especially ephemeral plants in deserts. In all of these annuals, the phases of the life cycle follow one another in uninterrupted sequence. The primary shoot appears immediately after germination, then a few leaves, after which the first flowers may already open. The shoot continues to grow, and at the same time vegetative and reproductive organs develop alternately. In species of determinate growth, flowers form only when intensive growth of the shoot is completed. Even while the fruits are ripening, signs of ageing appear in the vegetative parts of the the plant. Finally, the whole plant dies, leaving only the seeds; these remain in a dormant state until they are stimulated by conditions favourable for germination.

In regions where conditions are *favourable to growth all year*, such as the permanently wet tropics and mild-winter regions of the warm temperate zone, there are perennial plants that grow continuously and can become as tall as trees, e.g. tree ferns, cycads, palms, certain herbaceous phanerophytes (e.g. *Musa* spp.), as well as woody dicotyledons (for example, *Carica papaya*). In monocarpic plants the shoot grows without any marked interruption until flower formation exhausts the vegetative tip. Once the fruit has ripened, the entire plant dies (e.g. *Agave, Corypha*), although in some cases remnants of the vegetative part of the plant survive and produce new plants.

5.3.1.2 Plants with Intermittent Growth

There are many plants with a tendency to alternate between *activity* and *rest*. In species of this kind, growth and thickening of the shoot take place in spurts, old leaves are replaced with new in a burst of activity, and the storage organs fill and become depleted periodically. A striking example of *intermittent* growth is presented by some Fagaceae (Fig. 5.16: *Quercus* type) and conifers (*Pinus, Picea* and *Abies*). In these functional types, elongation of the new shoots is interrupted after the first phase of growth in spring, and is resumed later in a second burst. At the height of summer, shoots may again be produced ("lammas" shoots). Species of *Populus, Betula, Tilia* and *Robinia* com-

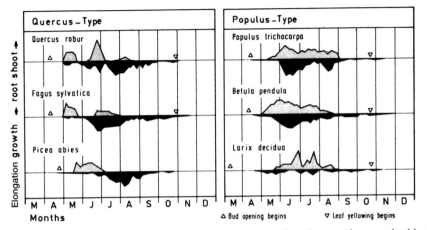

Fig. 5.16. Elongation growth of shoots (upwards, *stippled*) and roots (downwards, *black*) in several tree species in central Europe during the course of the year. (After Hoffmann 1972)

plete their growth without interruption (Fig. 5.16: *Populus* type). In regions with a seasonal climate, timing of the spurts of development is regulated by the phytochrome system and phytohormones, so that in a particular region most plants grow with the same timing. Root growth often begins before that of the shoot, and continues into the late autumn; it is to a large extent regulated by soil temperature, availability of water and distribution of nutrients.

In tropical and subtropical climates, woody plants respond to even small fluctuations in temperature and heavy rainfall with a growth spurt, or with other developmental processes. In the permanently wet tropics only 20% of all evergreen trees grow steadily throughout the year; in the rest, *growth is interrupted* at intervals. Although tropical forests are green all year, individual trees periodically replace old leaves with new and may even become temporarily bare. Usually the tree or single branches form new foliage within a few days, and shoot elongation also occurs during this period. In the permanently wet tropics it is not uncommon that these spurts of growth, which may take place several times each year, occur within one tree population, or even within the crown of a single individual *at different times*. The ecological significance of the production of new growth in multiple and staggered periods lies in the reduction of the danger that herbivores and parasites will strip the foliage and multiply disastrously, in a climate that favours reproduction all the year round.

5.3.1.3 Reproductive Cycles

When a plant has reached maturity, the reproductive cycle must also be accommodated in addition to the periodic growth of the shoot. Vegetative growth and reproductive development can proceed either concurrently or alternately, depending upon the plant species. In annuals and many tropical plants with continuous growth, the two processes run concurrently. Alternation of mainly

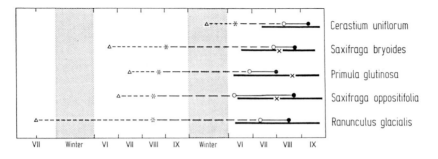

△ flower initiation * flower primordia ○ flowering ● ripe seeds x differentiated xylem

Fig. 5.17. Flower bud formation, flowering and seed ripening in rosette and cushion plants of the European Central Alps (2600–3200 m a.s.l.) in years with favourable weather. *Thick lines* Vegetative development, production and storage of material. (After Moser and Zachhuber, from Larcher 1980)

vegetative growth with flowering and fruit formation is typical for perennial plants of intermediate and higher latitudes and for dry regions, although it can also occur in the tropics. Particularly conspicuous is the alternation between vegetative and reproductive phases in biennial rosette plants and in geophytes.

In tropical and subtropical arid regions there are trees and shrubs that bear flowers only on the bare twigs *after leaf fall.* Examples are species of *Erythrina, Bombax* and *Tabebuia,* and many Caesalpiniaceae. If, in contrast, development of the fully differentiated flower primordia remains inhibited during the resting phase following leaf fall, the plant may produce flowers on the bare branches *before unfolding* of the new leaves; these are *early-flowering* types. Many deciduous forest trees, fruit trees and soft-fruit shrubs of the temperate zone are of this type. In spring geophytes (various Ranunculaceae and Liliaceae) the flower buds are fully differentiated during winter and are therefore ready to open when temperatures rise; thus the plants are able to use the brief period between the end of winter and full foliation of the tree storey for flowering and fruit set. Also, in many herbaceous species and dwarf shrubs of arctic and high mountain regions, the flowers have already been preformed in the preceding year and can therefore open immediately after snow melt (Fig. 5.17). Thus, the short summer is long enough for ripening of the fruits as well as for the subsequent accumulation of reserves before the plants are once more covered by snow.

5.3.2 Synchronization of Growth and Climatic Rhythms

The time course of the vegetative activity of plants is adjusted to the local duration of favourable growing conditions. In the dry tropics and subtropics the *growing season* is limited by increasing water deficiency once the dry period begins; activity of plants in the temperate and cold climate zones is synchronized with the seasons by seasonal photo- and thermoperiodicity. Not uncommonly, the change in day length is the pacemaker eliciting the switchover, which is reinforced by the changing temperature.

Above latitudes of 40° the days are longer than the nights during the entire growing season, and above 50° the difference is quite considerable (see Fig. 1.43). Taxa with centres of origin at high and intermediate latitudes are adapted to this periodicity, most of them behaving as long-day plants with respect to the production of new shoots, leaves and flowers, and as short-day plants with respect to the periodicity of their vegetative growth. In some cases, *ecotypic differences* have been observed; in spruce of subarctic origin, formation of the terminal bud, and thus the end of extension growth for a season, is induced when the days become shorter than the critical length of 20 h, as opposed to 14 hours in spruce originating in central Europe [49]. The northern photoperiodic ecotype of *Liquidambar styraciflua* in North America [275] is geared to a shorter growing season than the corresponding southern ecotype.

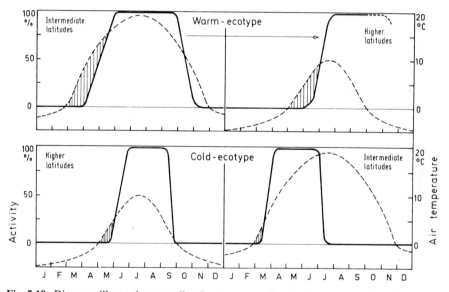

Fig. 5.18. Diagram illustrating coordination between climatic rhythm and the rhythm of growth in trees. The *left half* of the figure shows the synchronization of temperature (*dashed curve*) and physiological activity (*solid curve*) over the course of the year. The *right half* of the figure illustrates the effect of lack of appropriate timing. In warm-climate ecotypes introduced into higher latitudes (*upper right*), a considerable rise in temperature is necessary to break dormancy (wide-spaced shading): spring activation is delayed, and new growth cannot mature during the short growing season in the North. In cold-climate ecotypes transplanted to warmer latitudes (*lower right*) a slight rise in temperature is sufficient to break dormancy (*close-spaced shading*): the plants resume activity prematurely and thus are endangered by spring frosts; growth — which is adjusted to a short growing season — ends before full use can made of the favourable season. From transplanting experiments by Langlet as cited by Bünning (1953). For day length and temperature responses in northern deciduous trees: Heide (1993). For typology of dormancy patterns and predictive models for trees of boreal and temperate regions, see Hänninen (1990)

In cultivated strawberries the transition from the phase of asexual reproduction (by runners) to that of sexual reproduction (flowers) is controlled by temperature and day length; southern varieties cultivated in northern regions propagate by runners for a considerable time, and only late in the season produce a few flowers, whereas northern varieties at low latitudes flower too early and send out only a few runners [229].

A plant species, variety, or ecotype is well acclimatized if the growing season is utilized to the full, without risk of injury in the approaching unfavourable season. In woody plants this is ensured by the coupling of developmental processes with the level of frost resistance. Poorly adapted plants may sprout too late, continue to develop too slowly, and be damaged by the first winter frosts; conversely, the situation would be equally disadvantageous if they began to grow too early in the year (risking injury from late frosts) and stopped development too soon to make full use of the favourable season (Fig. 5.18). Poor synchronization between periods of plant activity and the rhythmicity of the climate thus restricts the spread of a species; such maladaptations can be overcome in evolution by the differentiation of ecotypes.

5.3.3 Winter Dormancy in Cold Regions

Woody plants adapt to winter cold by periodic changes in the state of the protoplasm, in metabolic activity, in developmental processes and in low-temperature resistance (Fig. 5.19).

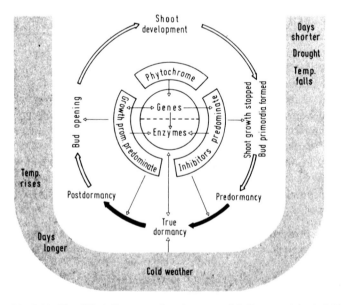

Fig. 5.19. Simplified diagram of environmental influences (*shaded U*) and endogenous interactions affecting the seasonal alternation of vegetative activity and dormancy in woody plants. (Original)

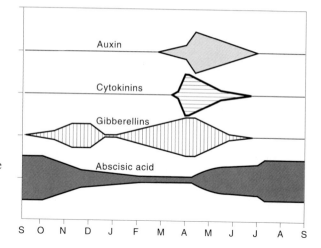

Fig. 5.20. Levels of phytohormones during the course of the year in fruit trees of the temperate zone. (After Luckwill et al., from Seeley 1990)

Towards the end of the summer, lateral buds are formed in the axils of the leaves, and the tips of the shoots become transformed into winter buds, or wither and die off. Before the leaves begin to turn colour, the buds are rendered dormant by hormone action (*correlated bud dormancy*) which prevents their opening if the weather should turn warm before the onset of winter (Fig. 5.20). Other parts of the plant also enter *winter dormancy*; the cambium, for example, and the other tissues of the shoot, can thus become hardened to frost and dehydration. Gene activity is selectively suppressed, translation processes are inhibited and mitotic activity in the meristems is greatly reduced or stops altogether. The nuclei remain in the G1 phase of the cell cycle, in readiness for the reduplication of DNA that occurs towards the end of winter. In this state the genome should be best protected against the effects of low temperature.

The transition to winter dormancy is also evident in the compartmentation and ultrastructure of the cells; for example, the endoplasmatic membrane systems and the plasmodesmata are condensed (protoplast isolation), the mitochondria become smaller, the thylakoid structures in the chloroplasts of some plants and tissues (cortical parenchyma in particular) are reduced, and vacuoles disperse into smaller units (Fig. 5.21). Metabolic activity declines, and the enzyme patterns and activities change (Fig. 5.22). This adjustment to the winter state is not an abrupt event, but takes place gradually, some alterations appearing earlier and others later; also, the process of transformation is not synchronized throughout the plant.

Dormancy is a temporary, endogenously regulated and environmentally influenced interruption of growth. Dormancy exists in many forms and degrees, depending on the growth form of the plant, the organ or tissue considered (usually buds and meristems) and the factors by which it is elicited (intracellular regulation, hormonal coordination, external factors). The typical series of events leading to winter dormancy begins with *predormancy,* which is initiated by decreasing day length and lower temperatures (around and below

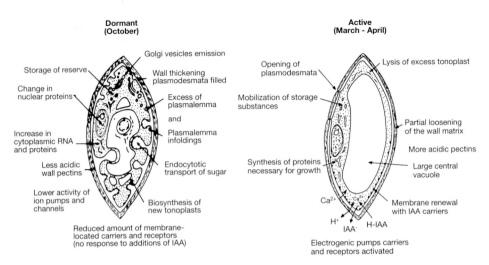

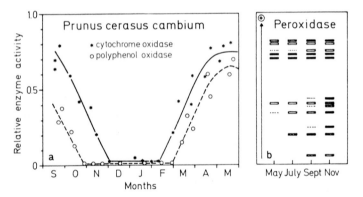

Fig. 5.21. Cytological differences between dormant and active cambium cells in woody plants of the temperate zone. After Lachaud (1989)

Fig. 5.22 a, b. Enzyme activities and enzyme patterns during the transition from vegetative activity to winter dormancy. **a** Relative activity of cytochrome oxidase and polyphenol oxidases in the cambium of cherry twigs in autumn, winter and spring. (Meyer 1968). **b** Seasonal changes in the isoenzyme pattern of peroxidases in spruce needles. (Esterbauer et al. 1978)

+5 °C). In intermediate and northern latitudes, complete dormancy (*true dormancy* or *endodormancy*) is attained by November or December (Fig. 5.23). At this point, the plants can no longer be activated by temporary warming. The inability of plants to emerge prematurely from the resting state once they have entered true dormancy is an important ecological factor in resistance, in view of the unpredictability of the weather. If plants of cold-winter regions were to respond to mild days in midwinter, they would suffer injury in the next cold wave.

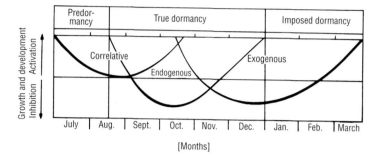

Fig. 5.23. Schematic representation of the progression of dormancy in woody plants of temperate regions. (Saure 1985). For further classifications of dormancy (para-, endo- and ecodormancy) see Lang et al. (1987)

In many woody plants the *termination of dormancy* depends on the fulfillment of certain cold requirements. In most cases, dormancy can only end after exposure for a number of weeks (up to 5 weeks for almond trees, up to 8 weeks for apple and pear trees, up to 10 weeks for cherry trees), to temperatures of 2 to 7 °C [210]. At the end of dormancy there is an increase in the concentration of the phytohormones (first a rise in GA, then IAA and CK; see Fig. 5.20) responsible for promoting the activity of genes and enzymes. Basal metabolism, mobilization of reserves and biosyntheses are set in motion, and cell division gradually begins. Once the plant is thus ready to develop, the appearance of new shoots and leaves is prevented only by bad weather conditions, mainly by cold (*imposed dormancy* or *ecodormancy*). With warmth, and the increasing hours of daylight, development proceeds rapidly. At intermediate latitudes, the endogenous dormancy ends at approximately the time of the winter solstice, after which initiation of growth is determined by weather conditions.

Dormancy also occurs in some *herbaceous* plants. Particularly *geophytes* overwinter in a state of dormancy, from which they emerge only after many weeks to months. Many of these geophytes require cold to accelerate their release from dormancy. Most perennial herbs and grasses do not enter an endogenously controlled state of dormancy; instead, in the event of severe cold they temporarily experience an enforced interruption of growth imposed by the environment. Such plants resume activity as soon as the temperature rises.

5.3.4 Phenology: an Indicator of Weather Characteristics and Changes of Climate

The onset and duration of particular phases of development vary from year to year, depending on the weather. Early in history, people living in close contact with nature began making observations about the time of occurrence of important events such as the unfolding of leaves, flowering and leaf senescence. Much of the traditional wisdom of farmers comes from a capacity for sharp observation and a deep insight into the relationship between weather phenomena and the development of the vegetation. In fact, a phenological calendar existed in China more than 2000 years ago. The time at which the cherry trees begin to flower has been recorded in Kyoto since 750 A.D., and since 1736 phenological data have been documented by succeeding generations in England [136]. Old series of phenological observations are attracting ever-increasing interest as important sources for historical studies of climate [6].

The correlative approach to phenology combines the questions of applied botany with those of meteorology. It is based on the time of onset and the duration of visible changes in the life cycle of plants and seeks statistical connections between climatic factors and defined stages of development of certain indicator plants. In the course of studies of this kind it has become clear that such correlations are valid for only a limited region and for the same type of weather pattern. It is no easy task to uncover the factors responsible for triggering phenological phenomena, because the effective impulse, such as the

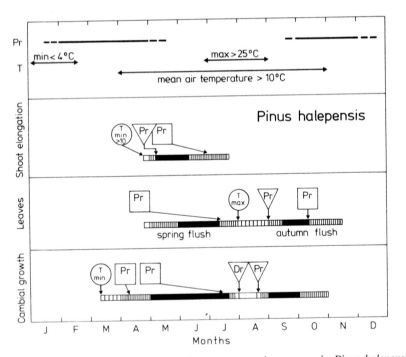

Fig. 5.24. The influence of climatic factors on growth processes in *Pinus halepensis* in the northern mediterranean region. *T* Temperature; *Pr* precipitation; *Dr* drought. *Triangles* indicate events with triggering effects (e.g. thunderstorms); the *squares*, amount-dependent effects (e.g. supply of water in the soil); *black bars* periods of greatest activity. The chief factors determining the time course and the amount of growth are the average minimum temperature in winter and spring, the quantity of rain during the main growth period, heat and dryness in summer, and the onset of equinoctial rains in autumn. The previous year's weather (not shown in the figure) also affects development. (Simplified from Serre 1976a, b)

passing of a temperature threshold, is modified by a large number of internal and external conditions; even the weather and the production yield of the previous year have their after-effects (Fig. 5.24). A means of obtaining suitable material for carrying out analytical studies of phenology would be provided by the establishment of phenological gardens throughout a continent, where clonal progeny of informative indicator plants (e.g. certain phenoecotypes of spruce, beech, oak, various species of polar and willow, linden 'and Rosaceae) could be observed over periods of years.

5.3.5.1 Phenophases and Phenological Dates

Knowledge of phenology, even today, is based on the observation of externally visible stages of development (*phenophases*), such as the germination of seed, bud burst, unfolding of leaves, flowering, leaf discoloration, and senescence of herbs. The observations would gain considerably in value if, in addition to the external events, also anatomical differentiation and morphogenesis (Fig. 5.25), as well as histochemical (e.g. storage patterns) and biochemical criteria (enzyme activities) could be taken into consideration.

The compilation of phenological dates provides ecologically valuable information about the average duration of the different phenophases of the characteristic species in an area, and about local and weather-determined differences in the dates of onset of these phases. The rhythmicity of different plant communities can be presented in the form of *symphenological diagrams*

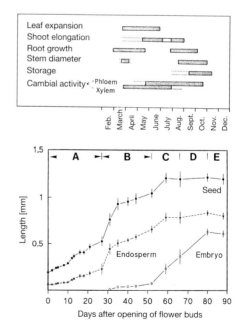

Fig. 5.25. The time-courses of different processes involved in growth and development. *Above* Periods of growth and storage in trees of the temperate zone (Wardlaw 1990). *Below* Seed development of the arctic-alpine cushion plant *Saxifraga oppositifolia. A* Zygote stage; *B* early embryonic development and rapid growth of endosperm; *C* formation of cotyledons, endosperm development almost complete; *D* elongation of the embryonic shoot; seed is ripe. (Wagner and Tengg 1993)

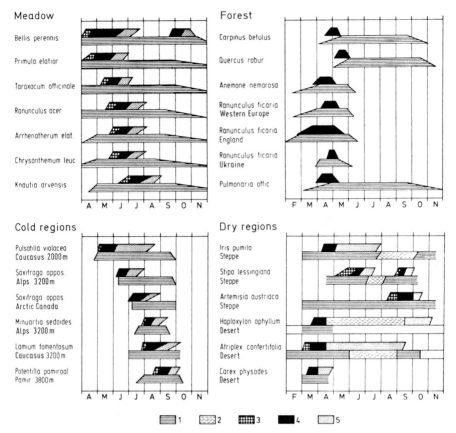

Fig. 5.26. Phenological diagrams for meadow plants (*Arrhenatheretum* in Poland), for trees and herbs of an oak mixed forest in northwest Germany (for comparison: *Ranunculus ficaria* in England and Russia), for plants from regions where the growing season is limited by cold (high mountains, arctic) and by dryness (steppe, desert). The strip diagrams of Schennikow (1932) are interpreted as follows: *1* growing season, the foliated period in the case of trees; *2* drought dormancy; *3* flower buds visible; *4* flowering period; *5* ripe fruits and seed scattering. (After diagrams and data of Ackerman and Bamberg, as cited in Lieth 1974; Borissovaya, as cited in Walter and Breckle 1986; Ellenberg 1939; Nakhutsrishvili and Gamtsemlidze 1984; Jankowska, as cited in Lieth 1970; Michelson and Togyzaev, as cited in Voznesenskii 1977; Moser et al. 1977; Salisbury 1916; Shalyt as cited in Beideman 1974; Shteshtenko 1969; Svoboda 1977)

(Fig. 5.26), showing the dates of the growing season and the component phenological stages. Large-scale events, such as leaf emergence ("green wave"), the ripening of grain, and the changing of leaf colour over entire continents of terrain can be recognized by spectral reflection analyses of the plant cover employing remote sensing. The spatial distribution of one and the same phenological phase (*isophanes*) and the local duration of growing season can then be represented by phenological maps (Fig. 5.27) or by altitudinal profiles (Fig. 5.28).

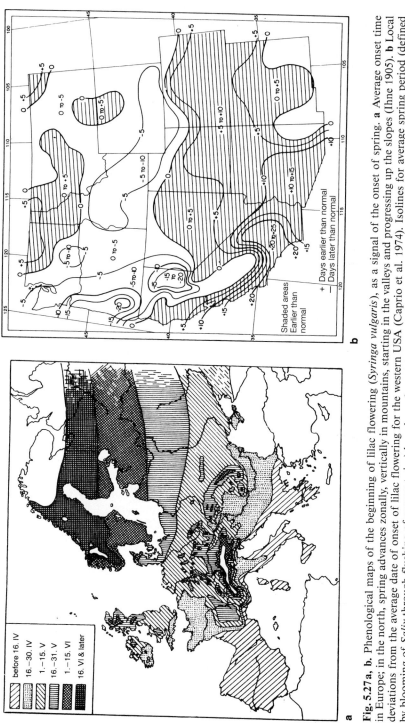

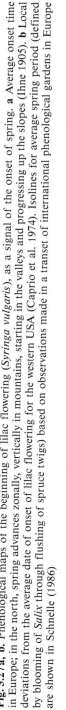

Fig. 5.27a, b. Phenological maps of the beginning of lilac flowering (*Syringa vulgaris*), as a signal of the onset of spring. **a** Average onset time in Europe; in the north, spring advances zonally, vertically in mountains, starting in the valleys and progressing up the slopes (Ihne 1905). **b** Local deviations from the average date of onset of lilac flowering for the western USA (Caprio et al. 1974). Isolines for average spring period (defined by blooming of *Salix* through flushing of spruce twigs) based on observations made in a transect of international phenological gardens in Europe are shown in Schnelle (1986)

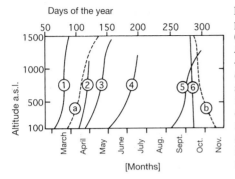

Days of the year

Fig. 5.28. Climatic effect of elevation on phenological dates in the Austrian Alps (see also Fig. 1.37). *1 Tussilago* in flower; *2* sweet cherry in flower; *3* lilac in flower; *4* cherries ripe; *5 Aesculus* fruits ripe; *6 Fagus* leaves turn colour; *a* thaw of soil surface in spring; *b* beginning of soil freezing in autumn. A considerable difference in phenological dates with increasing altitude shows that temperature plays a decisive role, whereas a small difference indicates that photoperiod is the dominant factor. (After Roller 1963)

The sequence of phenophases in the temperate zone: the time of onset of phenophases in the *first half of the year* depends primarily on certain *temperature thresholds.* This can be shown by comparing the temperature distribution over a certain terrain with the phenological dates. The opening of the buds, sprouting, the onset of flowering in trees and shrubs and the germination of seeds are possible only after the temperatures of both air and soil regularly exceed a threshold value characteristic for each stage. In general, the temperature threshold for the opening of the buds and flowering is 6–10 °C, although in spring flowering and mountain plants it is lower (0–6 °C), and in late-flowering plants higher (in many cycloporous trees between 10 and 15 °C). Poplar, birch, and some species of conifer begin shoot extension at just above 0 °C [213]. Flushing and flowering, however, can be elicited by warmth only if the plants are already in a state of readiness to develop, that is, if they have emerged from winter dormancy.

Phenological dates falling in the *second half of the year,* such as those for ripening of fruit, colour change of leaves, leaf fall, and the times for harvesting crops, can be affected by all those environmental conditions that delay or accelerate the processes of maturation and ageing. Once again, temperature is of the greatest significance, but in this case primarily with respect to its role in enhancing production. Thus threshold temperatures are of less concern than the *heat sum*, i.e. the time integral of average daily temperatures (degree-days [18, 89]) exceeding the specific temperature threshold levels. Other decisive factors are the supplies of nutrients and water and, above all, the influence of the diurnal photoperiod on times of flowering, leaf fall, and the onset of winter dormancy. In some species the plant is made ready for colour change and leaf fall by the decreasing day length (for example in birch, poplar, willow, beech, oak and maple). These closing phases of the phenological calendar become visible as soon as temperatures drop below threshold values between 5 and 10 °C.

Phenological events in the tropics and subtropics: in those parts of the tropics and subtropics where *rainy and dry seasons* occur, the phenophases are linked to this *hydroperiodic* alternation. The rainy season is the main growth period. In the dry season grasses and herbaceous plants dry up, deciduous trees shed their leaves, and in the first half of this period evergreen trees lose a large portion of their older leaves.

Even in the *humid* tropical regions with abundant rain throughout the year there are phenological events, but these are less conspicuous than those in regions with a marked seasonal rhythmicity of climate. In the equatorial zone itself a number of climatic factors change in the course of the year (Fig. 5.29). In some tropical countries these variations occur at irregular times and are only slight, while in others they are both considerable and predictable. This is well illustrated in the equatorial zone, in which rainfall is normally plentiful, but where dry periods occur regularly. It would be wrong to conclude from the almost constant mean monthly temperatures over the year, that the air temperature is always the same. The slight variability in these monthly averages masks considerable short-term fluctuation. Even the very slight change in day length over the year affects the process of development because tropical plants respond to even extremely weak photoperiodic stimuli.

Phenological events in evergreen forests of the wet tropics can only be described *statistically*, by noting when they occur in a greater than average number of species and individuals. The production of new shoots, unfolding of leaves and elongation growth often culminate near the equinoxes; that is, many tropical trees put out new shoots and leaves twice a year (Fig. 5.30). Leaf fall occurs throughout the year in tropical forests, but it appears to be greater

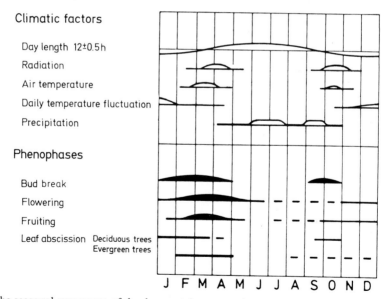

Fig. 5.29. The seasonal occurrence of developmental processes in trees of evergreen tropical forests in Ghana (6°N), and the changes in climatic factors over the year. At the equinoxes there is a striking concentration of bud opening, of abscission, flowering, and fruit formation. (After Longman and Jenik 1987). For induction of bud break by rainfall or irrigation in deciduous trees of tropical dry forests, see Borchert (1994)

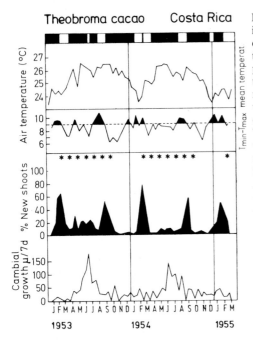

Fig. 5.30. Time courses of shoot flushing, cambial growth and flowering of cocoa trees in Costa Rica (10°N), with the changes in air temperature and in precipitation (*black regions* in the bar at the top of diagram indicate periods with more than 25 mm precipitation per week). Shoot growth is given as the percentage of branches with new shoots. The measure of cambial activity is the weekly increase in stem diameter. The time during which the trees are in flower is marked by *asterisks*. Shoot growth is greater at times when the diurnal fluctuation in temperature is greater, whereas cambial activity is promoted by high average temperatures; during the main elongation phase of the new shoots cambial activity is slowed. Precipitation is sufficient throughout the year and thus is not a controlling factor in this case. (After Alvim 1964). For periodicity of cambial growth of trees in a tropical wet forest monitored with dendrometers see Breitsprecher and Bethel (1990)

in connection with soil drying and with decreasing day length. Tropical plants can be classified according to the periodicity of flowering, into ever-blooming (e.g. *Hibiscus, Heliconia, Cocos, Carica papaya*), seasonally blooming (e.g. *Cassia fistula, Spathodea, Lagerstroemia*), and occasionally blooming species (e.g. *Dendrobium* species); an additional category comprises those species flowering at intervals of several years (e.g. species of bamboo). Characteristic for forests of the wet tropics is the absence of a distinct flowering season; there

Fig. 5.31. Seasonal patterns of fructification (trees bearing fruit) of the various growth forms in a semiarid deciduous forest in Panama. (After Foster, from Howe and Westley 1986). For seasonal patterns of fruiting in tropical, temperate and semiarid woodlands see Jordano (1992)

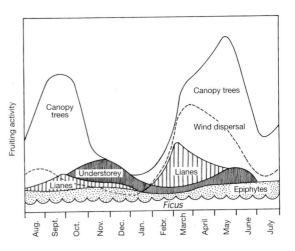

are always trees in flower, although the time of flowering can differ within a species or even from branch to branch on one and the same tree. The nearest approach to a flowering season is seen in regions with regularly occurring dry periods, where hydroperiodic induction [4] of flowering has been demonstrated for certain trees. Fruits can be found in growing and ripening stages throughout the year in tropical forests (Fig. 5.31), although fruit-bearing trees are more likely to be encountered in dry periods than in the rainy seasons.

5.3.4.2 Dendrochronology: Annual Growth Rings and Variability of Climate

A useful quantitative phenological method is provided by *annual ring phenometry*. The duration of cambial activity and the type of wood formed (the differentiation into early or late wood), are affected by environmental factors. In general, the formation of spring wood is promoted by all those factors that favour bud burst and elongation of the new shoots. The spring wood indicates not only the growth conditions prevailing in spring, it also indicates the nutritional status in the preceding year. All factors tending to slow down shoot growth and accelerate senescence of the foliage lead to differentiation of late wood (Fig. 5.32). The width of the annual rings, the thickness of the cell walls and the density of the cell wall structures in autumn wood depend on the supply of nutrients, and are thus an indication of the dry matter production of that year. Other factors exerting direct or indirect influences on the width and

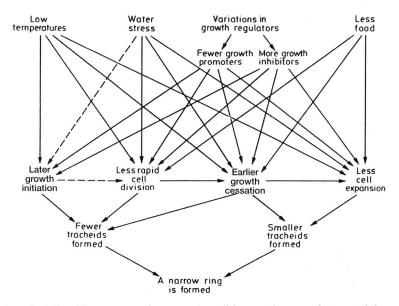

Fig. 5.32. Causal relationships among environmental conditions, endogenous factors and the growth of annual rings. (After Fritts 1976)

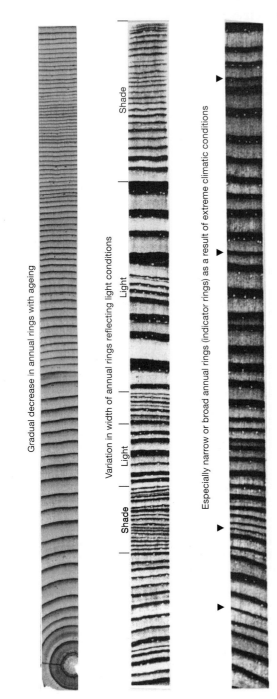

Gradual decrease in annual rings with ageing

Variation in width of annual rings reflecting light conditions

Shade | Light | Light | Shade

Especially narrow or broad annual rings (indicator rings) as a result of extreme climatic conditions

Fig. 5.33. Typical changes in the widths of the annual rings in the wood of conifer trunks as a response to both inherent developmental patterns and environmental conditions. (Schweingruber 1983; Schweingruber et al. 1983)

appearance of the annual rings (Fig. 5.33) include radiation, temperature, availability of mineral nutrients, water supply, and duration of the photoperiod, as well as all kinds of harmful environmental influences. Examples of the latter are attack by parasites, consumption by animals, excessive heat or frost, and absorption of pollutants. In cases where the particular influences to which the cambium is most subject are known in detail for a given species, the structure of the annual rings provides an important historical document of the growth-determining events of past years, and this in turn gives an idea of the climate and its extremes during the same period (Fig. 5.34).

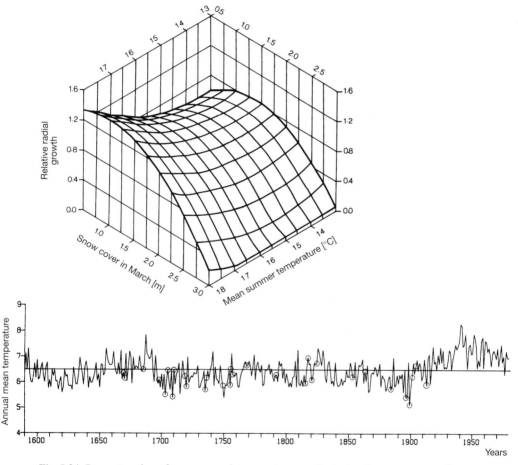

Fig. 5.34. Reconstruction of mean annual temperatures on the basis of measurements of annual rings of *Tsuga mertensiana* at the forest limit in the Cascade Mountains (Mount Rainier). *Above* Diagram linking the growth of annual rings, mean summer temperature (July to September) and the depth of snow in March. Relative radial growth indices are dimensionless units with mean equal to 1. *Below* Estimated mean annual temperatures for 1590–1913; the average is 6.3 °C (*line*). (Graumlich and Brubaker 1986)

6 Plants Under Stress

Wherever they grow, plants are subject to a great variety of stresses tending to restrict their chances of development and survival. There are large regions of the earth, such as the arid zones, regions with salty soils, the arctic and antarctic, and high mountains, where conditions favouring plant growth, if they occur at all, are of short duration. One tenth of the land surface has been transformed in such a way that many wild plants have been either exterminated or supplanted. Even in places where conditions are constantly favourable for the majority of the plants, the very exuberance of the vegetation represents a disadvantage for some members of the community. A good example of this is seen in dense forests, where the extreme deficiency of light beneath the crowns inhibits the undergrowth. In any closed stand of plants the weaker species and individuals in the competition for space are suppressed. Moreover, dense phytomass is particularly attractive to parasites and fungal diseases. The manifold unfavourable but not necessarily immediately lethal conditions occurring either permanently or sporadically in a locality are generally known as "stresses".

6.1 Stress: Disturbance and Syndrome

6.1.1 What is Stress?

Stress is in most definitions considered to be a significant deviation from the conditions optimal for life, and eliciting changes and responses at all functional levels of the organism which, although at first reversible, may also become permanent. Even if the stressful event is merely temporary, the vitality of the plant becomes weaker the longer the stress is maintained (Fig. 6.1). When the limit of the plant's ability to adjust is reached, hitherto latent damage develops into chronic disease or irreversible injury.

The literal meaning of the word "stress" is a constraining (derived from the Latin *stringere*) or impelling force. Physicists employ the term "stress" to denote the tension (with the dimension Pa) produced within a body by the action of an external force (with the dimension N). The resulting extension or compression, or relative difference in length is termed *strain*. The ratio of tension to strain (i.e. the elasticity module) and the transition from elastic (reversible) to plastic (irreversible) deformability are characteristic properties of a material.

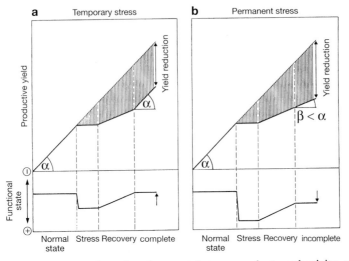

Fig. 6.1 a, b. The effect of environmental stress on plant productivity. **a** Temporary stress; **b** permanent stress. Brief exposure to stress not exceeding the plant's limits of tolerance causes only temporary changes in physiological processes, whereas prolonged exposure results in permanent disruption. After recovery from reversible disturbances the increase in dry matter returns to its previous rate (angle of slope α), but in the case of chronic stress the growth rate is permanently lowered (angle $\beta < \alpha$), and production losses are thus greater. (After Härtel 1976, from Larcher 1981)

In biology also, the "stress-strain" terminology is frequently employed in this sense [272].

A departure from the physical terminology took place as the term "stress" was gradually adopted in everday usage to denote everything connected with stressful situations, in fact to indicate the *event* as well as the *state* evoked within the organism. In order to avoid misunderstanding, the meaning of the term employed must in every case be unambiguous: *stress factor* (or *stressor*) indicates the stress stimulus, and *stress response* or *state of stress* denotes the response to the stimulus as well as the ensuing state of adaptation.

In recent decades the study of stress has gained increasing importance in many fields of biology, and a wide variety of approaches have been employed in the attempt to gain more insight into the processes involved.

The *stimulus-oriented* approach [140] is aimed at explaining the specific mechanisms of the stress response. Stress is regarded as a directional event, elicited by highly specific factors; investigations are oriented toward understanding the way in which the stress factor produces its effect. Often, the external factor does not reach the ultimate site of the stress reaction (the protoplasm) immediately or in its original intensity, because plants possess a variety of protective mechanisms to delay or even prevent disruption of the thermodynamic or chemical equilibrium between environment and cell interior. Only those components of the original disturbance that elicit a state of stress in the protoplasm need to be tolerated. Here also, within the protoplasm, processes can intervene to prevent or avoid a disturbance. All such protective and buffer-ing resistance mechanisms are aimed at *"avoidance"* of stress, whereas the ability of the protoplasm to resist stress represents *"tolerance"* (for example of salt resistance see Fig. 6.73). The insight into the nature of factor-specific mechanisms yielded by the analytical approach to stress effects and the pro-cesses leading to the acquisition of resistance are of particular value for molec-ular-biological, ecophysiological and applied studies.

Alternatively, stress can be regarded as a *functional state* or, in other words, as the dynamic response of the whole organism [219]. Studies are not confined to stressor-specific responses but are also concerned with non-specific effects. The reaction pattern evoked in response to stress is the characteristic *stress syndrome*, and also includes the processes leading to resistance. Such processes are of a homeostatic nature, aimed at normalizing the plant's vital functions and raising its powers of resistance. The *stress response* is a race between the effort to adapt and the potentially lethal processes in the protoplasm. Thus the *dynamics of stress* comprises a destabilizing, destructive component ("dis-tress"), as well as countermeasures promoting restabilization and resistance ("eustress"). Constraint, adaptation and resistance are interconnected parts of the whole event. The relative success of the harmful and protective reactions determines whether stress causes only slight and temporary deviations from the normal state, or severe and permanent injuries (see Fig. 6.34).

6.1.2 What Happens During Stress?

According to the dynamic concept of stress, the organism under stress passes through a succession of characteristic phases (Fig. 6.2).

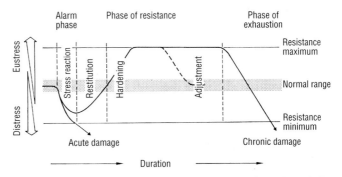

Fig. 6.2. A phase model of stress events and responses, after Selye (1936) and Stocker (1947). The impact of stress factors destabilizes vital structures and functions, inducing an "alarm phase", in which there are functional declines (*stress reaction*); these are offset by counter-reactions (*restitution*), which may lead to over-compensation (*hardening*). Under prolonged exposure to a constant stress, a higher degree of resistance is developed and this may result in restabilization (*adjustment*). If the organism is overtaxed either by acute or chronic stress (*exhaustion*), irreversible damage occurs. (After Larcher 1987; Arndt et al. 1987; Tesche 1989)

The *alarm phase:* the onset of a disturbance is followed by destabilization of the structural (e.g. proteins, biomembranes) and functional (biochemical processes, energy metabolism) conditions required for the normal pursuance of vital activities. Too rapid an intensification of the impairment results in acute collapse of cell integrity before defensive measures can take effect. The alarm phase begins with a *stress reaction* in which catabolism predominates over anabolism. If the intensity of the stimulus remains unchanged, *restitution* in the form of repair processes such as protein syntheses or de novo synthesis of protective substances is quickly initiated. This leads on to the *resistance phase,* in which, under continuing stress, the resistance increases (*hardening*). Due to the resulting improvement in stability, normalization can take place despite continued stress (*adaptation*). Resistance may remain elevated for some time after the disturbance has ceased.

If the state of stress lasts too long, or if the intensity of the stress factor increases, a *state of exhaustion* may set in during the *end phase,* thus rendering the plant susceptible to infections that occur as a consequence of impaired host defences (e.g. parasites causing opportunistic infections) and leading to premature collapse. However, if the impairment was merely temporary, the functional state is restored to its original level. If necessary, any damage incurred may be repaired in a *phase of regeneration.*

To sum up: from the botanical point of view, stress can be described as a state in which increasing demands made upon a plant lead to an initial destabilization of functions, followed by normalization and improved resis-

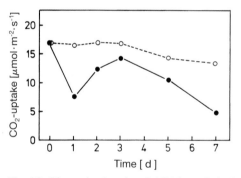

Fig. 6.3. Phases in the photoinhibition of shade-adapted *Oxyria digyna* after transfer to strong light. After measurement of their photosynthetic capacity, plants grown in weak light (100 μmol photons $m^{-2}s^{-1}$) were transferred to strong light (800 μmol photons $m^{-2}s^{-1}$), care being taken to avoid overheating and desiccation. The CO_2 uptake of the plants subjected to the stress of strong light (●) and of those left in weak light (○) was measured for short periods each day at 1000 μmol photons $m^{-2}s^{-1}$ and 20 °C. Under stress conditions the CO_2 uptake dropped sharply during the reaction phase (day 1), but recovered to some extent on the 2nd and 3rd days, in the phase of restitution. Under continuing stress, photosynthesis gradually declined in the phase of exhaustion. After 5 days the leaves in strong light began to turn yellow whereas those of the control plants in weak light were still green. (After Engel et al. 1986)

tance. If the limits of tolerance are exceeded and the adaptive capacity is over-taxed, permanent damge or even death many result [135].

The phase concept of the stress syndrome described above is still an abstract attempt to show up possible sequences and trends. It should not therefore be expected that the phenomena and trends discussed will occur in response to every stress. Nevertheless, sufficient evidence is available to justify the assumption of a sequence of phases (Fig. 6.3). Perhaps the most significant conclusion to be drawn from all these observations is that stress events are always *time-dependent*. Processes taking place under stress must therefore be considered to be subject to continuous change, which means that the laws governing the dynamics of equilibrium are not always applicable.

6.1.3 How to Recognize Stress

Organisms respond differently to a particular stressor, depending on their genetically determined *"reaction norm"*. Furthermore, the nature and intensity of the response of individual plants to a particular stress factor may vary considerably, depending upon age, degree of adaptation, and on seasonal and even diurnal activity. Although there is often good correlation between intensity of the stress factor and the elicited response, it cannot be assumed that the degree of impairment suffered by a plant is proportional to the strength of the stress factor. Nor need all stress-induced changes in the plant be either unequivocally harmful or protective. The question of whether a plant suffers stress in a given situation can only be answered by comparison with its normal behaviour.

Abnormal demands upon a plant can be recognized by a variety of *symptoms* or visible indications of a state of destabilization. Repair and resistance mechanisms also constitute **stress criteria**. A clear distinction between destructive and constructive processes is difficult to achieve; not only do they overlap, they may also occur simultaneously.

Stressor-specific effects, as a rule, involve a well-defined target within the plant. Intense radiation, for example, causes direct damage to the thylakoid membranes, or again, toxic concentrations of ions and heavy metals have an immediate effect on enzyme proteins. Consequently, in each case the *symptoms* evoked are highly specific. Specific mechanisms of resistance involve all functional levels; in many cases they are elicited by differential gene activation, as in the synthesis of stress proteins and special isoenzymes (Fig. 6.4).

Also characteristic for a state of stress are *non-specific* manifestations, which are primarily an expression of the *degree of severity* of the disturbance. A process is said to be non-specific if it follows a stereotypic pattern, whatever the nature of the stress factor. Examples of non-specific indications of a state of stress are changes in enzyme activities (especially of peroxidase, glutathione reductase, dehydroascorbate reductase); biosynthesis of polyamines; de novo synthesis and accumulation of antioxidants (ascorbic acid, tocopherol), of stress metabolites and compatible osmotically active substances (proline, betaine, polyols), as well as numerous secondary plant substances (polyphenols,

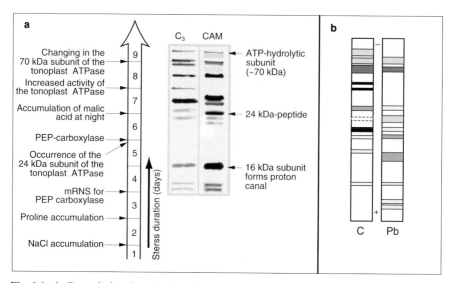

Fig. 6.4 a, b. Stress-induced synthesis of polypeptides and isoenzymes. **a** Sequence of occurrence of proteins of crassulacean acid metabolism in *Mesembryanthemum crystallinum* after the triggering of CAM induction by addition of 400 mM NaCl the root medium. (Heun et al. 1981; Michalowski et al. 1989, from Beck and Lüttge 1990). For a model of stress perception and molecular response in *Mesembryanthemum crystallinum* see Vernon et al. (1993). **b** Electrophoretic pattern of esterase from the roots of *Allium cepa*. *C* Control; *Pb* 10 days after a 5-day contamination with 200 ppm Pb. (Maier 1979)

anthocyanins), but above all the appearance of stress hormones (abscisic acid, jasmonic acid, ethylene). Further non-specific effects of stress are alterations in membrane properties (membrane potential, transport of substances), increased respiration, inhibition of photosynthesis, reduced dry matter production, growth disturbances, lower fertility and premature senescence.

Yet another early indication of a state of stress may be an intracellular decrease in the availability of energy (due to metabolic impairment), or an increase in energy consumption for repair syntheses (Fig. 6.5). Uncoupling of phosphorylation reactions means that less ATP is formed, even if respiration increases. An adenylate energy charge (AEC; [7])

$$AEC = \frac{[ATP] + 0.5\,[ADP]}{[ATP] + [ADP] + [AMP]} \tag{6.1}$$

below 0.6 due to the drop in the ATP/ADP ratio indicates a deterioration in the vitality of a plant. A plant under stress, however, must expend extra energy for maintaining normal functions, or if special metabolic pathways are necessary to regulate the internal milieu, or if breakdown processes set in prematurely. For example, in halophytes, a high energy expenditure to support the functioning of the membrane ATP-ases is required continuously for the transfer of salt ions into the vacuoles. It can therefore be deduced that

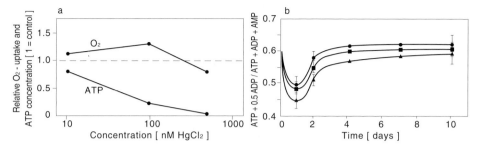

Fig. 6.5 a, b. Cellular energy supplies under heavy-metal stress. **a** Oxygen uptake and ATP production in the tissues of *Beta vulgaris* as functions of the concentration of mercury in the external medium (relative to the controls). (Lüttge et al. 1984). **b** Energy status of the adenylate system of a culture of of algae (*Euglena gracilis*) in continuous light, following the addition of toxic salts in sublethal concentrations. *Curves: top*, 50 μM ZnCl₂; *middle* 0.1 μM CdCl₂; *bottom* 0.01 μM HgCl₂. (De Filippis et al. 1981)

although halophytes are salt-resistant they are at the same time permanently under stress.

Reactions that indicate a state of stress make possible the employment of sensitive plant species as *bioindicators* of environmental stress, or the use of living plants, or parts thereof (e.g. cell or tissue cultures) as *biomonitors*. Caution is necessary, however, in connection with such methods, since many of the early diagnostic indications of stress can be elicited by a wide variety of situations, so that their non-specificity often makes it impossible to draw reliable conclusions about causal factors.

6.1.4 Stress and Plant Life

6.1.4.1 Stress Affects the Entire Organism

Ultimately, every organ of the plant is affected by stress even if only a limited part of the plant was involved initially. Coordination of the stress responses within the plant is carried out by *phytohormones*. As soon as part of a plant experiences a disturbance, a non-specific response is elicited in the form of characteristic changes in the hormone system. These changes bring about short-lived metabolic and longer-lasting morphogenetic processes aimed at minimizing the stress and preserving the life of the plant. Above all, disturbances in the root region, such as water shortage (see Fig. 6.59), or deficiencies of nutrients or oxygen (see Fig. 6.51) lead to adjustments in the distribution of assimilates, in the ratio of shoot to root growth, to premature flowering ("emergency flowering") and leaf abscission. Because the entire organism is affected by stress, caution is advisable in the application of test results obtained with simple systems to the behaviour of whole plants.

6.1.4.2 Tuning of the Organism in Response to Stress

Plants under stress respond to simultaneous or subsequent stressful events, and even to normal environmental influences, in an abnormal manner. Seldom in nature does one stress factor occur alone and uninfluenced by other phenomena. Frequently *multiple stresses* are involved, such as an obligatory combination of strong radiation, overheating and drought in open habitats.

Response to stress depends also upon the *background conditions*, such as the diurnal light rhythm, or the change of seasons. Responses under the variable climatic conditions prevailing in natural surroundings are often much stronger than under controlled laboratory conditions (e.g. proline accumulation in cold-stressed plants). Also the way in which a plant responds to climatic and many other types of stress alters fundamentally with the seasonal growth rhythm.

A combination of stress factors or a series of stressful events may *reinforce*, *weaken*, *mask*, or even *reverse* the response of a plant to a single stress factor. Enhancement of an effect in combination with other factors can often be observed: additional stress factors result in additional disturbances. Chemical strains have an especially strong tendency to mutual enhancement. Repair and resistance mechanisms triggered under stress depend on the particular situation; it is frequently unpredictable, whether the effects elicited with be synergic or antagonistic. Under the influence of frost, woody plants of winter-cold regions not only become freezing tolerant, they also become more resistant to

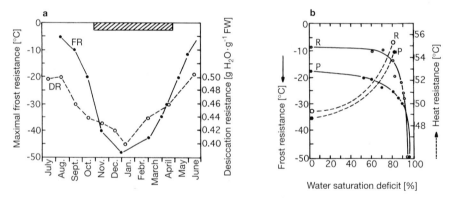

Fig. 6.6 a, b. The coupled acquisition of resistance to desiccation by plant species that develop tolerance to freezing during the course of the winter (see Sect. 6.2.2.3). **a** Synchronous course of the two resistances in needles of *Pinus cembra* growing at the alpine treeline. *FR* Frost resistance: temperature at which 5–10% of the needles exhibit freezing damage; *DR* desiccation resistance: water content of the needles with 1–2% damage to mesophyll; *hatched bar* period during which frost occurs on more than 15 days per month. (After Pisek and Larcher 1954). **b** Frost and heat resistance of the leaves of desiccation-resistant vascular plants; *R: Raymonda myconi* and *P: Polypodium vulgare* during dehydration in the winter condition. The measure of resistance is taken as the temperature that causes 10% damage to the leaves. (After Kappen 1965)

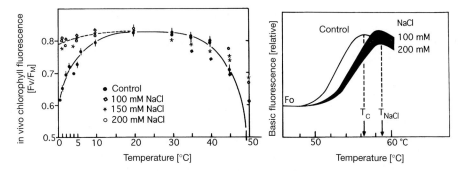

Fig. 6.7. Extension of the temperature limits of photosynthesis in *Vigna unguiculata* induced by salt treatment. *Left* Activity of the primary processes of photosynthesis as a function of temperature, expressed by the fluorescence ratio F_v/F_M, which is correlated with the quantum yield (see Fig. 6.11). For control leaves, the curve shows a marked impairment of photosynthesis below 6 °C and above 45 °C. Previous stress due to NaCl in the nutrient solution increases the stability of the protoplasm both at high and at low temperatures. *Right* Temperature curve of basic fluorescence (F_0) as an indicator for stability of the thylakoid membranes (see also Fig. 6.23). T_C, T_{NaCl} Threshold temperatures for irreversible damage to chloroplasts in the control plants and in salt-stressed plants. (Larcher et al. 1990). Adaptation to drought can also reduce the sensitivity of photosynthesis to heat (Havaux 1992)

desiccation (Fig. 6.6). Salt stress, depending on its intensity, duration and the phase of stress, can contribute toward increasing the stability of a plant to cold, heat and desiccation (Fig. 6.7), or may, on the other hand, lower the frost-hardening capacity.

The *coupled* acquisition of tolerance during adaptation to one factor is often due to the general stabilization of the protoplasm resulting from structural changes in biomembranes and proteins. Structural similarities between the proteins involved in salt stress, heat shock, and drought stress probably constitute the molecular basis of this kind of *cross protection*. Also nonspecific stress reactions such as the accumulation of flavonoids, or the activation of peroxidases, play a variety of protective roles, as against UV radiation, fungal attack and grazing.

6.1.4.3 The Costs of Overcoming Stress

Stress disrupts normal structures and the coordination of various processes at the molecular, cellular and whole organism levels. The restabilizing and reparative counter reactions, the achievement of readjusted, adapted states, and the maintenance of greater powers of resistance all call for additional energy and metabolites.

The "survival strategy" of plants in stress-dominated habitats is thus not directed at maximizing productivity, but rather at achieving a *compromise between yield and survival*. Species that are specialized to life on nutrient-poor, shallow soils, with a tendency to dryness, grow slowly and are often small. This

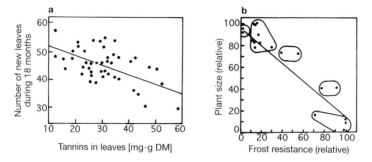

Fig. 6.8 a, b. Reduction of growth in favour of greater resistance. **a** Genotypes of the tropical tree species *Cecropia peltata* that are rich in tannins produce fewer leaves per unit time than those containing little tannin, but are better protected from consumption by insects. (Coley 1986). **b** Seedlings of frost-resistant genotypes of *Pseudotsuga menziesii* grow more slowly than those of frost-sensitive genotypes. (Braun and Scheumann 1989)

enables them to maintain adequate concentrations of mineral elements and a sufficiently high tissue water potential in spite of inadequate nutrients and water. Some partly lignified plants of arid regions (cacti, baobab) lower their expenditure for structural support by reducing the amount of lignified tissues and by a peripheral arrangement of supporting elements. Protection against herbivores and parasites, and the replacement of the plant parts removed by the animals also require energy (Fig. 6.8 a). The specific costs can be computed

Table 6.1. Construction costs (glucose units) for deterring grazing and renewing plant parts. (From calculations of Merino et al. 1984; Gulmon and Mooney 1986; Williams et al. 1987; Diamantoglou et al. 1989; Kull et al. 1992; unpub. data of U. Kull, pers. comm.)

Plant substance Plant part	Construction cost (g glucose/g dry matter)
Defence substances	
Tannins	1.55 – 1.6
Cyanogenic glycosides	1.9 – 2.1
Alkaloids	2.8 – 3.3
Monoterpenoids	2.8 – 3.5
Latex	3.3
Cell-wall substances	
Lignin (conifer wood)	2.44 – 2.49
Lignin (angiosperm wood)	2.48 – 2.52
Cost of replacing organs	
Soft leaves	
low in defence substances	1.3
rich in defence substances	1.8
Sclerophyllous leaves	1.35 – 1.55
Conifer needles	ca. 1.5
Shoots non-lignified	1.1 – 1.35
Shoots lignified	1.4 – 1.55

in $g\,CO_2$ required per g defence substance or material needed to replace the lost biomass. Metabolic costs for the biosynthesis of secondary substances can be quite considerable as well (Table 6.1). Also, the increase in resistance needed to survive unfavourable climatic seasons is often acquired at the cost of biomass production, growth and reproductive efficiency. This is because the recurrent transition to a resting period, a basic requirement for the development of maximum freezing tolerance and resistance to desiccation, shortens the period of time available for carbon assimilation and thus reduces the increase in biomass (Fig. 6.8 b).

6.1.4.4 Survival of Stress

The persistence of a species in sites endangered by stress is more likely if *stress prevention* is possible, if the *stress resistance* of the most vulnerable and indispensable part of the plant is adequate, and if the *recovery capacity* is sufficient for complete repair of any injuries incurred.

Survival = stress evasion, resistance, recovery   (6.2)

Stress can be excluded *spatially* by withdrawal to sheltered positions (e.g. underground organs), and *temporally* by exploiting the favourable seasons for growth (the rainy-season ephemerals, for example). Stress resistance includes both stress reduction (avoidance of stress) and stress tolerance. Recovery from injuries takes place through surviving organs such as the buds and regenerative tissues. On sites frequently exposed to harsh conditions, selection favours plant types that reproduce chiefly by vegetative means.

6.1.4.5 Stress and Evolution

The recognition that a moderate degree of stress contributes towards the healthy functioning and adaptability of the human organism also applies to every branch of biology. Thus in plants as well, exposure to *too little stress results in inadequate powers of defence.* Extreme stress, on the other hand, elicits the greatest degree of hardening. Stress as a constraint and as a stimulant, apart from affecting the individual, also promotes the development of better adapted genotypes, as is very clearly seen along stress gradients. In populations exposed to long-term stress, selection gradually brings about a transition from a *reproductive strategy* (r-strategy) to a *conservative survival strategy* (K-strategy): in abiotically propitious habitats, competitive pressure leads to crowding stress, whereas in marginal habitats the stress is due to abiotic factors. It seems that no place is totally free from stress: moderate stress must be regarded as a normal part of life and not as an exceptional state. Although every demand upon a plant contributes towards setting selective limits, it at the same time promotes the processes involved in performance and survival, and thus ensures the persistence and fitness of the species.

6.2 Natural Environmental Constraints

Environmental stress can be caused by too little or too much energy input, by too rapid or too slow a turnover of substrate, or it may be the result of unsuitable and unusual external influences (Fig. 6.9).

Among the *abiotic* environmental stressors a large number are climatic factors, exerting their effects in the atmosphere, in the soil and under water: high or low radiation stress; excessively high or low temperatures, the latter accompanied by frost, frozen ground, or coverings of snow and ice; deficient

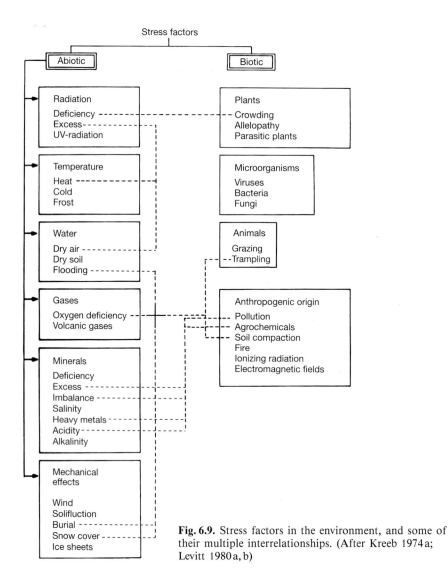

Fig. 6.9. Stress factors in the environment, and some of their multiple interrelationships. (After Kreeb 1974a; Levitt 1980a, b)

precipitation and drought; strong wind. In the soil, plants may be confronted with high concentrations of salt and minerals, or mineral deficiency; excessively acid or alkaline soils are unfavourable and stressful for most plants, and unstable soils, shifting sands and run-off water confront plants with mechanical stress; in dense or flooded soils and at the bottom of some lakes and ponds oxygen is deficient.

Biotic stress is particularly common in dense plant stands and wherever plants are intensively used by animals and microorganisms. In addition to the natural factors causing stress, *human beings* are responsible for introducing physical and mechanical stresses and, above all, chemical pollutants as stressors into the environment. Many of these factors are extremely hazardous, since they represent stresses to which plants are unable to develop any kind of defence mechanism.

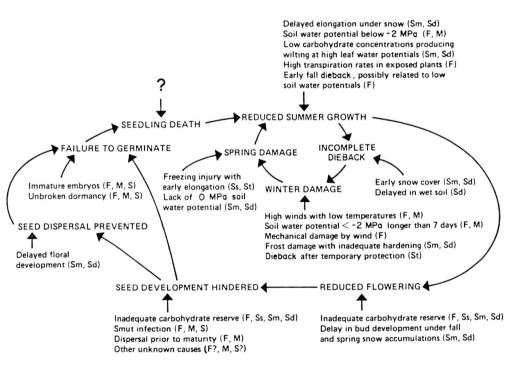

Fig. 6.10. Stress factors limiting the small-scale distribution of the alpine sedge *Kobresia myosuroides*. Microenvironments: *F* fellfield with little or no snow cover, abrasive wind effects, high summer evaporation rates and low summer soil moisture; *M Kobresia* meadow (control site); *Ss* shallow snow accumulation, early snow melt and premature spring growth accompanied by reduced frost resistance; *Sm* moderate snow accumulation; *Sd* deep snow accumulation associated with long duration of snow cover, restriction of the growing season and prevention of frost hardening of the snow-protected plants: *St* temporary snow accumulation dispersing before spring, freezing risk for unhardened plants if snow cover is removed. *S* all conditions connected with snow accumulation. (Bell and Bliss 1980)

Although it is necessary to discuss *individual* stressors in the following sections, this does not mean that they occur singly in nature. In habitats exposed to stress, the interplay of numerous stressors restricts the area on which a particular plant species can survive (Fig. 6.10). This is how climatic and edaphic ecotones and distributional limits (such as moisture-, polar- and altitudinal treelines) arise, and is also the explanation for the existence of islands of vegetation in arid, polar, saline, windy and stony types of desert.

6.2.1 Radiation Stress

Excessive quantities of photosynthetically active radiation and increased absorption of UV radiation produce radiation stress in plants. In both cases photoenergetic processes are involved.

6.2.1.1 Light Stress

The photosynthetic apparatus of plants is capable of optimally efficient absorption and utilization of visible light. However, strong light presents the leaf with more photochemical energy than can be utilized for photosynthesis; overloading of the photosynthetic process results in lower quantum utilization (Fig. 6.11), and a lower assimilation yield (*photoinhibition*). Extremely high irradiance destroys photosynthetic pigments and thylakoid structures (*photodamage*). Photodamage of single chloroplasts in the uppermost layer of the palisade parenchyma appears to be common, and is probably partly responsible for the declining photosynthetic capacity of ageing leaves. In addition, extensive injuries to chloroplast-containing tissues are not uncommon, and are recognizable as colourless patches in areas facing the sun.

Many cryptogams (algae, sciophytic lichens and mosses, some ferns), submersed cormophytes and all shade plants among the phanerogams, whether genetic or adapted, are sensitive to light (*photolabile* [158]) and may be damaged even by brief exposure to moderate irradiation. Phytoplankton respond to intense light by descending to deeper water; algae and seaweeds attached to the ground or to rocks gradually become photoinhibited before noon, so that their carbon assimilation begins to decrease early in the day (Fig. 6.12). Following a series of overcast days, direct sunlight inhibits the photosynthesis of lichens in their hydrated state; plants of the forest undergrowth suffer a *light shock* if they are suddenly exposed (treefall due to wind, or to clearing). Even sunflecks penetrating the crown canopy are a disturbing stimulus for tree saplings. As a rule, the more efficient the trapping of incoming light by the pigment complex, the more sensitive is a plant to strong irradiance.

Plants that grow on open ground, such as in high mountains, deserts, seacoasts, fallow land, and fields before the crop plants are high, are accustomed to strong light and better able to tolerate high levels of irradiance; they are usually *photostable*. However, even in such plants photoinhibition and photo-

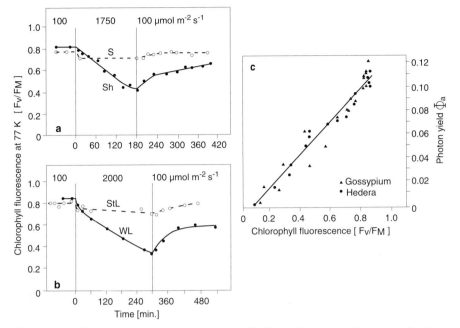

Fig. 6.11a–c. Time courses and extent of photoinhibition of photosynthesis and of subsequent recovery. The in vivo chlorophyll fluorescence (as a sensitive criterion for the functioning of photosystem II) was measured before and during exposure to strong light (1750 and 2000 μmol photons $m^{-2} s^{-1}$) and subsequently under weak light (100 μmol photons $m^{-2} s^{-1}$). **a** Comparison of sun leaves (*S*) and shade leaves (*Sh*) of *Hedera canariensis*. **b** Comparison of leaves of cotton plants grown in strong light (*StL*, at 1000 μmol photons $m^{-2} s^{-1}$) and in weak light (*WL*, 150 μmol photons $m^{-2} s^{-1}$). **c** Relationship between PSII antennal efficiency (i.e. the ratio of the variable and maximal fluorescence, F_V/F_M) and the photon yield (Φ_a) of the photosynthetic O_2 evolution for these two species. (Demmig and Björkman 1987)

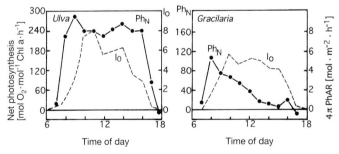

Fig. 6.12. Daily course of photoinhibition of photosynthesis of seaweeds in their natural habitats. Ph_N (*solid line*) Net photosynthesis at 70% of irradiance above water; I_0 (*broken line*) PhAR intensity, referred to the surface area of a sphere. In the green alga *Ulva curvata*, which is accustomed to high intensities of light in the eulittoral and upper sublittoral zones, only a slight depression of photosynthesis occurs at midday; the red alga *Gracilaria foliifera*, on the other hand, which inhabits the lower sublittoral zone is more photosensitive ("photolabile") and its photosynthesis is inhibited at an early hour, and only recovers during the night. (Ramus and Rosenberg 1980)

damage can occur if electron transfer to the Calvin cycle is prevented or delayed for some reason. Situations of this kind develop if the plant has been stressed previously, or if an additional stress occurs, especially from heat, cold, drought, excessive salt, poor supplies of mineral nutrients (particularly nitrogen and trace elements), or in cases of infection or exposure to toxic substances. In principle, it can be assumed that photoinhibition always occurs in the presence of strong radiation if the secondary processes of photosynthesis are unable to proceed at a normal rate. A photoinhibitory component is also involved in the commonly observed midday depression of photosynthesis [43], and in the reduced photosynthetic performance during dry periods and winter dormancy.

Photoinhibition – a Dynamic Stress

The processes involved in strong-light stress provide an excellent example of the interplay of destructive and reparative mechanisms, and of the resistance and decline that characterize a dynamic stress syndrome (see Fig. 6.3).

The primary site of attack in stress due to strong light is the reaction centre of photosystem II, where certain protein subunits (e.g. D 1, the 32-kDa protein) are rapidly broken down. Photosynthetic electron transport is thus interrupted

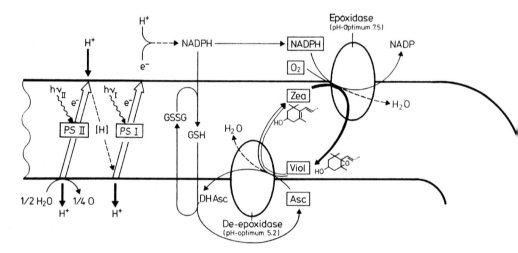

Fig. 6.13. A simplified scheme of the xanthophyll cycle in thylakoid membranes. During irradiation of chloroplasts the enzyme *de-epoxidase* is activated by the drop in pH within the thylakoid due to photosynthetic electron transport. Under the conditions of a proton gradient across the thylakoid membrane, violaxanthin (*Viol*) is reduced to zeaxanthin (*Zea*) with the participation of the redox systems glutathione/oxidized glutathione (*GSH/GSSG*) and ascorbic acid/dehydroascorbic acid (*Asc/DHAsc*). Reconversion of zeaxanthin to violaxanthin, which proceeds in the dark or in weak light, requires the uptake of oxygen and is catalyzed by *epoxidase. PSI, PSII* Photosystems I and II. (Hager 1975, 1980). For responses of plants to strong light, and photoprotection, see Powles (1984); Demmig-Adams and Adams (1992)

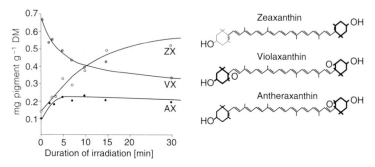

Fig. 6.14. Rapid conversion of violaxanthin (*VX*), via antheraxanthin (*AX*), to zeaxanthin (*ZX*) in *Chlorella pyrenoidosa* under high irradiance (approx. 2000 μmol photons m^{-2}s^{-1}). (Hager 1967)

and the efficiency of photosystem II is lessened (*PSII photoinactivation*). As an immediate protective measure, the superfluous radiation energy is diverted directly from the photosystems via fluorescence and, above all, as heat. The surplus reductive capacity in the chloroplasts is dissipated by the *xanthophyll cycle* (Fig. 6.13), in which, with the participation of ascorbate and NADPH$_2$, the di-epoxide violaxanthin is reduced to zeaxanthin via the mono-epoxide antheraxanthin. The conversion of violaxanthin to zeaxanthin takes place within a few minutes in the presence of strong irradiance (Fig. 6.14). Reconversion to violaxanthin, which again consumes reductants, also occurs rapidly in weak light or in darkness. Another protective mechanism by which energy can be diverted in a cyclical process is glycolate metabolism.

Under strong light, aggressive oxygen species accumulate; they can destroy chloroplast pigments and membrane lipids. Oxidoreductases (superoxide dismutase, peroxidases, catalases) function here as intercepting systems.

Adaptation to Stress from Strong Light
The effect of excessive irradiance can be mitigated by escape movements, such as positioning the leaves at an angle to the incoming light so as to receive less radiation, by rolling up the shoots (especially seen in mosses and pteridophytes), and by chloroplast movements in assimilatory tissue. Dense coverings of trichomes on the upper surface of the leaf or thickened walls in the epidermis and hypodermal tissue (e.g. in conifer needles, sclerophylls and cacti) act as diffusive filters and lessen the effect of strong radiation. Anthocyanin in unfolding leaves, particularly in the tropics, acts as a darkening filter and shields the mesophyll. Furthermore, in strong light the quantity of protective pigments such as carotene and lutein is known to increase in the chloroplasts.

6.2.1.2 Ultraviolet Radiation

After its passage through the atmosphere, sunlight contains ultraviolet up to about 290 nm, which includes the entire longwave UV-A (315–400 nm) and part of the UV-B (280–315 nm). In nature, UV-B occurs only at low intensities, but it could be expected to increase if the filtering effect of the stratospheric ozone layer were to be significantly weakened as a consequence of emission of oxides of nitrogen and halogenated hydrocarbons (Fig. 6.15). There is a large increase in the intensity of UV with increasing altitude and toward lower latitudes.

The greater part of the UV radiation penetrating cells is adsorbed and causes acute injuries on account of the high quantum energy. Longwave UV-A is chiefly photooxidative; UV-B, in addition to its photooxidative action, also causes photolesions, particularly in biomembranes. The molecular mechanism of UV damage to protoplasm consists in breaking down the disulfide bridges in protein molecules, and in dimerizing thymine groups of DNA, which results in defective transcription. Furthermore, UV inhibits the violaxanthin-deepoxidase, so that if at the same time the light is very strong the xanthophyll cycle cannot adequately fulfil its protective role. UV damage to a plant can be identified by changes in enzyme activity (increased peroxidase activity, inhibition of cytochrome oxidase), poor energy status of the cells (cessation of protoplasmic streaming), lower photosynthetic yield (Fig. 6.16), and by disturbed growth (reduced extension growth and pollen tube elongation).

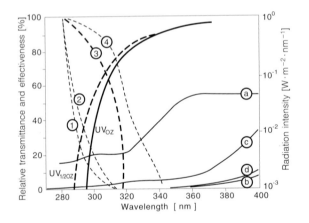

Fig. 6.15. Availability and effectiveness of ultraviolet radiation. UV_{oz} Short-wavelength limit of solar radiation with a steep angle of incidence; $UV_{1/2oz}$ UV radiation at sea level if stratospheric ozone density were reduced by half. *1* Inhibition of phytochrome-induced synthesis of anthocyanin in mustard cotyledons; *2* mutagenic effect on liverwort spores; *3* cessation of protoplasmic streaming in epidermal cells of the scales of onion bulbs, *Allium cepa*; *4* production of flavonoids in cell cultures and in maize coleoptiles. Spectral transmittance of *a* detached outer epidermal walls of *Sambucus coerulea*; *b* living epidermis of *Sambucus* with flavonoids in the cell sap; *c* cuticle of *Atriplex hastata*; *d* cuticle of *Eryngium maritimum*. (After Cappelletti 1961; Caldwell 1977; Robberecht and Caldwell 1978; Wellmann 1983)

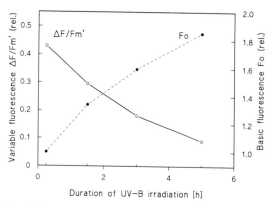

Fig. 6.16. Changes in the in vivo chlorophyll fluorescence of zoospores of *Haematococcus lacustris*, indicating progressive disturbance of photosynthesis during exposure to UV-B radiation. The irradiance employed (2.13 W m^{-2}) simulates the UV-B dose to be expected on a summer day in Sweden, assuming a 10% reduction in stratospheric ozone. (Hagen et al. 1992). The primary inhibition of photosynthesis by UV-B takes place at photosystem II. (Bornman 1989)

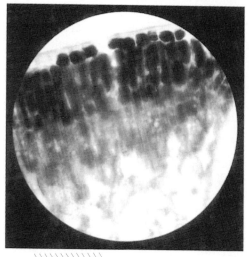

Fig. 6.17. Absorption of UV by plant cells. *Left* Microphotograph of a cross section of a fresh leaf of *Arbutus unedo* using the wavelength range 280–320 nm. Cell vacuoles containing flavonoids and tannins appear *black*. (Original: W. Larcher). *Right* UV is absorbed chiefly in the vacuoles of the epidermis and to some extent in the outer walls, thus protecting the cell nuclei which are near the inner walls. Only 5–10% of the incident UV radiation penetrates to the upper layer of the mesophyll. (After Wellmann from Caldwell et al. 1983). For comparative investigations on screening effectiveness among plant life forms, see Day et al. (1992), Day (1993)

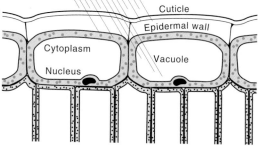

Due to the effective absorption of UV by epicuticular wax and by flavonoids dissolved in the cell sap (Fig. 6.17), the protoplasm in the cells of higher plants is largely protected from radiation injuries. Synthesis of the protective pigments is induced by UV, and they accumulate as stress increases.

6.2.2 Stress Due to Extreme Temperatures

Heat and cold are thermodynamic states characterized, respectively, by high and low kinetic energy of the molecules. Heat accelerates the movement of the molecules, bonds within macromolecules are loosened, and the lipid layers of biomembranes become more fluid. In contrast, at low temperatures biomembranes become more rigid, and the energy required for activating biochemical processes increases. Freezing is the transition of water to the solid state, and is therefore the most serious event for the vital processes of the plant.

Heat and cold, depending on their intensity and duration, impair the metabolic activity, growth and viability of plants and thus set limits to the distribution of a species. At the *activity limit* the vital processes are reversibly reduced to a minimal speed. Resting stages, such as dry spores or poikilohydric plants in the desiccated state, are insensitive, so that they can survive great extremes of temperature undamaged. At the *lethal limit*, which is characteristic not only for a species but also for different organs and tissues (Tables 6.2 and 6.3), permanent injuries occur. When critical temperature thresholds are crossed, cell structures and cellular functions may be damaged so suddenly that the protoplasm is killed immediately. In other cases, damage may develop gradually, with one or more processes thrown off balance and impaired, until eventually vitally important functions cease and the cell dies.

6.2.2.1 Extreme Temperatures and the Temperature Limits for Life

In few environments does the temperature always remain within the safest and most favourable temperature range (approximately $5-25\,°C$) for the vital functions of plants. Near to the soil surface on land, in the littoral zone, and in shallow water the temperature fluctuates between very low and very high values diurnally and, except in the equatorial zone, seasonally, and may even attain levels dangerous to life.

Overheating in a habitat is invariably the result of a large influx of absorbable energy combined with insufficient loss of heat. Under natural conditions the most important source of energy is solar radiation, in addition to heat imported by air currents. Particularly high *air temperatures* occur in the tropics, where the absolute maxima of $57-58\,°C$ have been recorded in North Africa, India, Mexico and California. On about 23% of the earth's land surface annual mean air temperatures of above $40\,°C$ can be expected [95], which means that under strong radiation plant temperatures of $50\,°C$ and more are a regular occurrence. High temperatures in water and soil may also arise in connection with *volcanic phenomena.* The hottest places on earth that are in-

Table 6.2. Maximal temperature resistance of poikilohydric plants and microorganisms in the hydrated and dehydrated states. Poikilohydric organisms in a desiccated state are all completely frost tolerant, surviving the temperature of liquid nitrogen (−196 °C). (Larcher 1973a; Brock 1978; Kappen 1981; Lüning 1984, 1990; Sakai and Larcher 1987; supplemented by data from Fujikawa and Miura 1986; Stetter et al. 1990)

Plant group	Cold injuries[a] below °C Hydrated state	Heat injuries[b] above °C Wet	Dry
Bacteria			
Archaebacteria		100–110	
Cyanobacteria and other photoautotrophic bacteria		55–75	
Saprophytic bacteria		60–70	
Thermophilic bacteria		to 95	
Bacteria spores		80–120	Up to 160
Fungi			
Phytopathogenic fungi	0 to below −10	45–65 (70)	
Saprophytic fungi	−5 to −10 (−30)	40–60 (80)	75–100
Fruiting bodies of fungi		50–60 (100)	>100
Fungus spores			
Algae			
Marine algae			
Tropical seas	+14 to +5 (−2)	32–35 (40)	
Temperate seas			
Eulittoral	−2 to −8	25–30	
Intertidal	−8 to −40	30–35	
Polar seas	−10 to −60	(15) 20–28	
Freshwater algae	−5 to −20 (−30)	35–45 (50)	
Aerial algae	−10 to −30	40–50	
Eukaryotic thermal algae	+20 to +15	45–50	
Lichens			
Polar regions	−80		
High mountains, deserts	−80		
Temperate regions	−50	33–46	70–100

Table 6.2 (continued)

Plant group	Cold injuries[a] below °C Hydrated state	Heat injuries[b] above °C	
		Wet	Dry
Mosses			
Humid tropics	−1 to −7		
Temperate zone			
Wet habitats	−5 to −15	40−45	80−95
Forest floor	−15 to −25	40−50	100−110
Epiphytic and epipetric mosses	−15 to −35		
Polar regions	−50 to −80		
Poikilohydric ferns	−20	47−50	60−100
Phanerogams			
Ramonda myconi	−9	48	56
Myrothamnus flabellifolia			80

[a] After at least 2 h exposure to cold.
[b] After exposure to heat for 1/2 h.

Table 6.3. Temperature resistance of the leaves of vascular plants of different climatic zones. Threshold temperatures at 50% damage (LT_{50} in °C) after 2 h or more exposure to cold, and 30 min heat treatment. (Larcher 1973a; Kappen 1981; Sakai and Larcher 1987; Nobel 1988; supplemented by data from Bannister and Smith 1983; Lösch and Kappen 1983; Larcher et al. 1989; Yoshie 1989)

Plant group	Threshold temperature for cold injury in the hardened state	Threshold temperature for heat injury during the growth season
Tropics		
Trees	+5 to −2	45−55
Forest understorey	+5 to −3	45−48
Plants of high mountains	−5 to −15 (−20)	ca. 45
Subtropics		
Evergreen woody plants	−8 to −12	50−60
Deciduous woody plants	(−10 to −15)[a]	
Subtropical palms	−5 to −14	55−60
Succulents	−5 to −10 (−15)	58−67
C_4 grasses	−1 to −5 (−8)	60−64
Desert winter annuals	−6 to −10	50−55
Temperate zone		
Evergreen woody plants of coastal regions with mild winter	−7 to −15 (−25)	46−50 (55)
Relict species of the tertiary tree flora	−8 to −20 (−15 to −30)[a]	
Dwarf shrubs of Atlantic heaths	−20 to −25	45−50
Deciduous trees and shrubs	(−25 to −35)[a]	ca. 50
Herbaceous species of		
sunny habitats	−10 to −20 (−30)	47−52
shady habitats	−10 to −20 (−30)	40−45
Graminoids of the steppes	(−30 to LN_2[b])[a]	60−65
Halophytes	−10 to −20	
Succulents	−10 to −25	(42) 55−62
Aquatic plants	−5 to −12	38−44
Homoiohydric ferns	−10 to −40	46−48
Regions with cold winters		
Evergreen conifers	−40 to −90	44−50
Boreal deciduous trees	(−30 to LN_2)[a]	42−45
Arctic-alpine dwarf shrubs	−30 to −70	48−54
Herbaceous plants of the high mountains and the Arctic	(−30 to LN_2)[a]	44−54

[a] Vegetative buds.
[b] LN_2 = Temperature of liquid nitrogen (−196 °C).

habited by living organisms are geysers and hot pools, in which water reaches the surface at temperatures of $92-95\,°C$. On volcanoes and within range of volcanic vents, soil is heated up to $40-70\,°C$ by subterranean magma and hot water.

Overheating is often experienced on calm days by plants growing close to the ground on open slopes and in depressions facing the sun. The boundary layer near the ground heats up particularly strongly over rubble and sandy ground, concrete and asphalt. In the temperate zone, soil surface temperatures of 60 and $70\,°C$, and as high as $80\,°C$ on deserts, have been recorded.

High temperatures from solar radiation usually last for only a few hours each day. The extreme temperatures measured on plant organs during this time are peak values; the temperature of leaves and flowers with good heat exchange and low heat capacity quickly equilibrates with that of the surrounding air, so that the peak temperatures usually only affect them for a few minutes. On the other hand, even in temperate latitudes, succulents, shoot apexes of grasses and sedges near the ground, prostrate shoots and runners, and the root collar in young plants and seedlings may be exposed to temperatures around $40\,°C$ for several hours or, for shorter periods, to as much as $50\,°C$. They are thus exposed to a temperature range in which vital processes are severely stressed, and which is dangerously close to the lethal limit (Figs. 6.18 and 6.19). In spite of this, there have been very few reports of heat necroses in wild plants growing in their natural surroundings. The explanation may be that this has received too little attention until now, and unequivocal evidence can only be obtained by continuous recording of the plant's temperature throughout the damaging event.

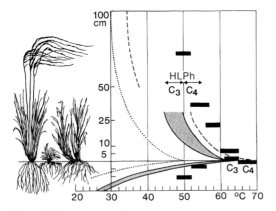

Fig. 6.18. Heat stress and threshold values for the heat resistance of steppe grasses. The temperature curves show the thermal gradients at different heights above the ground, for the hottest part of the day during a heat wave. *Dotted curves* Mean temperature maxima for air and soil on sunny days at the height of summer; *broken line* highest temperatures measured in the habitat; *solid lines with grey shading* range of the hottest temperatures measured frequently on leaves, stalks and organs on or in the soil. The *black bars* indicate average ranges of the heat resistance of inflorescences, leaf blades, shoot bases, and rhizomes and of roots from various depths; *HLPh* heat limit for photosynthesis of C_3 and C_4 grasses. (Larcher et al. 1989)

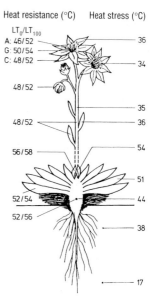

Heat resistance (°C)   Heat stress (°C)

LT$_0$/LT$_{100}$

A: 46/52 — 36
G: 50/54
C: 48/52 — 34

48/52 — 35

48/52 — 36

56/58 — 54

— 51

52/54 — 44

52/56 — 38

— 17

Fig. 6.19. Heat stress in *Sempervivum montanum* growing in a highly insolated, wind-sheltered mountain habitat, and the heat resistance of different organs in summer. *A* Anthers; *G* ovaries; *C* corolla; *LT$_0$* highest temperature survived; *LT$_{100}$* temperature at which only some isolated groups of cells survive. (Larcher and Wagner 1983)

A special kind of heat risk is presented by *fires*. The most common type of fire in dense forests and semiarid woodland (chapparal, macchia) is *surface* fires, which burn litter and often kill the understorey vegetation, sometimes starting *ground* fires which burn below the ground surface in the thick organic material. Surface fires may also spread upward via dry branches and lichens, starting *crown* fires which move from one crown to the next, developing temperatures of 500–700 °C and destroying most of the trees [251]. The surface fires so common in savannas, steppes and heaths reach somewhat lower temperatures (200–400 °C in the flames) as they spread at the speed of the wind over the dry grassland (Fig. 6.20).

Low temperatures result from a negative heat balance on the earth's surface. At high and medium latitudes the net *radiation balance* becomes increasingly negative as winter approaches, due to the low angle of the sun and the lengthening nocturnal phase of thermal reradiation. This leads to greater cooling and the accumulation of cold air masses, which are conveyed to lower latitudes by the atmospheric circulation. Thus cooling results both globally and locally when heat losses predominate, as well as from *advection of cold air*. Winter frosts in the temperate zone and exceptional cold in the subtropcis and tropics are the result of incoming polar air. Cold air of local origin (mountains, bare sites) moves into valley bottoms and hollows in the terrain, where it accumulates and becomes covered by layers of warmer air (temperature inversion). At the microclimatic level also, cold layers form as a result of heat loss by radiation from the ground and from the surface of plant stands, and due to the descent of cold air (Fig. 6.21). Therefore, in estimating the danger of frost for plants with different growth forms, the existence of vertical temperature gradients must always be taken into consideration.

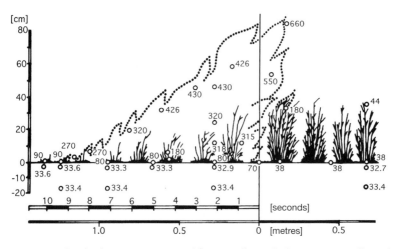

Fig. 6.20. The rise in temperatures and heat gradients during a savanna fire. The *numbers* indicate the temperatures in °C; a comparison of the time scale with the metre scale shows the speed at which the fire spreads. (Vareschi 1962)

The *lowest air temperatures* on the earth, almost $-90\,°C$, have been measured in the Antarctic. In valleys and lowlands in eastern Siberia, minimum air temperatures between -66 and $-68\,°C$ have been recorded. Relatively severe frost (mean annual minimum air temperature below $-20\,°C$), can be expected to occur on 42% of the earth's surface; only one third of the total land area is absolutely safe from frost (Fig. 6.22).

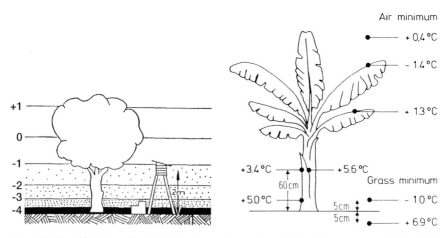

Fig. 6.21. Stratification of cold air during a clear, calm night with strong reradiation. *Left* The minimum temperatures of the plants depends on their growth form (trees, grasses, rosette plants) and may deviate considerably from the minimum temperatures registered at weather stations. (Hofmann, from Burckhardt 1963). *Right* Minimum temperatures of various parts of a banana plant, and air and soil temperatures during radiation cooling. (Shmueli 1960)

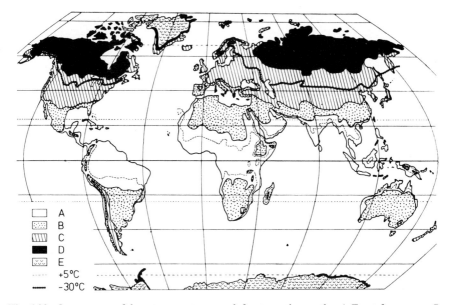

Fig. 6.22. Occurrence of low temperatures and frost on the earth. *A* Frost-free zone; *B* episodic frosts down to −10 °C; *C* regions with a cold winter, and average annual minimum temperature between −10 and −40 °C; *D* average annual minimum below −40 °C; *E* polar ice and permafrost; ⎯⎯⎯ −30 °C minimum isotherm; -------- +5 °C minimum isotherm. The above zones correspond to the areas of distribution of species with different types of frost resistance. *Chilling-sensitive plants* Equatorial zone with minima not below +5 °C; *freezing-sensitive plants* zone A; plants protected by *freezing point depression and effective supercooling* zone B; plants with *limited freezing tolerance* and trees with wood capable of *deep supercooling* zone C; completely *freezing-tolerant* plants: zone D. (Larcher and Bauer 1981)

Whether frost occurs *periodically* in the course of the seasons, or is episodic, is of considerable significance. Plants prepare for the annual recurrence of winter cold by terminating their growth activities and gradually becoming hardened. Only in extremely cold winters, such as may occur at intervals of decades in the northern hemisphere, does the natural vegetation suffer freezing damage. Cultivated plants, on the other hand, which are often planted up to the extreme limits of survival (e.g. fruit trees, vines, and particularly citrus and exotic fruits) frequently suffer quite extensive damage from frost. During brief spells of cold weather, *episodic frosts* may occur in temperate zones as late frosts in spring or early frosts in autumn, and as summer frosts in higher latitudes and in mountains, usually reaching minimum temperatures not below −5 to −8 °C. This type of frost can be dangerous for the native plants which are caught unprepared at a sensitive, unhardened phase in their life. In equatorial uplands and mountains, isolated night frosts as low as −10 to −12 °C may occur at all times of year, but as they last only a few hours the native species are not endangered.

Table 6.4. Thresholds (°C) for functional disturbances (10% inhibition) after 10 min heat exposure of the leaves of the C_4 desert plants *Atriplex sabulosa* and *Tidestromia oblongifolia*. (Björkman et al. from Berry and Raison 1981)

Function	*Atriplex*	*Tidestromia*
Leaf functions		
Net photosynthesis	43	51
Dark respiration	50	55
Semipermeability	52	56
Chloroplast functions		
Photosystem I	>55	>55
Photosystem II	42	49
RuBP carboxylase	49	56
PEP carboxylase	48	54
3-PGA kinase	51	51
Adenylate kinase	47	49
Phosphohexose isomerase	52	55
Ru5P kinase	44	52

6.2.2.2 Heat

Functional Disturbances and Injury Patterns
High temperatures cause reversible alterations in the physicochemical state of biomembranes and the conformation of protein molecules. Because thylakoid membranes are especially sensitive to heat, disturbances in photosynthesis are among the first indications of heat stress. At first, photosystem II is inhibited, after which carbon metabolism is gradually thrown out of balance (Table 6.4). As a result of damage to the chloroplasts, photosynthesis is subsequently depressed, and eventually this results in death of the cell. Since photosystem II is also affected by photoinhibition, the combination of heat and irradiation has an additive effect. In the leaves of tropical herbaceous Fabaceae (*Macroptilium atropurpureum*, *Vigna unguiculata* [149]), heat-dependent photoinhibition begins at 42 °C, whereas in the dark they suffer damage only at temperatures from 48 °C upwards. In the presence of an additional stressor (e.g. drought), there are already signs of incipient inhibition of photosynthesis from 30 °C upwards.

Critical heat thresholds for reversible and irreversible inactivation of photosynthesis, which correlate with lethal injuries that only later become apparent, can be usefully determined by the method of in vitro chlorophyll fluorometry (Fig. 6.23). If, finally, a number of thermolabile enzymes are inactivated, nucleic acid and protein metabolism are disrupted, the fine structure of the biomembranes breaks down, selective membrane transport and mitochondrial respiration cease, and the cell dies.

Fig. 6.23. The temperature curve for basic chlorophyll fluorescence (F_0-T diagram) during gradually increased heating of the leaves of *Atriplex sabulosa* and *Tidestromia oblongifolia*, two species differing in heat resistance. The *lower break in the curve* corresponds to the inactivation temperature of photosystem II (see Table 6.5), the *upper break* marks the temperature at which the thylakoid membranes are irreversibly damaged. (Schreiber and Berry 1977). For fluorescence-heating curves of different crop plants from tropical and temperate zones, see Smillie and Nott (1979)

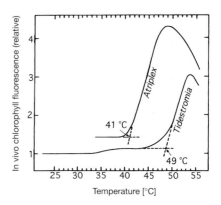

Survival Capacity Under Heat Stress

Plants can survive high temperatures by prevention and mitigation of heating, and due to the capacity of the protoplasm to tolerate high temperatures.

Prevention of dangerous overheating of the leaves is achieved by some plants through evasion of strong sunlight. Protection against fire is afforded by heat-insulating bark (thick fibrous bark of the trunks of *Sequoia* and *Sequoiadendron*; rough, suberized bark of many trees, particularly in semiarid regions), by dense leaf sheaths covering the basal buds (graminous plants) and by withdrawal to underground organs (bulbs and tubers). Monocotyledons such as palms and grass trees (*Xanthorrhoea*) are merely singed by fire, possibly due to some fire-checking action of the silica in their cell walls; also, external burns on their stems cause less harm to monocotyledonous phanerophytes than to the dicotyledons, with a peripheral cambium layer. Overheating of leaves can be effectively prevented by *transpirational cooling*, as long as sufficient water is available; in this way, the leaves of desert and steppe plants remain 4–6 K, in extreme cases even 10–15 K, cooler than the air (Fig. 6.24).

Heat tolerance of protoplasm is a highly specific property; closely related species of the same genus may differ quite markedly in this respect, and even different organs and tissues of one and the same plant are unequally resistant to heat (see Figs. 6.18 and 6.19). Typical differences in resistance, obviously connected with the conditions in the area of distribution and the geographical origin of a species, have evolved in the course of time. Plants of cold regions (tundra, high mountains), are distinctly more heat-sensitive with respect to function (cessation of plasmatic streaming as symptom of an energy crisis and denaturation of the cytoskeleton: Fig. 6.25) and lethal thresholds (see Tables 6.2 and 6.3) than those of temperate regions, which in turn are more sensitive than tropical and desert plants. Some sedges and grasses, among them particularly C_4 plants, seem to be especially heat-tolerant.

The effect of heat depends on its duration; in other words, it obeys the heat *dose law* [15], which says that less heat for a longer time causes as much damage as more heat for a shorter time. It is therefore customary for resistance

Citrullus colocynthis

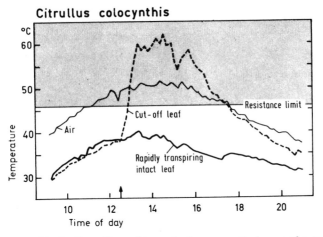

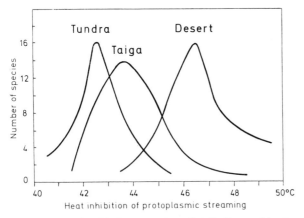

Fig. 6.24. Cooling effect of transpiration upon the leaves of a well-watered *Citrullus colocynthis* plant under desert conditions. Strongly transpiring leaves, despite intense insolation, are much cooler than the air. If a leaf is cut off (*arrow*) so as to make transpiration impossible, the leaf temperature rises rapidly above that of the air, becoming so high that signs of heat injury appear (shown in grey). Plants like *Citrullus*, which maintain a temperature lower than that of the air, can survive in hot habitats only if they are able to transpire at high rates. (After Lange 1959)

Fig. 6.25. Relationship between plant distribution and heat resistance. Here the measure of resistance is the thermostability of protoplasmic streaming after a 5-min exposure to the indicated temperatures. Plants of cooler regions are usually more sensitive to heat than those from dry habitats exposed to higher temperatures. Individual species, however, can deviate considerably from the statistical mean with respect to heat resistance. For example, among the tundra plants most resistant to heat, protoplasmic streaming ceases only at temperatures high enough to produce the same effect in the most sensitive third of the hot-semidesert species that were studied. (Kislyuk et al. 1977). Critical heat thresholds for leaf respiration are 35–40 °C for arctic plants, 40–45 °C for temperate herbaceons species, and up to 53 °C for hot desert plants. (Semikhatova et al. 1992)

data to refer to a heat duration of half an hour. The injury threshold for one hour's exposure to heat would be $1-2$ K lower.

Three **types of heat resistance** can be distinguished:

Heat-sensitive species: this group comprises all species injured at $30-40\,°C$, or at the most $45\,°C$: eukaryotic algae and submerged cormophytes, lichens in a hydrated state (these, however, soon dry out in strong sunlight and are then completely resistant), and most of the soft-leaved terrestrial plants. In addition, bacteria pathogenic to plants, as well as viruses, are destroyed even at relatively low temperatures (e.g. tomato wilt virus is killed at $40-45\,°C$). All these sensitive species can colonize only habitats in which they are not exposed to overheating, unless they can effectively keep their own temperature down by means of transpirational cooling.

Relatively heat-resistant eukaryotes. The plants of sunny and dry localities are, as a rule, able to acquire hardiness to heat; they can survive heating to $50-60\,°C$ for half an hour. However, between 60 and $70\,°C$ there seems to be an absolute limit for survival of highly differentiated cells and organisms.

Heat-tolerant prokaryotes. Some thermophilic prokaryotes can endure exceedingly high temperatures. In the waters of volcanic vents and geysers, cyanobacteria form colonies in hot zones up to $75\,°C$, bacteria up to $90\,°C$, while hyperthermophilic archaebacteria from the depths of the oceans (e.g. *Pyrobaculum*, *Pyrococcus*, *Pyrodictium*) live at temperatures up to $110\,°C$ [235]. All these organisms have especially resistant cell membranes, nucleic acids and proteins.

Acclimatization to heat takes place rapidly in response to heat stress, the shift to higher limiting temperatures being accomplished within hours; on hot days, heat resistance is greater in the afternoon than in the morning. The process of dehardening, or loss of heat resistance in cooler weather, takes place more slowly, although it is completed within a few days. In order to produce hardening, the temperature must be high enough to elicit a stress reaction in the protoplasm. This is usually the case at temperatures exceeding $35\,°C$ for most terrestrial plants, whereas grasses require temperatures higher than $38-40\,°C$, and succulents harden best at high nocturnal temperatures (Fig. 6.26).

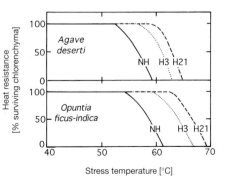

Fig. 6.26. Adjustment of succulents to heat. *NH* Heat tolerance of unhardened plants (daytime temperature $30\,°C$/nocturnal temperature $20\,°C$); *H3* increase in resistance after 3 days at $50/40\,°C$ (day/night); *H21* extent of hardening after 3 weeks at $50/40\,°C$. Measure of resistance: proportion of intact chlorenchyma cells after exposure to the stress temperature for 1 h. (Nobel 1988)

Heat resistance [% surviving chlorenchyma]

Agave deserti

NH   H3 H21

Opuntia ficus-indica

NH   H3 H21

Stress temperature [°C]

A most effective form of heat protection is provided by specific *heat-shock proteins* (HSPs, proteins with molecular weights ranging between 15 and 110 kDa, chiefly HSP90, HSP70, HSP60, HSP20 and ubiquitine [260]); they are rapidly encoded by the cell nucleus, synthesized in the cytosol and transferred to the chloroplasts and mitochondria [160]. The role played by HSPs in heat tolerance is apparently that of stabilizing chromatin structures and membranes and promoting repair mechanisms. They disappear gradually a few hours after the heat stress.

In many plants, heat resistance follows an annual cycle (Fig. 6.27), since it is coordinated with the processes of development, as well as the temperature in the open air. During their most active period of growth all plants are very sensitive to heat. The heat resistance of freshwater and marine algae adjusts to the temperature of the surrounding water, and is thus highest in late summer

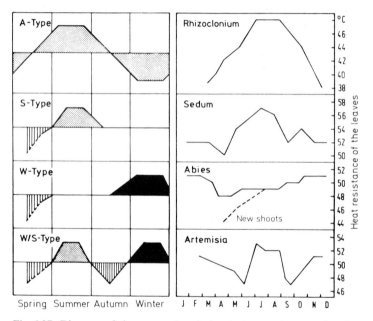

Fig. 6.27. Diagram of the seasonal variation in heat resistance of various types of plants (*left*) and the measured levels of heat resistance of the leaves of various species throughout the year (*right*). In the diagram, resistance changes occurring as an adaptation to the prevailing temperatures in the habitat are identified by *stippling*, the decline in resistance during the period of intensive growth is *hatched*, and the rise in resistance during winter dormancy is shown in *black*. *A-type* Species that continually adapt their heat resistance to the local temperature (example: the alga *Rhizoclonium* sp.); *S-type* species with increased heat resistance during the warm season (example: *Sedum montanum*); *W-type* species with increased resistance associated with protoplasmic alterations during winter dormancy (example: *Abies alba*); *S/W-type* species with increased resistance in summer and winter and with two resistance minima associated with increased growth activity in spring and autumn (example: *Artemisia campestris*). Apart from these types, there are plants with no detectable seasonal variation in heat resistance (e.g. *Asplenium ruta muraria*). After data of various authors, from Lange (1967), Larcher (1973a) and Kappen (1981)

and lowest in winter. The amplitude of this annual oscillation is larger, the greater the difference between the winter and summer water temperatures. Among land plants there are species in which resistance increases only in summer, while others acquire the greatest heat resistance at the time of winter dormancy. The latter pattern, an ecological paradox, is governed primarily by the stage of development rather than by the ambient temperature. Finally, there are also plant species with no detectable seasonal fluctuations in heat resistance.

The ability to survive fires depends on whether new shoots can be produced by the residual stems and trunks, by perennating buds at the base of the old shoot, or by underground organs. Shrubs of the mediterranean macchia (e.g. *Arbutus unedo, Quercus coccifera, Erica arborea* [253]), chaparral (e.g. *Adenostoma fasciculatum, Ceanothus megacarpus* in North America [159]) and heath plants (*Calluna vulgaris*), as well as certain trees (birch, poplar) typically produce new shoots from stems and root stocks. After exposure to fire, *Pinus canariensis* regenerates by prolific growth from surviving twigs and branches; eucalypts send out shoots from lignotubers. In tropical forests, especially palms, some Fabales (*Inga* species) and Lecithidaceae exhibit very good powers of recovery [113]. A special kind of adaptation to regularly recurring bush and grass fires is seen in the fire-resistant seeds and fruits of *pyrophytes*. The seeds of such plants, which include various species of pine and Cupressaceae, *Eucalyptus, Protea*, certain palms, many shrubs, subshrubs (*Cistus*) and herbs of the savannas (*Lantana camara* and some grasses) germinate better, or only germinate at all after exposure to heat.

Although sporadic fires destroy part of the plant cover and influence the process of selection among species, fire also has positive effects on the development of the vegetation and on the ecosystem as a whole. Not only are the accumulations of necromass and the overlying layers of litter completely remineralized, the thinning-out of dense stands promotes rejuvenation. Thus every larger fire initiates a cycle of successsions ("fire cycles") [252].

6.2.2.3 Cold and Frost

Impairment of Function and Onset of Injury
As the temperature decreases, so also does the speed of chemical reactions, with the result that equilibrium reactions shift increasingly in the direction of energy release (Le Chatelier principle). Thus, in the cold, less metabolic energy is available, the uptake of water and nutrients is restricted, less biosynthesis takes place, assimilation is reduced and growth stops. The more frequent, the longer, and the colder the periods of low temperatures, the more serious are the consequences for the plant.

The various cellular functions are not equally sensitive to temperature (Fig. 6.28). The first detectable result of low temperatures is the cessation of cytoplasmic streaming, which is a phenomenon directly dependent on energy supplied by respiratory processes and on the availability of high-energy phosphate. Impairment of photosynthesis soon follows; this is detectable at an ear-

ly stage by gas-exchange measurements and by in vivo chlorophyll fluorometry
(Fig. 6.28), both of which can be regarded as early-warning methods. Only
rarely do plants recover immediately from cold stress when the temperature
rises. Strong irradiation during or immediately after the cold spell enhances
the damage to chloroplasts and delays or even prevents recovery; in plants that
are especially sensitive to cold there is the danger of chlorophyll destruction
by photooxidation. In some cases, the temperature coefficient of metabolic activity deviates from the normal in the critical temperature range; this is marked

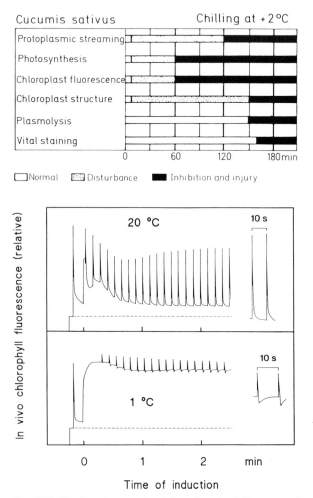

Fig. 6.28. The impairment of cell functions in chilling-sensitive plants with increasing duration of exposure to cold. *Above* Differences in the sensitivity of various functions in
cucumber leaves at +2 °C. (After Kislyuk 1964). *Below* Chlorophyll fluorescence transients
measured on soybean leaves at 20 and 1 °C. Abnormal deviations in the fluorescence transient during chilling (*undershoots*, see enlarged detail) indicate the beginning of a reversible
disturbance of photosynthesis. (Larcher and Neuner 1989)

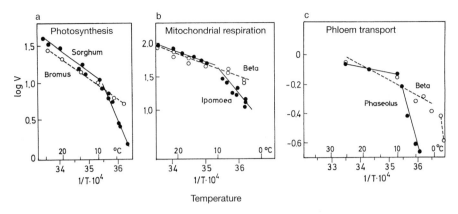

Fig. 6.29a–c. Arrhenius diagrams for the temperature dependence of metabolic functions of chilling-sensitive (*solid line*) and chilling-tolerant plants (*broken line*). If the reaction velocity (V) is plotted logarithmically against the reciprocal of the Kelvin temperature ($1/T$), an inflection in the regression line indicates an abnormal deviation from the optimal temperature coefficient [see formula (2.7)]. **a** Net photosynthesis of leaf disks of *Sorghum bicolor* and *Bromus unioloides*. (McWilliam and Ferrar 1974). **b** Oxidative activity of mitochondria from tubers of *Ipomoea batatas* and from roots of *Beta vulgaris*. (Lyons and Raison 1970). **c** Velocity of phloem transport in petioles of *Phaseolus vulgaris* and *Beta vulgaris*. (Giaquinta and Geiger 1973)

by a change of slope in the Arrhenius diagram (Fig. 6.29). After cold stress, sometimes a temporary increase in respiration may occur.

Low temperatures may be the direct cause of injuries to plant cells: *chilling-sensitive* plants suffer lethal injuries at temperatures a few degrees *above* the freezing point. Just as with heat, death due to cold is the consequence of lesions in biomembranes and interruption of the cellular energy supply. Plants that are insensitive to cold *above* their freezing point are damaged at temperatures *below* this point as a result of ice formation. While *freezing-sensitive* tissues are killed as soon as ice forms within them, the tissues of *freezing-tolerant* plants can survive considerable ice formation before they, too, gradually die.

Injury Patterns in Chilling-Sensitive Plants

Certain plants of tropical origin, as well as the fruits of some species with chilling-resistant vegetative organs, are lethally damaged at temperatures between about 10 and 0 °C. The progress and the extent of chilling injuries within a plant depend on the degree of cooling, its duration, and the speed with which the temperature changes during cooling and rewarming. A sudden temperature change (*temperature shock*) is particularly detrimental. It can be seen from Fig. 6.30 that the lower the temperature and the longer the duration of stress, the more extensive is the damage.

Saintpaulia leaves

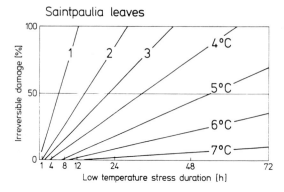

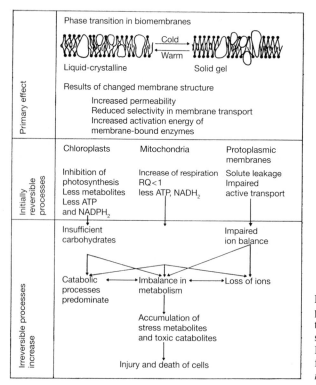

Fig. 6.30. Dose dependence of chilling damage to *Saintpaulia* leaves. The extent of the tissue necroses at each of the stress temperatures increases with decreasing temperature and the duration of exposure. At temperatures above 8 °C no damage occurs. (Larcher and Bodner 1980)

Chilling damage to the protoplasm develops stepwise in sensitive plants (Fig. 6.31). At first, only certain isolated functions are temporarily impaired or arrested; later, this is followed by irreversible disturbances in permeability and their unfavourable consequences. As a rule, it takes some days or even weeks for the full extent of the damage to develop. The primary effect is the transition of the lipid components of the biomembranes from a fluid-crystalline to a gel-like state, and lateral separation of membrane proteins. The selectivity of the permeation processes is lowered, uncontrolled exchange of metabolites and

Fig. 6.31. Events and pathways of events leading to damage in chilling-sensitive plant cells. (After Lyons 1973; Levitt 1980a, from Larcher 1985, *modified*)

ions between the various cell compartments occurs, and cellular contents leak out. Because the metabolic processes are no longer in harmony with one another, and the energy yield is reduced on account of shifts favouring anaerobic respiration, there is an accumulation of toxic intermediary and end products. With this loss of cell compartmentation the cell deteriorates and dies.

Among **chilling-sensitive** plants two types can be distinguished: the *totally* sensitive species, in which every part of the plant is susceptible to damage, and species which are only *partially* damaged by cold (e.g. their flower primordia or fruits). There are also remarkable differences in sensitivity between the organs and tissues of one and the same plant (Fig. 6.32), and also certain stages in the life cycle of a plant (such as seed imbibition, germination and senescence) may be more sensitive to cold than others; in some varieties of rice and millet the pollen is sterile if the flower primordia have been exposed to temperatures below about 10 °C during the tetrad stage.

Freezing Patterns and Freezing Injury

In plants, ice forms first in parts that cool down soonest and freeze most readily, usually the peripheral vascular bundles and water arising from condensa-

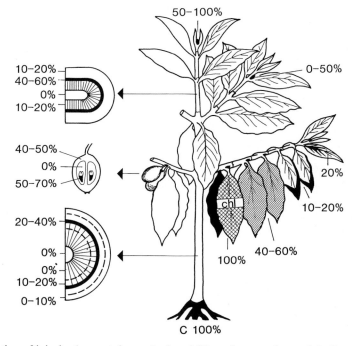

Fig. 6.32. Distribution of injuries (percent damage) after chilling of young plants of *Coffea arabica* at +1 °C for 36 h. The most susceptible organs and tissues are the roots, the cambial zone, chlorotic (*chl*) and senescent leaves, opening buds and the embryos in the seeds. (Sakai and Larcher 1987)

tion in the intercellular spaces. Ice-nucleation active (INA) bacteria (species of *Erwinia* and *Pseudomonas* [143]) might play a role in ice formation in plant tissues; due to the arrangement of the water molecules adhering to the surface proteins of the bacteria, freezing occurs slightly below 0 °C. Ice formation spreads rapidly through vascular bundles and homogeneous tissues, but its progress is hindered by discontinuities such as air spaces, and tissues with densely lignified or cutinized cell walls.

The protoplast freezes *intracellularly* if it contains a high proportion of water, if it is not hardened, or if it has been previously deeply undercooled. Ice crystals form suddenly within the cell (Fig. 6.33) and the cytoplasm is destroyed. Quite often, on the other hand, ice formation occurs first in the intercellular spaces, and then between the cell wall and the protoplast; this is known as *extracellular ice formation*. While ice is crystallizing out, it has same effect as dry air because the vapour pressure above ice is lower than above a supercooled solution. Thus water is withdrawn from the protoplasts, which may shrink to two thirds of their volume, and the concentration of the solutes rises accordingly. Water movements and the process of freezing continue until

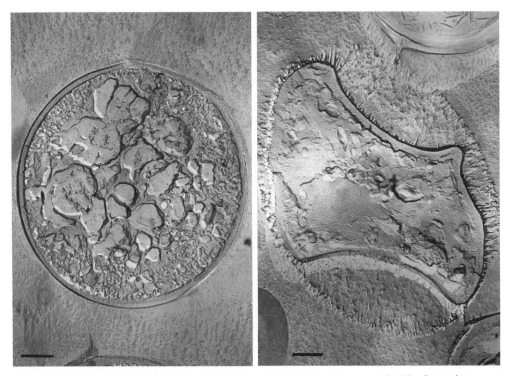

Fig. 6.33. Intracellular (*left*) and extracellular (*right*) freezing of yeast cells. The formation of intracellular ice crystals causes total destruction of the protoplasmic structures. The extracellularly frozen yeast cell has shrunk due to withdrawal of water and is surrounded by ice crystals which delineate the original margins of the cell; bars = 1 μm. (Moor 1964)

thermodynamic equilibrium between the ice layer and the cell sap is reached. The position of equilibrium is temperature-dependent: at a temperature of $-5\,°C$ it is at roughly $-6\,MPa$, at $-10\,°C$ roughly $-12\,MPa$.

Subzero temperatures therefore have the same effect on protoplasm as desiccation. During cold stress and dehydration of the protoplasm, salt ions and organic acids in the remaining unfrozen solutions reach abnormally high concentrations; they exert a toxic effect and inactive enzymes. Under these conditions, the biomembranes are overtaxed both osmotically and by the volume reduction; their lipids are broken down (Fig. 6.34), proteins dissociate, and ATPase activity declines. An indication of membrane destruction caused by the stress of freezing and thawing is the escape of plastocyanin from thylakoid membranes (see Fig. 6.37a). In the end, so much water is withdrawn from the cell that, at a specific degree of dehydration (given by the ratio of liquid to solid phase), the fine structure of the protoplasm is irreversible destroyed.

Surviving Frost Stress

Plants have developed a wide variety of mechanisms for surviving in habitats exposed to episodic or prolonged periods of frost (Fig. 6.35). In addition to measures for reducing freezing stress, plants have also evolved ways of meeting other stresses connected with cold winters, such as photoinhibition, winter drought and snow. Frost damage can be avoided by protective and evasive mechanisms, so that tissue freezing can be delayed or even prevented. Ultimately, in cold climates the survival of a plant is determined by the degree of its freezing tolerance.

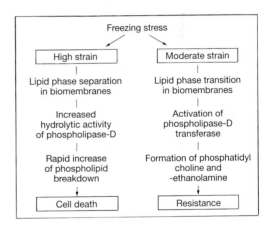

Fig. 6.34. Processes involved in the freezing of leaves of rape plants. The first effect of low temperatures is (in the alarm phase) destabilization of the biomembranes. Severe stress leads to acute damage, in the course of which the membranes are destroyed, predominantly by the hydrolytic activity of phospholipase-D. Under moderate stress, if the cells survive the critical phase, increased activity of phospholipase-D transferase results in the formation of membrane-stabilizing phosphatidyl choline and phosphatidyl ethanolamine, which promote the frost resistance of the cells. (After Sikorska and Kacperska 1982)

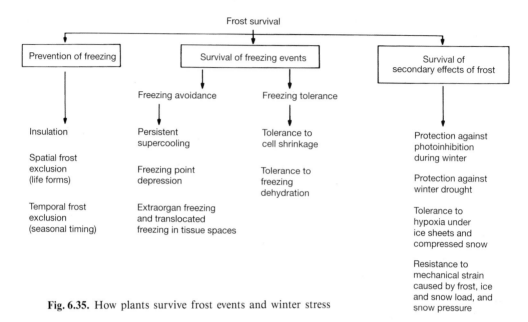

Fig. 6.35. How plants survive frost events and winter stress

Frost prevention is based chiefly on thermal insulation to reduce heat losses. This can be achieved by dense growth surrounding the regenerative buds (cushion forms), by withdrawal of the perennating organs beneath a covering of leaves or layers of litter or in the ground (geophytes) and by the abscission of sensitive organs before severe frost occurs (timely leaf fall in deciduous woody plants). In the tropical high mountains the leaves of "giant rosette" plants close over the shoot tips at night and thus protect them from freezing.

Depression of the freezing point and supercooling can delay the freezing of water in plant tissues. Because the freezing point is lowered by the presence of solutes, the temperature at which cell sap freezes (between -1 and $-5\,^{\circ}C$) depends on the concentration of its solutes. This *depression of the freezing point* provides only moderate, but reliable protection against frost. In addition to this, the cell fluids may be *supercooled*, which means that they can be cooled to a temperature lower than the freezing point without immediately freezing. Both depression of the freezing point and supercooling can be detected by monitoring the temperature of the plant. Initially the tissue temperature drops parallel to the cooling of the outside air; this is followed by a sudden rise in temperature which indicates the release of heat of crystallization. This "exotherm" marks the beginning of the freezing process (Fig. 6.36).

In parenchyma, with large, well-hydrated cells, and in xylem vessels, the supercooled state is extremely labile (*transient* supercooling) and can seldom be maintained for more than a few hours; it can at best provide some degree of temporary protection against the intensity and duration of a radiation frost. In tissues in which dense and thick cell walls form barriers to nucleation, a more *persistent* supercooling can be maintained until the temperature drops

Fig. 6.36a–c. Freezing patterns in a liquid medium and in various plant organs. **a** Temperature and phase changes during freezing and thawing of an aqueous solution: ($\varDelta$) depression of freezing point; *SC* transient super-cooling; *NT* nucleation temperature; *ET* exotherm; *FP* freezing point; *MP* melting point.
b Typical freezing pattern of organs with a high pro-portion of parenchyma, e.g. roots. The broad exo-therm in the region of the freezing threshold tempera-ture (T_f) is preceded by a transient high exotherm peak caused by freezing of the interstitial water. *Abscissa* time course.
c Freezing pattern of a branch, showing supercool-ing in woody tissues, mea-sured by differential ther-moanalysis. The various stages in freezing are recognizable by an increase in the temperature dif-ference (*ordinate*) between the freezing sample and a reference containing no water. *Abscissa* reference temperature. The first exo-therm peak at $-5\,°C$ marks the freezing of the water in the dead elements of the wood; the high tem-perature exotherm (*HTE*) which follows at about $-10\,°C$ is caused by freez-ing of the cortical paren-chyma, and the low tem-perature exotherm (*LTE*) indicates ice nucleation in the supercooled cells of the xylem parenchyma. (Larcher 1985)

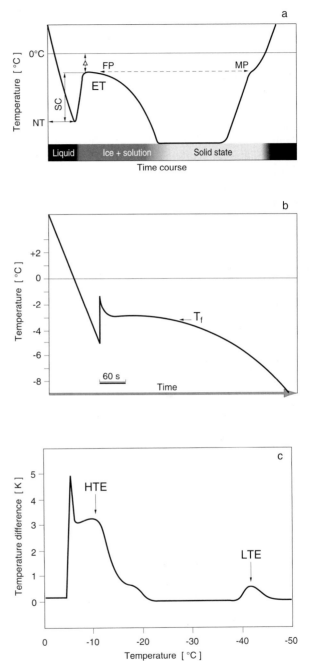

below a characteristic threshold value. Persistent deep supercooling down to -10 and $-12\,°C$ is possible in some leaves, and even down to -30 to $-50\,°C$ in the buds and wood of various trees and shrubs of the temperate zone (many forest and fruit trees). When the supercooling threshold is crossed, however, the metastable state collapses spontaneously and ice crystals form inside the cells (low temperature exotherms: see Fig. 6.36c).

Some seeds, buds and bark tissue have a third protective mechanism known as *translocated ice formation* or "extratissue freezing". This involves the movement of water from the tissues into intercellular lacunae or other spaces, where it freezes to form extensive masses of ice. The cell sap is thereby concentrated and intracellular freezing is delayed. Water can also be bound by hydrophilic mucopolysaccharides, which are known to accumulate in winter, for example in the shoots of *Opuntia humifusa* [145].

For plants that have to withstand severe frost it is essential for their protoplasm to be **freezing-tolerant.** This condition of hardiness or freezing tolerance is achieved by increased incorporation of cold-stable phospholipids into the biomembranes (see Fig. 6.35) and by the accumulation of soluble carbohydrates (sugar and oligosaccharides; Fig. 6.37b), polyols, low-molecular nitrogen-containing substances (amino acids, polyamines) and water-soluble proteins. The protective action of these solutes is due partly to the fact that, as a result of the increased concentration of particles in the cytoplasm, thermodynamic equilibrium is established at a higher intracellular water content, thus lessening the risk of dehydration stress. In addition, the accumulation of low-molecular, membrane-neutral substances reduces the concentration of toxic ions and causes the displacement of electrolytes from structural proteins.

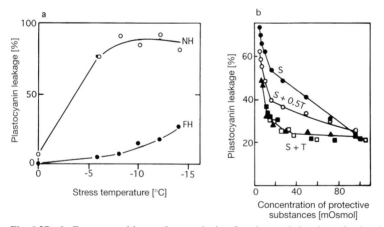

Fig. 6.37a, b. Damage to biomembranes during freezing and thawing of spinach leaves, and the protective effect of soluble carbohydrates. **a** Leakage of plastocyanin from the thylakoid membranes immediately after freezing and thawing of non-hardened (*NH*) and frost-hardened (*FH*) leaves. **b** The protective effect of soluble carbohydrates during a freezing cycle. *S* Sucrose alone, *S+0.5 T* addition of 0.5 mM trehalose; *S+T* addition of 1–10 mM trehalose. (Hincha 1989; Hincha et al. 1989)

A plant is not able to acquire freezing tolerance at every stage in its phenological cycle, and when it does, then not always to the same extent. During the most intensive phase of elongation growth, most plants can scarcely become hardened at all, and are therefore extremely sensitive to low temperatures. In regions with a seasonal climate, land plants acquire in autumn the capacity to survive a substantial degree of extracellular ice formation (Fig. 6.38). The first prerequisite for transition to the state of *readiness to harden* is the termination (in woody plants) or interruption (in herbaceous species) of growth. Once readiness to harden has been attained the **process of hardening** can begin. This is a stepwise process [254], each stage of which prepares for the transition to the next.

In *woody plants* hardening is begun by exposure to temperatures slightly above zero for several days to weeks (Fig. 6.39). In this prehardening stage sugars and other protective substances are accumulated, the cells become less turgid and the central vacuole splits up into many small vacuoles. The protoplasm is now ready for the next phase, which in this type of plant takes place under the influence of moderate frosts between −3 and 5 °C. In this phase biomembranes and enzymes are modified in such a way that the cells are able

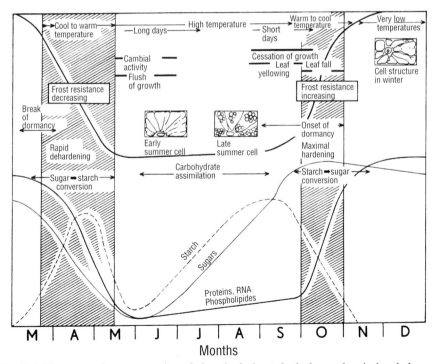

Fig. 6.38. Diagrammatic representation of phenological, cytological, cytochemical and physiological events and processes associated with the acquisition and loss of frost hardening, based on the results of studies on cortical tissues of *Robinia*. The *hatched areas* indicate periods of maximum change in hardiness. (Siminovich 1981)

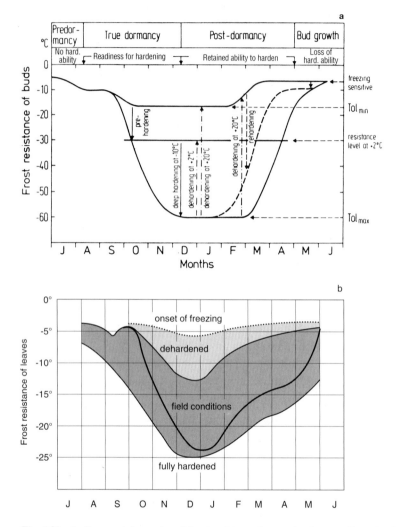

Fig. 6.39a, b. Seasonal dynamics of frost resistance in woody plants. **a** Stepwise frost hardening of the vegetative buds of apple trees (cv. Antonovka). *Upper curve* (Tol$_{min}$) lowest frost-
tolerance level of trees protected from frost in a greenhouse; *lower curve* (Tol$_{max}$) highest
frost tolerance attained after progressive hardening and uninterrupted freezing stress at
$-10\,°C$; *broken line* reduced frost tolerance in late winter and in spring after dehardening
and rehardening. The different stages of winter dormancy are described in Chapter 5.3.3.
(After Tyurina and Gogoleva 1975). **b** Annual trends of frost resistance and the onset of
freezing in the leaves of *Rhododendron ferrugineum*. The *shaded area* indicates the range
between *minimal resistance* after several days in a warm room and *potential resistance*
achieved by stepwise hardening to frost. The *actual freezing resistance* of the plants in their
natural surroundings (*bold line*) lies between these limits. It depends upon the preceding
weather conditions. (After Pisek and Schiessl 1947)

to tolerate the loss of water incurred by ice formation. The highest level of hardening is achieved by uninterrupted exposure to at least -5 to $-15\,°C$, although the effective temperature range differs from species to species. Thus each level of cold produces the degree of hardening necessary for the plant to survive its effects. When frost becomes less severe, the protoplasm reverts to the first level of hardening, but as long as the plant remains dormant, it can repeatedly be raised to a higher level by cold periods. As soon as growth commences at the end of winter, the ability to harden is rapidly lost.

Superimposed on the seasonal pattern of frost hardening in winter are short-lasting *induced adaptations* by means of which the level of resistance can adjust promptly to changes in weather. Frosts are especially effective in promoting hardening in the pre-winter season; at this time resistance can be raised to a maximum within only a few days. Conversely, above-zero temperatures, particularly in late winter, cause a rapid loss of resistance, although even in mid-winter exposure to $+10$ to $+20\,°C$ may suffice for plants to lose most of their hardening within a couple of days.

Most *herbaceous plants* acquire freezing tolerance more simply [109]. The hardening process is triggered by temperatures of $+5$ to $-2\,°C$, without a preceding stage of readiness, the day length apparently being of little or no significance. The essential condition for acquiring higher resistance is the accumulation of carbohydrates. This is possible because herbaceous species usually continue to photosynthesize in winter, but at this time their yield of assimilates is not invested in growth. Thus, as the periods of frost (especially when the ground freezes) gradually become longer, resistance to cold also increases. Characteristic of this *one-phase* pattern of hardening is the ready adjustment to winter cold. The process starts up quickly, and within a short time can exert its maximum effect, but when the plant emerges from winter dormancy, hardening is also lost quickly (Fig. 6.40).

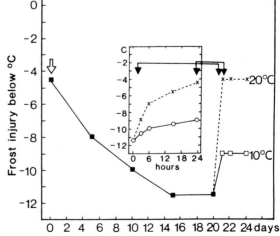

Fig. 6.40. Time course of cold hardening and dehardening of leaves of the tuber-bearing *Solanum commersonii* from the Andes. Plants were cold-acclimated at $2\,°C$ (*white arrow*) for 20 days and then deacclimated (*black arrows*) at either 10 or $20\,°C$. The *inset* demonstrates the dehardening rate at intervals over 24 hours. (Chen and Li 1980)

Categories of Cold Resistance

Not all plants are able to survive low temperatures or ice formation, and not all freezing-tolerant species can go through all phases of hardening. On the basis of their maximum resistance (see Tables 6.2 and 6.3) plants form ecologically significant groups that reflect the role of specific resistance in determining plant distribution (Fig. 6.41). The resistance of a plant is usually expressed as the temperature at which half of the samples are killed (lethal temperature TL_{50}).

Chilling-sensitive plants: this group includes all plants seriously damaged even at temperatures above the freezing point: algae of warm oceans, certain fungi and some, but not all, vascular plants of the tropics.

Freezing-sensitive plants: these plants are protected from injuries solely by measures that delay freezing. In the colder seasons there is an increase in the concentration of osmotically active substances in the cell sap and protoplasm. Algae living at great depths in cold oceans and some freshwater algae, tropical and subtropical vascular plants and a variety of species of the warm temperate regions are sensitive throughout the entire year.

Freezing-tolerant plants: the intertidal algae and some freshwater algae, aerial algae, mosses of all climate zones (including the tropics), and perennial terrestrial species in regions with a cold winter are seasonally freezing-tolerant. Some algae, many lichens and various woody species can harden sufficiently to tolerate extremely low temperatures; they are not damaged by lengthy periods of severe frost, and some survive cooling down to the temperature of

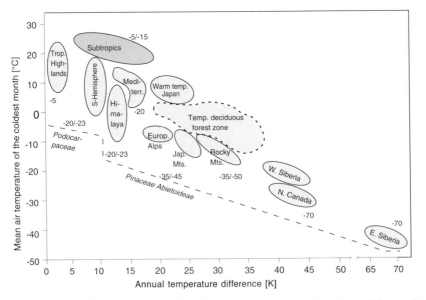

Fig. 6.41. Potential frost resistance (in °C) of winter buds of conifers from different climatic zones. The annual temperature amplitude is an indication of a maritime (small difference) or a continental climate (large difference). (After Sakai 1983, from Sakai and Larcher 1987)

liquid nitrogen. Some species of the high mountains and of arctic regions are able to acquire freezing tolerance within a few days at all times of the year, and so are able to survive extracellular ice formation also in summer.

Variations in Frost Resistance and Survival Capacity
The various organs and tissues of a plant may differ considerably with respect to temperature resistance (Fig. 6.42). *Reproductive organs*, e.g. the flower primordia in winter buds or the ovaries in flowers, are often especially sensitive to frost (Fig. 6.43). *Underground organs* are also quite sensitive, and therefore the frost resistance acquired by bulbs, tubers and rhizomes is vitally important for the survival of geophytes. In woody plants, the ability of the whole plant to survive severe winters depends upon the resistance of the root system, above all the root collar. If these parts die due to inadequate resistance, the shoot will also perish.

The *aboveground shoot* is the least sensitive part of the plant; in its fully hardened state, the cambium of the shoot is the most resistant of all tissues. The resistance of regenerative buds increases in proportion to their exposure to frost (Fig. 6.44). Even buds that overwinter without any special protection usually acquire the same degree of resistance to frost as the shoot that bears

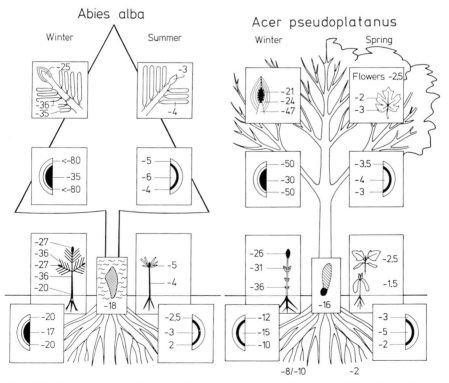

Fig. 6.42. Frost resistance of various life stages, organs and tissues of *Abies alba* and *Acer pseudoplatanus* in winter and during the growing season. (Larcher 1985)

them, and are in any case more resistant than the leaves. By comparison, buds that overwinter near the ground develop only a moderate degree of resistance. For survival of winter constraints, the resistance of buds is of primary impor-

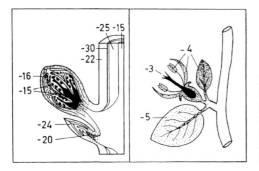

Fig. 6.43. Frost resistance of flower buds of the almond tree in winter (*left*) and of apple blossoms in spring (*right*). The temperatures (°C) are those causing 50% damage. *Black* Most sensitive regions. (After Pisek 1958; Larcher 1970)

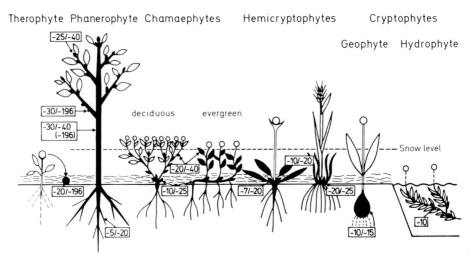

Fig. 6.44. Cold resistance of winter-surviving ecomorphs of cold-winter regions. The degree of cold resistance is indicated by the range of temperatures below which the plants are damaged; the smaller number applies to the more sensitive species and the larger to those more resistant. Depending on their specific physiological constitution some individual species may deviate widely from these limits. The terminology for winter-survival ecomorphs corresponds to the system of "life forms" of Raunkiaer (1910). The parts of the plants that overwinter are shown in *black*; those left *unshaded* die off at the end of the growing season. *Phanerophytes* Trees and shrubs with perennating buds above the snow; *Chamaephytes* small shrubs regularly protected by the winter snow cover; *Hemicryptophytes* perennial herbs with perennating buds just above the soil surface, under litter or enclosed in the remnants of dead leaves and leaf sheaths still attached to the plant; *Cryptophytes* perennial herbs with persistent organs under ground (geophytes) or under water (hydrophytes); *Therophytes* annual plants that complete their life cycles during the growing season and overwinter as seeds. These ecomorphs represent an evolutive adaptation not only to winter cold but also to winter drought (cf. Chap. 6.2.2.4) and other kinds of winter stress. The resistance data are taken from the measurements of many authors (cf. Table 6.3). For frost resistance of fern rhizomes in relationship to habitat distribution see Sato (1982)

tance; as long as the buds remain healthy, losing its leaves need not affect a plant too seriously, whereas the loss of buds as well as leaves can be disastrous. Fortunately, even under such circumstances many plants can still put out new leaves from reserve buds. If trees need to resort to this type of regeneration too often, they become bushy and acquire a stunted form.

The different **life stages** of plants are unequally resistant, and vary in the risk involved. For distribution, the young individuals are of special importance, since this is the most sensitive life stage; basically it sets the limits for the persistence and spread of the species. This is illustrated in Fig. 6.45 by the example of a mediterranean community of evergreen oak. The entire progeny of one year can be destroyed by a single frost event no colder than $-4\,°C$. The regular occurrence of frosts between -8 and $-10\,°C$ in successive winters excludes the possibility that young plants will survive, whereas the mature shrubs and trees are not damaged in any way at these temperatures. For a stand of *Quercus ilex*, frost has catastrophic consequences at temperatures as low as -20 to $-25\,°C$ only if they last sufficiently long to penetrate the thicker trunks. In forests, the outer layer of the crowns is most exposed to danger from radiation frost; that these layers develop greater frost resistance than the juvenile stages growing in their shade is a good example of acclimatization of the different components of a forest according to individual exposure.

There are also differences in hardening capacity and patterns of resistance among the individuals of a *population*. A wide spectrum of resistance is the basis upon which *climatic ecotypes* with above-average resistance can evolve. Such types are able to migrate and survive in harsher environments if long-term changes in climate produce unsuitable conditions for the species within its original area of distribution.

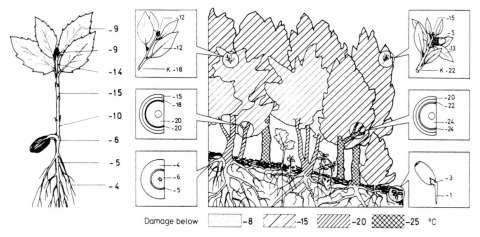

Damage below ☐ -8 ▨ -15 ▨ -20 ▩ -25 °C

Fig. 6.45. Frost resistance of the various strata and age groups in a stand of *Quercus ilex*. *Numbers in the marginal pictures* represent the threshold temperatures below which first frost damage is to be expected. (Larcher and Mair 1969)

Evolution of Frost Hardening in Vascular Plants

The appearance of ecotypes resistant to low temperatures, as well as fine gradations in the resistance of plants from different climatic zones indicate that the ability to become hardened to freezing has been developed in a step-wise process of evolution along a series of plant types [134]. The first step in the evolution of cold adaptation must have been the adaptation of the enzymes and biomembranes of initially *chilling-sensitive* plants of the warm, humid tropics to tolerate lower temperatures (Fig. 6.46). The next step could have been adaptation to temperatures even a little below zero, due to an improved capacity to supercool and to lower the tissue freezing point. This would allow the plants to survive moderate episodic frosts (down to about −10 °C) without injury. In regions with this type of climate, especially in the subtropics, avoidance of freezing is the best compromise between safe survival and an adequate yield of photosynthates during the cold season. With sufficient *capacity to supercool*, leaves and shoots as well as the metabolic processes are safe, because moderate subzero temperatures do not result in cell dehydration. The ability of the protoplasm to acquire *tolerance to freezing* in a stepwise process of hardening makes possible an existence under extreme cold stress, albeit at the price of a shorter growth season. Therefore the highest freezing tolerance was most likely to be achieved by plants whose life cycle included a genetically determined alternation of periods of activity and dormancy, or at least an interruption in growth (see Chap. 5.3.2).

The freezing resistance of phanerogames could have developed along two initially separate *evolutionary lines*. It is striking that there is an enhancement

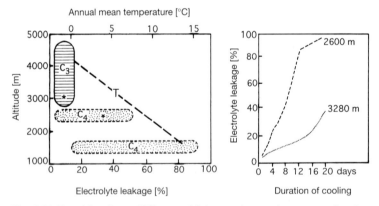

Fig. 6.46. Transition from chilling sensitivity to tolerance in grass species along an altitudinal transect on Mt. Wilhelm in Papua New Guinea. *Left* Up to 2600 m a.s.l. chilling-sensitive C_4 grasses are predominant (11 species were studied); after 6 days' exposure to 0 °C they suffer moderate to severe damage, as indicated by leakage of electrolytes. At higher altitudes only the chilling-tolerant C_3 grasses remain (also 11 species studied). *Asterisks* indicate values for *Miscanthus floridulus* (see diagram on the right). *T* Altitudinal gradient of annual mean temperature. *Right* Populations of the C_4 grass *Miscanthus floridulus* from an altitude of 3280 m are clearly more resistant to cold than the population at 2600 m. (After Earnshaw et al. 1990)

Table 6.5. Frost resistance (°C) of giant rosette plants exposed to night frosts in high altitudes of tropical mountains; Venezuelan Andes, Mt. Kenya. (Beck et al. 1982, 1984; Goldstein et al. 1985)

Species	Provenance		Threshold for 50% leaf damage (°C)
Espeletia atropurpurea	Andes	2850 m	− 6.1
		3100 m	− 8.1
Espeletia schultzii	Andes	3560 m	− 10.0
		4200 m	− 11.2
Dendrosenecio brassica	Mt. Kenya 4100 m		Down to − 10
Dendrosenecio keniodendron	Mt. Kenya 4200 m		− 14
Lobelia telekii	Mt. Kenya up to 4500 m		Down to − 20

of the trend towards improved avoidance of freezing with increasing altitude, and of the trend towards greater freezing tolerance with increasing latitude. In the tropical highlands, which are considered to be a reservoir for early angiosperm evolution, the large diurnal fluctuations in temperature characteristic of tropical mountains may have been influential in establishing and selecting the metabolic adaptations and *mechanisms for avoidance of freezing* which represent the first step in the phasewise process of hardening (Table 6.5), and which still to-day are effective throughout the year in mountain plants.

In regions with a monsoon climate (alternation of dry and rainy periods), the activity rhythm which is a prerequisite for *freezing tolerance* could have been imposed upon woody plants as they radiated to the interior of the continents, the onset of dormancy being coupled with an increase in drought resistance. If freezing tolerance of the protoplasm is regarded as the ability to survive the withdrawal of water resulting from extracellular ice formation in tissues, it might safely be inferred that plants capable of tolerating severe desiccation during the dormant phase are resistant to water loss due to freezing. The prevailing opinion among phylogeneticists is that the evolution of vascular plants began under humid tropical conditions, later proceeding along mountain ranges into subtropical regions with alternating rainy and dry seasons, and eventually into the temperate zone. On the way across the alternately wet and dry subtropics with episodic frosts, the evolution of plants that had already undergone preadaptation in tropical highlands and mountains may have led to the degree of perfection of frost resistance that has enabled perennial species to colonize regions with cold winters.

6.2.2.4 Frozen Soil, Snow and Ice

Winter cold not only endangers plants by its *direct* effects on vital functions and by the formation of ice in the tissues, it also has secondary consequences due to water freezing on and in the soil, to snowfall and ice rain, and the snow cover itself. Winter damage is enhanced by the effects of deep-reaching frost in the absence of an adequate covering of snow, by the presence of long-lasting snow cover and encasement by ice, as well as by winter-specific disturbances of parasitic or anthropogenic origin. To reduce the danger of injuries resulting from this *complex stress pattern*, plants have developed a variety of resistance mechanisms (see Fig. 6.35). The most significant consequence of the secondary effects of cold and frost is curtailment of the period of growth and production, especially by prolonged snow cover.

Winter Desiccation

If the soil freezes, plants can no longer take up enough water to cover their requirements, even if evaporative losses are small. For plants, frozen ground means dry ground. The strain imposed on the water balance by winter conditions may result in damage due to desiccation; this is known as "frost drying" or "winter desiccation".

At low but positive soil temperatures the uptake of water and nutrients by the roots is already impaired due to increased transfer resistances between soil and roots, and due to the absence or root growth. The temperature influence on water uptake seems to adjust to the soil temperature in the plant's habitat. Plants of the arctic tundras and high mountains, and trees of northern forests can take up water from soil at temperatures barely above zero, and even from partially frozen soils, whereas plants from warmer regions have difficulty in doing so at temperatures below 10°C, and in fact symptoms of winter desiccation have been observed on unfrozen ground, for example in *Citrus* species [54, 119].

Winter desiccation is most commonly encountered on sites with little snow cover, when deep soil freezing occurs (Fig. 6.47). Characteristic features are the gradually increasing water deficit and the occurrence of injuries in late winter before the soil thaws, but when the sun is already warming the twigs more strongly and stimulating transpiration (Fig. 6.48). Winter desiccation can proceed according to one of two patterns:

Acute collapse of the water balance occurs in plants that readily open their stomata even in winter. Thus more water is lost by stomatal transpiration than can be withdrawn from the cold or frozen soil, or shifted from water reserves. The pull exerted by transpiration promotes rupture of the water columns in the conducting vessels, and causes cavitation; since the embolisms spread and completely block the passage of water through the xylem, the leaves and tips of the twigs are very soon damaged.

Fig. 6.48 a, b. Progressive desiccation of conifer needles in the course of the winter. **a** *Pinus cembra* at and above the timberline in the Central Alps. ● Mature trees; ○ trees 1−2 m in height with scant water reserves in the trunk; * isolated stunted individuals at the tree line. **b** *Picea abies* from different altitudes. ◆ Foothills at 330 m; only minimum values are shown here, joined with a *thin line*; ● mountain forests at 1000 m; ○ alpine timberline at 1950 m; ★ tree line at 2100 m. The *triangles* and the *continuous line* show the water content at 10% damage to the needle mesophyll. (After Larcher 1963 b; Michael 1967; Tranquillini 1982)

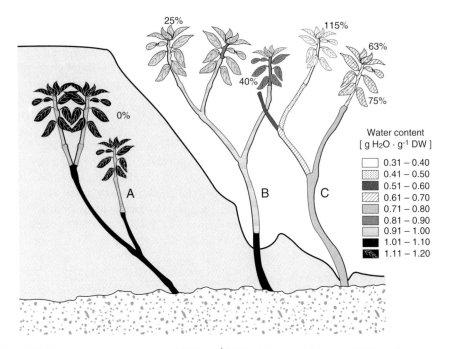

Fig. 6.47. Decrease in water content (gH_2O g^{-1} DW) of leaves and stems of *Rhododendron ferrugineum* at the alpine timberline during winter. *A* Permanently covered by snow; *B* terminal twigs mostly without snow cover in February; *C* plants without snow cover on frozen soil in March. *Percentages* indicate relative drought indices [*RDI*; for explanation see formula (6.4)]. (After Larcher 1963c)

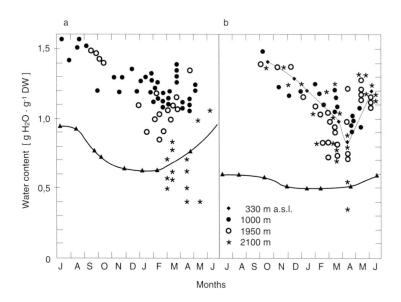

Chronic damage is typical of plants in which the stomata open only slightly, if at all, during winter dormancy, and therefore transpire almost solely via the cuticle or periderm, and lose water very slowly. Coniferous and deciduous tree species are of this type. In general, it can be said that the higher the transpiration of a plant and the lower its water-storage capacity, the greater its susceptibility to winter desiccation. Other factors increasing the risk of desiccation are abrasion of the cuticle due to wind, browsing animals and parasitic fungi (Fig. 6.49).

For plants not covered by snow, winter is not only a cold, but also a dry season. In the extreme north, just as in the mountains, the critical selective factors determining the limits for tree survival are the duration of snow cover, depth of snow, and frozen soil. Wherever the snow is blown away, the young trees perish, and where the rooted soil strata are frozen too long, or where permafrost rises too high, no trees can grow. The forests becomes sparse and gradually give way to tundra in subarctic regions, and in the high mountains

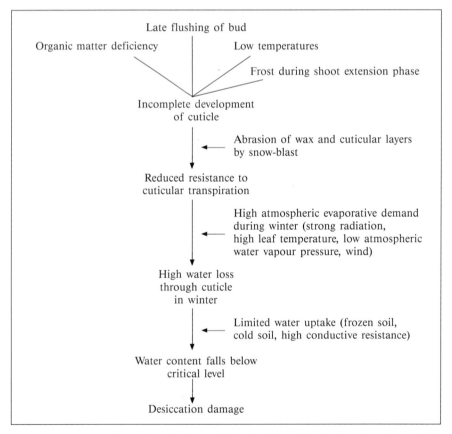

Fig. 6.49. Factors leading to chronic winter desiccation in conifers in the Alps. (Tranquillini 1979)

they are gradually replaced by dwarf shrubs and alpine grass and herb communities.

Harmful Effects of Long Periods Beneath Ice or Snow

Although a covering of snow affords protection from very low temperatures, wind and winter desiccation, the plants beneath it have to support its weight, and they are also deprived of light. Depending on the packing density, only 1–15% of daylight penetrates a 20 cm covering of snow [76].

A special danger arises from the low CO_2 and O_2 permeability of *ice sheets* and compact snow, as a result of which the gas exchange of the enclosed plants is greatly impeded. The respiratory CO_2 of the plants and microorganisms attains high concentrations, at the same time as the oxygen concentration drops to abnormal levels. Under such conditions, comparable to flooding-induced hypoxia, toxic substances accumulate via abnormal metabolic pathways. Especially injurious is the combination of ethanol and elevated CO_2 concentrations. Cytologically, the abnormal state is expressed in proliferation of endoplasmic membranes and the formation of concentric bundles of biomembranes. Such debilitated plants are less resistant to frost and are prone to attack by psychrophilic fungi (e.g. snow moulds).

6.2.3 Oxygen Deficiency in the Soil

Oxygen deficiency within the rhizosphere is a widespread condition, and its causes are manifold. Extensive areas of land are temporarily inundated by the flood waters of large rivers, and even smaller rivers or streams repeatedly overflow their banks; the plant cover of valley soils is often buried for long periods of time beneath mud and landslides; tundras, mires, swamps and other depressions are waterlogged (especially during snowmelt); and soils are compacted and become impermeable as a result of construction activities.

The soil atmosphere is low in oxygen in any case, due to its consumption by roots, soil animals and aerobic microorganisms (see Chap. 1.1.4). Atmospheric oxygen diffuses so slowly into dense, wet and flooded soils that it can drop to a few volumes percent, or disappear completely, within a few hours. As soon as the soil is oxygen-free, anaerobic microorganisms take over, creating a strongly reducing milieu in which Fe^{2+}, Mn^{2+}, H_2S, sulphides, lactic acid, butyric acid, among others, are present in toxic concentrations. Nitrogen turnover in the soil is also noticeably impaired. Clay soils with their fine pores are especially prone to oxygen deficiency, whereas gravel and sandy soils bordering fast-flowing rivers and streams are well supplied with oxygen even at maximum water flow.

Only certain plant species are found on permanently wet soils, e.g. herbaceous helophytes (swamp plants), a few tree species including poplar, willow, alder, *Taxodium distichum*, *Nyssa spp.*, mangroves, palms such as *Nypa fruticans*. For many other plants, inundation represents a risk that may lead to death within days or weeks. As a rule, the majority of plants succumb more quickly to flooding than to depletion of soil moisture.

6.2.3.1 Functional Disturbances and Patterns of Injury

Although roots are capable of respiring anaerobically, if this continues for some hours irregularities in metabolism occur. When the partial pressure of oxygen drops to 1–5 kPa (*hypoxia*), the respiratory quotient (RQ = mol CO_2/mol O_2) becomes greater than 1; alternative respiratory pathways, in which the oxygen affinity of the key enzymes is lower than that of the cytochrome oxidases, are activated, and the energy status of the adenylate system drops substantially. At this point, root growth stops (Fig. 6.50), root tips entering the low-oxygen zones die off, and adventitious roots develop. Older parts of the root systems often develop corky intumescences and swollen lenticels.

In the event of total and near-total oxygen deficiency (*anoxia*), respiration switches to anaerobic dissimilation. In the absence of terminal oxidation, acetaldehyde and ethanol accumulate. Increased ethanol content is the characteristic symptom of oxygen deficiency. Abscisic acid, ethylene and ethylene precursors are formed in larger amounts, evoking in the leaves partial stomatal closure, epinasty, and often abscission (Fig. 6.51). Cellular membrane systems break down, mitochondria and microbodies disintegrate and their enzymes are partially inhibited.

Surviving Oxygen Deficiency
Many plants can germinate, root and grow in oxygen-deficient soils because they have developed certain adaptations to meet conditions in an anoxic environment.

A **functional adaptation** to lack of oxygen is the absence of an increase in alcohol dehydrogenase (ADH) during anaerobiosis. At the same time, protein metabolism is adjusted within a few hours; after gene activation, for example,

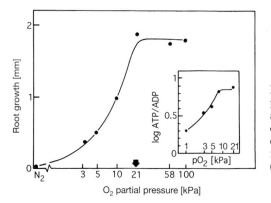

Fig. 6.50. The elongation rate of primary roots of maize seedlings as a function of the partial pressure of oxygen in the substrate. *Inset* cellular energy status at the different partial pressures of O_2. (Saglio et al. 1984)

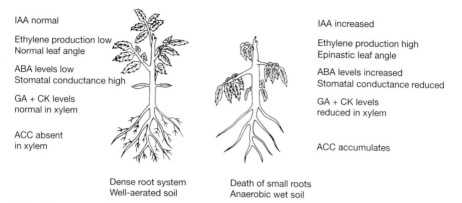

IAA normal

Ethylene production low
Normal leaf angle

ABA levels low
Stomatal conductance high

GA + CK levels
normal in xylem

ACC absent
in xylem

IAA increased

Ethylene production high
Epinastic leaf angle

ABA levels increased
Stomatal conductance reduced

GA + CK levels
reduced in xylem

ACC accumulates

Dense root system   Death of small roots
Well-aerated soil   Anaerobic wet soil

Fig. 6.51. A comparison of tomato plants with oxygen deficiency in the rhizosphere as a result of 24–48 h flooding (*right*), and plants growing on well aerated soil (*left*). The wilted appearance under anaerobic conditions is due to pathological downward growth of the leaves (epinasty), elicited by the production of ethylene; because the water potential remains high, the leaves still retain their turgescence. *ABA* Abscisic acid; *ACC* 1-aminocyclopropan-1-carboxylic acid (ethylene precursor); *CK* cytokinin; *GA* gibberellic acid; *IAA* indoleacetic acid. (After Bradford and Yang 1981)

roughly 20 different *anaerobiosis polypeptides* are produced in young maize plants [88] to function as isoenzymes replacing the original enzymes (e.g. aerobic ADH and phosphate-transfer enzymes). The end product is thus not the toxic ethanol, but lactic acid, shikimic acid or malic acid. The ADH activity in hypoxia provides a quantitative indication of the specific sensitivity of a plant species to oxygen deficiency (Fig. 6.52).

Morphological adaptations to a hypoxic milieu consist in the development of ventilating tissue (*aerenchyma*) with a continuous system of intercellular spaces through which oxygen can pass down the shoot to the underground or submerged organs (Fig. 6.53). The volume of the intercellular system in the root parenchyma of swamp plants amounts to between 20 and 60%, whereas in plants on well-aerated soils it makes up less than 10% of the parenchyma [37]. Well-aerated roots may even lose oxygen to the surrounding soil, where it can detoxify harmful reducing substances, e.g. by precipitating Fe^{2+} as Fe-III-oxide. Aeration is also furthered by temperature gradients. In order to obtain oxygen, plants growing on very dense and poorly aerated soils develop a system of laterally spreading roots near the surface; in flooded regions submerged parts of trunks and branches put out dense bundles of water roots (e.g. poplar, willow, alder, ash, *Acer rubrum*, *Eucalyptus* spp. and some conifers). Extreme types of specialization are seen in certain mangrove plants, in the form of lenticel-covered respiratory roots (pneumatophores) with a large volume of aerenchyma, and in swamp Cupressaceae and Taxodiaceae which produce "knee roots" that protrude above the surface of the soil and standing water.

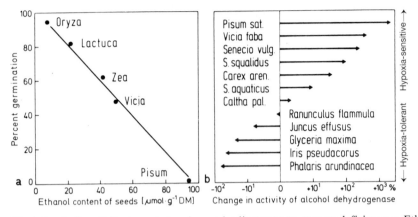

Fig. 6.52 a, b. Sensitivity to waterlogging and adjustment to oxygen deficiency. **a** Ethanol content and percent germination (as a measure of vitality) of seeds differing in sensitivity to oxygen deficiency. The seeds were examined after being submerged under water for 3 days. (Crawford 1977). **b** Increase (*positive values*) and decrease (*negative values*) in alcohol dehydrogenase activity of the roots of plants with different oxygen requirements, after growing for 1 month in waterlogged soil with inadequate oxygen supply. The activity of alcohol dehydrogenase is greatly enhanced in sensitive species, leading to ethanol accumulation and cell damage. In hypoxia-tolerant species there is a shift to alternative metabolic pathways, and the activity of alcohol dehydrogenase actually decreases. (McManmon and Crawford 1978)

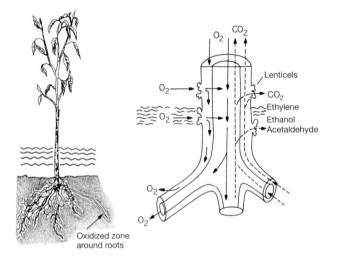

Fig. 6.53. Aeration system in the stem and roots of tupelo seedlings (*Nyssa aquatica* or *N. sylvatica*) and dual diffusion pathway of oxygen from the atmosphere to the rhizosphere and of volatile compounds from the root to the atmosphere. The O_2 escaping from the root surface creates an oxidizing milieu in the rhizosphere. (After Hook and Scholtens 1978)

6.2.4 Drought

Drought causes stress to a plant if too little water in a suitable thermodynamic state is available. This situation can occur for a variety of reasons, such as intense evaporation, osmotic binding of water in saline soils, or with frozen soils. A gradually increasing drought stress can also be the result of inadequate water uptake by plants growing in soil too shallow for the development of an adequate root system. Stress due to drought, in contrast to many other stressful events, does not occur abruptly, but develops slowly and increases in intensity the longer it lasts. Thus the dimension of time plays an important role in the survival of drought stress.

6.2.4.1 Drought as a Stress Factor

The term "drought" denotes a period without appreciable precipitation, during which the water content of the soil is reduced to such an extent that plants suffer from lack of water. Frequently, but not invariably, dryness of the soil is coupled with strong evaporation caused by dryness of the air and high levels of radiation. Shortage of precipitation alone, however, is not enough to cause aridity. The cold polar regions, for example, despite their low precipitation, are saved from aridity by low evaporation; and in dry regions the roots of the sparse vegetation reach permanently moist horizons in the vicinity of groundwater or rivers (e.g. gallery forests, and riparian vegetation).

On a large scale, **dryness** results from the combination of low precipitation and high evaporation. In dry regions drought is of such regular and prolonged occurrence that annual evaporation may exceed total annual precipitation. Such a climate is called *arid*, as opposed to the *humid* climate in regions with surplus precipitation [249]. About one third of the earth's continental area has a rain deficit, and half of this (about 12% of the land area) is so dry that annual precipitation is less than 250 mm, which is not even a quarter of the potential evaporation (Fig. 6.54). Extensive arid regions are found mainly between the latitudes of 15° and 30° north and south (Fig. 6.55), and on the lee sides of high mountain ranges which intercept rain-bearing winds. The greater the distance from the ocean, or in the rain shadow of a mountain range, the drier is the climate. There is a gradual transition from a humid climate, through a semiarid intermediate region with occasional or periodic dry periods, to an arid region characterized by permanent drought and increased salinity of the soil (Fig. 6.56).

The relationship between annual precipitation and annual evaporation gives only a rough indication of the humid or arid nature of an area. As far as the plants growing there are concerned, what matters is that there should be an assured water supply at the time of greatest need, which is the growing season. A picture of the relatively humid and relatively arid seasons can be obtained from climatic diagrams constructed from the records of meterological stations

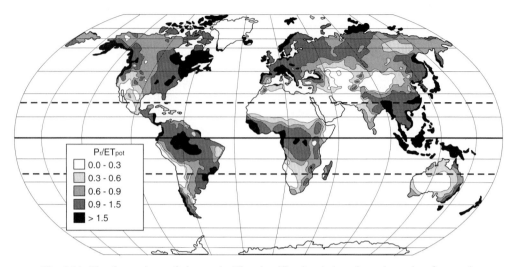

Fig. 6.54. The dry regions of the earth. The classification is based on the ratio of annual precipitation to regional potential evapotranspiration (Pr/ET_{pot}). *Arid climate* below 0.3; *semiarid* 0.3–0.6; *subhumid* 0.6–0.9; *humid* 0.9–1.5; *very humid* above 1.5. (UNESCO 1979; Box and Meentemeyer 1991)

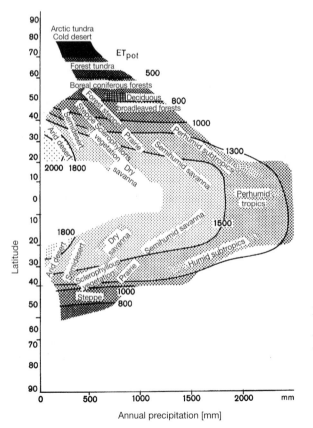

Fig. 6.55. Latitudinal distribution of annual precipitation and potential evapotranspiration (ET_{pot} in mm a^{-1}), and of characteristic plant formations. (After Schultz 1988)

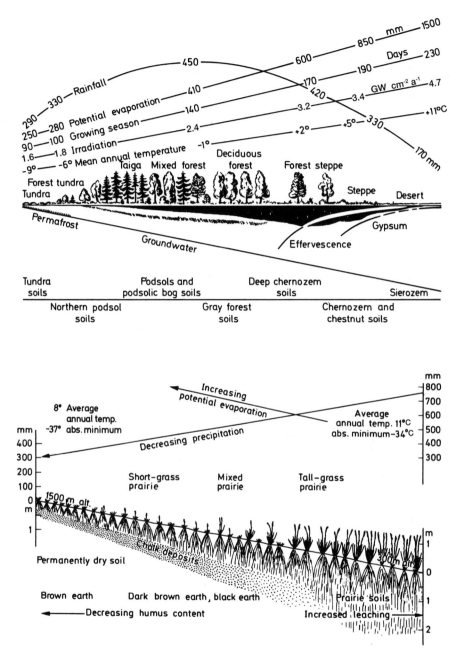

Fig. 6.56. Changes in climate, vegetation and soil along aridity gradients. The point of intersection of the precipitation curve and the curve for potential evaporation marks the boundary between humid and arid climates. *Above* Profile through eastern Europe from northwest to southeast, to the Caspian lowlands. (Vysotskom and Morozov, from Walter 1968). *Below* East-west profile throught the prairie region in the North America, rising from 300 m to the Great Plains at 1500 m above sea level. (After Walter 1968)

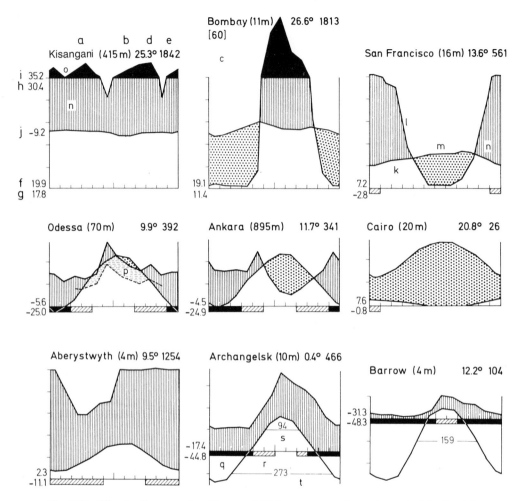

Fig. 6.57. Climate diagrams for *Kisangani* (Zaire, permanently wet equatorial climate); *Bombay* (India, tropical summer-rain climate); *San Francisco* (California, winter-rain region with summer drought); *Odessa* (Black Sea coast, semi-arid steppe climate); *Ankara* (Turkey, Mediterranean climate type with equinoctial rain); *Cairo* (Egypt, subtropical desert climate); *Aberystwyth* (Wales, maritime-temperate climate); *Archangelsk* (taiga zone on the White Sea, cold-temperate climate); *Barrow* (Alaska, arctic tundra climate). Interpretation of the climate diagrams: *Abscissa* In the northern hemisphere the months from January to December, in the southern hemisphere from July to June (the warm season is always in the middle of the diagram); *ordinate* one subdivision represents 10 °C or 20 mm precipitation. The labels denote: *a* station; *b* altitude above sea level; *c* number of years of observation; *d* mean annual temperature; *e* mean annual precipitation; *f* mean daily minimum in the coldest month; *g* absolute temperature minimum; *h* mean daily maximum in the warmest month; *i* absolute temperature maximum ; *j* mean daily temperature fluctuation (tropical stations with diurnal rather than seasonal variation); *k* curve of the mean monthly temperatures; *l* curve of mean monthly precipitation; *m* season of relative drought (*stippled*); *n* relatively humid season (*vertical shading*); *o* very humid season, mean monthly precipitation > 100 mm (scale reduced to 1/10, *black area*); *p* relatively dry season (precipitation curve shifted downward, 1 scale division represents 30 mm); *q* cold season, months with mean

(Fig. 6.57 [75, 120, 265]). In such a plot, the temperature curve serves as an indicator of the progressive change in evaporative power of the atmosphere during the year. That part of the year during which the precipitation curve lies below the temperature curve is a time of *drought* for the majority of plants that are not irrigated and are unable to utilize groundwater. This procedure offers the advantage that it can be used at stations for which no data on evaporation or radiation are available. However, climate diagrams cannot be applied everywhere; for example, the increase in evaporation with increasing altitude and the surface runoff of precipitation are not taken into consideration.

To assess the actual degree of drought stress experienced by plants in their environment (especially important in agriculture and forestry), various criteria based on measurements of soil moisture and analyses of the distribution of precipitation can be used. From the standpoint of ecophysiology, knowledge of external factors alone provides an insufficient basis for drawing accurate conclusions about the degree of drought. Only the plant itself can show reliably where and when lack of water becomes a stress, and this is best revealed by indicators of the state of plant's water balance. The extent to which individual plants, in their habitats, suffer from dry conditions depends not only upon their drought resistance, but also on the conditions prevailing in the habitat. Both of these factors are included in the **relative drought index** (RDI [96]). This index compares the actual water saturation deficit (WSD_{act}) with the critical threshold value for water saturation deficit (WSD_{crit}) for the species concerned.

$$RDI = \frac{WSD_{act}}{WSD_{crit}} \qquad (6.3)$$

The *critical threshold* can refer either to the first visible signs of drought injuries, or to the beginning of disturbance to a particular function, depending on the question to be answered. Instead of water saturation deficit, other indicators for water balance can be used, such as relative water content, water potential (with transpiration suppressed), or the osmotic potential. A low decimal value (RDI may also be expressed as a percentage) is an indication that drought stress in this plant species is slight. Information about the spatial and temporal differences in the severity of drought can be obtained by comparing several individuals of the same species, in different locations, with one another. The RDI values of the characteristic or most common species of a particular region also give an idea of the stress patterns typical of a particular vegetation or habitat (Table 6.6).

---

daily minimum below 0 °C (*black bar*); *r* period when daily minimum is below 0 °C and daily maximum above 0 °C, months with absolute minimum below 0 °C (*cross-hatched bar*); *s* number of days with mean temperatures above +10 °C; *t* number of days with mean temperatures above −10 °C. *These measures are representative of the climatic data important in plant ecology.* (After Walter and Lieth 1967)

Table 6.6. Drought stress in plants of different types of vegetation, expressed as relative drought index (RDI %). (Data of numerous authors: Larcher 1973b; Bannister 1976; Sveshnikova 1979; Bobrovskaya 1985)

Type of vegetation	Provenance	Average maximal water saturation deficit WSD_{act} at the natural growing site in % of the WSD_{crit} at $5-10\%$ damage
Mediterranean macchia	Dry limit	$80-85$
	Northern limit	$40-70$
Semishrubs	Mediterranean region	$90-105$
Ericaceous dwarf shrubs	Atlantic heaths	$15-10\ (88)^{a}$
Deciduous broadleaved trees and shrubs	Northern Europe	$10-40\ (50)$
Herbs of the forest understorey	Northern Europe	$6-25\ (50)$
Dicotyledonous herbs of xerothermic habitats	Central Europe	$40-85$
Meadow grasses	Central Europe	$20-40\ (75)$
Steppe plants	Central Europe	$50-90\ (108)$
	Central Asia	$60-80$
Desert plants		
Trees and shrubs	Karakorum	$30-50$
Semishrubs and forbs	Central Asia	$50-65$

[a] In brackets: rarely occurring extreme values.

6.2.4.2 Functional Disturbances and Patterns of Injury

Water deficiency results in a decrease in cell volume, an increase in concentration of the cell sap, and progressive dehydration of the protoplasm. There is no living process that is unaffected by a decline in water potential. Figure 6.58 provides an overview of a number of cellular functions that are altered by a decreasing water potential. Since drought stress in homoiohydric plants develops *gradually*, the relative stress sensitivity of the affected functions is revealed by the *sequence* of events.

The first and most sensitive response to **water deficit** is a decrease in turgor and a slowing down of growth processes (particularly elongation growth) in connection with this. Protein metabolism and synthesis of amino acids are both very soon impaired. One of the enzymes most strongly inhibited as a result of water deficiency is nitrate reductase. Even a brief period of negative water balance causes nitrate reductase values to drop by 20% or, after longer periods of water deficit, even by 50% [98]. This is why drought leads to a rise in the nitrate content of plants that have been treated with fertilizers. Also

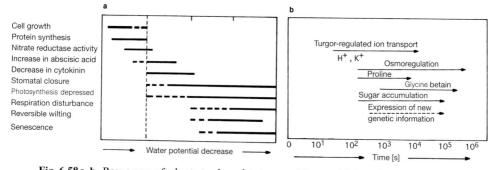

Fig. 6.58 a, b. Responses of plants to drought stress. **a** The sensitivity of cell functions and cellular processes during water deficiency. The *horizontal lines* show the ranges in which a clear response occurs in most plant species; *broken vertical line* the onset of stomatal closure. (After Hsiao 1973; Bradford and Hsiao 1982). **b** Time course of cellular responses to turgor perturbation. (After Wyn Jones and Pritchard 1989). For molecular responses to water deficit and drought-induced gene expression see Bray (1993)

nitrogen fixation in the root nodules of Fabales is very sensitive to drought, although the effect can be lessened by the presence of symbiotic VA mycorrhiza [177]. The reduction of the syntheses in protein metabolism supresses cell division; even moderate drought stress slows down mitosis, the S-phase being affected most. If water stress occurs during pollen development, the meioses exhibit chromosome anomalies, especially disturbances in the metaphase and anaphase. Thus drought lowers pollen fertility [164].

Even a moderate water deficit is sufficient to trigger the synthesis of abscisic acid from carotinoids in the roots. It is then transported as a "root signal" to different parts of the plant, where it elicits a variety of effects (Fig. 6.59); in the leaves it initiates stomatal closure. Under the influence of hormones synthesized in the leaves and roots in response to drought, changes occur in the allocation of assimilates, the ratio of shoot to root growth is altered, characteristic morphogenetic features develop and, as a rule, reproductive processes begin prematurely. If the degree of dehydration increases, catabolic processes become predominant, senescence is accelerated and older leaves dry out and are shed.

Many of these processes are at first completely reversible. Observations on the timing of the different events have revealed that protective responses and destructive processes succeed one another, and to some extent overlap (see Fig. 6.58). When turgor begins to decrease, *osmoregulatory measures* are initiated. A combination of syntheses of organic nitrogen compounds, and conversion of starch to soluble carbohydrates leads to the accumulation of low molecular organic substances in the cell compartments and the cytosol, thereby promoting the osmotic influx of water. This helps to maintain cell volumes, and also delays loss of turgor in the mesophyll as well as in the stomatal guard cells (Fig. 6.60), which means that the stomata remain open longer, leaving more time for carbon assimilation.

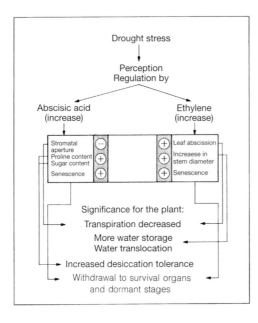

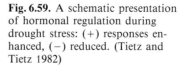

Fig. 6.59. A schematic presentation of hormonal regulation during drought stress: (+) responses enhanced, (−) reduced. (Tietz and Tietz 1982)

If the plant wilts its cells contract (*cytorrhysis*; Fig. 6.61), with the various cell compartments shrinking in proportion to the reduction in cell volume. Increasing concentrations of the intracellular solutes, especially ions, impair chiefly the secondary reactions of photosynthesis and mitochondrial respiration. In the final phase preceding cellular disruption, the central vacuole splits up into small fragmentary vacuoles, the thylakoids in the chloroplasts and the mitochondrial cristae first of all swell and are later broken down, the nuclear membrane becomes distended and the polyribosomes disintegrate [259]. If rehydration occurs before this series of changes has progressed too far, some

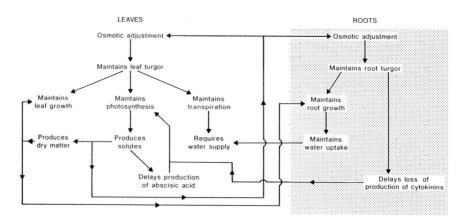

Fig. 6.60. Effects produced by osmotic adjustment in roots and leaves. (Turner 1986)

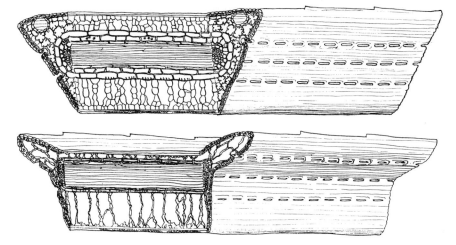

Fig. 6.61. Cell shrinkage in a needle of *Pinus strobus*. *Above* Water-saturated condition. *Below* After drying to 55% of its fresh weight. (Parker 1952)

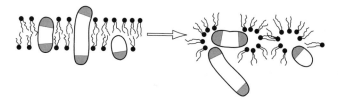

Fig. 6.62. Possible orientation of lipid and protein components in membranes in relation to their degree of hydration. Hydrated state (*left*) orientation of the hydrophilic poles (*heads*) of the phospholipids and of the membrane proteins (*stippled areas*) toward the external aqueous medium. Dehydrated state (*right*) reversed orientation of the polar ends of phospholipids and proteins toward interior water channels of the membrane. (Bewley and Krochko 1982)

plants (particularly drought-resistant species, and embryos of seeds) are able to restore their membrane structures and compartmentation. Thus it can be said that the fate of a cell after dehydration is determined by the degree of destruction of its membranes (Fig. 6.62) and its capacity to repair the damage incurred. Adhesions and disintegration of the biomembranes are the events leading ultimately to cell death.

6.2.4.3 Survival of Drought

Drought resistance is the capacity of a plant to withstand periods of dryness, and is a complex characteristic. The prospects for survival of a plant under extreme stress due to drought are better, the longer a harmful decrease in the water potential of the protoplasm can be delayed (*avoidance of desiccation*) and the more the protoplasm can dry out without becoming damaged (*desiccation tolerance*). Among homoiohydric plants desiccation tolerance is very slight, so that interspecific differences in drought resistance are chiefly due to differences in desiccation avoidance. A plant need not necessarily be drought resistant in order to live in dry regions. There are species which escape drought by timing their growth and reproduction to occur in the brief period when sufficient water is available. Figure 6.63 summarizes the possible ways for plants to survive in dry regions (*xerophytes*).

Drought-Escaping Xerophytes
For these species, which as a rule are not truly drought-resistant, survival of dry periods requires only the appropriate timing of the production of desiccation-resistant seeds or perennating organs specially protected from desiccation. The *pluviotherophytes* are ephemeral vascular plants that germinate after fairly heavy rainfall and rapidly complete their cycles of development. The majority of these plants are winter annuals. In order to be able to flower as soon as water is available, rain-ephemeral species must be neutral with respect to day length; their adaptive advantage lies in a prompt response of the ontogenetic cycle to the availability of water. Most plants of this kind survive the dry period in the desiccation-tolerant seed stage. The *geophytes*, on the other hand, have succulent underground organs such as rhizomes, tubers or bulbs. In the rainy period they can put out shoots at once by utilizing stored carbohydrates, thus enabling them to flower and develop fruits within a very short space of time.

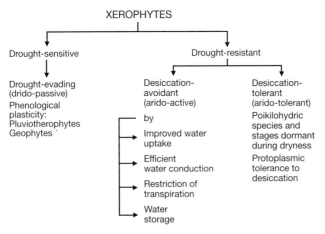

Fig. 6.63. Survival mechanisms of plants in dry regions. (After Shantz 1927; Evenari et al. 1975; Turner 1979; Ludlow 1989)

Desiccation Avoidance

Desiccation is delayed by all those mechanisms that enable the plant to maintain a favourable tissue water content as long as possible despite dryness of air and soil. This may be achieved by improved uptake of water from the soil, by reduced loss of water (early and effective increase in diffusion resistance and reduction of the transpiring surface), by a high water-conducting capacity, or by the storage of water. These functional measures for avoiding desiccation are also reflected in the morphology of the plant (Fig. 6.64).

Water uptake by an extensive root system with a large active surface area is improved further by rapid growth into deeper soil layers. The roots of plants of the steppes and deserts reach down, on an average, to a depth of $2-5$ m [247]. There the roots reach horizons containing water on which the plants can draw for some time. The seedlings of woody plants in dry regions may have tap roots ten times as long as the shoot, while grasses in such places develop a dense root system and send their threadlike roots to depths of some metres. Considerable portions of root systems of this kind are suberized or can store water. The shoot/root ratio is shifted further in favour of the roots, the greater the exposure to drought. The situation becomes critical when there is not

Fig. 6.64. Some examples of different growth forms allowing plants to survive drought. *a* Deciduous "bottle" trees with water-storing trunks (*Adansonia/Chorisia* type); *b* succulents storing water in the stem (Cacti/*Euphorbia* type); *c* succulents with water-storing leaves (*Agave*/Crassulaceae type); *d* evergreen trees and shrubs with deep tap root systems (sclerophyll type); *e* deciduous, often thorny shrubs (*Capparis* type); *f* chlorophyllous-stemmed shrubs (*Retama* type); *g* tussock grasses with renewal buds protected by leaf sheaths, and with wide-ranging root system (*Aristida* type); *h* cushion plants (*Anabasis* type); *i* geophytes with storage roots (*Citrullus* type); *j* bulb and tuber geophytes; *k* pluviotherophytes (annual plants); *l* desiccation-tolerant plants (poikilohydric type). (Compiled and expanded after Troll 1960; Stocker 1970, 1971, 1974a; Sen 1982)

enough room for the root system to expand. Plants with extensive root systems (woody plants in particular) growing on shallow soils are especially endangered by drought. Analogous problems arise in attempting to establish plants on and over flat roofs, tunnels and other underground structures. Even in humid regions only drought-resistant plants are suitable for such purposes, for they can manage with the small amount of water stored in the limited volume of soil.

Water conducting capacity is increased by enlarging the area of the conducting system (more xylem, dense leaf venation) and reducing the transport distance (shorter internodes). If the transpiring surface is reduced at the same time, the *relative* area of the conducting system is increased even though the *absolute* area (conducting cross section) is unchanged.

Reduction of transpiration helps conserve the available water. As a modulative adaptation, this takes place by timely closure of the stomata. A modificative change is seen when leaves growing under conditions of water deficiency develop smaller but more densely distributed stomata (Fig. 6.65). This modification makes a leaf able to reduce transpiration by a quicker onset of stomatal regulation. The leaves of genotypically adapted plants have more densely cutinized epidermal walls and are covered with thicker layers of wax. Stomata are present only on the under side of the leaf, are smaller and are often hidden beneath dense hairs or in depressions (grooves or crypts: Fig. 6.66). In this way the boundary layer resistance is increased and the air outside the stomata becomes moister. An effective reduction of water losses also can be achieved by rolling of the leaves (see Fig. 4.41) to reduce the transpiring surface. Further, leaves developed under conditions of poor water supply are usually correspondingly smaller or more deeply laciniated, and have a smaller *specific leaf area* (SLA).

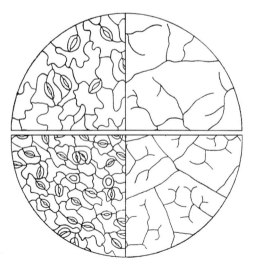

Fig. 6.65. Density of the stomata and venation on leaves of *Phaseolus vulgaris* grown with sufficient water (*top*) and with limiting soil moisture (*below*). (After Tumanov 1927). Morphological features of this kind can be the result of high endogenous levels of abscisic acid in drought stressed plants, or of treatment with ABA (Quarrie and Jones 1977; Ristic and Cass 1991)

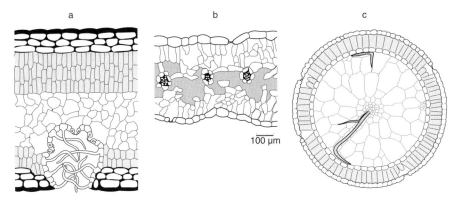

Fig. 6.66a–c. Examples of leaf structures of xerophytes. **a** Cross section of a scleromorphic leaf of *Nerium oleander* with thickened hypodermis and multilayered palisade parenchyma (*stippled*); the stomata are protected from the effects of wind. **b** Leaf of the Californian Asteraceae *Hemizonia luzulifolia* ssp. *rudis*, which stores water in a pectinaceous intercellular substance (*stippled*). **c** Cylindrical leaf of *Zygophyllum simplex*, a succulent plant of the north African desert, with external chlorenchyma (*stippled*) which surrounds the extensive, centrally situated water-storing tissue. (After Stocker 1952; Morse 1990). Cacti contain water-binding mucilage: Nobel et al. (1992b)

$$\text{SLA} = \frac{\text{surface area}}{\text{leaf mass}} \quad (\text{cm}^2\,\text{g}^{-1}) \qquad (6.4)$$

A particularly effective reduction of the transpiring surface of a plant can take place by the partial or complete *abscission of the leaves*. A number of woody plants in dry regions shed their leaves regularly during the drought season. Through peridermal transpiration the trunks and branches of large trees lose only a small fraction of the amount of water evaporated from the leaves when the water supply is good. Chlorophyllous-stemmed shrubs and some stem-succulents (e.g. Euphorbiaceae, Pachypodiaceae) shed their leaves as required, reducing the surface area to one third to one fifth of its usual value.

With *water storage*, mechanisms for desiccation avoidance reach their peak of perfection, especially when they are coupled with surface reduction and high transpiration resistance of the epidermis. A measure of storage capacity is given by the degree of succulence [42]:

$$\text{Degree of succulence} = \frac{\text{water content at saturation (g)}}{\text{surface area (dm}^2)} \qquad (6.5)$$

Under certain circumstances succulent plants can survive years of drought; the water reserves after the last rain may be sufficient for several weeks to pass before the stomata remain permanently closed (Fig. 6.67). As a rule, succulents accommodate their water reserves in relatively capacious specialized tissue. The *water-storage tissue* (see Fig. 6.66c) is usually situated in the interior of leaves and shoots. A special form of water conservation is the binding of water

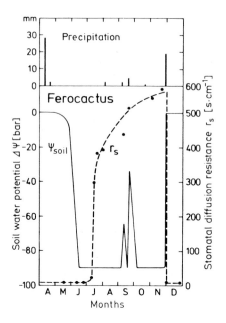

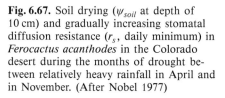

Fig. 6.67. Soil drying (ψ_{soil} at depth of 10 cm) and gradually increasing stomatal diffusion resistance (r_s, daily minimum) in *Ferocactus acanthodes* in the Colorado desert during the months of drought between relatively heavy rainfall in April and in November. (After Nobel 1977)

to strongly hydratable carbohydrates (mucilage) in cells and ducts, and in intercellular cavities; water reserves of this kind can protect the plant from too sudden wilting and severe shrinkage of the leaves.

The **movement of water** from storage tissue and massive organs (trunks and larger branches of trees, and the underground storage organs of herbaceous plants) becomes very significant during prolonged drought. Some tree species in dry regions store in the wood and cortex considerable quantities of water which can be redistributed to twigs and leaves in periodic (diurnal) rhythms. In baobab trees (*Adansonia digitata*) in Kenya, measurements showed that the trunk provided the leaves with 400 liters of water in one day [61]. Succulent trees and cacti contain two to six times as much water in their trunks and stems as other trees in the same habitat [168]. Even in humid regions the water stored in wood and bark provides a buffer for maintaining the water balance. It is known that in summer, 30 – 50% of the water lost during midday transpiration in 40-year old pine trees comes from the trunks and branches, which reduces the strain on the water status of the needles [270]. During dry periods the water stored at the base of the trunk is used up first, then gradually that further up the trunk and in the branches. The movement of water from stems and roots to the leaves is more effective because it can continue after the stomata have closed and transpiration is very slow. For evergreen woody plants exposed to the risk of winter drought the redistribution of water from the trunks to the leaves or needles is an important mechanism for survival.

Desiccation Tolerance

Desiccation tolerance is the species-specific and adaptable capacity of proto-plasm to tolerate severe loss of water. The lowest limiting values for the state of hydration of the plant or its tissues, in terms of RWC, WSD or water-poten-tial at which the first irreversible effects ("critical limit") or necrotic injuries appear (sublethal at $5-10\%$ injury, 50% injury is desiccation lethality, DL_{50}) are employed as the *measure of tolerance*. Although data based on water con-tent are informative in assessing the reserves available in the event of drought, they should not be used for comparisons of different species, since their absolute values are influenced by the anatomical peculiarities of the sample investigated (e.g. different proportions of supporting structures). The most suitable data for giving a measure of *specific* desiccation tolerance of plants are osmolality and water potential of cells or tissues.

Within the plant kingdom, as Table 6.7 shows, the desiccation tolerance of protoplasm varies over a wide range.

Desiccation-sensitive species: the protoplasm of most plants is highly sen-sitive to loss of water. The cells in the leaves of *homoiohydric* vascular plants suffer lethal injury if they are exposed directly to a relative air humidity be-tween 95 and 96% (corresponding to -11 MPa and -5.5 MPa) for several hours; root cells are even more sensitive, whereas bud tissues are more resistant. The level of resistance can be raised by hardening to dry conditions (Fig. 6.68). During periods of growth the cells are especially sensitive to desiccation; dur-ing dormant periods they are somewhat more resistant.

Among the *thallophytes* the most sensitive are the planktonic algae and those seaweeds attached so far below the surface that they are normally always covered by water. Algae of the intertidal zone, which regularly dry out at low

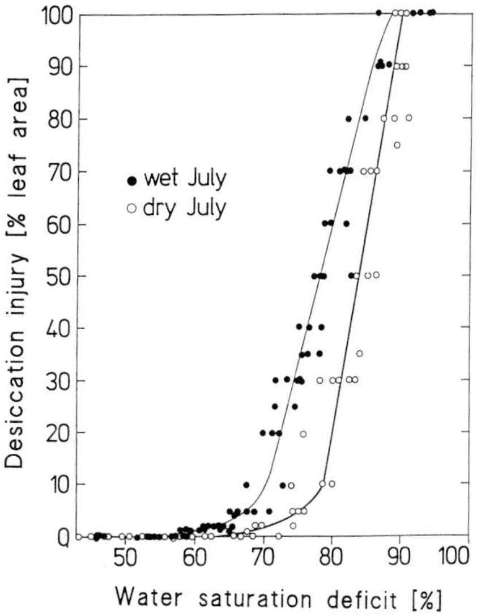

Fig. 6.68. Desiccation resistance of leaves of *Olea europea* in response to wet and dry summer conditions. In the rainy July 1953, the leaves were less tolerant to water saturation deficits than in the dry July 1952. (After Larcher 1963b)

394   Plants Under Stress

Table 6.7. Desiccation tolerance of plant cells after 12–48 h equilibration in vapour chambers with different relative air humidities. (After Iljin, Biebl, Höfler, Härtel, Abel, Kappen and Parker as cited in Larcher 1973b; Gaff 1980)

Plant group	Tolerated without injury	
	Relative air humidity (%)	Water potential (MPa)
Marine algae		
Lower sublittoral algae	99 – 97	– 1.4 to – 4
Sublittoral algae	95 – 86	– 7 to – 20
Eulittoral (intertidal) algae	86 – 83	– 20 to – 25
Liverworts		
Hygrophytes	95 – 90	– 7 to – 14
Mesophytes	92 – 50	– 11 to – 94
Xerophytes	(36) – 0	(– 140) to ∞
Mosses		
Hygrophytes	95 – 90	– 7 to – 14
Mesophytes	90 – 50 (10)	– 14 to – 93 (– 310)
Xerophytes	5 – 10	– 400 to ∞
Hymenophyllaceae	90 – 75	– 14 to – 38
Fern prothalli		
Pteridium aquilinum	>90	Up to – 14
Cystopteris fragilis	>90	Up to – 14
Asplenium ruta-muraria	40 – 60	– 70 to – 120
Homoiohydric cormophytes (tissue sections)		
Leaf epidermis	96	– 6
Mesophyll	95	– 7
Root cortex	97	– 4
Poikilohydric cormophytes		
Borya nitida	85 – 90	– 14 to – 22
Xerophyta villosa	66	– 56
Myrothamnus flabellifolia		
at natural site	11	– 298
in the laboratory	11 – 0	– 298 to ∞

tide, are more tolerant to desiccation than those living in deep water. Many bryophytes (especially liverworts) are adapted to a narrow range of water potential (there are differences even between species of a single genus), so that the various species are of value as indicators of the humidity of a habitat. It should be mentioned that in mosses hardening also can considerably increase resistance to desiccation.

Desiccation-tolerant plants: among *thallophytes* many species are capable of tolerating complete desiccation. Most lichens can withstand a state of dryness (in equilibrium with the surrounding air) for months and even years, resuming metabolic activity as soon as they are moistened again. Also completely desiccation-tolerant are some cyanobacteria, fungal mycelia, various mosses, and certain species of *Selaginella* and ferns.

Completely desiccation-tolerant species are also found among the *cormophytes* (vascular plants): e.g. *Ramondia serbica* and some species of *Haberlea* (both Gesneriaceae) in the Balkan region, and many so-called "resurrection plants" among the Myrothamnaceae, Scrophulariaceae, Lamiaceae, Cyperaceae, Poaceae, Liliaceae and Velloziaceae of the dry regions of Central Asia, Australia, South America and (in the greatest numbers) South Africa [71]. Such plants are as a rule perennial hemicryptophytes; their leaves are small and often rolled up; their growth is slow and their reproductive capacity poor. As they have no effective means of delaying desiccation (late stomatal closure in C_4 grasses, and lack of water storage) they very quickly become dehydrated and their water potential drops to extremely low values.

All plants capable of tolerating desiccation survive the extreme degree of cellular dehydration by transition to an *anabiotic state*, in which metabolism is almost at a standstill. As dehydration proceeds, the anabiotic state is brought about by synthesis of drought-stable proteins and incorporation of phospholipid-stabilizing carbohydrates (raffinose, trehalose) into the membranes. In some species, cell shrinkage is slowed down by gelation of the cell sap. Yet another factor contributing to desiccation tolerance is the capacity of the protoplasm of such species to rehydrate in a coordinated way when water becomes available (Fig. 6.69). Step by step, by reconstruction of the cell components, the conditions necessary for reactivating energy metabolism (first respiration, then photosynthesis) are again established.

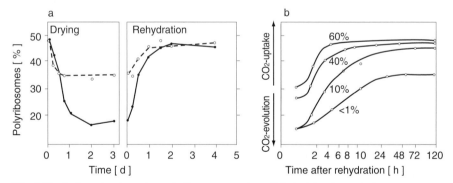

Fig. 6.69 a, b. Inactivation of metabolic processes in poikilohydric plants during desiccation and their reactivation during rehydration. **a** Breakdown and recovery on rehydration of polyribosomes during rapid (*broken line*) and slow changes in hydration (*solid lines, dots*) in the moss *Tortula ruralis*. Changes in polyribosome density give an indication of the rate of protein synthesis. (After Bewley 1981). **b** Time required for recovery of photosynthesis in the lichen *Verrucaria elaeomelaena* following desiccation at different air humidities (%). (After Ried 1953). Recovery of ultrastructure and photosynthetic function in a poikilohydric vascular plant (Velloziacea): Tuba et al. (1993, 1994)

6.2.5 Salt Stress

Salt stress may have been the first chemical stress factor encountered during the evolution of life on earth. The earliest living organisms were marine forms, and even today brackish biotopes are colonized from the marine side rather than from the freshwater. Thus, from the very beginning, organisms must have evolved effective mechanisms for ion regulation and for the stabilization of protoplasmic structures. After developing the capacity to cope with salt in the cell, and other adjustments involving the whole organism, halophytes were well adapted for survival in strongly saline habitats. Saline soils cover roughly 6% of the continental surface.

6.2.5.1 Saline Habitats

Saline habitats are defined by the presence of an abnormally high content of readily soluble salts. The oceans, salt lakes and saline ponds are *aquatic* saline habitats; on land there are *saline soils* under both humid and arid climatic conditions. In humid regions, it is possible for the soil to become salty in the spray region of the tidal zone, on dunes and in marshes; saline environments can also be found in the vicinity of sources that have been in contact with salt deposits. In aerosol form, some oceanic salt is carried inland up to 100 km or more by wind and clouds. Moreover, the salt content of the soil can be increased by spray, dust, and run-off water along roads that are salted in winter to keep them free of ice.

Only in the open ocean does the salt concentration remain constant (on an average 480 mM Na^+, 560 mM Cl^{-1}). In the intertidal zone the salinity fluctuates over a very wide range (290–810 mM Na^+ in the mangrove belt), and in salts marshes the Na^+ concentration can increase at times to as much as 600–1000 mM [9, 66].

During the growing season salts accumulate in the *plant canopy* as residues from evaporation and transpiration; after the leaves and other parts have died and fallen to the ground the salts are leached out by rain and returned to the soil. A survey of salt turnover and movement in halophytic habitats is given in Fig. 6.70. *Soil salinity* is greatly increased in arid regions where evaporation from the soil is greater over the course of the year than the amount of precipitation infiltrating the soil. Especially large amounts of salt accumulate in hollows where the groundwater table is high, in depressions with no drainage (salt flats), and in intensively irrigated areas where there is appreciable salt content in the irrigation water and insufficient drainage.

Saline soils in humid regions contain predominantly NaCl. Neutral saline soils of this sort also occur in dry regions. More often, the soils of steppe and desert contain sulphates and carbonates of Na, Mg, and Ca, which render them more alkaline. The high pH values (8.5–11) of the sodium soils (solonetz) occurring in steppes are due to the presence of basic salts like $NaHCO_3$ and Na_2CO_3, and NaOH; they arise during the wet season from Na-

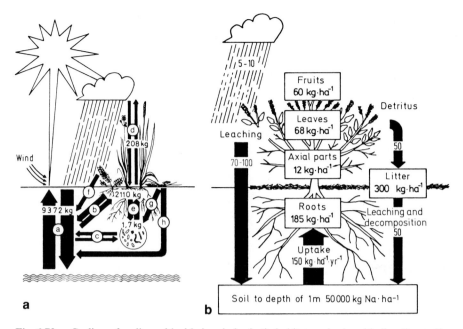

Fig. 6.70. a Cycling of sodium chloride in a halophytic habitat under humid climatic conditions. Below the soil surface is a saline groundwater horizon; during the growing season the soil solution contains 1.9% NaCl. *a* NaCl transport in the soil by capillary ascent during evaporation and by percolating water after precipitation; *b* NaCl uptake by vascular plants (*Triglochin maritima, Juncus gerardi, Glaux maritima*) and NaCl release after the roots died in autumn and winter; *c* NaCl uptake by bacteria in the soil and release by dying bacteria; *d* distribution and accumulation in the aboveground parts of the plants in spring and summer; *e* salt uptake and release by bacteria in association with halophytes; *f* leaching from living plants; *g* NaCl uptake by bacteria during decomposition of dead parts of plants; *h* leaching from plant detritus and litter. All quantities in kg NaCl per ha in the top 10 cm of soil. (Steubing and Dapper 1964). **b** Sodium turnover and content of Na^+ in a halophytic habitat under arid climatic conditions. Turnover is given in kg Na^+ ha^{-1} a^{-1}, content in kg Na^+ ha^{-1}. (After Breckle 1976, 1982)

humus-clay complexes. The highly saline solonchak soils in semiarid areas are usually recognizable by the salt efflorescences on the surface; these consist of chlorides, sulphates, and (bi)-carbonates of sodium, magnesium and calcium. Such soils have a high content of gypsum, are pasty, and, when dry, the surface is hard and cracked.

Salinity, especially in soil science and agronomy, is usually given in terms of the electrical conductivity of the aqueous saturation extract (EC_e) in $S\,m^{-1}$, formerly also in mmhos/cm (equivalent to $0.1\,S\,m^{-1}$ or $1\,mS\,cm^{-1}$). The EC_e and osmotic potential are linearly proportional ($1\,mS\,cm^{-1} = -0.036\,MPa$) and the salt concentration of the soil can be read off from a nomogram. This can vary with the amounts of salts other than NaCl which are present. The yield of plants sensitive to salt stress is reduced if the EC_e values exceed $4\,mS\,cm^{-1}$ and for this reason irrigation water should not have

EC_e values above $2 \, mS \, cm^{-1}$. The electrical conductivity of seawater is at least $44 \, mS \, cm^{-1}$ [57].

6.2.5.2 Effects of High Salt Concentrations on Plants

The burden of high salt concentrations for plants is due to the osmotic retention of water and to specific ionic effects on the protoplasm. Water is *osmotically* held in salt solutions, so that as the concentration of salt increases water becomes less and less accessible to the plant (see Fig. 6.71). An excess of Na^+ and, to an even greater extent, an excess Cl^- in the protoplasm leads to disturbances in the ionic balance (K^+ and Ca^{2+} to Na^+) as well as *ion-specific* effects on enzyme proteins and membranes. As a result, too little energy is produced by photophosphorylation, and by phosphorylation in the respiratory chain, nitrogen assimilation is impaired, protein metabolism is disturbed and accumulation of diamines such as putrescine and cadaverine, and of polyamines occurs [65].

If the adverse osmotic and ion-specific effects of salt uptake exceed the level tolerable to the plant, **functional disturbances** and injuries occur. Photosynthesis is impaired, not only by closure of the stomata, but also by the effect of salt on the chloroplasts, in particular on electron transport and the secondary processes. Respiration, especially in the roots, can be either increased or decreased by salt. It seems that the enzyme systems of glycolysis and the tricarboxylic acid cycle are more sensitive than those of the alternative metabolic pathways. When the NaCl content of the soil is high the uptake of mineral nutrients, particularly NO_3^-, K^+ and Ca^{2+}, is reduced.

Growth processes are especially sensitive to the effects of salt, so that growth rates and biomass production provide reliable criteria for assessing the degree of salt stress and the ability of a plant to withstand it. In *obligatory halophytes* (euhalophytes, e.g. *Salicornia, Salsola, Suaeda, Halocnemum*), which are confined to saline habitats, growth is promoted by the uptake of moderate amounts of salt (Fig. 6.72); only if the salt reaches high levels is

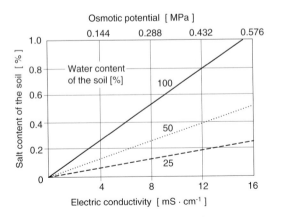

Fig. 6.71. Relationship between salt content, relative water content (100% = water saturation) and electrical conductivity (EC_e) of the saturation extract of the soil. (Marschner 1986)

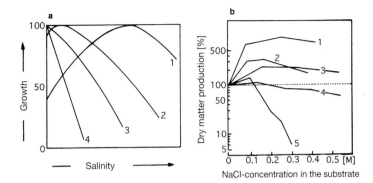

Fig. 6.72 a, b. The influence of salinity on growth. **a** Schematic representation of the effect of increasing NaCl concentration in the substrate on the growth of *1* euhalophytes; *2* facultative halophytes; *3* slightly salt-tolerant non-halophytes; *4* halophobic plants. **b** Dry matter production of various halophytes in relation to the NaCl concentration in the nutrient medium (as compared with NaCl-free controls). *1: Salicornia europaea*; *2: Aster tripolium*; *3: Suaeda maritima*; *4: Spartina foliasa*; *5: Puccinellia peisonis*. (After Baumeister and Schmidt, Flowers, Phleger, Stelzer and Läuchli, as cited in Albert 1982). Comparative study of growth and photosynthesis of seashore halophytes: Rozema and Van Diggelen (1991)

growth impaired and the plant shows signs of stress, such as the production of anthocyanin or the breakdown of chlorophyll. *Facultative* halophytes (e.g. many species of Poaceae, Cyperaceae, Juncaceae; *Glaux maritima, Plantago maritima, Aster tripolium*) occur on slightly saline soil, but their growth is soon inhibited if the salt level rises. Other species of plants are *indifferent* to salt; although they are usually found on non-saline soils they also tolerate slight salinity (e.g. ecotypes of *Festuca rubra, Agrostis stolonifera, Phragmites communis*, species or subspecies of *Puccinellia, Lotus* and *Atriplex hastata*, and crop plants such as *Vigna unguiculata* and sugar beet). Uptake of salt by non-halophytic or *halophobic* species leads to marked or even severe impairment of growth.

The colonization of saline habitats by halophytes depends on the *germination requirements* and salt *resistance of the seedlings*. Both seedlings and young plants are more sensitive to salt than the mature halophyte. Furthermore, the young plants are exposed to greater hazard since their roots are confined to the upper soil layers which contain the highest salt concentration. Germination is most successful in salt-free or low-saline environments, and there are only a few halophytes that can germinate at a salinity of $30-40$ mS cm^{-1} (with an osmotic potential of roughly $1-2$ MPa); as a rule, the upper limit for germination and subsequent growth is $15-20$ mS cm^{-1} (about $0.5-0.7$ MPa) and below [202].

Extreme salt stress leads to dwarfism and to inhibition of root growth. Bud opening is delayed, shoots are stunted, and the leaves are small; cells die and necroses appear in roots, buds, leaf margins and shoot tips; the leaves become yellow and dry before the growing season has ended, and finally whole portions of the shoot dry out. Lower levels of cytokinin and increased amounts of abscisic acid and ethylene are involved in this premature onset of senescence.

6.2.5.3 Survival in Saline Habitats

Plants growing in saline habitats cannot evade the effects of salt and must therefore develop at least some degree of resistance to it. *Salt resistance* is the ability of a plant either to avoid, by means of salt regulation, excessive amounts of salt from reaching the protoplasm, or, alternatively, to tolerate the toxic and osmotic effects associated with the increased ion concentration (Fig. 6.73).

Regulation of the Salt Content
Halophytes can regulate their salt content in a variety of ways:
   Salt exclusion: in some mangrove species ultrafiltration by *transport barriers in the roots* prevents the salinity of the water in the conducting system from becoming too high. In *Prosopis farcta* almost no salt reaches the leaves although salt ions, Na$^+$ in particular, are taken up by the roots. *Interruption of salt transport* also takes place in various crop plants, and especially in halophobic species excess ions are held back in roots, in the upper stems, and in leaf- and flower stalks, thus lessening the amount of salt reaching meristems, developing leaves and young fruits [107].
   Salt elimination: a plant can rid itself of excess salt by releasing volatile methyl halides, through exudation by glands and excretion of salt at the surface of the shoot, and by shedding parts heavily loaded with salt. Marine phytoplankton, macroalgae and fungi, but also *Mesembryanthemum crystallinum* (through an S-adenosyl methionine transfer mechanism), produce methyl

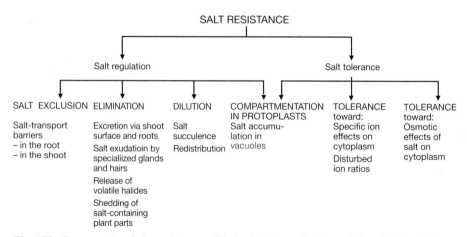

Fig. 6.73. Components of the resistance of halophytes to salt stress. (After Steiner 1934; Waisel 1972; Kreeb 1974b; Flowers et al. 1977)

chloride, bromide and iodide, with an estimated annual global emission into the atmosphere of 10^6 tons [278]. Salt-secreting glands actively eliminate salts, thus keeping accumulation in the leaves within certain limits (Fig. 6.74). Glands of this kind are found in various mangrove plants (for example, *Avicennia*), species of *Tamarix*, *Glaux maritima*, various Plumbaginaceae, and halophilic grasses such as *Spartina*, *Distichlis*, etc. The vesicular hairs of some *Atriplex* species in arid regions and of *Halimione* accumulate chlorides in the cell sap, then die off and are replaced by new hairs. Another means of desalination is the *abscission of older leaves* that have accumulated considerable amounts of salt. Meanwhile, young leaves capable of accumulating salt while performing their normal function grow to replace those that fall. This is a characteristic feature of halophytic rosette plants such as *Plantago maritima*, *Triglochin maritimum* and *Aster tripolium*.

Salt succulence: the stressful factor in the action of salts on protoplasm is not their total amount but rather their concentration. If the storage volume of the cells increases in step with the uptake of salt (as the cells steadily take up water) the salt concentration can be kept reasonably constant for extended periods. This kind of succulence, for which chloride ions are responsible, is widespread among halophytes of wet saline environments (*Salicornia* and other coastal plants of the family Chenopodiaceae, and the mangrove *Laguncularia*; Fig. 6.75) as well as in xerohalophytes of dry regions. The latter type of plant exhibits in addition features typical of drought-adapted succulence.

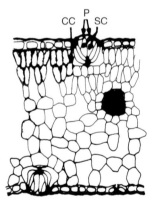

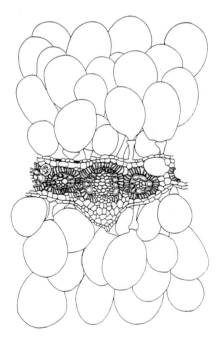

Fig. 6.74. Salt elimination by exudation and vesicular hairs. *Left* Multicellular salt gland complex in the leaf epidermis of *Limonium gmelinii*. *P* Pores; *SC* secretory cells; *CC* collecting cells. (Ruhland 1915). *Right* Salt-accumulating, pediculate vesicular hairs on leaves of *Atriplex mollis*. (Berger-Landefeldt 1959)

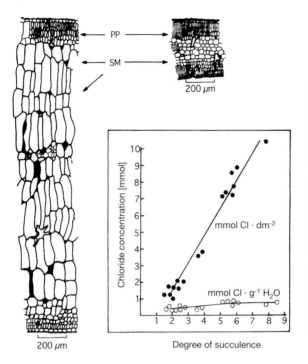

Fig. 6.75. The salt-induced increase in succulence during leaf development in mangroves. Tissue structure of young (*right*) and mature leaves (*left*) of *Sonneratia alba*. *PP* Palisade parenchyma; *SM* succulent storage mesophyll. Magnification is the same in both drawings. (Walter and Steiner 1936). *Graph* The chloride concentration in leaves of *Laguncularia racemosa* related to the leaf surface (*dots*) and to the saturation water content (*circles*). (After Biebl and Kinzel 1965)

Salt redistribution: Na^+ and Cl^- can be readily translocated in the phloem, so that the high concentrations arising in actively transpiring leaves can be diluted by redistribution throughout the plant.

Salt Accumulation and Intracellular Compartmentation

Halophytes compensate for the osmotic potential in the saline environment by accumulating salt in the cell sap. Marine algae maintain their turgor by active absorption of salt to lower their internal osmotic potentials to values of about -2.6 to -3 MPa, which are below that of sea water (-2.5 MPa). Algae of the intertidal zone, where the salt concentrations are subject to diurnal fluctuations, regulate their osmotic potential by rapid transport of salts in either direction according to need. In the brackish water of river estuaries ecotypes adapted to different salt contents can be found among the brown and red algae.

Terrestrial plants growing on saline soils accumulate salt in their cell sap up to a level at which their osmotic potentials are lower in the soil solution. Water can then move along an osmotic gradient (Fig. 6.76). In addition to the osmotic effect exerted by the accumulated salt, osmotic potential is also lowered by organic acids and soluble carbohydrates. In halophytic monocotyledons, especially grasses, which contain less salt than halophytic dicotyledons, the accumulation of soluble carbohydrates plays a substantial part in main-a

b

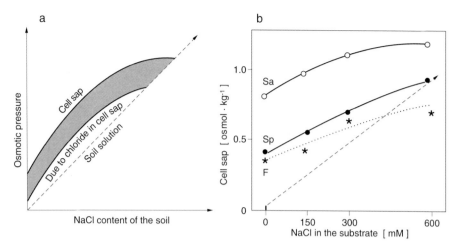

Fig. 6.76 a, b. The relationship between increasing salt content of the soil and the osmotic pressure in the cell sap of halophytes. **a** Diagrammatic representation: with increasing NaCl concentration in the soil, the osmotic pressure in the soil solution increases linearly (*broken line*). Halophytes overcome the correspondingly greater difficulty in absorbing water from the soil by taking up salt in excess and accumulating it in the cell sap up to a certain limiting value given by the intersection of the curve for chloride in the cell sap with the straight line for the soil. Added to the osmotic effect of NaCl storage is that of the other cell-sap components (the osmotic effectiveness of non-chlorides in the cell sap is indicated by the *hatched region*). Only when the salt concentration in the soil is very high are the halophytes no longer able to compensate for the osmotic pressure of the soil. (Walter 1960). **b** Types of response in cell sap osmolarity to increasing salt concentration in the substrate: *Sa: Salicornia europaea* (euhalophyte); *Sp: Spartina anglica* (salt-regulating halophyte); *F: Festuca rubra* ssp. *litoralis* (slightly salt-tolerant, salt excluder). (After Rozema et al. 1985)

taining a low osmotic potential in the cell sap (Fig. 6.77). The *ion ratios* Na^+/K^+ and Cl^-/SO_4^{2-} in the cell sap of halophytes are very often specific to the family and species of plant. Whereas many halophytic dicotyledons accumulate more Na^+ than K^+, in most halophytic monocotyledons the amount of K^+ equals or exceeds that of Na^+. The anion most commonly accumulated is Cl^- (for example, in species of *Salicornia* and *Suaeda*, *Atriplex confertifolia* and many grasses and Juncaceae), although SO_4^{2-} predominates in Plumbaginaceae, *Plantago maritima*, *Lepidium crassifolium*, species of *Tamarix* and various other dicotyledons. Different species of a single genus can be of different types with respect to ion accumulation: *Salsola kali* takes up $K^+ > Na^+$ and $Cl^- > SO_4^{2-}$, *Salsola turcmanica* $Na^+ > K^+$ and $Cl^- > SO_4^{2-}$, and *Salsola rigida* also accumulates $Na^+ > K^+$, but $SO_4^{2-} > Cl^-$.

The accumulation of salt in the plant nevertheless endangers metabolic processes. With the exception of the membrane-bound ATPases, most enzymes of halophytes appear to be as sensitive to salt as those of halophobic species. The ability of the protoplasm to tolerate high concentrations of salt depends on

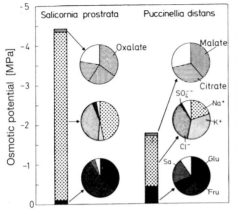

Fig. 6.77. Total osmotic potential of the cell sap of *Salicornia prostrata* (Na/Cl halophyte) and of *Puccinellia distans* (K/Cl glycohalophyte), with component potentials as indicated. *Blocks* Relative contribution to the osmotic potential by inorganic ions (*stippled*), organic ions (*hatched*) and soluble carbohydrates (*black*). *Circles* Proportions of various ions and sugars involved in each component potential; *Glu* glucose; *Fru* fructose; *S* sucrose; *white sectors* remaining ions and sugars. (Albert and Popp 1977, 1978)

selective **compartmentalization** of ions entering the cell. The majority of the salt ions accumulate in the vacuoles (includer mechanisms); this reduces the concentrations to which the cytoplasm and, most important, the chloroplasts are exposed (Table 6.8) and so protects their enzyme systems from direct salt stress. Compartmentalization of NaCl into the vacuole is achieved by the membrane ATPases. In order to provide the energy continually required for the transport of ions and the maintenance of their unequal distribution within the cell, the plant must divert part of its assimilates which would otherwise be invested in production, growth and competitive capacity. Halophytes can therefore be said to exist in a state of permanent stress.

Salt Tolerance

Salt tolerance is the *protoplasmic component* of resistance to salt stress. This involves the degree to which the protoplasm, depending on the plant species, tissue type and vigour, can tolerate the ionic imbalance associated with salt stress, and the toxic and osmotic effects of increased ion concentrations.

Some organisms are extraordinarily salt tolerant. The photoautotrophic flagellate *Dunaliella salina* lives in concentrated salt pools; the enzymes of the

Table 6.8. Distribution of salt ions between the different cell compartments in leaves of *Suaeda maritima* grown on a substrate containing 340 mol m^{-3} NaCl. (Flowers 1985)

Cell compartment	Proportion of cell volume (%)	Ion concentration (mol m^{-3})	
		Na$^+$	Cl$^-$
Cytoplasm	11	116	60
Chloroplasts	1.4	104	98
Vacuole	81.6	494	352
Cell wall	6	194	138

halophilic bacterium *Pseudomonas salinarum* and the yeast *Debaryomyces hansenii* remain functional even at concentrations of 20–24% NaCl. The salt tolerance of marine algae is exactly matched to the concentration amplitude of their particular habitats; this is especially well illustrated by the marked differences betweeen the euryhaline marine algae of the intertidal zone and the stenohaline algae of the sublittoral (Fig. 6.78).

Halophytic vascular plants can thrive on soils with salt concentrations of 2–6% and more (in extreme cases up to 20%), and can accumulate NaCl in their cell sap NaCl up to a 10% solution, whereas cells of halophobic plants are killed by solutions of only 1% (171 mM) NaCl; callus cultures of their more resistant mutants tolerate up to 300 mM.

Salt tolerance of the protoplasmic components (proteins, biomembranes) is attained by means of **stress proteins and compatible cytoplasmic osmotic solutes.** In the event of increasing salt concentrations, stress proteins are induced within 3–6 h after activation of certain DNA sequences [36]; similarities

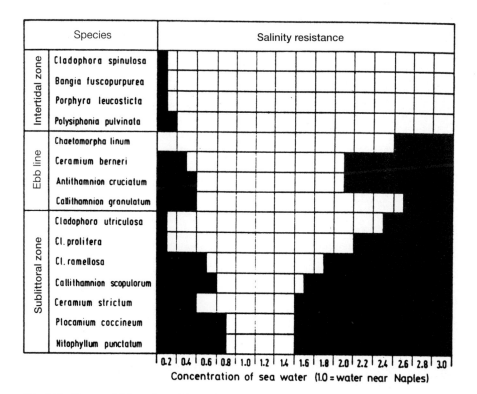

Fig. 6.78. The sensitivity of maritime algae from different depths to high salinity. *White areas* surviving; *shaded* dead after 24 h. Algae from the intertidal zone, which in their habitat are exposed to particularly large fluctuations of salt concentration, tolerate both hypotonic and hypertonic conditions better than the constantly wet algae of the ebb line, and even more so than the sublittoral algae. (After Biebl 1938)

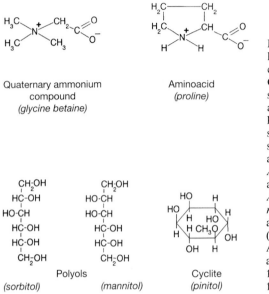

Quaternary ammonium
compound
(glycine betaine)

Aminoacid
(proline)

Polyols

(sorbitol)

(mannitol)

Cyclite
(pinitol)

Fig. 6.79. Examples of osmotically active, compatible organic compounds in halophytes. Glycine betaine: in *Atriplex* species, *Salicornia, Suaeda,* and among *Avicennia* mangroves. Proline: in *Armeria, Artemisia* species, *Salicornia, Suaeda,* and some mangroves (in *Aegialitis* and *Xylocarpus*). Polyols: in *Aster* species, *Juncus, Puccinellia,* and some mangroves (in *Aegiceras, Luminitzera* and *Sonneratia*). Pinitol: in *Spergularia* and in all Rhizophoraceae (mangroves). (After Popp and Albert 1980; Popp 1984; Popp et al. 1984; Aspinall and Paleg 1981; Wyn Jones and Storey 1981; Jeffrey 1987)

to drought- and heat-shock proteins suggest the possibility of cross protection. Osmotic equilibria between the cytoplasm and the different cell compartments are maintained by compatible (non-toxic) organic compounds (Fig. 6.79). Without compensation by the synthesis of innocuous osmotically active substances, the active accumulation of salt in the vacuoles would necessarily result in large concentration gradients toward the cytoplasm. In fact many plants, as a non-specific reaction to stress from excess salt, drought or frost, produce and accumulate amino acids and amides (proline, alanine, glutamine, asparagine), quarternary ammonium bases (betaines), and various sugars and polyols (e.g. mannitol, sorbitol). Certain plant groups even have their own characteristic patterns of stress metabolites (e.g. mannitol and pinitol are especially prominent in mangroves). Moreover, soluble carbohydrates and amino acids contribute towards protection of proteins and biomembranes from the detrimental effects of high ionic concentrations.

6.2.5.4 Salt Sensitivity of Halophobic Plants

Plants that normally are not encountered in saline habitats, and cannot therefore be called halophytes, are not necessarily completely sensitive to salt. In the different zones of vegetation along seashores, the gradual transitions indicate different degrees of salt resistance in the various plant species and even the existence of ecotypes (e.g. of *Plantago maritima, Juncus bufonius* [105]).

The salt resistance of some food and fodder plants is shown in Fig. 6.80. Not only are resistant species less vulnerable to direct damage by salt, but their

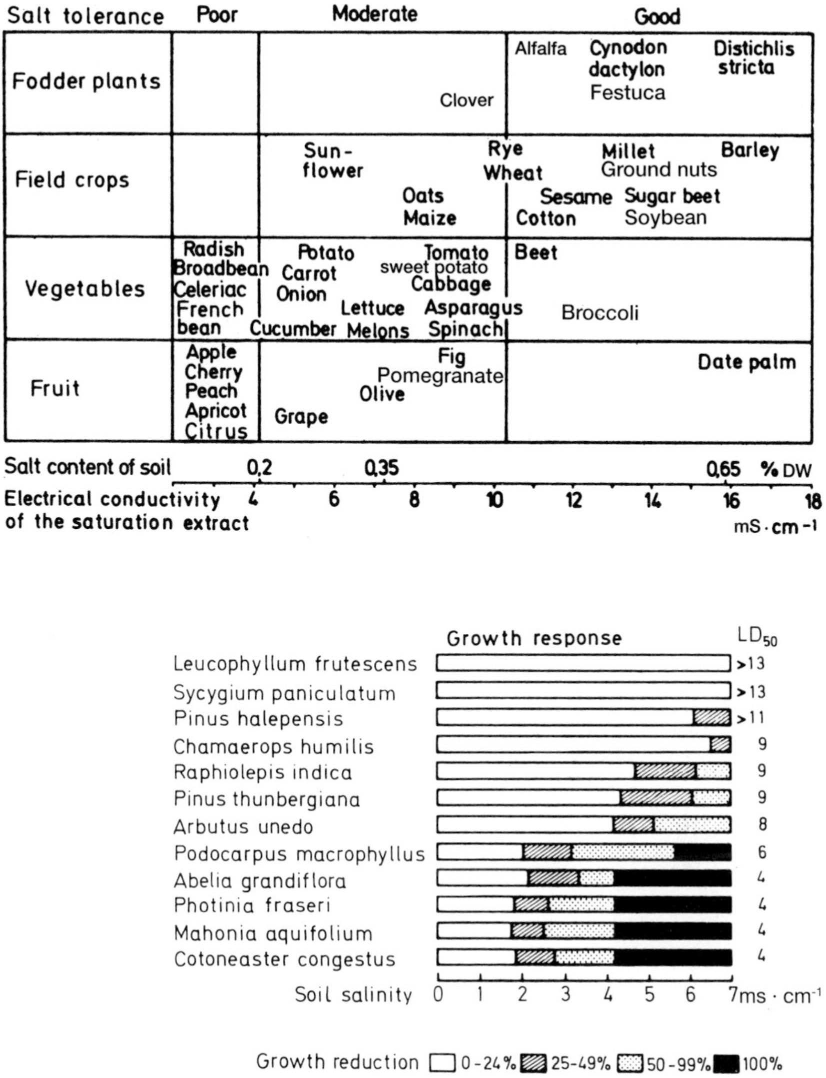

Salt tolerance	Poor	Moderate		Good		
Fodder plants			Clover	Alfalfa Festuca	Cynodon dactylon	Distichlis stricta
Field crops		Sun- flower	Rye Wheat Oats Maize	Millet Ground nuts Sesame Sugar beet Cotton Soybean		Barley
Vegetables	Radish Broadbean Celeriac French bean	Potato Carrot Onion Cucumber Melons	Tomato sweet potato Cabbage Lettuce Asparagus Spinach	Beet Broccoli		
Fruit	Apple Cherry Peach Apricot Citrus	Grape	Fig Pomegranate Olive	Date palm		

Salt content of soil	0,2	0,35		0,65	% DW
Electrical conductivity of the saturation extract	4 6	8 10	12 14	16 18	mS·cm⁻¹

Growth response   LD$_{50}$

Leucophyllum frutescens   >13
Sycygium paniculatum   >13
Pinus halepensis   >11
Chamaerops humilis   9
Raphiolepis indica   9
Pinus thunbergiana   9
Arbutus unedo   8
Podocarpus macrophyllus   6
Abelia grandiflora   4
Photinia fraseri   4
Mahonia aquifolium   4
Cotoneaster congestus   4

Soil salinity   0   1   2   3   4   5   6   7 ms·cm⁻¹

Growth reduction ☐ 0-24% ▨ 25-49% ▥ 50-99% ■ 100%

Fig. 6.80. Salt resistance of various economic plants. *Above* Agricultural, horticultural and orchard crops. The different species are placed in the table above the soil salt content at which a 50% reduction in yield is to be expected. Data of many authors. (Kreeb 1965; Cox and Atkins 1979). *Below* Salt resistance of ornamental shrubs and trees (indicated on the *right* by the dose for 50% lethality) and growth reduction on saline soils (*bars*). Soil salinity and LD$_{50}$ are given in terms of the electrical conductivity (mS cm⁻¹) of the saturation extract (Francois and Clark 1978). Salt-tolerant plants for landscape architecture: Barrick (1978)

growth and yield are also less affected than those of more sensitive species. For agronomists, foresters and gardeners, even a moderate degree of *salt resistance in crop plants* is useful in their attempts to utilize potentially arable saline soils in dry regions, especially in the subtropics. Great efforts are being made by plant breeders – particularly with the help of tissue culture and gene technology – to develop more resistant varieties, primarily for use in poor developing countries.

In temperate or cold regions, protection and proper management of the environment requires a knowledge of the resistance of trees, shrubs and grasses to the de-icing salts (NaCl, $MgCl_2$, $CaCl_2$) spread in winter on streets and highways. The salt reaches the aerial parts of the vegetation as spray, and often also its roots via percolation into the soil. Table 6.9 summarizes the salt tolerance of trees and shrubs commonly planted along roadsides and sidewalks. An alternative is to employ chloride-free de-icing salts, such as ammonium sulfate or calcium-magnesium acetate.

Table 6.9. Relative sensitivity of non-halophytic trees and shrubs to saline soils and to salt spray. Toxicity threshold values for the chloride content of leaves in early summer in salt-sensitive broadleaved trees and shrubs: 0.3 to 0.5% Cl in the dry matter; in sensitive conifers: $0.2-0.4\%$; in salt-resistant broadleaved trees and shrubs: $0.8-1.6\%$; in resistant conifers: about 0.6%. (Sucoff 1975; Meyer 1978; Carter 1982; Däßler 1991)

Salt-sensitive	Relatively resistant to salt
Deciduous trees and shrubs	
Acer platanoides	*Acer negundo*
Aesculus	*Ailantus*
Carpinus	*Elaeagnus*
Euonymus	*Fraxinus*
Fagus	*Gleditsia*
Juglans	*Hippophae*
Ligustrum vulgare	*Lycium*
Platanus	*Potentilla fruticosa*
Prunus serotina	*Quercus* species
Rosa rugosa, some species	*Rosa rugosa*, some species
Syringa	*Robinia*
Tilia	*Sophora*
Evergreen broadleaved woody plants	
Ilex species	*Coccoloba*
Ligustrum lucidum	*Ficus* species
Mahonia	*Magnolia grandiflora*
Trachelospermum	*Nerium*
Conifers	
Abies, many species	*Juniperus chinensis*
Picea, many species	*Picea mariana*
Pinus strobus	*Picea parryana*
Pinus silvestris	*Pinus halepensis*
Pseudotsuga	*Pinus nigra*
Taxus	*Pinus ponderosa*

The *recovery of roadside trees* following exposure to deicing salt in winter presents a serious problem. Even in regions with plentiful rainfall, the salts in the soil are not as a rule adequately leached out by the precipitation. The remaining salt is then taken up by plants and transported to the branches, buds and leaves, where it causes visible damage, as well as lessening the degree of frost hardening. Effective countermeasures are, in the first place, a reduction in the amount of salt used, and the best possible protection of the plants from accumulations of melt water. Covering the soil surface with a layer of absorbant peat can be helpful, but these layers and the covering litter must be removed at the end of the winter, and may on no account be used for compost. A further possibility is to treat the soil with ion exchange solutions (containing K^+, Mg^{2+}, Ca^{2+}, NO_3^-, and SO_4^{2-}) for several years. Nevertheless, recovery of damaged trees can be expected only in habitats that have not been exposed to more than moderate additions of salt.

6.3 Anthropogenic Stress

6.3.1 Man-Made Pollutants and Their Impact on the Phytosphere

Sun, air, water and soil together supply plants with the energy, nutrients and milieu necessary for their survival. At the same time, however, the surroundings also contain *phytotoxic* substances in *harmful concentrations*. In addition, some of the chemical elements and compounds normally present in nature are potentially dangerous for plants, such as SO_2 from volcanic emissions, H_2S in swampy areas, oxides of nitrogen arising from microbial denitrification processes in the soil and from electrical discharges in thunderstorms, as well as salts and airborne particles of dust. Today, as a result of human activities, plants are exposed to far greater amounts of harmful substances than before; the situation is the more serious since these are chiefly substances foreign to plants (*xenobiotics*) and to which they have not (as yet) become accustomed (Table 6.10). A great variety of phytotoxic substances enter the atmosphere, water and soil as a result of industrial processes, traffic, the use of chemical agents in agriculture and in households, and especially the excessive consumption of fossil fuels. In addition to hazards of this nature, the vegetation may be exposed to excess applications of biocides and fertilizers, to eutrophication of soils and water, and to NH_3 and methane entering the air as a result of intensive, large-scale livestock farming. The explosive increase in commercial traffic, the escalation of recreational activities, catastrophic accidents (nuclear reactors, oil tankers), conflagrations and wars all have detrimental consequences far greater than their immediate and local effects. Each contributes to widespread, long-term environmental stress affecting ecosystems, continents, and even the entire globe.

Table 6.10. Typical concentrations and residence times of atmospheric pollutions. (Freedman 1989; Lahmann 1990; Legge and Krupa 1990; Kuttler 1991). Highly divergent values for residence times are encountered in the literature

Chemical compound	Typical concentration (ppm)		Average residence times in the atmosphere
	Clean air	Polluted air	
CO_2	340	400	2 – 6 years
CO	0.1	40 – 70	2 – 6 months
SO_2	0.0002	0.2	1 – 10 days
H_2S	0.0002		0.5 – 2 days
NH_3	0.01	0.1	2 – 14 days
N_2O	0.25		4 – 10 years
NO	<0.002	1 – 2	3 – 6 days
NO_2	<0.004	0.2	5 – 10 days
O_3	0.02	0.5	days-months
CH_4	1 – 1.7	3	4 – 10 years
Other hydrocarbons	<0.02	0.3	ca. 2 days

Single highly concentrated toxic substances causing acute damage to vegetation are mostly spatially restricted, like SO_2 and halogens within range of fossil-fuelled power plants and metallurgical and ceramic factories; heavy metals and metalloids in sewage sludge, in the surroundings of waste deposits and on rubbish dumps; de-icing salt, herbicides and exhaust gases from motor traffic along roadsides; toxic chemicals in waste water, rivers, lakes and in the oceans.

Far more common, however, is the *combined* occurrence of a number of pollutants, such as, e.g. the photooxidant complex and SO_2. Also, the atmosphere is not the only medium by which gaseous pollutants exert their effects, and the influence of liquid pollutants is not confined to the hydrosphere; examples of this are the acidification of soil and water by acid deposition, and the movement of heavy metals and fertilizer salts from the soil into groundwater, lakes and rivers. For many years investigations focussed on the effects of individual pollutants; currently, however, the centre of interest has progressed to the interactions between combinations of pollutants and the interrelationships between the different spheres of the environment.

The pollutants occurring at a particular place and at a particular time are called *immissions*; the pollutants released from a source are called the *emission*. The concentration of a pollutant in the environment is the immission concentration. For gaseous pollutants the concentrations are given either as dilutions (ppm = $1:10^6$ or as ppb = $1:10^9$) or as weight per volume (mg m^{-3}). In studies concerning the biochemical and cytophysiological (and phytotoxic) activity of pollutants, data must be converted into dimensions of molar concentration since chemical compounds act in the cell as molecules or ions. Table 6.11 shows conversion factors for the most common pollutants. *Injury threshold doses* (%) for the most frequently occurring atmospheric pollutants are laid down by international organizations (WHO, the World

Table 6.11. Conversion factors between volumetric $(1\ \text{ppb} = 1 : 10^9)$ and gravimetric $(\mu\text{g m}^{-3})$ concentrations of pollutant gases at $20\,^\circ\text{C}$ and $101.3\ \text{kPa}$ atmospheric pressure. (Lendzian and Unsworth 1983; Däßler 1991)

Pollutant	Chemical formula	To convert from ppb to $\mu\text{g m}^{-3}$, multiply ppb by	To convert from $\mu\text{g m}^{-3}$ to ppb, multiply $\mu\text{g m}^{-3}$ by
Sulphur dioxide	SO_2	2.67	0.38
Hydrogen sulphide	H_2S	1.42	0.70
Ammonia	NH_3	0.71	0.42
Nitric monoxide	NO	1.25	0.80
Nitric dioxide	NO_2	1.91	0.52
Ozone	O_3	2.00	0.50
Peroxyacetyl nitrate	$CH_3COO - O - NO_2$	4.37	0.23
Ethylene	C_2H_4	1.16	0.86
Hydrogen fluoride	HF	0.83	1.20
Hydrogen chloride	HCl	1.51	0.66

Table 6.12. Reference values for maximal permissible concentrations of different pollutants $(\mu\text{g m}^{-3})$. Ranges for sensitive to less sensitive plants. (Jäger et al. 1989)

Pollutant	Mean values of peak concentrations $(30-60\ \text{min})$	Mean values for permanent load	
		For 1 day $(8-24\ \text{h})$	For the growing period or for 1 year
HF	3	$1-2$	$0.2-0.4$
SO_2	400	$70-100$	$20-80$
NO_2	$200-6000$	$70-100$	$30-60$
O_3	$300-500$	$80-300$	$50-60$
NH_3	10000	600	

Health Organization; IUFRO, the International Union of Forest Research Organizations) and by governmental bodies in individual countries. However, owing to differences in sensitivity of the various species and varieties of plants, the official threshold values are useful only as rough guides. The recommended values (see Table 6.12) are *maximal (permissible) immission concentrations* for short concentration peaks and for longer-lasting immissions; this allows for the fact that a high input of pollutants causes acute damage within a short time, whereas chronic damage results from exposure to lower concentrations for a longer period.

6.3.2 Pollution Injury

The extent to which vital functions are affected by pollutants, and whether there is visible damage depends on many factors, both biotic and abiotic. The most important of these are the species, growth form, age, phase of activity and general vigour of the plant (Fig. 6.81), climatic and edaphic conditions, and the chemical nature, concentration, time and duration of action of the pollutants. In some cases, the effects of pollution are proportional to the product of their concentration and the duration of exposure (Fig. 6.82), although the relationship is linear only over a certain range. The lower limit is set by the concentration *threshold* below which there are no observable changes even after prolonged exposure. At the other end of the range, when a certain high concentration is exceeded, even very brief exposure causes damage. The effect of immissions also depends on the *time* at which their concentration is highest. Peak concentrations of atmospheric pollutions occurring before noon, when the stomata are usually fully open, are more harmful than if they occur during the night. On the other hand, if the plants have only been exposed to injurious immissions (such as photooxidants) for a few hours during the day, the night can be a time for recovery.

90 – 100
80 – 89
70 – 79
60 – 69
50 – 59
40 – 49
30 – 39
20 – 29
10 – 19
 0 – 9

Fig. 6.81. Distribution of necroses (% of leaf surface damaged) at different stages of development and at different ages, on leaves of a tobacco plant after treatment with photochemical smog. (After Glater, Solberg and Scott, as cited in Guderian et al. 1985). An unequal distribution of functional disturbances within the leaf due to SO_2 uptake can be detected by means of image instrumentation systems. (Omasa et al. 1987)

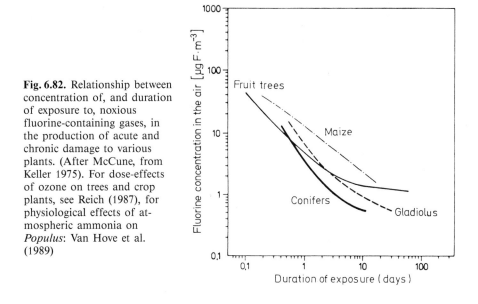

Fig. 6.82. Relationship between concentration of, and duration of exposure to, noxious fluorine-containing gases, in the production of acute and chronic damage to various plants. (After McCune, from Keller 1975). For dose-effects of ozone on trees and crop plants, see Reich (1987), for physiological effects of atmospheric ammonia on *Populus*: Van Hove et al. (1989)

6.3.2.1 Airborne Pollutants

The most deleterious atmospheric pollutants for plants are SO_2, nitric oxides, ozone, peroxyacetyl nitrate (PAN) and halides of hydrogen. Other less harmful airborne substances are ammonia, hydrocarbons, tar fumes, soot, and dust. The **symptoms of damage** are varied and usually nonspecific; the same substance may elicit quite different effects in different species, and on the other hand the same symptom may be produced by several different substances. Also, the nature of and the damage caused by an individual substance are modified by the other environmental and stress factors affecting the plant at the same time; these often enhance the injurious effect of the main pollutant. Trees subjected to the stress of noxious immissions have been observed to suffer greater damage from drought and frost than healthy individuals.

Criteria for *early recognition* of incipient pollution damage are: accumulation of the toxic substance or its derivatives in the plant (Fig. 6.83), reduction of the buffering capacity of the tissues, erosion of the epicuticular wax of needles by acid immissions, diminished or increased activity of certain enzymes (Table 6.13), qualitative and quantitative shifts among metabolites, the appearance of stress hormones (especially ethylene), increased or decreased respiration (see Fig. 6.5), disturbances in photosynthesis (see Fig. 6.88), alterations in stomatal opening and closure, and diminished allocation of photosynthates to the root system (Fig. 6.84). In any case it is inadvisable to draw conclusions from one symptom alone; *response patterns* involving several criteria provide a more reliable basis for a diagnosis concerning pollution stress.

Acute lethal damage is recognizable by the occurrence of chlorosis, leaf discoloration, necrosis of areas of tissues and organs, or death of the entire

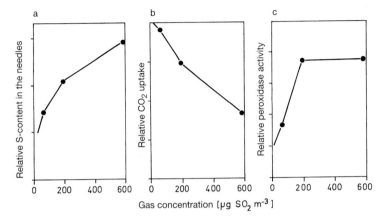

Abb. 6.83 a–c. Stress responses in young spruce plants after fumigation for 3 months with different concentrations of SO_2. **a** Accumulation of sulfur in the needles. **b** Decreased photosynthetic uptake of CO_2. **c** Increased peroxidase activity. (Keller 1982). Effects of SO_2, O_3 and NO_2 measured by in vivo chlorophyll fluorescence technique: Schmidt et al. (1990)

Table 6.13. Biochemical and physiological indications of stress from atmospheric pollutants. (Härtel 1972; Horsman and Wellburn 1976; Jäger 1982; Darrall and Jäger 1984; Weigel et al. 1989; Grill et al. 1989; Cape and Vogt 1991)

Indicator	Pollutant	Increase	Decrease
Enzymes			
Peroxidase	F_2, HF, SO_2	×	
Polyphenol oxidase	SO_2, NO_2, hydrocarbons	×	
Glutamate dehydrogenase	SO_2, NO_x	×	
RuBP-carboxylase	SO_2		×
Nitrite reductase	SO_2, NO_x		×
Superoxide dismutase	Acid precipitations, O_3	×	
Stress metabolites			
Ascorbic acid	Non-specific	×	
Glutathione	SO_2	×	
Polyamine	Non-specific	×	
Ethylene	Non-specific	×	
Metabolism			
Adenylate status	Non-specific		×
Photosynthesis	Non-specific		×
Optical reflectance	O_3, SO_2, acid precipitations		×
Turbidity test[a]	Acid precipitations	×	

[a] Turbidity of the hot-water eluate from conifer needles.

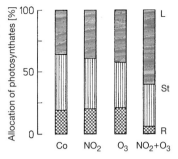

Fig. 6.84. Allocation of photosynthates in young bean plants after fumigation with O_3 or NO_2, and after exposure to a combination of the two gases. *Co* Control, not exposed to the gases. The amounts of photosynthates translocated to the roots (*R*), stem (*St*) and newly formed leaves (*L*) are given as relative proportions of the carbon compounds exported from the lowest leaf. (Okano et al. 1984). For synergistic responses of trees to pollutant mixtures see Freer-Smith (1984)

plant. Damage of this kind is in general seen only in the immediate vicinity of the pollution source. *Chronic injury,* however, often without distinctive symptoms, leads to reduced productivity and defective fertility (e.g. pollen sterility). In trees, growth, especially cambial growth, is less vigorous. These changes in the structure of the wood (Table 6.14) and the analysis of annual rings can give a clear record of the progress of chronic pollution injury (Fig. 6.85). Foliage becomes sparser from year to year, water transport to the top of the crown is interfered with, branches dry out, and gradually the tree dies.

In the following sections, the specific processes leading to plant injury and the mechanisms by which plants can resist exposure to noxious immissions will be discussed, with particular reference to the most widely occurring gaseous pollutants, sulphur dioxide and photooxidants.

Table 6.14. Possible alterations in the wood of trees exposed to pollution stress. (+) Increase, (−) decrease, (×) no apparent change. (After Liese et al. 1975; Keller 1980; Halbwachs and Wimmer 1987; Wimmer and Halbwachs 1992)

Feature	Conifer wood	Angiosperm wood
Cambial growth	−	−
Formation of late wood	±	
Proportion of sclerenchymatic tissue	+	−
Thickness and density of cell walls	−	±
Fibre length		−
Tracheids or tracheary elements per unit area	+	+
Length of tracheids or tracheary elements	−	
Diameter of tracheids of tracheary elements	−	−
Number of bordered pits per tracheid	+	
Diameter of bordered pits	−	
Number of rays per unit cross area	+	+
Number of resin ducts	+	
Storage substances	−	×
Proportion of cellulose	×	×
Proportion of lignin	−	×
Other cell-wall incrustations	+	×

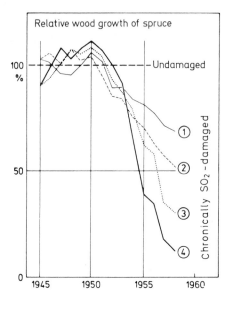

Fig. 6.85. Reduced wood production in stands of spruce exposed to strong SO_2 pollution. The relative annual ring growth in stands exposed to pollution is expressed as a percentage of the thickness in healthy stands, and takes into consideration age and weather-dependent effects. Levels of damage: *1* beginning of needle loss; *2* advancing needle loss; *3* severe damage, with dieback of twigs and branches; *4* dying trees, drying at tops. (After Vinš 1962). Effect of SO_2 emissions on a boreal forest in North America: Amundson et al. (1990)

Serious forest damage is also caused by halides of hydrogen, but these gases are as a rule confined to smaller areas. Hydrogen fluoride is extremely toxic; fluorides enter the plant through the stomata and are distributed via the apoplastic route. The first visible sign of damage is chlorosis of the leaves, followed by the characteristic necroses of leaf tips and margins.

Sulphur Dioxide

Of the toxic gases, SO_2 causes the most damage to plant life, and this even preceded the early smelting of sulphur containing ores more than 4000 years ago; as long as life has existed on earth, volcanic emissions have been a source of SO_2. Not only is more known about the effects of sulphur dioxide on plants than for any other toxic gas, it is also the gas to which plants have been able to adapt genotypically.

Entry of sulphur dioxide into the plant: SO_2 can enter the leaf through the open stomata as readily as CO_2 ($r_s^{SO_2} = 0.84\ r_s^{CO_2}$ [246]). If the stomata are closed the SO_2 can (like oxygen) overcome the cuticular resistance and enter the leaf via this route. At low external concentrations (about 45 µg SO_2 m^{-3}), SO_2 can trigger a loss of turgor in the epidermal cells surrounding the guard cells, thus causing the stomata to open. Higher ambient concentrations of SO_2 (above 1300 µg m^{-3}) cause stomatal closure. This is probably the explanation for conflicting observations, sometimes of an increase in transpiration, and in other cases a reduction [173].

The diffusion pathway of SO_2 is the same as that of CO_2 (see Fig. 2.14), and also the concentration gradient between the atmosphere and the other end of the transport route, the chloroplasts, is similar. In the cell walls the SO_2 is first dissolved in water [$SO_2 \cdot H_2O$], and its reaction products, hydrogen sulphite HSO_3 and sulphite SO_3^{2-}, are then distributed inside the cell between chloroplasts, cytosol and vacuole, in proportions which have been reported to be $96:3:1$ [179]. The chloroplasts, which, when illuminated, have a pH of 8, function as an ion sink.

Intracellular processes: sulphur compounds, including SO_2 and H_2S, are not foreign substances for plants. In the earliest phases of phylogeny pro-karyotes adapted to life in an acidic, sulphur-containing milieu; therefore sulphur taken up by a plant can be incorporated via existing metabolic pathways. The first step, the oxidation of sulphite to sulphate, is started in the cell walls by peroxidases, but further incorporation is mainly effected by photosynthetic sulphur metabolism, the end-products of which are the S-con-taining amino acids cysteine and methionine (Fig. 6.86). If too much sulphur is taken up and the concentration of thiols continues to rise, sulphur can be accumulated in the form of glutathione. Surplus SH groups and sulphite may be converted into sulphide and lost as H_2S by gas exchange, thus relieving the plant's metabolism. The effectiveness of the detoxification reactions is never-theless limited. With continued uptake of SO_2 and increasing acidification, the buffering capacity of the protoplasm is overtaxed, the sulphite level in the

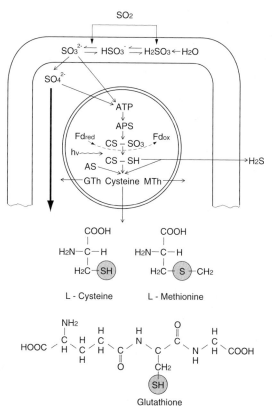

Fig. 6.86. Simplified diagram of the processes taking place when SO_2 enters the cell, and its detoxi-fication via sulphur metabolism in the chloroplasts. *APS* Adeno-sine phosphosulphate is the first product of the reaction of SO_4^{2-} with ATP. The activated sulphate (phospho-APS) is bound to a sulphur-containing carrier protein (*CS*), and the resulting protein-sulphite complex (*CS-SO₃*) is reduced to the sulphide (*CS-SH*) via ferredoxin (*Fd_red*, *Fd_ox*). By transfer on o-acetyl serine (*AS*), cysteine, methionine and glutathione may be formed. If the reducing power exceeds the avail-able sources of carbon, sulphur can be reduced to H_2S. *hv* Photons. (After Rennenberg 1984; Garsed 1985; Wellburn 1985; Lendzian 1987)

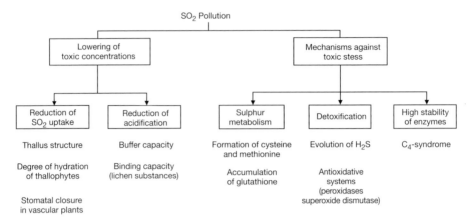

Fig. 6.87. Mechanisms of resistance to SO_2 pollution. (After Wirth and Türk 1975; Tesche 1989)

chloroplasts rises, and SO_2 occupies the CO_2-binding sites on the RuBP carboxylases. This results in inhibition of the secondary processes of photosynthesis (which also causes stomatal closure via the CO_2 control circuit), and in addition the tertiary structure of enzymes is disturbed. Photooxidation of sulphite to sulphate in the chloroplasts gives rise to superoxide radicals which, if they are not quickly rendered harmless by superoxide dismutase (SOD), destroy the chlorophyll.

Mechanisms of resistance to SO_2 stress: most plants are able, by various passive and active processes, to endure moderate SO_2 toxicity (Fig. 6.87). Some characteristics and functional traits of plants, although not causally related to the pollutant (i.e. non-specific), do in fact provide means of protection from SO_2 stress. The regular development of new leaves with a short functional life span means that herbaceous species are less endangered than, for example, evergreen trees with needles surviving for several years. Since they shed their leaves in autumn, deciduous woody plants escape the strongest emissions of SO_2, which occur during the winter from heating with fossil fuels. In thallophytes, structural, chemical and ecological characteristics hinder the entry of SO_2, exemplifying the usefulness of quite fortuitous features which influence the degree of sensitivity of different plant types.

Stressor-specific mechanisms for avoiding toxic effects and acidification are of a chemical and biochemical nature: they include a high buffering capacity from an increased uptake of alkali- and alkali-earth cations, the binding of SO_2 to secondary products of metabolism (such as lichen substances), and metabolic use of sulphur and the detoxifying oxidative reactions previously mentioned. Another adaptation to the environment (developed independently of immissions), the C_4 syndrome, promotes resistance to moderate SO_2 stress. The reason is that PEP-carboxylase is less sensitive to SO_2 than RuBP-carboxylase, and due to the CO_2 concentrating mechanism there is less competitive

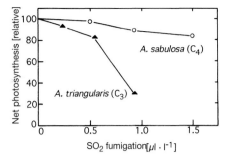

Fig. 6.88. Decrease in net photosynthesis (given in percent of the clean air control) in *Atriplex triangularis* (C$_3$ plant) and *Atriplex sabulosa* (C$_4$ plants) after 8-h exposure to various concentrations of SO$_2$. (Winner and Mooney 1980)

inhibition of RuBP carboxylase. Thus C$_4$ plants are less sensitive to SO$_2$ than C$_3$ plants (Fig. 6.88), which accounts for the occurrence of C$_4$ grasses (*Miscanthus sinensis* [244], *Andropogon virginicus* [276]) in the vicinity of volcanic vents, along with the specifically SO$_2$ resistant C$_3$ plants as Polygonaceae on volcanoes of various continents and *Metrosideros collina* on Hawaii.

Species-specific sensitivity to immissions: different species, individual varieties and ecotypes, as well as different life stages, vary characteristically in their sensitivity to pollutants. Knowledge of specific resistance to pollutants and the ability to adapt is of practical significance when plants are to be established in areas with concentrated population and industry (Table 6.15).

Table 6.15. Trees and shrubs suitable for localities exposed to heavy loads of SO$_2$. (Krüssmann 1970; Däßler 1991)

Well-suited	Moderately suited
Broadleaved trees	
Acer platanoides	
Buxus sempervirens	*Castanea sativa*
Celtis australis	*Ginkgo biloba*
Gleditsia triacanthos	*Magnolia hypoleuca*
Platanus x *hybrida*	*Populus candicans*
Quercus species	*Robinia pseudoacacia*
Sophora japonica	
Conifers	
Juniperus species	*Chamaecyparis* species
Picea omorika	*Picea pungens f. glauca*
Taxus baccata	*Pinus mugo, P. nigra, Thuja* species
Shrubs	
Calluna vulgaris	*Berberis* species
Erica carnea	*Forsythia intermedia*
Gaultheria shallon	*Prunus laurocerasus*
Ligustrum species	*Rosa rugosa*
Sambucus nigra	*Weigelia florida*

Table 6.16. SO$_2$ pollution and lichen growth on trees and rocks. (After Kershaw 1985; Arndt et al. 1987, based on data of various authors)

Average SO$_2$ concentration (μg m^{-3})	Epiphytic lichens[a]		Epipetric lichens[b]	
	Eutrophic bark	Non-eutrophic	Substrate alkaline	Substrate acid
>125	*Lecanora conizaeoides* *Lecanora expallans*	*Lecanora conizaeoides* *Lepraria incana*	Lecanorion dispersae	Conizaeoidion
ca. 70	*Buellia canescens* *Physcia adscendens*	*Hypogymnia physodes* *Lecidea scalaris*		
ca. 60	*Buellia canescens* *Xanthoria parietina* *Physcia orbicularis* *Ramalina farinacea*	*Hypogymnia physodes* *Evernia prunastri*	Xanthorion	Conizaeoidion *Acarospora fuscata*
ca. 50	*Pertusaria albescens* *Physconia pulverulenta* *Xanthoria polycarpa* *Lecania cyrtella*	*Parmelia caperata* *Graphis elegans* *Pseudevernia furfuracea*		
ca. 40	*Physcia aipolia* *Ramalina fastigiata* *Candelaria concolor*	*Parmelia caperata* *Usnea subfloridana* *Pertusaria hemisphaerica*	Xanthorion (increasing diversity)	*Cladonia* spp.
<30	*Ramalina calicaris* *Caloplaca aurantiaca*	*Lobaria pulmonaria* *Usnea florida* *Teloschistes flavicans*	up to 20 spp. of *Xanthoria*	Increasing diversity, no *Lecanora conizaeoides*

[a] Lichens on tree trunks.
[b] Lichens on rocks and walls.

However, even the careful choice of species, optimal fertilization and other measures can at best do no more than lessen the effects of environmental pollution; only by a drastic reduction of immissions at their source would it be possible to eliminate the danger to the plants altogether.

Plants showing extraordinarily high sensitivity to SO_2 (and to HF and HCl) are certain mosses and lichens (Table 6.16), and many phytopathogenic fungi (e.g. *Rhytisma acerinum, Diplocarpon rosae, Gymnosporangium* species, *Puccinia graminis*). As a rule the gelatinous lichens are particularly sensitive, and the crustose and fruticose lichens slightly less so. Certain lichens are even tolerant to SO_2 (e.g. species of *Stereocaulon* in volcanic craters), but in many lichens as little as one-hundredth of the SO_2 concentration injurious to higher plants is sufficient to cause disturbances in respiration, loss of chlorophyll and inhibition of growth.

Atmospheric Oxidants: Nitrogen Oxides, Ozone and Secondary Photooxidants

Nitric oxides (nitrogen monoxide, NO, and nitrogen dioxide, NO_2) come to the atmosphere from hot combustion processes. In the light, after absorption of UV radiation in the 300–400 nm range, NO_2 splits into NO and atomic (reactive) oxygen, which then combines with atmospheric molecular oxygen to form *ozone* (O_3). The ozone again oxidizes the photolytically produced NO to NO_2. The three oxidants exist in dynamic equilibrium, until a further component enters the photochemical cycle, i.e. hydrocarbons emitted chiefly by motor vehicle exhaust or by industries. Beside O_3, peroxy radicals are another secondary immission; these react further to produce substances such as peroxides of acetyl-, butyl- and benzyl nitrate. Due to their dependence on sunlight, *photooxidants* exhibit marked diurnal and annual weather-dependent fluctuations over a wide range of concentrations (Fig. 6.89). In the Alps, the long-term ozone concentrations increase with rising altitude, as the exhaust gases ascending from the valleys encounter more and more UV light.

Uptake by the plant: all of these gases can enter the leaf through the open stomata just as readily as SO_2; NO_2 can diffuse through the cuticle much faster than SO_2 (this is why it is readily taken up from wet depositions); and most of the O_3 dissociates to oxygen in the outer wall of the epidermis. When nitric oxides come into contact with water, which occurs at the latest in the cell walls, HNO_2 and HNO_3 are formed and dissociate to nitrite and nitrate ions, which can be taken up actively by the living cells.

Events within the cell: the *nitrite* ions are assimilated into amino acids in a reaction catalyzed by nitrite reductase, and here chloroplasts also play an important role in detoxification. The atmospheric nitric oxides are in fact an additional source of nitrogen for plants, but may also have negative effects, such as acidification of the leaves. In addition, SO_2 increases their toxic effects since it inhibits nitrite reductases and thus hinders the breakdown of nitric oxide. *Ozone* dissociates rapidly in plant tissues to form atmospheric oxygen and peroxides. These peroxides first adversely affect the plasmalemma, and then all other biomembranes, so that transfer processes are impaired. Soon necroses

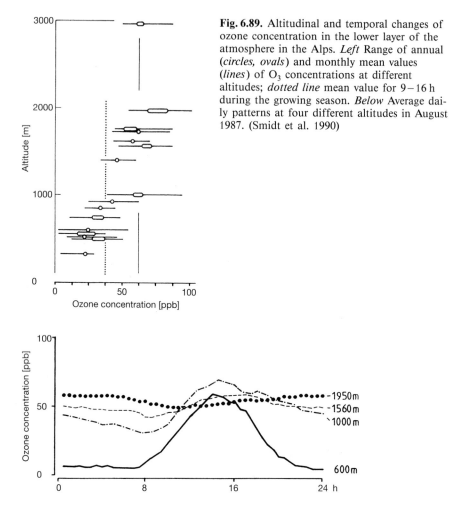

Fig. 6.89. Altitudinal and temporal changes of ozone concentration in the lower layer of the atmosphere in the Alps. *Left* Range of annual (*circles, ovals*) and monthly mean values (*lines*) of O_3 concentrations at different altitudes; *dotted line* mean value for 9–16 h during the growing season. *Below* Average daily patterns at four different altitudes in August 1987. (Smidt et al. 1990)

develop and spread out from wherever the gas enters the leaf. Prolonged experimental exposure of trees [150], wild herbs [195] and grasses to ozone concentrations similar to those repeatedly measured in the atmosphere ($0.05-0.1 \mu l \ l^{-1}$) has been found to cause a significant reduction in growth and yields. *Peroxyacetyl nitrate* oxidizes many substances, especially those containing SH groups, and also lipids and the growth hormone indoleacetic acid, which can be converted to the growth inhibiting hydroxymethyloxindol.

Specific sensitivity to immissions: most of the data concerning the specific sensitivity of various plants to photooxidants (Table 6.17) is based on experimental treatment with individual gases. In nature, however, this group of pollutants always occurs in combination but with varying composition, so that some discrepancies between the various experimental observations are unavoidable. In obvious cases of damage it is also extremely difficult, on the

Table 6.17. Relative sensitivity of various vascular plants to nitrogen oxides, ozone and PAN. (Data of different authors, from Ormrod 1978; Guderian et al. 1985; Däßler 1991; Treshow and Anderson 1991)

Species	NO_x	O_3	PAN
Crop plants and vegetables			
Most cereals	+	+	
Rice		−	
Forage grasses	+	+	
Fabaceae	+	+ / ×	+
Brassicaceae	−	×	−
Apiaceae	+		−
Beta		− / ×	+
Spinach		+	+
Solanaceae	×	+	+
Asteraceae		−	
Cichoriaceae		+ / ×	+
Ornamental shrubs			
Cornus		−	−
Cotoneaster		+	
Gardenia	×		
Hibiscus	+		
Ilex		−	
Ligustrum	×	+	
Pyracantha		−	
Rhododendron	+	+ / ×	× / −
Syringa		+	× / −
Viburnum		−	
Broadleaved trees			
Acer	×	−	−
Betula	+	−	−
Carpinus	−		
Fagus	−	−	
Fraxinus		+ / −	−
Populus		+ / −	
Quercus	−	+ / −	−
Robinia	−	−	
Tilia	×		−
Conifers			
Abies	×	× / −	−
Juniperus	−	−	
Larix	+	+ / ×	−
Picea		−	
Pinus	−	+ / −	−
Pseudotsuga		−	−
Sequoia		−	
Taxus	−	−	
Thuja	−	−	

+ Sensitive; − less sensitive; × intermediate.

basis of chemical analyses of leaves or by stress physiological procedures, to decide with certainty whether or not photooxidants are chiefly responsible.

6.3.2.2 Heavy Metal Contamination of Soil and Waters

Among the great variety of substances entering the soil, inland waters and the ocean as waste products (whether deliberately or through carelessness), heavy metals especially create long-term problems. Not only do they accumulate in organisms, and thus circulate in food chains, they also remain in the ecosystem in dangerous concentrations for longer periods in sediments. The soil covering ore-bearing rock or near slag heaps contains heavy metals (especially Zn, Pb, Ni, Co, Cr, Cu) and metalloids (Mn, Cd, Se, As) in amounts toxic to most plants. *Heavy metal contamination* also occurs in industrial zones, where the sources include heavy vehicular traffic, refuse dumps and sewage sludge. Emissions of dust from the metal processing industries contain every kind of heavy metal; waste water from factories contains Cd, Zn, Fe, Pb, Cu, Cr, Hg, sewage sludge Cd, Zn, Fe, Cu, Cr, Ni, Hg; along roads and in aerosols there is Pb, and waters often contain metal contaminants coming from inflows and seepage, as well as those settling out from the atmosphere.

The uptake of heavy metals by plants and their toxic effects: the uptake of metallic elements by plant cells, especially in the roots, is facilitated by appropriate mechanisms for their transport and accumulation, since several heavy metals are in fact required by plants as microelements. The plant cannot, however, prevent toxic elements from entering by the same mechanisms. The toxicity of heavy-metal ions is due chiefly to their interference with electron transport in respiration and photosynthesis, the inactivation of vital enzymes (Fig. 6.90), as a result of which the energy status is lowered (see Fig. 6.5), decreased uptake of mineral nutrients, and reductions in growth.

Resistance to heavy metals: most plants are sensitive to heavy metals in excess of minimal concentrations. However, certain plant species can grow on contaminated habitats because they have developed a variety of avoidance mechanisms by which the excess of heavy metals can be rendered harmless (Fig. 6.91). These mechanisms include

- *reduced uptake*;
- *immobilization* of the toxic ions in the cell walls (Table 6.18), thus preventing contact with the protoplasts as well as further transport through the apoplasts;
- *impeded permeation* across the boundary layers of the protoplasm;
- *chelation* in the cytoplasm to sulphur-containing polypeptides (glutathione and glutamylcysteine derivatives), to SH-containing proteins and to induced stress proteins giving protection from metal toxicity;
- *compartmentalization* and the formation of complexes with organic and inorganic acids, phenol derivatives and glycosides in the vacuole; and lastly,
- *retranslocation*.

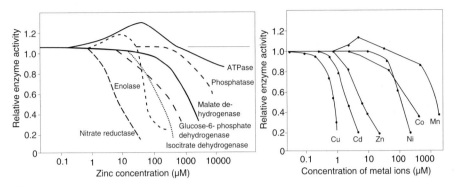

Fig. 6.90. Response of enzymes from leaves of *Silene cucubalus* to increasing stress from heavy metals (relative to controls = 1). *Left* Effects of increasing zinc concentration on the activity of various enzymes. *Right* Effects of some heavy metals on nitrate reductase activity. (Ernst 1967; Ernst and Joosse-van Damme 1983)

Fig. 6.91. Possible mechanisms of resistance to heavy metals. *1* Immobilization of metal ions in the cell wall, especially by pectins; *2* impeded permeation across the cell membrane; *3* formation of chelates by metal-binding proteins and polypeptides (phytochelatins) in the cytoplasm; *4* compartmentalization in the vacuoles; *5* active export; *CW* cell wall; *CYT* cytoplasm; *VAC* cell vacuole. (After Ernst 1976; Tomsett and Thurman 1988; Grill 1989; Cumming and Taylor 1990)

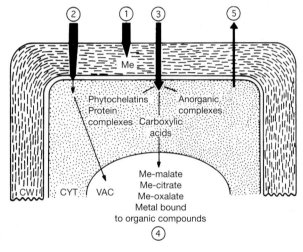

Even without human agency plants have been exposed to heavy metals throughout their existence, for example on outcrops of ores, on serpentine soils, or on strongly acid soils with toxic contents of aluminium, to mention only a few possibilities. In such places *chemo-ecotypes* have evolved, and among them even "specialized" species (e.g. "ore" species [233], or species characteristic of the serpentine flora [262] on different continents; see Table 6.22). The ability to develop heavy-metal resistance is genetically based but can be modified by adaptation [46]. Chemo-ecotypes display characteristic patterns of isoenzymes; they show element-specific increases in the ability of the protoplasm to resist the high concentrations of heavy metals in their tissues when they grow on soils rich in these elements. The greater the exposure to a

Table 6.18. Distribution[a] of zinc in organs and cell compartments of metallophytes on soils rich in zinc. (Ernst 1976)

Species, organ	Vacuole, cytoplasm (%)	Cell organelles (%)	Cell wall (%)
Cardaminopsis halleri			
leaves	**82**	*6*	12
roots	38	*5*	**57**
Silene vulgaris			
leaves	**64**	*10*	26
roots	18	*10*	**72**
Agrostis tenuis			
leaves	**48**	*11*	41
roots	38	*10*	**52**
Minuartia verna			
leaves	46	*8*	46
roots	20	*8*	**72**

[a] Boldface type indicates largest percentage in respective line; italics indicates smallest value.

certain element, the more tolerance is developed toward that element. There are also species with such a high degree of genetic plasticity that they have evolved a number of specialized ecotypes (with several resistance genes) resistant to several heavy metals and to other elements. Examples are seen in the resistance of *Agrostis tenuis*, *A. canina*, *Festuca ovina*, *Plantago lanceolata* and *Silene vulgaris* to Zn, Cu, Cd, Ni, and some of them also to As, Al, Fe and NaCl. An understanding of the genetic and physiological basis of heavy-metal resistance is an essential prerequisite in selecting appropriate species and breeding suitable varieties for the revegetation of strongly contaminated areas (slag heaps, dumps), and of course also for selecting bioindicator plants.

6.3.2.3 Bioindicators of Pollution Impact

As sedentary organisms, plants are continuously exposed to the stress of local immissions, and in greater intensity than humans and animals. For this reason they can be of great use as **bioindicators** for a variety of toxic substances. Bioindicators are organisms or communities of organisms that are sensitive to pollution stress and respond by alterations in their vital processes (*response indicators*) or by accumulation of the pollutant (*accumulation indicators*). Both types are useful as indicator plants in different ways.

Indicator organisms respond to changes in their surroundings, depending on their specific requirements, by decline, disappearance or, conversely, by abundant growth and reproduction. Changes in floristic composition, and the decline of certain species are good indications of long-term stress. A well-known application of this is the mapping of lichen growth (Fig. 6.92) from which, knowing the species composition of the lichen communities, it is possible to identify different zones of SO_2 pollution. It is also important of course, to take into account the specific SO_2 sensitivity of the different species of lichen, the prevailing climatic (irradiation, air humidity) and substrate conditions (pH and degree of eutrophication of the substrate).

Test organisms are characterized by a high degree of sensitivity to certain pollutants. Under standardized conditions and on the basis of biochemical, physiological or morphological criteria, they can indicate the presence of phytotoxic substances and in certain cases even the amounts present. In addition to microorganisms and animals, algae are frequently employed as test organisms, as well as submerged aquatic plants (Table 6.19), seedlings, and tissue- and cell cultures.

Monitor organisms are species which on account of their specific responses to pollutants can be employed for the qualitative or quantitative detection of stress situations. This involves observation or analysis (*passive* monitoring) of the plant in its own natural environment, or of standard plant material exposed in the study area (*active* monitoring). A number of highly sensitive response indicators can be found among both cryptogams and phanerogams (Table 6.20). Accumulation indicators able to accommodate large amounts of pollutants without undergoing damage are useful for both passive and active monitoring ("trapping plants"). Such species usually accumulate a particular element; chemical analysis then permits assessment of the degree of pollution in a locality. Mosses, with their especially high adsorption capacity, take up large quantities of ions (Table 6.21); explants of lichens (e.g. of *Hypogymnia physodes*) not only reveal damage due to gaseous pollutants (principally SO_2) on the basis of chlorophyll destruction, but also accumulate mineral pollutants (Pb, Zn, Fe, Mn in urban areas). Some fungi and bacteria accumulate large amounts of heavy metals; bacteria are therefore employed to retrieve heavy metals from sewage sludge and waste water ("microbial leaching").

Following the accident in the reactor of the Chernobyl nuclear power station in April 1986, many fungi (e.g. *Amanita muscaria, Xerocomus badius, Lactarius deliciosus* and *L. chrysorrheus*, but not *L. volemus* and *L. vellereus*), lichens (e.g. *Cetraria islandica, Cladonia* spp.) and mosses (e.g. species of *Bryum, Thuidium* and *Dicranum*) were found to be highly **radioactive** due to accumulation of ^{137}Cs, ^{134}Cs, ^{90}Sr and ^{105}Ru [56, 90]. Also in connection with Chernobyl, vascular plants proved to be important indicators also for the degree of local radioactive contamination and for the uptake of radionuclides from the soil. The specific activity of the radionuclides in a plant (A_{plant}, referred to fresh- or dry weight) can be compared with that in the soil (A_{soil}) by using the *transfer factor* (TF), which corresponds to the concentration factor for the bioaccumulation within a nutrition chain [84]

$$TF = \frac{A_{plant}\ (Bq\ kg^{-1})}{A_{soil}\ (Bq\ kg^{-1})} \tag{6.6}$$

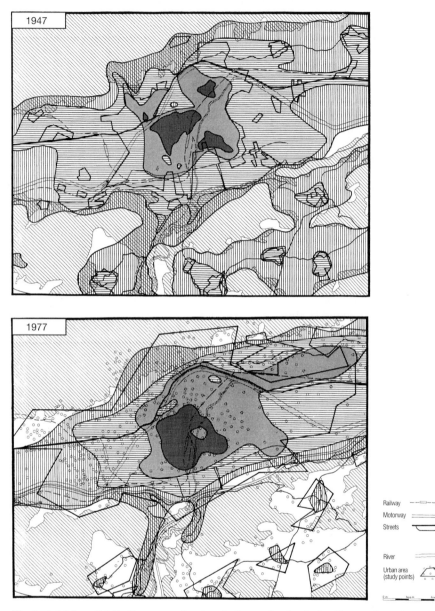

Fig. 6.92. Lichen distribution maps showing the increasing load of pollutants in and around Innsbruck. With this method it is possible to compare pollution in a particular area over longer periods of time (1947, 1977, 1987). *Zone I* Normal abundance and luxuriance of species-rich lichen populations on trees; *Zone II* abundant growth but changes in species composition, indicating a slight degree of pollution; *Zone III* neutrophilic epiphytic lichens and *Xanthoria parietina* predominate; lichens still abundant; *Zone IV* lichen vegetation with poor species diversity and low density; foliose lichens poor and in some cases deformed; *Zone V* lichen growth on trees almost non-existent, and only crustaceous lichens ("lichen desert") on walls. (After Beschel 1958; Bortenschlager and Schwarzer 1988)

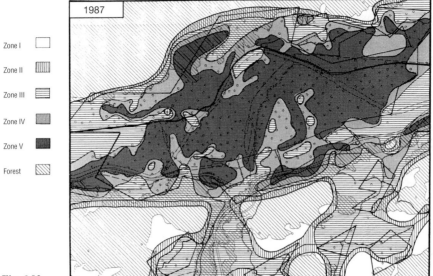

Zone I
Zone II
Zone III
Zone IV
Zone V
Forest

Fig. 6.92

Table 6.19. Submersed macrophytes as test organisms for pollutants. (Nobel et al. 1983)

Pollutant	Test plant	Threshold concentration for injury (ppm)	50% Lethality (ppm)
Phenol	*Potamogeton lucens*	0.2	0.6
	Potamogeton coloratus	0.6	
	Potamogeton crispus	0.6	
o-Cresol	*Potamogeton lucens*	0.2	0.7
	Potamogeton coloratus	0.6	1.1
	Potamogeton crispus	1.1	
KH_2PO_4	*Potamogeton alpinus*	0.2	2.0
	Elodea canadensis	0.5	>5.0
NH_4Cl	*Potamogeton coloratus*	<5	15
	Potamogeton crispus	<5	15
	Ranunculus fluitans	25	
H_3BO_3	*Elodea canadensis*	<1.0	10
	Myriophyllum alterniflorum	<2.0	5.0
	Ranunculus penicillatus	<1.0	10
Lead	*Potamogeton crispus*	2.1	
	Elodea canadensis	10.4	
	Potamogeton lucens	10.4	
Cadmium	*Elodea canadensis*	0.01	0.6
	Potamogeton crispus	0.01	0.6
	Potamogeton lucens	0.01	0.6
Copper	*Potamogeton crispus*	<0.03	0.06
	Elodea canadensis	0.006	0.3
	Potamogeton lucens	0.06	>0.3
Zinc	*Elodea canadensis*	<0.7	3.3
	Potamogeton lucens	0.7	6.5
	Potamogeton crispus	4.9	6.5

Table 6.20. Terrestrial vascular plants as selective pollution-sensitive indicators (examples). (Data of numerous authors, from Steubing 1976; Ernst and Joosse-van Damme 1983; Arndt et al. 1987; Rabe 1990; Schulze and Stix 1990; Schubert 1991)

Pollutant	Plant species	Most sensitive variety
SO_2	*Populus tremula* *Medicago sativa*	
H_2S	*Pseudotsuga menziesii* *Spinacia oleracea*	
HF, F_2	*Prunus armeniaca* *Gladiolus communis*	Snow Princess, Shirley Temple
HCl	*Syringa vulgaris* *Fragaria vesca*	
NH_3	*Taxus baccata* *Brassica oleracea*	Cauliflower Le Cerf
NO_x	*Apium graveolens* *Petunia* x *hybrida*	
O_3	*Nicotiana tabacum* *Phaseolus vulgaris*	Bel W 3 Sanilac, Pinto III, Tempo
PAN	*Petunia* x *hybrida* *Phaseolus vulgaris* *Poa annua*	Blue magic, Red magic Provider, Astro
Ethylene	*Petunia* x *hybrida*	White Joy

Table 6.21. Mosses tolerant to heavy metals, as examples of accumulation indicators. (Arndt et al. 1987; Tyler 1990)

Element	Species
Lead	*Bryum pseudotriquetrum* (also Zn) *Dicranella varia* (also Zn) *Philonotis fontana* *Fontinalis squamosum* *Scapania undulata*
Copper	*Calypogeia muelleriana* *Merceya ligulata* *Mielichhoferia elongata* (also Fe, Cr) *Mielichhoferia nitida* (also Fe, Pb, Zn) *Nardia scalaris* *Pleuroclada albescens*
Nickel	*Oligotrichum hercynicum* (also Cu)
Zinc	*Cephalozia bicuspidata* (also Cu) *Pohlia nutans* (also Cu)

Table 6.22. Examples of high mineral contents in metallophytes and toxicophytes (As, Se). Concentrations are shown in mg kg^{-1} dry matter. (After various authors: Duvigneaud and Denaeyer-De Smet 1973; Ernst 1976, 1990; Baumeister and Ernst 1978; Steubing et al. 1989)

Species	Occurrence	Element	Concentration	Degree of accumulation [*]
Eichhornia crassipes	Tropical waters	Fe	14400	10
Minuartia verna	Central Europe	Cu[a]		
leaves			1030	147
roots			1850	109
Thlaspi coerulescens	British Isles	Zn[b]		
leaves			25000	208
roots			11300	140
Minuartia verna	SE Europe	Pb[c]		
leaves			11400	950
roots			26300	970
Minuartia verna	Central Europe	Cd[d]		
leaves			348	3480
roots			382	3820
Jasione montana	British Isles	As[e]	31000	
leaves				
Mechovia grandiflora	Congo basin	Mn	7000	7
leaves				
Acrocephalus robertii	Congo basin	Co	1490	50
leaves				
Psychotria douarrei	New Caledonia	Ni[f]		
leaves			45000	
roots			92000	
Pearsonia metallifera	E. Africa	Cr[g]		
leaves			490	98
roots			1620	162
Astragalus preussi	N. America	U[h]	70	116
leaves, roots				
Astragalus racemosus	N. America	Se[i]	15000	
leaves				

[*] Degree of accumulation according to Duvigneaud = M_c/M_o.
M_c: Mineral concentration in plants from contaminated soils. M_o: Mineral concentration in plants on normal soils.
[a] Other cuprophytes: ecotypes of *Silene vulgaris* in Europe, *Haumaniastrum robertii* in Congo basin, *Becium homblei* and *Indigofera dyeri* in SE Africa, *Polycarpea spirostylis* in Australia, *Gypsophila patrini* in Central Asia, *Gladiolus* species in Africa.
[b] Other zinc-accumulating species in Europe: *Minuartia verna*, *Viola calaminaria*, ecotypes of *Silene vulgaris* and *Armeria maritima*.
[c] Other lead-accumulating species: *Agrostis tenuis*, *Festuca ovina*, *Erianthus giganteus*, *Cerastium holosteoides*, also *Calluna vulgaris*.
[d] Other cadmium-accumulating species: *Thlaspi coerulescens*.
[e] Other arsenic-accumulating species: *Calluna vulgaris*, *Agrostis tenuis*.
[f] Other nickel-accumulating species: *Hybanthus austrocaledonicus*, *H. floribundus*, *H. caledonicus*, *Sebertia acuminata* in Australia and New Caledonia.
[g] Other chromium-accumulating species: *Sutera fondina*, *Dicoma niccolifera*, *Convolvolus ocellatus* in E. Africa.
[h] Uranium-accumulating species in Central Europe: *Sambucus nigra*.
[i] Other selenophytes: *Aster xylorrhiza*, *Stanleya* species, various *Astragalus* species in N. America, *Neptunia amplexicaule*, *Acacia cana*.

It could be shown that the degree of accumulation of caesium and strontium isotopes was influenced by the nature of the plants, as well as by the many meteorological, edaphic and other habitat-related factors. The growth form and especially the rooting pattern (high concentration of radioactivity in grasses), the developmental stage, and chemotaxonomic and physiotypical features (*Ribes nigrum* was more strongly contaminated than *Ribes rubrum*) were influential.

There are certain plant species that are distinctly **heavy-metal indicators** (*metallophytes*; Table 6.22). Metallophytes take up large amounts of heavy-metal ions and store them at concentrations up to $0.5-0.8$ g kg^{-1} dry matter (in extreme cases up to 25 g kg^{-1}). This is $100-1000$ times the normal concentration of trace elements in a plant; it even approaches the magnitude of the macronutrients phosphorus and sulphur. The great capacity of the tropical floating plant *Eichhornia crassipes* to accumulate metal ions is exploited for detoxification of lakes and rivers.

6.3.3 The Effects of Atmospheric Pollutants on the Ecosystem and at the Global Level

The atmosphere is the most susceptible part of the global environment. Most people tend to overestimate the extent of the atmosphere; in fact, the layer of air in which living organisms can exist without artificial means extends only to an altitude of 6000 m. This can best be envisioned by mental comparison with a horizontal distance of 6 km. In this relatively thin coat of air are deposited all of the gaseous waste products resulting from human activities. Even if the whole troposphere, extending to an altitude of maximally 15 km, is considered to be storage space, the capacity of the atmosphere for exhaust gases, vapours, droplets and dust is still far less than that of any other sphere on our planet. The economic growth and prosperity of the industrialized countries, with the accompanying effects of abundance and mobility, as well as the exponential growth in population in poorer regions of the earth, have led to an enormous worldwide rise in atmospheric immissions (and of course also to much destruction of vegetation and soil, as well as pollution of the oceans).

Many immissions are washed out of the air by rain, snow and fog, and then precipitated. As *acid precipitation* they wet the surfaces of plants, and enter the oceans, lakes and rivers, and the soil. This is the origin of the "*new type of forest decline*" (or "novel forest decline") in large parts of Europe, the "new" aspect being the greater spread and the chronic nature of the damage, the extremely variable symptoms, and the involvement of a variety of tree species. The decline of the forests is a *complex disease* which would not have occurred in the absence of anthropogenic pollution of the air.

The increasing use of fossil fuels, clearcutting of forests and the growth of agriculture and animal farming are the principal causes of the rise in carbon dioxide and methane in the atmosphere. These inherently innocuous trace gases are chiefly responsible for the anthropogenic component of the *green-house effect*.

6.3.3.1 Continental Forest Decline: an Ecosystem Stress Syndrome

Widespread forest disease first became noticeable toward the end of the 1960s, in the form of a limited degree of previously unknown damage in central Euro-pean fir forests, and roughly a decade later in the spruce forests of northern and central Europe. From 1980 onward large areas of forest were affected, first of all conifer forests, later deciduous forests as well (Fig. 6.93). After an initial phase in which the condition of the forests underwent spectacular deteriora-tion, from about 1985 to 1988 the condition of many forests appeared to have stabilized, and in places there are even signs of recovery [212].

Forests have always been subject to decline and local damage, but formerly the causes were quickly recognized and could be remedied; these included age-ing of the stand, episodic damage by pests, extremes of climate, inappropriate management, interruption of mineral recycling in the ecosystem as a result of litter gathering, exhaustion of soil nutrients by monocultures, and toxicity caused by identifiable local emitters. The "new type of forest decline" could initially not be attributed to any one particular cause, but all evidence pointed to the participation of atmospheric pollutants. Not until the very intensively pursued investigations of forest decline underwent a change from the mono-causal approach to ecosystem-oriented and dynamic concepts of stress was it possible to gain insight into the innumerable causes and complexity of this multiple disease.

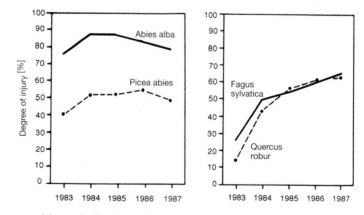

Fig. 6.93. Development of forest decline in Germany between 1983 and 1987. Damage is ex-pressed as percentage of the stand area of the particular tree species that were affected. (Schmidt-Vogt 1989). For various aspects of forest decline and possible interpretations see Kandler (1992); Huettl and Mueller-Dombois (1993)

The Symptoms of Forest Decline

The outward signs of forest decline may take a great variety of forms, depending on the tree species, growth form (e.g. the different branching form of local spruce populations), the site (valley or slope, altitude, exposure), the type of soil, geological origin, and the superimposition of various stressors (toxicity and additional attack by parasites or climatic stress). Even experienced forestry pathologists often have difficulty in deciding whether observed damage can be safely termed "forest decline". In general, it has to be carefully established whether the damage can be attributed to one particular factor, e.g. a virus infection, specific fungal attack, animal pests, acute toxicity (for example, nearby emitter, high salt concentrations along roads), or even to the possible influence of microwave fields. Even if the *differential diagnosis* suggests one particular cause, the damage may nevertheless be augmented by a latent predisposition arising from previous stress.

Despite some uncertainty concerning details, there are indeed **typical symptoms** which occur with characteristic frequency in the "new type of forest decline": these include anomalous growth (such as shortened internodes in the top of the crown, e.g. of *Abies* ("stork's nest"), discoloration of needles and leaves (with Mg deficiency the older needles turn yellow, with Fe and Mn deficiency the younger ones are affected), necroses of isolated areas of needles, leaves and branches, shedding of needles (thinning of crowns, bareness of the hanging branches of spruce: "lametta syndrome"), dieback of leader and branch tips and increasing shallowness of the root system. For *grading the condition of the crowns* in conifer forests, categories have been fixed on the basis of discoloration (0 none, 3 strong discoloration) and loss of needles (Fig. 6.94). Surveys of forest damage are unfortunately not carried out uniformly in all countries. In Germany, conifer forests are classified as slightly damaged (damage level 1) if the loss of needles is estimated at $11-25\%$, moderately damaged (2) at $26-60\%$ loss, severely damaged (3) at $61-99\%$ loss of needles, and level 4 indicates forest death. In Austria a $5-6$ grade scale is used. It is clear that evaluations based on appearance can result only in approximate assessments, especially where different tree species occur. The surveys can be supported by *aerial photography* employing false colour films, which show up the damaged tree canopies in a contrasting colour.

A typical feature of forest decline is the very different degree of damage observable over short distances. Thus a tree growing on poor, stony ground may be severely damaged, whereas another growing in a nearby hollow providing more nutritive conditions may be apparently healthy. The same applies to the changing weather conditions over the years; at a particular site the extent of damage may show a substantial increase or decrease from year to year. Surveys and research on forest decline must therefore include careful observation of trends in order to allow for variability of individual habitats.

Fig. 6.94. Appearance of crowns of *Picea abies* suffering from decline. *Percentages* Estimates of the degree of needle losses. (Müller and Stierlin 1990)

Causes, Development and Pattern of Forest Decline

Soon after forest decline began to spread it became clear that there must be a connection with large-scale atmospheric immissions. Suspicion increasingly concentrated on the **acidic effect of precipitations** deposited on plant surfaces, and falling on the ground, and then reaching soils and water. Acidic components of the atmosphere are acid-forming gases (SO_2, NO/NO_2), free acids (H_2SO_4, HNO_3, HCl) and H^+. Acid precipitation has been spreading over Europe (Fig. 6.95) and other industrialized regions (N. America, Japan) since the mid-20th century.

The *acidity* of precipitation depends on the local level and the origin of immissions. Since about 1970 pH values in Europe have been between 4 and 5, and occasionally between 3 and 4. Acidity is greater where fog and low clouds are frequent. At the base of tree trunks pH values may be up to 2 units lower than beneath the crown due to acid runoff from the trunks.

Apart from the toxic effects of their chemical components, depositions may cause **direct acid damage** to the assimilatory organs, such as necroses of the margins of leaves, and destruction of the cuticle and cuticular waxes of conifer

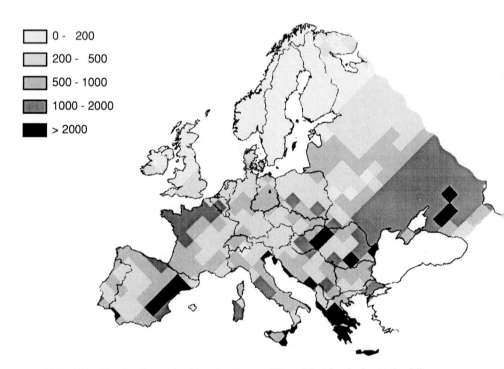

0 - 200
200 - 500
500 - 1000
1000 - 2000
> 2000

Fig. 6.95. Critical loads of actual acidity for Europe. The critical load of actual acidity was computed on the basis of weathering of base cations, hydrogen leaching and aluminium leaching. *Units* Acidity equivalents per ha and year. (Hettelingh et al. 1991). For pollution loads on other continents, see Kuttler (1984), Koziol and Whatley (1984)

needles. Acidification of the apoplast can also affect the distribution of phyto-hormones. In the fine roots chromosome anomalies may occur during cell division, the cells are damaged, and dissolution of the cell walls leads to tissue disruption.

An important role in the development of forest decline is played by **acidification of the soil** (Fig. 6.96). One of the first effects is that the soluble $Ca(HCO_3)_2$ is leached out at a pH value of about 7.0. In soils overlying limestone, despite this loss of calcium, the pH value does not shift to harmful levels of acidity, although deficiencies of potassium and iron may occur. On silicate soils, mainly Ca^{2+} and Mg^{2+} are leached out in exchange reactions; together with anions they are lost from the soil in drainage water. At pH values between 6 and 4 buffering of the soil is taken over by clay-humus complexes in the role of cation exchangers. Below pH 4.0 clay minerals are broken down, metal hydrous oxides are brought into solution, and increasing amounts of free ions of aluminium and heavy metals are detectable in the soil solution. As a result of soil acidification and its side effects, the breakdown of organic matter is slowed down and nitrification is inhibited. The depletion of basic cations and nitrate, and the fixation of phosphate also are important features of soil acidification.

The variability in the **complex of symptoms** of regional forest decline arises from the combination of the underlying stresses from acid immissions, and

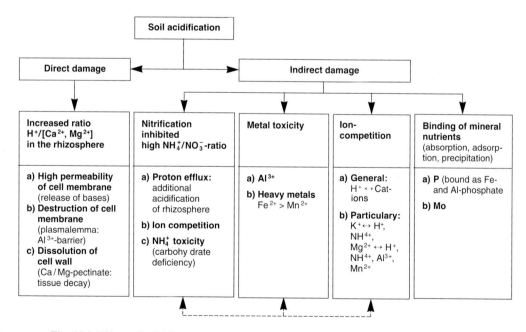

Fig. 6.96. Effects of acidification of the soil on its properties and on plant roots. (After Isermann 1983)

additional disturbances elicited by climatic, edaphic and biotic stressors (Fig. 6.97).

The physiological processes involved in the individual course taken by the disease are at present understood only in some cases. Acidification of the soil plays an important role. In spruce forests on silicate soil the fine roots are confined mainly to the upper layers of soil which, as a result of litter breakdown and leaching from the canopy, are more basic than the deeper, leached horizons. This means that the trees are more sensitive to surface drying of the soil. As a consequence of the restricted range of the root system and weakened mycorrhizal symbiosis, the tree receives insufficient mineral nutrients. If the deficient uptake of certain ions (especially Mg^+, K^+ and Ca^{2+}, possibly also Mn^{2+}) coincides with the same ions being leached from the leaves in substantial amounts by acid precipitations (Fig. 6.98) then *conditions of deficiency and ionic imbalance* arise.

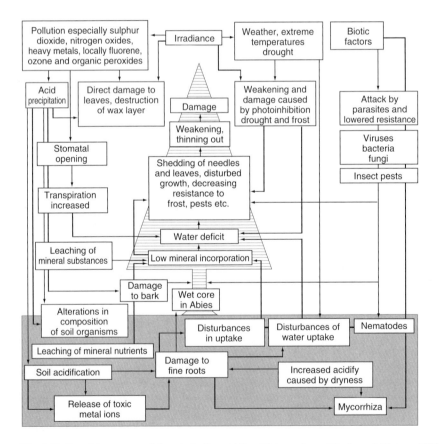

Fig. 6.97. Possible causes of forest decline. (After Elstner, from Hock and Elstner 1988)

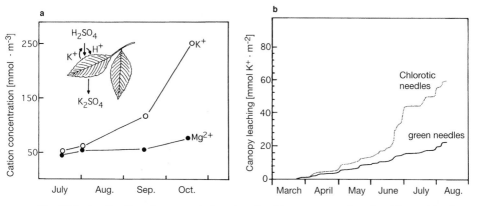

Fig. 6.98 a, b. Results of augmented cation leaching as a result of acid precipitation. **a** Neutralization mechanism on the leaf surface (*inset*) and cation concentration in the canopy throughfall in a beech-maple-birch mixed forest. (After Cronan and Reiners 1983). **b** Leakage of potassium from needles (referred to ground area) of a severely damaged spruce (*chlorotic needles*) and a less affected tree (*green needles*)). (After Glatzel et al. 1987)

If the atmospheric precipitation contain a high proportion of nitrogen compounds (NO_3^- combined with NH_3 and NH_4^-) shoot growth is at first promoted to such an extent that a dilution effect occurs (see Chap. 3.3.2). Mg^{2+} and K^+, which are readily translocatable elements, are then withdrawn from the stores accumulated in mature parts of the plant, with the result that the *older* needles of spruce turn yellow. Chlorosis in conifers can be treated successfully by the administration of carefully balanced mixtures of dolomite, magnesium sulphate and $K+Mg$ fertilizers. However, for promoting forest recovery, the application of chemical fertilizers can be considered only as a form of first aid; lasting effects can be achieved only if measures are aimed at combatting the primary cause, which is atmospheric pollution.

The nutrient balance in forests is influenced by many variables and is extremely labile (Fig. 6.99). Small alterations in nitrogen input, in the uptake of mineral substances from the soil or their loss by leaching, affect growth and vitality, and determine whether further damage or recovery ensue; each succeeding year can bring unforeseeable changes.

6.3.3.2 Global Increase in Infrared-Absorbing Trace Gases

From the middle of the 18th century onward, the CO_2 content of the atmosphere has been rising, at first slowly, but since the middle of the 20th century with increasing rapidity (on an average by $1.3 \, \mu l \, l^{-1}$ per year; Fig. 6.100). During this span of time vast forest regions in North America have been converted into arable land by immigrants, in all parts of the world hitherto unpopulated regions have become inhabited, transition to the industrial age has taken place,

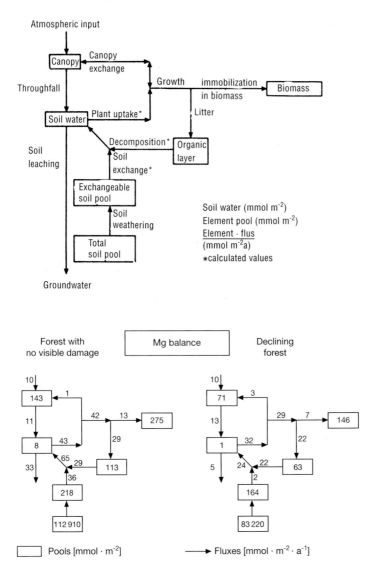

Fig. 6.99. Magnesium pools and fluxes of spruce stands with no visible damage, and of a declining forest. *Above:* Compartments and fluxes in the ecosystem. The *numbers in boxes* indicate the amount present (pools) in the stand, and the numbers *next to the arrows* indicate the annual fluxes. In the damaged stand the turnover of magnesium and the amount present in the phytomass and soil are lower than in the spruce forest showing no visible damage. (Horn et al. 1989)

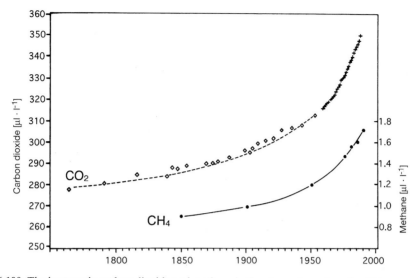

Fig. 6.100. The increase in carbon dioxide and methane in the atmosphere since the 18th century. Since 1958 the measurements were made at the Mauna Loa Observatory (Hawaii); values prior to 1958 are from assays of ice cores taken in the Antarctic (Siple station). (After Keeling et al., from IGBP 1990 and WMO 1990)

and enormous quantities of material have been destroyed in wars. All these events have greatly reduced the carbon reserves in biomass and soil, and have loaded the atmosphere with additional carbon dioxide. At the same time there has been a parallel rise in the methane content of the air.

Carbon dioxide in the atmosphere, and also water vapour, methane, ozone and nitrous oxide (N_2O) exert an influence on climate. These gases allow the unhindered passage of short-wave solar radiation to the earth's surface, but they absorb roughly 90% of the infrared energy radiated back from the earth, and it is this energy that contributes to the warming of the atmosphere. The natural *"greenhouse effect"* provides the temperatures necessary to support life on the earth. To the naturally occurring greenhouse gases in the atmosphere, human activities have now added halogenated hydrocarbons and other industrial trace gases. These, and the increase in the original components, contribute toward the warming up the air. There are indications that the mean temperature of the earth has risen by roughly 0.7 K since the middle of the last century, but as temperature patterns fluctuate continuously (see Fig. 1.44) calculations of this kind may be subject to error. It is important to remember that temperature changes averaged over the entire earth give no information about trends in individual regions. Such changes are strongly influenced by the zonal radiation balance and especially by the movement of air masses and by oceanic currents.

In connection with the accumulation of CO_2 in the atmosphere the following aspects should be borne in mind: (1) carbon dioxide is a nutrient for photoautotrophic organisms, which could therefore help counteract the increase in CO_2; (2) carbon dioxide, as an infrared-absorbing trace gas, plays an important role in climatic changes that could alter the vegetation pattern of the earth; in the event of a dramatic change in the vegetation the global carbon balance would also alter.

Interplay Between the Increasing CO_2 in the Air and CO_2 Uptake by Plants
In general, the biomass production of plants is limited by the natural supply of CO_2. Many experiments have shown that photosynthesis can be increased by CO_2 concentrations up to three times its current ambient level (about 350 µl l^{-1}, which is equivalent to a partial pressure of 35 Pa): treating plants with 100 Pa CO_2 has been shown to produce substantial increases in growth. Doubling the CO_2 supply above the normal leads to faster growth and to greater biomass production (Fig. 6.101). At elevated levels of CO_2 the stomata open less widely and therefore less water is lost by evaporation, so that drought-stressed plants are able to take up CO_2 for longer periods (Table 6.23). Opposite effects have also been observed, however, and results may differ from species to species [161]. Negative effects of CO_2 "fertilizing" include overfilling of the chloroplasts with starch (feedback inhibition) or hypertrophy of leaf tissues (leaves are softer and the ratio of leaf surface to

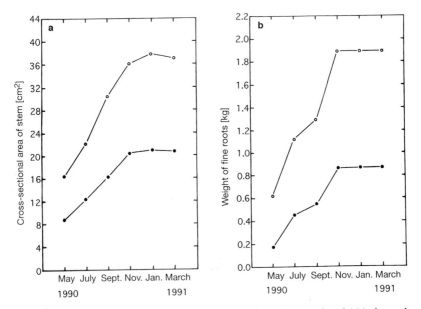

Fig. 6.101 a, b. Growth of *Citrus aurantium* trees with ($\bullet$) normal and ($\circ$) elevated concentrations of CO_2. **a** Increase in the cross-sectional area of the stem at a height of 45 cm above the ground. **b** Amount of fine roots at a soil depth of 40 cm. The young plants were grown as a plantation in open-top chambers treated with additional CO_2 in situ. (Idso and Kimball 1992)

Table 6.23. Possible responses of individual plants and plant communities to a doubling of CO_2 in the air. (Strain and Cure 1985; Farrar and Williams 1991; Hunt et al. 1991; Nobel 1991c; Woodward et al. 1991; Körner and Arnone 1992; Weigel et al. 1992; Coleman et al. 1993; Overdieck 1993; Thomas et al. 1993)

Morphogenesis and development	
Mesophyll thickness	$0/+$
Leaf area/leaf dry weight	$0/-$
LAI	$+/\times$
Shoot growth	$+$
Branching density	$+/-$
Root growth	$+$
Root/shoot ratio	$+/-$
Flower formation	$-/+$
Seed size	$+$
Life span and ageing	$+/\times$
Metabolism, dry matter production and water balance of individual plants	
Net photosynthesis	$+ \ (C_3 > C_4)$
RuBP-Carboxylase	$0/-$
CAM function	$+$
Dark respiration	$+/-$
Sugar transport	$+/\times$
Dry matter production	$+$
Transpiration	$- \ (C_4 > C_3)$
Water use efficiency (WUE)	$+$
Nitrogen use efficiency (NUE)	0
Mineral concentration in dry matter	$-$
Dry matter production under heat, drought and saline conditions	$+$
Plant communities	
Light utilization	$+/-$
Water utilization	$+/0$
Mineral utilization	$+/0$
Competitivity	$+/0 \ (C_3 > C_4)$

$+$ Increase; $-$ decrease; 0 no change; $\times$ not clear.

dry weight is lower). In consequence of the excess of carbohydrates more assimilates are diverted to the roots; this results in alterations in the shoot/root ratio, elevated root respiration and loss of photosynthetic products via the roots. If correspondingly more mineral substances, particularly nitrogen, are not taken up, the plant produces fewer flowers and fruits (Fig. 6.102). In the event of imbalance in the interplay of the bioelements it can be expected that dilution effects will lead to weaker growth and will certainly negatively affect the development of stress resistance. However, the overall impression is that the advantages predominate; but it must be remembered that these data have been obtained chiefly on crop plants, usually under experimental and therefore un-natural conditions, and over short periods of time. It would therefore be especially valuable to carry out studies on plants accustomed to elevated CO_2 in habitats where the ambient CO_2 concentration is regularly higher than the

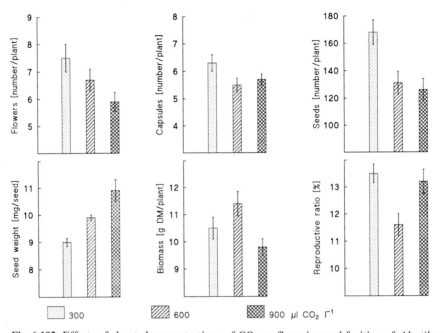

Fig. 6.102. Effects of elevated concentrations of CO_2 on flowering and fruiting of *Abutilon theophrasti*. The number of flowers, fruit capsules and seeds, and the total biomass are given per plant; the investment in reproduction ("reproductive ratio") is expressed as the percentage contribution of the fruits to the total mass. In this case, the number of flowers decreases but the weight of seeds increases. (Bazzaz et al. 1985). For differing effects on other plant species see Garbutt and Bazzaz (1984)

normal atmospheric CO_2 concentration (e.g. herbs of the forest floor and rosette plants beneath the soil surface).

Some idea of how plants would react to a gradual rise in atmospheric CO_2 over decades would be given by comparisons with growth during bygone periods of time. Examination of fossilized and preserved leaves (e.g. herbarium material, archaeological finds and the contents of graves [13]) suggest a connection between high levels of CO_2 and lower stomatal density. Analyses of growth rings of trees have proved to be contradictory since the CO_2 effect is masked by too many other influences. It is possible that the physiological adjustments to a higher intake of CO_2 (lowering of RuBP-carboxylase activity) and also the morphological changes (lower specific leaf area) might in the long run suppress the initially large increase in productivity. Also to be expected would be shifts in the competition between different functional groups of plants such as C_3 versus C_4 plants, and fast-growing herbs versus perennial types and woody plants. In contrast, the productivity of marine plankton (with CO_2-concentrating mechanisms) would probably not change appreciably.

It is impossible to predict whether in *natural ecosystems* with their many and highly varied interrelationships, a gradual rise in the CO_2 supply would result in a substantial and sustainable increase in biomass. It seems more likely that control circuits such as are active in every complex system would play a compensatory role. Considered on a global scale, the carbon turnover between biosphere and atmosphere is in any case strongly buffered in the vast storage capacity of hydrosphere and lithosphere (Fig. 6.103).

Possible Effects of Changes in Climate

Climatic changes connected with the accumulation of infrared-absorbing gases in the atmosphere may affect plant life much more seriously than the inevitable direct effects of CO_2 on metabolism and growth. The great unknown factor is the possible *extent* of the climatic changes. It is perhaps easier, due to the present state of knowledge concerning the influence of climatic factors on plant life, to foresee the *consequences* of a change in climate (that could proceed much faster than any previously known fluctuations in climate).

Intensive research on climate changes being carried on throughout the world under the guidance of meteorologists, oceanologists and glaciologists has produced a number of predictive models leading to quantitatively differing expectations. Calculations based on General Circulation Models (GCMs) predict that, by 2050 A.D., global warming of the earth by about 2.5 K with doubling of the atmospheric CO_2 concentration can be expected [101]. Most models foresee warming in the higher latitudes (Scandinavia, Siberia, Canada), small thermal alterations in the tropics, and an expansion of the arid zones. Using extensive model calculations, climatological data can be related

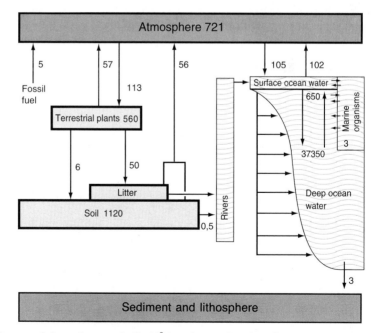

Fig. 6.103. Estimates of the carbon pools (in 10^9 t) and annual carbon fluxes in the global carbon cycle. (After King, De Angelis and Post, from Jarvis 1989)

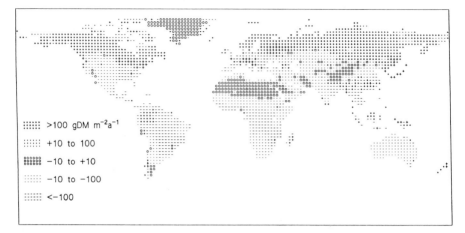

Fig. 6.104. Simulation of the effects of a rise in global temperature of 2 K on the function of the phytosphere as a CO_2 sink. Positive values indicate that more carbon would be assimilated as a result of the temperature increase. The vegetation in higher latitudes and in moist habitats would respond to a rise in temperature by an especially large increase in carbon uptake. In the tropics and subtropics, plants of highly humid regions would take up more, those of dry regions considerably less carbon (Esser 1987). Model calculations for wetlands can be found in Masing et al. (1990); for the effects on agriculture, see Carter et al. (1992)

to ecological parameters and represented as scenarios. Figure 6.104 shows a simulation from which the regional changes in carbon fixation by the vegetation for a global warming of 2 K can be read. Basic data for predictive calculations is also provided by results of analytical vegetation science. For example, correlation diagrams for *potential climatic limits of distribution* provide useful information about limiting factors in the global climate (Fig. 6.105).

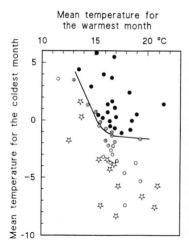

Fig. 6.105. Temperature thresholds limiting the area of distribution of *Hedera helix* in Europe. (●) The plants develop normally within their natural area of distribution; (☆) data from meteorological stations where the plants are missing. The limits to distribution of the species are set by the temperature conditions under which the fruit ripens only in favorable years (⊙), or not at all (○). (After Iversen 1944). Diagrams of this kind make it possible to recognize migrations due to environmental changes. Climate indices for determining potential limitations to distribution can be found in Hintikka (1963), Brisse and Grandjouan (1974), Tuhkanen (1980)

The response of plants to increasing warmth, dry periods and inundations, such as might be caused by a change in climate, is fairly obvious. Warmer summer temperatures in the zones of *tundra and boreal forests*, where at present temperatures are mostly suboptimal for plant growth, would promote dry matter production. The increased yield of photosynthates would bring the disadvantage of a greater release of respiratory CO_2. A rise in temperature in such regions would also accelerate bacterial breakdown of organic matter accumulated in rawhumus and in the soil, and thus release considerable amounts of CO_2 into the air. In *temperate latitudes* the assimilation period would become longer. Evergreen trees might achieve dominance if the period of foliation of the deciduous trees were to remain unchanged (provided that synchronization of flushing and leaffall depended more on photoperiod than temperature). Mild winters might delay the onset of winter dormancy, and might not meet the cold-requirements of the buds. In regions that become *drier*, the effects not only of longer drought but also of greater heat and increased salinity would mean additional stress for the plants. In the *oceans* the currents would shift and the composition of the phytoplankton would change.

Various species and functional types of plants, due to their specific climatic requirements and their tolerance toward climatic stress, give an indication of changes in the relevant climate. Visible changes in the composition of the vegetation, such as alterations in the vegetation limits, immigration or suppression of certain species, are preceded by changes in the metabolic and developmental processes of the individual plants. If sufficient is known about the climatic requirements and tolerance of different plant species, it is possible to find suitable indicator plants to serve as *climate monitors*.

Acclimation indicators are plants with a broad modulative amplitude for acclimation (physiological response), a wide range of modificative differences in form, and a high degree of genetic plasticity (a strong tendency to ecotype differentiation). Such plants will remain in the same habitat for a long time. Stress accelerates their evolutive development; their state of acclimation reflects the changed conditions in the habitat.

Migration indicators are plant groups characterized by a high degree of phenotypical and genotypical stability over many generations. A shift in their area of distribution or a change of habitat is thus conspicuous and provides an indication of climatic trends. Tree species in particular, and some herbaceous plants, are of this type, responding to climatic changes by migration rather than by ready acclimation [101]. A knowledge of vegetation history and of palaeoecology can provide valuable clues in this respect; but since migration usually proceeds slowly, it is not easy to predict/estimate the way in which different plants would respond to an unprecedently rapid change in climate.

The changing climatic conditions of the global environment provide the field of ecophysiology with a unique challenge. Investigators are faced with the urgent need for comparative studies aimed at identifying and analyzing interactions between environmental factors and plant distribution, and elucidating their underlying mechanisms. An important task will be the systematic search

References in the Text

[1] Ajtay GL, Ketner P, Duvigneaud P (1979) Terrestrial primary production and phyto-mass. In: Bolin B, Degens ET, Kempe S, Ketner P (eds) The global carbon cycle. Wiley, Chichester, pp 129−181
[2] Alexeev VA (1975) Svetovoi rezhim lesa. Nauka, Leningrad
[3] Altman PhL, Dittmer DS (1972) Biology data book, vol I; 2nd edn. Federation of American Societies for Experimental Biology, Bethesda
[4] Alvim PD (1960) Moisture stress as a requirement for flowering of coffee. Science 132:354
[5] André M, Massimino D, Daguenet A (1978) Daily patterns under the life cycle of a maize crop. I. Photosynthesis, transpiration, respiration. Physiol Plant 43:397−403
[6] Aono Y, Omoto Y (1993) Variation in the March mean temperature deduced from cherry blossom in Kyoto since the 14th century. J Agric Meteorol 48:635−638
[7] Atkinson DE, Walton GM (1967) ATP conservation in metabolic regulation. J Biol Chem 242:3239−3241
[8] Badger MR, Pfanz H, Büdel B, Heber U, Lange OL (1993) Evidence for the functioning of photosynthetic CO_2-concentrating mechanisms in lichens containing green algal and cyanobacterial photobionts. Planta 191:57−70
[9] Ball MC (1988) Ecophysiology of mangroves. Trees 2:129−142
[10] Bauer H, Martha P, Kirchner-Heiss B, Mairhofer I (1983) The CO_2 compensation point of C_3 plants − a re-examination. II. Intraspecific variability. Z Pflanzenphysiol 109:143−154
[11] Bazzaz FA, Ackerly DD (1992) Reproductive allocation and reproductive effort in plants. In: Fenner M (ed) Seeds. The ecology of regneration in plant communities. CAB International, Wallingford, pp 1−26
[12] Beadle CL, Long SP (1985) Photosynthesis − is it limiting to biomass production? Biomass 8:119−168
[13] Beerling DJ, Chaloner WG (1993) Stomatal density responses of Egyptian *Olea europaea* L. leaves to CO_2 change since 1327 BC. Ann Bot 71:431−435
[14] Belaya GA (1978) Ekologiya dominantov kamchatskogo krypnotraviya. Izdat Nauka, Moskva
[15] Bĕlehrádek J (1957) Physiological aspects of heat and cold. Annu Rev Physiol 19:59−82
[16] Berendse F, Aerts R (1987) Nitrogen-use-efficiency: a biologically meaningful definition? Funct Ecol 1:293−296
[17] Bierhuizen JF (1973) The effect of temperature on plant growth, development and yield. In: Slatyer RO (ed) Plant response to climatic factors. UNESCO, Paris, pp 89−98
[18] Bierhuizen JF, Wagenvoort WA (1974) Some aspects of seed germination in vegetables. 1. The determination and application of heat sums and minimum temperature for germination. Sci Hortic 2:213−219
[19] Björkman O, Demmig B (1987) Photon yield of O_2 evolution and chlorophyll fluorescence characteristics at 77 K among vascular plants of diverse origins. Planta 170:489−504
[20] Blackman FF (1905) Optima and limiting factors. Ann Bot 19:281−295
[21] Bothe H, Yates MG, Cannon FC (1983) Physiology, biochemistry and genetics of dinitrogen fixation. In: Läuchli A, Bieleski RL (eds) Inorganic plant nutrition. Ency-

clopedia of plant physiology, vol 15 A. Springer, Berlin Heidelberg New York, pp 241–285

[22] Boysen-Jensen P (1932) Die Stoffproduktion der Pflanzen. Fischer, Jena

[23] Briggs GE, Kidd R, West C (1920a) Quantitative analysis of plant growth. Ann Appl Biol 7:103–123

[24] Briggs GE, Kidd R, West C (1920b) Quantitative analysis of plant growth. Ann Appl Biol 7:202–223

[25] Brix H (1990) Uptake and photosynthetic utilization of sediment-derived carbon by *Phragmites australis* (Cav.) Trin. ex Steudel. Aquat Bot 38:377–389

[26] Brown RH, Hattersley PW (1989) Leaf anatomy of C_3-C_4 species as related to evolution of C_4 photosynthesis. Plant Physiol 91:1543–1550

[27] Bryant JP, Chapin-III FS, Reichhardt PB, Clausen TP (1987) Response of winter chemical defense in Alaska paper birch and green alder to manipulation of plant carbon/nutrient balance. Oecologia 72:510–514

[28] Buesa RJ (1977) Photosynthesis and respiration of some tropical marine plants. Aquat Bot 3:203–216

[29] von Caemmerer S, Evans JR (1991) Determination of the average partial pressure of CO_2 in chloroplasts from leaves of several C_3 plants. Aust J Plant Physiol 18:287–305

[30] von Caemmerer S, Farquhar GD (1981) Some relationships between the biochemistry of photosynthesis and the gas exchange of leaves. Planta 153:376–387

[31] Carpenter EJ (1992) Nitrogen fixation in the epiphyllae and root nodules of trees in the lowland tropical rainforest of Costa Rica. Acta Oecol 13:153–160

[32] Čermák J (1989) Solar equivalent leaf area: an efficient biometrical parameter of individual leaves, trees and stands. Tree Physiol 5:269–289

[33] Chapin DM, Bliss LC, Bledsoe LJ (1991) Environmental regulation of nitrogen fixation in a high arctic lowland ecosystem. Can J Bot 69:2744–2755

[34] Chapin FSt, Schulze ED, Mooney HA (1990) The ecology and economics of storage in plants. Annu Rev Ecol Syst 21:423–447

[35] Chiu ST, Ewers FW (1992) Xylem structure and water transport in a twiner, a scrambler, and a shrub of *Lonicera* (Caprifoliaceae). Trees 6:216–224

[36] Claes B, Dekeyser R, Villarroel R, Van den Bulcke M, Van Montagu M, Caplan A (1990) Characterization of a rice gene showing organ-specific expression in response to salt stress and drought. Plant Cell 2:19–27

[37] Crawford RMM (1982) Physiological responses to flooding. In: Lange OL, Nobel PS, Osmond CB, Ziegler H (eds) Encyclopedia of plant physiology, vol 12 B. Springer, Berlin Heidelberg New York, pp 453–477

[38] Crespo HM, Frean M, Cresswell CF, Tew J (1979) The occurrence of both C_3 and C_4 photosynthetic characteristics in a single *Zea mays* plant. Planta 147:257–263

[39] Dacey JWH (1981) Pressurized ventilation in the yellow waterlily. Ecology 62:1137–1147

[40] Dainty J (1976) Water relations of plant cells. In: Lüttge U, Pitman MG (eds) Encyclopedia of plant physiology, vol 2. Springer, Berlin Heidelberg New York, pp 12–35

[41] Dawson TE, Ehleringer JR (1993) Gender-specific physiology, carbon isotope discrimination, and habitat distribution in box elder, *Acer negundo*. Ecology 74:798–815

[42] Delf EM (1911) Transpiration and behaviour of stomata in halophytes. Ann Bot 25:485–505

[43] Demmig-Adams B, Adams WW, Winter K, Meyer A, Schreiber U, Pereira JS, Krüger A, Czygan FC, Lange OL (1989) Photochemical efficiency of photosystem II, photon yield of O_2 evolution, photosynthetic capacity, and carotenoid composition during the midday depression of net CO_2 uptake in *Arbutus unedo* growing in Portugal. Planta 177:377–387

[44] Dicke M, Dijkman H (1992) Induced defence in detached uninfested plant leaves: effects on behaviour of herbivores and their predators. Oecologia 91:554–560

[45] Dicke M, Sabelis MW (1988) Infochemical terminology: based on cost-benefit analysis rather than origin of compounds? Funct Ecol 2:131–139

[46] Dickinson NM, Turner AP, Lepp NW (1991) How do trees and other long-lived plants survive in polluted environments? Funct Ecol 5:5–11

[47] Dixon HH (1914) Transpiration and the ascent of sap in plants. Macmillan, London

[48] Doley D (1981) Tropical and subtropical forests and woodlands. In: Kozlowski TT (ed) Water deficits and plant growth. Vol VI: Woody plant communities. Academic Press, New York, pp 209–323

[49] Dormling I (1989) The role of photoperiod and temperature in the induction and the release of dormancy in *Pinus sylvestris* L. seedlings. Ann Sci For 46:228 s–232 s

[50] Downs RJ, Hellmers H (1975) Environment and the experimental control of plant growth. Academic Press, London

[51] Downton WJS (1975) The occurrence of C_4 photosynthesis among plants. Photosynthetica 9:96–105

[52] Dufrêne E, Dubos B, Rey H, Quencez P, Saugier B (1992) Changes in evapotranspiration from an oil palm stand (*Elaeis guineensis* Jacq.) exposed to seasonal soil water deficits. Acta Oecol 13:299–314

[53] Elakovich StD (1989) Allelopathic aquatic plants for aquatic weed management. Biol Plant 31:479–486

[54] Elfving DC, Kaufmann MR, Hall AE (1972) Interpreting leaf water potential measurements with a model of a soil-plant-atmosphere continuum. Physiol Plant 27:161–168

[55] Eliáš P (1983) Water relations pattern of understorey species influenced by sunflecks. Biol Plant 25:68–74

[56] Elstner EF, Fink R, Höll W, Lengfelder E, Ziegler H (1987) Natural and Chernobyl-caused radioactivity in mushrooms, mosses and soil-samples of defined biotops in SW Bavaria. Oecologia 73:553–558

[57] Epstein E (1983) Crops tolerant of salinity and other mineral stresses. In: Ciba Found Symp 97: better crops for food. Pitman, London, pp 61–76

[58] Erickson RO, Michelini FJ (1957) The plastochron index. Am J Bot 44:297–305

[59] Ernst WHO (1983) Ökologische Anpassungsstrategien an Bodenfaktoren. Ber Dtsch Bot Ges 96:49–71

[60] Evans JR (1989) Photosynthesis and nitrogen relationships in leaves of C_3 plants. Oecologia 78:9–19

[61] Fenner M (1980) Some measurements on the water relations of baobab trees. Biotropica 12:205–209

[62] Finke RL, Harper JE, Hageman RH (1982) Efficiency of nitrogen assimilation by N_2-fixing and nitrate-grown soybean plants [*Glycine max* (L.) Merr]. Plant Physiol 70:1178–1184

[63] Firbas F (1931) Untersuchungen über den Wasserhaushalt der Hochmoorpflanzen. Jahrb Wiss Bot 74:459–696

[64] Fischer RA, Turner NC (1978) Plant productivity in the arid and semiarid zones. Annu Rev Plant Physiol 29:277–317

[65] Flores HE (1990) Polyamines and plant stress. In: Alscher RG, Cumming JR (eds) Stress responses in plants: adaptation and acclimation mechanisms. Wiley-Liss, New York, pp 217–239

[66] Flowers TJ (1985) Physiology of halophytes. Plant Soil 89:41–56

[67] Francis D, Barlow PW (1988) Temperature and the cell cycle. In: Long SP, Woodward FI (eds) Plants and temperature. Company of Biologists, Cambridge, pp 181–202

[68] Fromm J, Eschrich W (1993) Electric signals released from roots of willow (*Salix viminalis* L.) change transpiration and photosynthesis. J Plant Physiol 141:673–680

[69] Frost-Christensen H, Sand-Jensen K (1992) The quantum efficiency of photosynthesis in macroalgae and submerged angiosperms. Oecologia 91:377–384

[70] Gaastra P (1959) Photosynthesis of crop plants as influenced by light, carbon dioxide, temperature and stomatal diffusion resistance. Meded Landbouwhogesch Wageningen 59:1–68

[71] Gaff DF (1989) Responses of desiccation tolerant "resurrection" plants to water stress. In: Kreeb KH, Richter H, Hinckley TM (eds) Structural and functional responses to environmental stresses: Water shortage. SPB Acad Publ, The Hague

[72] Gardner WR (1968) Availability and measurement of soil water. In: Kozlowski TT (ed) Water deficits and plant growth, vol I. Academic Press, New York, pp 107–135

[73] Gates DM (1973) Plant temperatures and energy budget. In: Precht H, Christophersen J, Hensel H, Larcher W (eds) Temperature and life. Springer, Berlin Heidelberg New York, pp 87–101

[74] Gausman HW (1985) Plant leaf optical properties in visible and near-infrared light. In: Grad Studies, Texas Tech Univ No 29. Texas Tech Press, Lubbock, pp 78

[75] Gaussen H (1954) Théories et classification des climats et microclimats. Rapp et Comm, Paris, Bot Cgr, Sect 7 b, vol 8:125–130

[76] Geiger R (1961) Das Klima der bodennahen Luftschicht, 4. Aufl. Vieweg, Braunschweig

[77] Gibson AH, Jordan DC (1983) Ecophysiology of nitrogen-fixing systems. In: Lange OL, Nobel PS, Osmond CB, Ziegler H (eds) Encyclopedia of plant physiology, vol 12 C. Springer, Berlin Heidelberg New York, pp 301–390

[78] Gimmler H, Weis U, Weiss C (1989) pH-Regulation and membrane potential of extremely acid resistant green alga *Dunaliella acidiphila*. In: Dainty J, De Michelis M, Marré E, Rasi-Caldogno F (eds) Plant membrane transport. Elsevier, Amsterdam, pp 389–390

[79] Golley FB, McGinnis JT, Clements RG, Child GI, Duever MJ (1975) Mineral cycling in a tropical moist forest ecosystem. University of Georgia Press, Athens

[80] Granhall U, Lid-Torsvik V (1975) Nitrogen fixation by bacteria and free-living blue-green algae in Tundra areas. In: Wiegolaski FE, Kallio P, Rosswall T (eds) Fennoscandian tundra ecosystems. 1. Plants and microorganisms. Springer, Berlin Heidelberg New York, pp 305–315

[81] Gregory FG (1926) The effect of climatic conditions on the growth of barley. Ann Bot 40:1–26

[82] Griffiths H (1989) Carbon dioxide concentrating mechanisms and the evolution of CAM in vascular epiphytes. In: Lüttge U (ed) Vascular plants as epiphytes. Evolution and ecophysiology. Springer, Berlin Heidelberg New York, pp 42–86

[83] Grime JP (1979) Plant strategies and vegetation processes. Wiley, Chichester

[84] Haisch A, Capriel P, Forster S (1985) Cäsiumverfügbarkeit für Pflanzen auf drei verschiedenen Böden. Transferfaktoren Boden – Pflanze. Landwirtsch Forsch 38:229–235

[85] Halevy AH (1976) Treatments to improve water balance of cut flowers. Acta Hortic 54:223–230

[86] Halevy AH (1985) Handbook of flowering. CRC Press, Boca Raton

[87] Hall DO, Coombs J (1979) The prospect of a biological-photochemical approach for the utilisation of solar energy. Watt Committee Energy Rep 6:2–15

[88] Hanson AD (1980) Interpreting the metabolic responses of plants to water stress. Hortic Sci 15:623–629

[89] Hayhoe HN (1975) Crop indices. In: Smith LP (ed) Progress in biometereology, vol 1, Div C. Swets & Zeitlinger, Amsterdam, pp 268–276

[90] Heinrich G, Müller HJ, Oswald K, Gries A (1989) Natural and artificial radionuclides in selected Styrian soils and plants before and after the reactor accident in Chernobyl. Biochem Physiol Pflanz 185:55–67

[91] Hellriegel F (1883) Beiträge zu den naturwissenschaftlichen Grundlagen des Ackerbaues. Vieweg, Braunschweig

[92] Hendry GAF, Houghton JD, Brown SB (1987) The degradation of chlorophyll – a biological enigma. New Phytol 107:255–302

[93] Hew CS, Lee GL, Wong SC (1980) Occurrence of non-functional stomata in the flowers of tropical orchids. Ann Bot 46:195–201

[94] Hirose T, Werger MJA (1987) Nitrogen use efficiency in instantaneous and daily photosynthesis of leaves in the canopy of a *Solidago altissima* stand. Physiol Plant 70:215–222

[95] Hoffmann G (1963) Die höchsten und tiefsten Temperaturen auf der Erde. Umschau 1963:16–18

[96] Höfler K, Migsch H, Rottenburg W (1941) Über die Austrocknungsresistenz landwirt-
      schaftlicher Kulturpflanzen. Forschungsdienst 12:50−61
[97] Holmgren P, Jarvis PG, Jarvis MS (1965) Resistances to carbon dioxide and water
      vapour transfer in leaves of different plant species. Physiol Plant 18:557−573
[98] Hsiao ThC (1973) Plant responses to water stress. Annu Rev Plant Physiol
      24:519−570
[99] Huber B (1928) Weitere quantitative Untersuchungen über das Wasserleitungssystem
      der Pflanzen. Jahrb Wiss Bot 67:877−959
[100] Humphreys MO (1991) Genetic control of physiological response − a necessary rela-
      tionship. Funct Ecol 5:213−221
[101] Huntley B (1991) How plants respond to climate change: migration rates, individual-
      ism and the consequences for plant communities. Ann Bot 67, Suppl 1:15−22
[102] Ingestadt T, Ågren GI (1988) Nutrient uptake and allocation at steady-state nutrition.
      Physiol Plant 72:450−459
[103] Inoue E, Uchijima Z, Udagawa T, Horie T, Kobayashi K (1968) CO_2-Environment
      and CO_2-exchange within a corn canopy. In: Monsi M (ed) Photosynthesis and
      utilization of solar energy. Level III Experiments. Jpn Natl Subcommittee PP, Tokyo,
      pp 1−8
[104] Ivanov LA (1953) O sosushschem apparate kornya drevesnykh porod sovetskovo
      soyuza. Dokl Akad Nauk SSSR 93:713−716
[105] Jefferies RL, Davy AJ (eds) (1979) Ecological processes in coastal environments.
      Blackwell, Oxford
[106] Jenny H, Gessel SP, Bingham FT (1949) Comparative study of decomposition rates
      of organic matter in temperate and tropical regions. Soil Sci 68:419−432
[107] Jeschke WD, Wolf O, Hartung W (1992) Effect of NaCl salinity on flows and parti-
      tioning of C, N, and mineral ions in whole plants of white lupin, *Lupinus albus* L.
      J Exp Bot 43:777−788
[108] Jurik ThW, Chabot BF (1986) Leaf dynamics and profitability in wild strawberries.
      Oecologia 69:296−304
[109] Kacperska A (1985) Biochemical and physiological aspects of frost hardening in her-
      baceous plants. In: Kaurin A, Junttila O, Nilsen J (eds) Plant production in the North.
      Norwegian University Press, Tromsø, pp 99−115
[110] Karabourniotis G, Papadopoulos K, Papamarkou M, Manetas Y (1992) Ultraviolet-B
      radiation absorbing capacity of leaf hairs. Physiol Plant 86:414−418
[111] Karlsson PS, Nordell KO, Eirefelt S, Svensson A (1987) Trapping efficiency of three
      carnivorous Pinguicula species. Oecologia 73:518−521
[112] Kasanaga H, Monsi M (1954) On the light transmission of leaves and its meaning for
      the production of matter in plant communities. Jpn J Bot 14:304−324
[113] Kauffman JB (1991) Survival by sprouting following fire in tropical forests of the
      Eastern Amazon. Biotropica 23:219−224
[114] Keeley JE, Sandquist DR (1992) Carbon: freshwater plants. Plant Cell Environ
      15:1021−1035
[115] Kelly GJ, Latzko E (1984) Photosynthesis. Carbon metabolism: on land and at sea.
      Prog Bot 46:68−93
[116] Kimball StL, Bennett BD, Salisbury FB (1973) The growth and development of mon-
      tane species at near-freezing temperatures. Ecology 54:168−173
[117] Kinzel H (1983) Influence of limestone, silicates and soil pH on vegetation. In: Lange
      OL, Nobel PS, Osmond CB, Ziegler H (eds) Encyclopedia of plant physiology,
      vol 12 C. Springer, Berlin Heidelberg New York, pp 201−244
[118] Kira T (1991) Forest ecosystems of East and Southeast Asia in a global perspective.
      Ecol Res 6:185−200
[119] Konakahara M (1975) Experimental studies on the mechanisms of cold damage and
      its production methods in *Citrus* trees. Shizuoka Pref Citrus Exp Sta Spec Bull
      3:1−164
[120] Köppen W (1931) Grundriß der Klimakunde. de Gruyter, Berlin

[121] Körner Ch (1993) Scaling from species to vegetation: the usefulness of functional groups. In: Schulze ED, Mooney HA (eds) Biodiversity and ecosystem function. Springer, Berlin Heidelberg New York, pp 117–140

[122] Körner Ch (1994) Leaf diffuse conductance in the major vegetation types of the globe. In: Schulze ED, Caldwell MM (eds) Ecophysiology of photosynthesis. Springer, Berlin Heidelberg New York, pp 463–490

[123] Kramer PJ (1983) Water relations of plants. Academic Press, New York

[124] Kullenberg B, Bergström G (1975) Kommunikation zwischen Lebewesen auf chemischer Basis. Endeavour 34:59–66

[125] Kummerow J, Krause D, Jow W (1978) Seasonal changes of fine root density in the southern Californian chaparral. Oecologia 37:201–212

[126] Küppers M (1987) Hecken. Ein Modellfall für die Partnerschaft von Physiologie und Morphologie bei der pflanzlichen Produktion in Konkurrenzsituationen. Naturwissenschaften 74:536–547

[127] Lambers H (1982) Cyanide-resistant respiration: a non-phosphorylating electron transport pathway acting as an energy overflow. Physiol Plant 55:478–485

[128] Lambers H (1985) Respiration in intact plants and tissues: its regulation and dependence on environmental factors, metabolism and invaded organisms. In: Douce R, Day AD (eds) Encyclopedia of plant physiology, vol 18. Springer, Berlin Heidelberg New York, pp 418–473

[129] Lange OL, Matthes U (1981) Moisture-dependent CO_2 exchange of lichens. Photosynthetica 15:555–574

[130] Lange OL, Zuber M (1977) *Frerea indica*, a stem succulent CAM plant with deciduous C_3 leaves. Oecologia 31:67–72

[131] Lange OL, Green TGA, Ziegler H (1988) Water status related photosynthesis and carbon isotope discrimination in species of the lichen genus *Pseudocyphellaria* with green or blue-green photobionts and in photosymbiodemes. Oecologia 75:494–501

[132] Larcher W (1961) Jahresgang des Assimilations- und Respirationsvermögens von *Olea europaea* L. ssp. *sativa* Hoff. et Link., *Quercus ilex* L. und *Quercus pubescens* Willd. aus dem nördlichen Gardaseegebiet. Planta 56:575–606

[133] Larcher W (1969) Physiological approaches to the measurement of photosynthesis in relation to dry matter production by trees. Photosynthetica 3:150–166

[134] Larcher W (1981) Resistenzphysiologische Grundlagen der evolutiven Kälteakklimatisation von Sproßpflanzen. Plant Syst Evol 137:145–180

[135] Larcher W (1987) Streß bei Pflanzen. Naturwissenschaften 74:158–167

[136] Lauscher F (1978) Neue Analyse ältester und neuerer phänologischer Reihen. Arch Meteorol Geophys Bioklimatol Ser B 26:373–385

[137] Lavrenko EM, Borisova IV, Popova TA (eds) (1981) Pushtynnye stepi i severnye pushtyni mongolskoi narodnoi republiki. Nauka, Leningrad

[138] Lenz F (1978) Sink-source relationships in fruit trees. In: Scott TK (ed) Plant regulation and world agriculture. Plenum Press, New York, pp 141–153

[139] Leonardi S, Rapp M, Denes A (1992) Organic matter distribution and fluxes within a holm oak (*Quercus ilex* L.) stand in the Etna volcano. A synthesis. Vegetatio 99–100:219–224

[140] Levitt J (1980) Responses of plants to environmental stresses. I. Chilling, freezing and high temperature stresses, 2nd edn. Academic Press, New York

[141] von Liebig J (1862) Die Naturgesetze des Feldbaues. Vieweg, Braunschweig

[142] Likens GE (1975) Primary productivity of inland aquatic ecosystems. In: Lieth H, Whittaker RH (eds) Primary productivity of the biosphere. Springer, Berlin Heidelberg New York, pp 185–202

[143] Lindow SE (1983) The role of bacterial ice nucleation in frost injury to plants. Annu Rev Phytopathol 21:363–384

[144] Lockhart JA (1965) An analysis of irreversible plant cell elongation. J Theor Biol 8:264–275

[145] Loik ME, Nobel PS (1991) Water relations and mucopolysaccharide increases for a winter-hardy cactus during acclimation to subzero temperatures. Oecologia 88:340–346

[146] Long SP, Postl WF, Bolhár-Nordenkampf HR (1993) Quantum yields for uptake of carbon dioxide in C_3 vascular plants of contrasting habitats. Planta 189:226–234

[147] Loveless AR (1961) A nutritional interpretation of sclerophylly based on differences in the chemical composition of sclerophyllous and mesophytic leaves. Ann Bot 25:168–184

[148] Ludlow MM (1985) Photosynthesis and dry matter production in C_3 and C_4 pasture plants, with special emphasis on tropical C_3 legumes and C_4 grasses. Aust J Plant Physiol 12:557–572

[149] Ludlow MM (1987) Light stress at high temperature. In: Kyle DJ, Osmond CB, Arntzen CJ (eds) Photoinhibition. Elsevier, Amsterdam, pp 89–108

[150] Matyssek R, Günthardt-Goerg MS, Saurer M, Keller Th (1992) Seasonal growth, $\delta 13C$ in leaves and stem, and phloem structure of birch (Betula pendula) under low ozone concentrations. Trees 6:69–76

[151] Maximov NA (1923) Physiologisch-ökologische Untersuchungen über die Dürreresistenz der Xerophyten. Jahrb Wiss Bot 62:128–144

[152] Milburn JA (1966) The conduction of sap. I. Water conduction and cavitation in water stressed leaves. Planta 69:34–42

[153] Mitscherlich G (1971) Wald, Wachstum und Umwelt. Eine Einführung in die ökologischen Grundlagen des Waldwachstums. Bd II: Waldklima und Wasserhaushalt. Sauerländer, Frankfurt

[154] Molisch H (1937) Der Einfluß einer Pflanze auf die andere – Allelopathie. Fischer, Jena

[155] Monsi M, Saeki T (1953) Über den Lichtfaktor in den Pflanzengesellschaften, seine Bedeutung für die Stoffproduktion. Jpn J Bot 14:22–52

[156] Monteith JL (1973) Principles of environmental physics. Arnolds, London

[157] Monteith JL (1978) Reassessment of maximum growth rates for C_3 and C_4 crops. Exp Agric 14:1–5

[158] Montfort C (1953) Hochsommerliche Veränderungen des Chlorophyll-Spiegels bei Laubbäumen von gegensätzlichem photochemischem Reaktionstypus. Planta 42:461–464

[159] Moreno JM, Oechel WC (1991) Fire intensitsy and herbivory effects on postfire resprouting of Adenostoma fasciculatum in southern California chaparral. Oecologia 85:429–433

[160] Morimoto RI (1993) Cells in stress: transcriptional activation of heat shock genes. Science 259:1409–1410

[161] Mousseau M, Enoch HZ (1989) Carbon dioxide enrichment reduces shoot growth in sweet chestnut seedlings (Castanea sativa Mill.). Plant Cell Environ 12:927–934

[162] Müller D (1928) Die Kohlensäureassimilation bei arktischen Pflanzen und die Abhängigkeit der Assimilation von der Temperatur. Planta 6:22–39

[163] Münch E (1930) Die Stoffbewegungen in der Pflanze. Fischer, Jena

[164] Namuco OS, O'Toole JC (1986) Reproductive stage, water stress and sterility. I. Effect of stress during meiosis. Crop Sci 26:317–321

[165] Napp-Zinn K (1973) Low temperature effect on flower formation: vernalization. In: Precht H, Christophersen J, Hensel H, Larcher W (eds) Temperature and life. Springer, Berlin Heidelberg New York, pp 171–194

[166] Nelson T, Langdale JA (1992) Developmental genetics of C_4 photosynthesis. Annu Rev Plant Physiol Plant Mol Biol 43:25–47

[167] Nilsen ET, Sharifi MR, Rundel PW, Virginia RA (1986) Influences of microclimatic conditions and water relations on seasonal leaf dimorphism of Prosopis glandulosa var. torreyana in the Sonoran Desert, California. Oecologia 69:95–100

[168] Nilsen ET, Sharifi MR, Rundel PW, Forseth IN, Ehleringer JR (1990) Water relations of stem succulent trees in northcentral Baja California. Oecologia 82:299–303

[169] Ochi H (1962) Autecological study of mosses in respect to water economy. I. On the minimum hydrability within which mosses are able to survive. Bot Mag 65:112

[170] Odum EP (1969) The strategy of ecosystem development. Science 164:262–270

[171] Odum HT, Pigeon RF (1970) A tropical rain forest. A study of irradiation and ecology at El Verde/Puerto Rico. Office Inf Serv, US Atomic Energy Comm, Oak Ridge, Tennessee

[172] Oertli JJ, Lips SH, Agami M (1990) The strength of sclerophyllous cells to resist collapse due to negative turgor pressure. Acta Oecol 11:281–289

[173] Omasa K, Hashimoto Y, Kramer PJ, Strain BR, Aiga I, Kondo J (1985) Direct observation of reversible and irreversible stomatal responses of attached sunflower leaves to SO_2. Plant Physiol 79:153–158

[174] Ovington JD (1954) A comparison of rainfall in different woodlands. Forestry 27:41–53

[175] Pandey U, Singh JS (1981) A quantitative study of the forest floor, litterfall and nutrient return in an oak-conifer forest in Himalaya. II. Pattern of litter fall and nutrient return. Acta Oecol Oecol Gen 2:83–99

[176] Pate JS (1983) Patterns of nitrogen metabolism in higher plants and their ecological significance. In: Lee JA, McNeill S, Rorison IH (eds) Nitrogen as an ecological factor. Blackwell, Oxford, pp 225–255

[177] Peña JI, Sánchez-Díaz M, Aguirreolea J, Becana M (1988) Increased stress tolerance of nodule activity in the *Medicago-Rhizobium-Glomus* symbiosis under drought. J Plant Physiol 133:79–83

[178] Penning de Vries FWT (1983) Modeling of growth and production. In: Lange OL, Nobel PS, Osmond CB, Ziegler H (eds) Encyclopedia of plant physiology, vol 12 D. Springer, Berlin Heidelberg New York, pp 117–150

[179] Pfanz H, Martinoia E, Lange OL, Heber U (1987) Flux of SO_2 into leaf cells and cellular acidification by SO_2. Plant Physiol 85:928–933

[180] Pigott CD, Huntley JP (1981) Factors controlling the distribution of *Tilia cordata* at the northern limits of its geographical range. III. Nature and causes of seed sterility. New Phytol 87:817–839

[181] Pisek A, Winkler E (1953) Die Schließbewegung der Stomata bei ökologisch verschiedenen Pflanzentypen in Abhängigkeit vom Wassersättigungszustand der Blätter und vom Licht. Planta 42:253–278

[182] Pisek A, Larcher W, Unterholzner R (1967) Kardinale Temperaturbereiche der Photosynthese und Grenztemperaturen des Lebens der Blätter verschiedener Spermatophyten. I. Temperaturminimum der Netto-Assimilation, Gefrier- und Frostschadensbereiche der Blätter. Flora Abt B 157:239–264

[183] Pitt JI, Christian JHB (1968) Water relations of xerophilic fungi isolated from prunes. Appl Microbiol 16:1853–1858

[184] Pócs T (1980) The epiphytic biomass and its effect on the water balance of two rain forest types in the Uluguru mountains (Tanzania, East Africa). Bot Acad Sci Hung 26:143–167

[185] Pons TL, Pearcy RW (1992) Photosynthesis in flashing light in soybean leaves grown in different conditions. II. Lightfleck utilization efficiency. Plant Cell Environ 15:577–584

[186] Potvin C, Strain BR, Goeschl JD (1985) Low night temperature effect on photosynthate translocation of two C_4 grasses. Oecologia 67:305–309

[187] Proctor MCF (1982) Physiological ecology: water relations, light and temperature responses, carbon balance. In: Smith AJE (ed) Bryophyte ecology. Chapman & Hall, London, pp 333–381

[188] Pukacki PM, Giertych M (1982) Seasonal changes in light transmission by bud scales of spruce and pine. Planta 154:381–383

[189] Rabotnov TA (1992) Fitotsenologiya, 3rd edn. Izd Moskva Univ, Moskva

[190] Raghavendra AS, Das VSR (1978) The occurrence of C_4 photosynthesis: a supplementary list of C_4-plants reported during late 1974–mid 1977. Photosynthetica 12:200–208

[191] Rascio N, Mariani P, Tommasini E, Bodner M, Larcher W (1991) Photosynthetic strategies in leaves and stems of *Egeria densa*. Planta 185:297–303

[192] Raskin I (1992) Role of salicylic acid in plants. Annu Rev Plant Physiol 43:439–463

[193] Raven JA (1990) Sensing pH? Plant Cell Environ 13:721–729
[194] Raven JA, Smith FA, Glidewell SM (1979) Photosynthetic capacities and biological strategies of giant-celled and small-celled macroalgae. New Phytol 83:299–309
[195] Reiling K, Davison AW (1992) The response of native, herbaceous species to ozone: growth and fluorescence screening. New Phytol 120:29–37
[196] Remmert H (1991) The mosaic-cycle concept of ecosystems. Springer, Berlin Heidelberg New York
[197] Richter H (1972) Wie entstehen Saugspannungsgradienten in Bäumen? Ber Dtsch Bot Ges 85:341–351
[198] Roberts DM, Harmon AC (1992) Calcium-modulated proteins: targets of intracellular calcium in higher plants. Annu Rev Plant Physiol 43:375–414
[199] Roberts StW, Strain BR, Knoerr KR (1981) Seasonal variation of leaf tissue elasticity in four forest tree species. Physiol Plant 52:245–250
[200] Rosnitschek-Schimmel I (1983) Biomass and nitrogen partitioning in a perennial and an annual nitrophilic species of *Urtica*. Z Pflanzenphysiol 109:215–225
[201] Roth LE, Jeon K, Stacey G (1988) Homology in endosymbiotic systems: the term "symbiosome". In: Palacios R, Verma DPS (eds) Molecular genetics of plant-microbe interactions. American Phytopathological Society Press, St Paul, pp 220–225
[202] Rozema J, Bijwaard P, Prast G, Broekman R (1985) Ecophysiological adaptations of coastal halophytes from foredunes and salt marshes. Vegetatio 62:499–521
[203] Rundel PW (1988) Water relations. In: Galun M (ed) Handbook of lichenology, vol II. CRC Press, Boca Raton, pp 17–36
[204] Rundel PW, Ehleringer JR, Nagy KA (1988) Stable isotopes in ecological research. Springer, Berlin Heidelberg New York
[205] Rychnovská-Soudková M (1966) Wasserhaushalt einiger *Stipa*-Arten am natürlichen Standort. Rozpr Česk Akad Ved Řada Math Přír Věd 76:1–32
[206] Salisbury FB (1963) The flowering process. Pergamon Press, Oxford
[207] Sánchez F, Padilla JE, Pérez H, Lara M (1991) Control of nodulin genes in root-nodule development and metabolism. Annu Rev Plant Physiol Plant Mol Biol 42:507–528
[208] Sánchez-Diaz M, Pardo M, Antolín M, Peña J, Aguirreolea J (1990) Effect of water stress on photosynthetic activity in the *Medicago-Rhizobium-Glomus* symbiosis. Plant Sci 71:215–221
[209] Sassa T, Oohata S, Wakabayasi Y, Negisi K (1984) Bark respiration rates of deciduous broad leaved trees in the Tokyo University Forest at Chichibu. Misc Inf Tokyo Univ For 23:117–129
[210] Saure MC (1985) Dormancy release in deciduous fruit trees. Hortic Rev 7:239–300
[211] Sauter JJ, Kloth S (1986) Plasmodesmatal frequency and radial translocation rates in ray cells of poplar (*Populus×canadensis* Moench *robusta*). Planta 168:377–380
[212] Schmidt-Vogt H (1989) Die Fichte. Bd II/2: Krankheiten, Schäden, Fichtensterben. Parey, Hamburg
[213] Schnelle F (1955) Pflanzenphänologie. Akademische Verlagsgesellschaft, Leipzig
[214] Schröder P (1989) Characterization of a thermo-osmotic gas transport mechanism in *Alnus glutinosa* (L.) Gaertn. Trees 3:38–44
[215] Schulze ED, Fuchs MI, Fuchs M (1977) Spatial distribution of photosynthetic capacity and performance in a mountain spruce forest of northern Germany. I. Biomass distribution and daily CO_2 uptake in different crown layers. Oecologia 29:43–61
[216] Schulze ED, Gebauer G, Schulze W, Pate JS (1991 a) The utilization of nitrogen from insect capture by different growth forms of *Drosera* from Southwest Australia. Oecologia 87:240–246
[217] Schulze ED, Gebauer G, Ziegler H, Lange OL (1991 b) Estimates of nitrogen fixation by trees on an aridity gradient in Namibia. Oecologia 88:451–455
[218] Schuster WS, Monson RK (1990) An examination of the advantages of $C_3–C_4$ intermediate photosynthesis in warm environments. Plant Cell Environ 13:903–912
[219] Selye H (1973) The evolution of the stress concept. Am Sci 61:693–699

[220] Semikhatova OA, Gorbacheva (1962) O dykhanii vysokogornykh rastenii zapadnogo Kavkaza. Trudy Teberdinsk gosydarstv zapovednika III:155–171

[221] Seybold A (1933) Zur Klärung des Begriffes Transpirationswiderstand. Planta 21: 353–367

[222] Shah N, Smirnoff N, Stewart GR (1987) Photosynthesis and stomatal characteristics of *Striga hermonthica* in relation to its parasitic habit. Physiol Plant 69:699–703

[223] Sharkey TD (1988) Estimating the rate of photorespiration in leaves. Physiol Plant 73:147–152

[224] Shearer G, Kohl DH, Virginia RA, Bryan BA, Skeeters JL, Nilsen ET, Sharifi MR, Rundel PW (1983) Estimates of N_2-fixation from the natural abundance of ^{15}N in Soronan desert ecosystems. Oecologia 56:365–373

[225] Singh JS, Gupta SR (1977) Plant decomposition and soil respiration in terrestrial ecosystems. Bot Rev 43:449–528

[226] Skre O, Oechel WC (1981) Moss functioning in different Taiga ecosystems in interior Alaska. I. Seasonal, phenotypic and drought effects on photosynthesis and response patterns. Oecologia 48:50–59

[227] Slatyer RO, Taylor SA (1960) Terminology in plant and soil water relations. Nature 187:922–924

[228] Slavíková J (1965) Die maximale Wurzelsaugkraft als ökologischer Faktor. Preslia (Praha) 37:419–428

[229] Smeets L (1980) Effect of temperature and daylength on flower initiation and runner formation in two everbearing strawberry cultivars. Sci Hortic 12:19–26

[230] Smirnova ES (1965) Die Samenstruktur der Blütenpflanzen. Biol Zentralbl 84:299–307

[231] Smith H (1982) Light quality, photoperception, and plant strategy. Annu Rev Plant Physiol 33:481–518

[232] Smith H, Casal JJ, Jackson GM (1990) Reflection signals and the perception by phytochrome of the proximity of neighbouring vegetation. Plant Cell Environ 13:73–78

[233] Smith RAH, Bradshaw AD (1979) The use of metal tolerant plant populations for the reclamation of metalliferous wastes. J Appl Ecol 16:595–612

[234] Stålfelt MG (1939) Vom System der Wasserversorgung abhängige Stoffwechselcharaktere. Bot Not (Lund) 1939:176–192

[235] Stetter KO, Fiala G, Huber G, Huber R, Segerer A (1990) Hyperthermophilic microorganisms. FEMS Microbiol Rev 75:117–124

[236] Stewart WDP (1982) Nitrogen fixation – its current relevance and future potential. Isr J Bot 31:5–44

[237] Stocker O (1929) Das Wasserdefizit von Gefäßpflanzen in verschiedenen Klimazonen. Planta 7:382–387

[238] Stocker O (1931) Transpiration und Wasserhaushalt in verschiedenen Klimazonen. I. Untersuchungen an der arktischen Baumgrenze in Schwedisch-Lappland. Jahrb Wiss Bot 75:494–549

[239] Stocker O (1935) Assimilation und Atmung westjavanischer Tropenbäume. Planta 24:402–445

[240] Stocker O (1956) Wasseraufnahme und Wasserspeicherung bei Thallophyten. In: Ruhland W (ed) Handbuch der Pflanzenphysiologie, Bd 3. Springer, Berlin Heidelberg New York, pp 160–172

[241] Sveshnikova VM (1979) Dominanty kasakhstanskikh stepei. Nauka, Leningrad

[242] Sweeney FC, Hopkinson JM (1975) Vegetative growth of nineteen tropical and subtropical pasture grasses and legumes in relation to temperature. Trop Grassl 9:209–217

[243] Szarek SR, Ting IP (1977) The occurrence of Crassulacean acid metabolism among plants. Photosynthetica 11:330–342

[244] Tagawa H (1965) A study of the volcanic vegetation in Sakurajima, southwest Japan. II. Distributional pattern and succession. Jpn J Bot 19:127–148

[245] Tani N (1978) Agricultural crop damage by weather hazards and the countermeasures in Japan. In: Takahashi K, Yoshino MM (eds) Climatic change and food production. Tokyo University Press, Tokyo, pp 197–215

[246] Taylor GE, Tingey DT (1983) Sulfur dioxide flux into leaves of *Geranium carolinianum* L. Plant Physiol 72:237–244

[247] Taylor HM, Terrell EE (1982) Rooting pattern and plant productivity. In: Rechcigl M (ed) CRC Handbook of agricultural productivity, vol I. CRC Press, Boca Raton, pp 185–200

[248] Terry N, Waldron LJ, Taylor SE (1983) Environmental influences on leaf expansion. In: Dale JE, Milthorpe FL (eds) The growth and functioning of leaves. Cambridge University Press, Cambridge, pp 179–205

[249] Thornthwaite CW (1948) An approach towards a rational classification of climate. Geogr Rev 38:55–94

[250] Tomlinson PB, Soderholm PK (1975) The flowering and fruiting of *Corypha elata* in South Florida. Principes 19:83–99

[251] Trabaud L (1979) Etude du comportement du feu dans la garrigue de chêne kermes à partir des températures et vitesses de propagation. Ann Sci For 36:13–38

[252] Trabaud L (ed) (1987) The role of fire in ecological systems. SPB Acad Publ, The Hague

[253] Trabaud L, Méthy M (1988) Modifications dans le système photosynthétique de repousses apparaissant après feu de deux espèces ligneuses dominantes des garrigues méditerranéennes. Acta Oecol 9:229–243

[254] Tumanov II (1979) Fiziologiya zakalivaniya i morozostoikosti rastenii. Izdat Nauka, Moskva

[255] Turesson G (1925) The plant species in relation to habitat and climate. Contributions to the knowledge of genecological units. Hereditas 4:147–236

[256] Tyree MT, Hammel HT (1972) The measurement of the turgor pressure and water relations of plants by the pressure-bomb technique. J Exp Bot 23:267–282

[257] Ulrich B (1968) Ausmaß und Selektivität der Nährelementaufnahme in Fichten- und Buchenbeständen. Allg Forsztg 47 (Sonderdruck)

[258] Ulrich B (1987) Stability, elasticity and resilience of terrestrial ecosystems with respect to matter balance. In: Schulze ED, Zwölfer H (eds) Potentials and limitations of ecosystem analysis. Springer, Berlin Heidelberg New York, pp 11–49

[259] Vieira da Silva J (1976) Water stress, ultrastructure and enzymatic activity. In: Lange OL, Kappen L, Schulze ED (eds) Water and plant life. Problems and modern approaches. Springer, Berlin Heidelberg New York, pp 207–224

[260] Vierling E (1991) The roles of heat shock proteins in plants. Annu Rev Plant Physiol Plant Mol Biol 42:579–620

[261] Vogelmann TC, Björn LO (1986) Plants as light traps. Physiol Plant 68:704–708

[262] Walker RB (1954) The ecology of serpentine soils; factors affecting plant growth on serpentine soils. Ecology 35:259–266

[263] Wallace CS, Rundel PhW (1979) Sexual dimorphism and resource allocation in male and female shrubs of *Simmondsia chinensis*. Oecologia 44:34–39

[264] Walter H (1931) Die Hydratur der Pflanze und ihre physiologisch-ökologische Bedeutung. Fischer, Jena

[265] Walter H (1955) Die Klimadiagramme als Mittel zur Beurteilung der Klimaverhältnisse für ökologische, vegetationskundliche und landwirtschaftliche Zwecke. Ber Dtsch Bot Ges 69:331–344

[266] Walter H, Walter E (1953) Das Gesetz der relativen Standortskontanz: Das Wesen der Pflanzengesellschaften. Ber Dtsch Bot Ges 66:228–236

[267] Wang J, Ives NE, Lechowicz MJ (1992) The relation of foliar phenology to xylem embolism in trees. Funct Ecol 6:469–475

[268] Warembourg FR, Montange D, Bardin R (1982) The simultaneous use of $^{14}CO_2$ and $^{15}N_2$ labelling techniques to study the carbon and nitrogen economy of legumes grown under natural conditions. Physiol Plant 56:46–55

[269] Waring RH, Cleary BD (1967) Plant moisture stress: evaluation by pressure bomb. Science 155:1248–1254

[270] Waring RH, Whitehead D, Jarvis PG (1979) The contribution of stored water to transpiration in Scots pine. Plant Cell Environ 2:309–317

[271] Watson DJ (1947) Comparative physiological studies on the growth of field crops. I. Variation in net assimilation rate and leaf area between species and varieties, and within and between years. Ann Bot 11:41–76

[272] Weatherley PE (1965) Some investigations on water deficit and transpiration under controlled conditions. In: Slavik B (ed) Water stress in plants. Czechoslovak Academy of Sciences, Praha, pp 63–71

[273] Webb T, Bartlein PJ (1992) Global changes during the last 3 million years: climatic controls and biotic responses. Annu Rev Ecol Syst 23:141–173

[274] Went FW (1957) The experimental control of plant growth. Waltham, Mass

[275] Williams GJ, McMillan C (1971) Frost tolerance of *Liquidambar styraciflua* native to the United States, Mexico, and Central America. Can J Bot 49:1551–1558

[276] Winner WE, Mooney HA (1985) Ecology of SO_2 resistance. V. Effects of volcanic SO_2 on native Hawaiian plants. Oecologia 66:387–393

[277] Woolhouse HW (1983) The effects of stress on photosynthesis. In: Marcelle R, Clijsters H, van-Poucke M (eds) Effects of stress on photosynthesis. Nijhoff-Junk, The Hague, pp 1–28

[278] Wuosmaa AM, Hager LP (1990) Methyl chloride transferase: a carbocation route for biosynthesis of halometabolites. Science 249:160–162

[279] Yoda K (1978) Light climate within the forest. In: Kira T, Ono Y, Hosokawa T (eds) Biological production in a warm-temperate evergreen oak forest of Japan. JIBP Synthesis 18, Jpn Committee for the Int Biol Program, Science Council of Japan, Univ Press, Tokyo, pp 46–54

[280] Zelitch I (1975) Improving the efficiency of photosynthesis. Science 188:626–633

[281] Zimmermann MH, Brown CL (1974) Trees, structure and function, 2nd edn. Springer, Berlin Heidelberg New York

References in the Tables and Figures

Addicott FT (1968) Environmental factors in the physiology of abscission. Plant Physiol 43:1471–1479

Aho N, Daudet FA, Vartanian N (1979) Evolution de la photosynthèse nette et de l'efficience de la transpiration au cours d'un cycle de dessèchement du sol. CR Acad Sci Paris Ser D 288:501–504

Ajtay GL, Ketner P, Duvigneaud P (1979) Terrestrial primary production and phytomass. In: Bolin B, Degens ET, Kempe S, Ketner P (eds) The global carbon cycle. Wiley, Chichester, pp 129–181

Albert R (1982) Halophyten. In: Kinzel H (ed) Pflanzenökologie und Mineralstoffwechsel. Ulmer, Stuttgart, pp 33–215

Albert R, Popp M (1977) Chemical composition of halophytes from the Neusiedler Lake region in Austria. Oecologia 27:157–170

Albert R, Popp M (1978) Zur Rolle der löslichen Kohlenhydrate in Halophyten des Neusiedlersee-Gebietes (Österreich). Oecol Plant 13:27–42

Alexeev VA (1975) Svetovoi rezhim lesa. Nauka, Leningrad

Allen LH, Yocum CS, Lemon ER (1964) Photosynthesis under field conditions. VI. Radiant energy exchanges within a corn crop canopy and implications in water use efficiency. Agron J 56:253–259

Altenburger R, Matile Ph (1990) Further observations on rhythmic emission of fragrance in flowers. Planta 180:194–197

Altman PhL, Dittmer DS (1972) Biology data book, vol I; 2nd edn. Fed Amer Soc Exp Biol, Bethesda

Altman PhL, Dittmer DS (1973) Biology data book, vol II; 2nd edn. Federation of American Societies for Experimental Biology, Bethesda

Alvim de P (1964) Tree growth periodicity in tropical climates. In: Zimmermann MH (ed) The formation of wood in forest trees. Academic Press, London, pp 479–495

Amundson RG, Walker RB, Schellhase HU, Legge AH (1990) Sulphur gas emissions in the boreal forest: the West Whitecourt Case Study. VIII. Pine tree mineral nutrition. Water Air Soil Pollut 50:219–232

Anderson JM, Osmond CB (1987) Shade-sun responses: compromises between acclimation and photoinhibition. In: Kyle DJ, Osmond CB, Arntzen CJ (eds) Photoinhibition. Elsevier, Amsterdam, pp 2–38

Anderson MC (1964) Light relations of terrestrial plant communities and their measurement. Biol Rev 39:425–486

Anfodillo T, Sigalotti GB, Tomasi M, Semenzato P, Valentini R (1993) Applications of a thermal imaging technique in the study of the ascent of sap in woody species. Plant Cell Environ 16:997–1001

Aoki M, Yabuki K, Koyama H (1975) Micrometeorology and assessment of primary production of a tropical rain forest in West Malaysia. J Agric Meteorol 31:115–124

Arndt U, Nobel W, Schweizer B (1987) Bioindikatoren. Möglichkeiten, Grenzen und neue Erkenntnisse. Ulmer, Stuttgart

Aspinall D, Paleg LG (1981) Proline accumulation: physiological aspects. In: Paleg LG, Aspinall D (eds) Physiology and biochemistry of drought resistance in plants. Academic Press, Sydney, pp 205–241

Aulitzky H (1963) Grundlagen und Anwendung des vorläufigen Wind-Schnee-Ökogrammes. Mitt Forstl Bundes-Versuchsanst Mariabrunn 60:765–834

462   References in Tables and Figures

Aulitzky H (1968) Über die Ursachen der Unwetterkatastrophen und den Grad ihrer Beein-flußbarkeit. Centralbl Gesamte Forstwes 85:2–32
Aulitzky H, Turner H, Mayer H (1982) Bioklimatische Grundlagen einer standortsgemäßen Bewirtschaftung des subalpinen Lärchen-Arvenwaldes. Mitt Eidg Forstl Versuchswes 58:327–580
Aussenac G (1973) Effets de conditions microclimatiques différentes sur la morphologie et la structure anatomique des aiguilles de quelques résineux. Ann Sci For 30:375–392
Baeumer K (1991) Allgemeiner Pflanzenbau, 3. Aufl. Ulmer, Stuttgart
Baker NR, Hardwick K (1976) Development of the photosynthetic apparatus in cocoa leaves. Photosynthetica 10:361–366
Bannister P (1976) Introduction to physiological plant ecology. Blackwell, Oxford
Bannister P, Smith PJM (1983) The heat resistance of some New Zealand plants. Flora 173:399–414
Barrick WE (1978) Salt-tolerant plants for Florida landscapes. Proc Fla State Hortic Soc 91:82–84
Barrs HD (1968) Determination of water deficits in plant tissues. In: Kozlowski T (ed) Water deficits and plant growth, vol I. Academic Press, New York, pp 235–368
Bartkov BI, Zvereva G (1974) Raspredelenie assimiliatov v period plodonosheniya bobovykh rastenii. O printsipe dublirovaniya v fitosistemakh. Fiziol Biokhim Kult Rast 6:502–505
Batanouny KH, Stichler W, Ziegler H (1988) Photosynthetic pathways and ecological distri-bution of grass species in Egypt. Oecologia 75:539–548
Bates TE (1971) Factors affecting critical nutrient concentrations in plants and their evalua-tion: a review. Soil Sci 112:116–130
Baumeister W, Ernst W (1978) Mineralstoffe und Pflanzenwachstum, 3. Aufl. Fischer, Stutt-gart
Bazilevich NI, Rodin LY (1971) Geographical regularities in productivity and the circulation of chemical elements in the earth's main vegetation types. In: Soviet Geography (Rev and Translation). American Geographical Society, New York
Bazzaz FA, Garbutt K, Williams WE (1985) Effect of increased atmospheric carbon dioxide concentration on plant communities. In: Strain BR, Cure JD (eds) Direct effects of in-creasing carbon dioxide on vegetation. US Dept Energy, Carbon Dioxide Res Div, DOE/ER0238, Washington, pp 155–170
Beadle CL, Long SP (1985) Photosynthesis – is it limiting to biomass production? Biomass 8:119–168
Beck E, Lüttge U (1990) Streß bei Pflanzen. Biol unserer Zeit 20:237–244
Beck E, Senser M, Scheibe R, Steiger HM, Pongratz P (1982) Frost avoidance and freezing tolerance in Afroalpine "giant rosette" plants. Plant Cell Environ 5:215–222
Beck E, Schulze ED, Senser M, Scheibe R (1984) Equilibrium freezing of leaf water and ex-tracellular ice formation in Afroalpine "giant rosette" plants. Planta 162:276–282
Beevers L (1976) Nitrogen metabolism in plants. Edward Arnold, London
Begg JE, Turner NC (1970) Water potential gradients in field tobacco. Plant Physiol 46:343–346
Beideman IN (1974) Metodika izucheniya fenologii rastenii i rastitelnykh soobshchestv. Nauka, Novosibirsk
Bell KL, Bliss LC (1980) Autecology of *Kobresia bellardii*: why winter snow accumulation limits local distribution. Ecol Monogr 49:377–402
Bennett RN, Wallsgrove RM (1994) Secondary metabolites in plant defence mechanisms. New Phytol 127:617–633
Berger W (1931) Das Wasserleitungssystem von krautigen Pflanzen, Zwergsträuchern und Lianen in quantitativer Betrachtung. Beih Bot Centralbl 48:364–390
Berger-Landefeldt U (1959) Beiträge zur Ökologie der Pflanzen nordafrikanischer Salzpfan-nen. Vegetatio 9:1–47
Bergmann W (1983) Ernährungsstörungen bei Kulturpflanzen – Entstehung und Diagnose. Fischer, Jena
Berry JA, Raison JK (1981) Responses of macrophytes to temperature. In: Lange OL, Nobel PS, Osmond CB, Ziegler H (eds) Encyclopedia of plant physiology, vol 12A. Springer, Berlin Heidelberg New York, pp 277–338

Beschel R (1958) Flechtenvereine der Städte, Stadtflechten und ihr Wachstum. Ber Naturwiss Med Ver Innsb 52:7–158

Bewley JD (1981) Protein synthesis. In: Paleg LG, Aspinall D (eds) Physiology and biochemistry of drought resistance in plants. Academic Press, Sydney, pp 261–282

Bewley JD, Krochko JE (1982) Desiccation-tolerance. In: Lange OL, Nobel PS, Osmond CB, Ziegler H (eds) Encyclopedia of plant physiology, vol 12B. Springer, Berlin Heidelberg New York, pp 325–378

Beyschlag W, Lange OL, Tenhunen JD (1990) Photosynthese und Wasserhaushalt der immergrünen mediterranen Hartlaubpflanze *Arbutus unedo* L. im Jahresverlauf am Freilandstandort in Portugal. III. Einzelfaktorenanalyse zur Licht-, Temperatur- und CO_2-Abhängigkeit der Nettophotosynthese. Flora 184:271–289

Biebl R (1938) Trockenresistenz und osmotische Empfindlichkeit der Meeresalgen verschieden tiefer Standorte. Jahrb Wiss Bot 86:350–386

Biebl R, Kinzel H (1965) Blattbau und Salzhaushalt von *Laguncularia racemosa* (L.) Gaertn. und anderer Mangrovebäume auf Puerto Rico. Oesterr Bot Z 112:56–93

Billings WD, Godfrey PJ, Chabot BF, Bourque DP (1971) Metabolic acclimation to temperature in arctic and alpine ecotypes of *Oxyria digyna*. Arct Alp Res 3:277–290

Björkman O, Pearcy RW, Harrison AT, Mooney HA (1972) Photosynthetic adaptation to high temperatures: a field study in Dealth Valley, California. Science 175:786–789

Black CC (1973) Photosynthetic carbon fixation in relation to net CO_2 uptake. Annu Rev Plant Physiol 24:253–286

Bliss LC (1975) Devon Island, Canada. In: Rosswall T, Heal OW (eds) Structure and function of tundra ecosystems. Ecological Bulletin 20. Natl Sci Foundation Editorial Service, Stockholm, pp 17–60

Bliss LC, Heal OW, Moore JJ (1981) Tundra ecosystems: a comparative analysis. Cambridge University Press, Cambridge

Bobrovskaya NI (1985) Wasserhaushalt der Wüsten- und Wüstensteppenpflanzen in der Nord- und der Trans-Altai-Gobi. Feddes Repert 96:425–432

Bolhàr-Nordenkampf HR, Draxler G (1993) Functional leaf anatomy. In: Hall DO, Scurlock JMO, Bolhàr-Nordenkampf HR, Leegood RC, Long SP (eds) Photosynthesis and production in a changing environment: a field and laboratory manual. Chapman & Hall, London, pp 91–112

Borchert R (1994) Induction of rehydration and bud break by irrigation or rain in deciduos trees of a tropical dry forest in Costa Rica. Trees 8:198–204

Bornman JF (1989) Target sites of UV-B radiation in photosynthesis of higher plants. J Photochem Photobiol B4:145–158

Bortenschlager S, Schwarzer Ch (1988) Flechtenkartierung im Raum Innsbruck – Vergleich 1977/87. In: Zustand der Tiroler Wälder. Landesforstdirektion, Innsbruck

Boukhris M, Lossaint P (1975) Aspects écologiques de la nutrition minérale des plantes gypsicoles de Tunisie. Rev Ecol Biol Sol 12:329–348

Bowen HJM (1979) Environmental chemistry of the elements. Academic Press, London

Box EO, Meentemeyer V (1991) Geographic modeling and modern ecology. In: Esser G, Overdieck D (eds) Modern ecology: basic and applied aspects. Elsevier, Amsterdam

Boyer JS (1970) Leaf enlargement and metabolic rates in corn, soybean, and sunflower at various leaf water potentials. Plant Physiol 46:233–235

Boyer JS (1971) Nonstomatal inhibition of photosynthesis in sunflower at low leaf water potentials and high light intensities. Plant Physiol 48:532–536

Boyer JS (1974) Water transport in plants: mechanism of apparent changes in resistance during absorption. Planta 117:187–207

Boyer JS, Bowen BL (1970) Inhibition of oxygen evolution in chloroplasts isolated from leaves with low water potentials. Plant Physiol 45:612–615

Bradford KJ, Hsiao TC (1982) Physiological responses to moderate water stress. In: Lange OL, Nobel PS, Osmond CB, Ziegler H (eds) Encyclopedia of plant physiology, vol 12B. Springer, Berlin Heidelberg New York, pp 263–324

Bradford KJ, Yang ShF (1981) Physiological responses of plants to waterlogging. Hort Science 16:25–30

Braun H, Scheumann W (1989) Erste Prüfungsergebnisse von Douglasienbestandesnach-
kommenschaften unter besonderer Berücksichtigung der Frostresistenz. Beitr Forstwirt
23:4−11
Braun HJ (1970) Funktionelle Histologie der sekundären Sproßachse. I. Das Holz. In:
Handbuch der Pflanzenanatomie. Spezieller Teil, Bd IX/1. Bornträger, Berlin
Braun HJ (1974) Rhythmus und Größe von Wachstum, Wasserverbrauch und Produktivität
des Wasserverbrauchs bei Holzpflanzen. I. *Alnus glutinosa* (L.) Gaertn. und *Salix alba*
(L.) „Liempde". Allg Forst-Jagdztg 145:81−86
Braun HJ (1977) Zum Wachstum und zur Produktivität des Wasserverbrauchs der Baumar-
ten *Acer platanoides* L., *Acer pseudoplatanus* L. und *Fraxinus excelsior* L. Z Pflanzen-
physiol 84:459−462
Braune W, Leman A, Taubert H (1987) Pflanzenanatomisches Praktikum I, 5. Aufl. Fischer,
Stuttgart
Bray EA (1993) Molecular responses to water deficit. Plant Physiol 103:1035−1040
Breckle SW (1976) Zur Ökologie und zu den Mineralstoffverhältnissen absalzender und
nichtabsalzender Xerohalophyten. In: Diss Bot, Bd 35. Cramer, Vaduz
Breckle SW (1982) The significance of salinity. In: Spooner B, Mann HS (eds) Desertifica-
tion and development: dryland ecology in social perspective. Academic Press, London,
pp 277−292
Breitsprecher A, Bethel JS (1990) Stem-growth periodicity of trees in a tropical wet forest
of Costa Rica. Ecology 71:1156−1164
Brisse H, Grandjouan G (1974) Classification climatique des plantes. Oecol Plant 9:51−80
Brock ThD (1978) Thermophilic microorganisms and life at high temperatures. Springer,
Berlin Heidelberg New York
Brown HT, Escombe F (1900) Static diffusion of gases and liquids in relation to the assimila-
tion of carbon and translocation in plants. Philos Trans R Soc Lond Ser B Biol Sci
193:223−291
Brunstein FC, Yamaguchi DK (1992) The oldest known Rocky Mountain bristlecone pines
(*Pinus aristata* Engelm.). Arctic and Alpine Res 24:253−256
Brünig EF (1987) The forest ecosystem: tropical and boreal. Ambio 16:67−79
Bünning E (1953) Entwicklungs- und Bewegungsphysiologie der Pflanzen, 3. Aufl. Springer,
Berlin Heidelberg New York
Burckhardt H (1963) Meteorologische Voraussetzungen der Nachtfröste. In: Schnelle F (ed)
Frostschutz im Pflanzenbau, Bd 1. Bayerischer Landwirtschafts-Verlag, München, pp
13−81
Burian K (1973) *Phragmites communis* Trin. Im Röhricht des Neusiedler Sees. Wachstum,
Produktion und Wasserverbrauch. In: Ellenberg H (ed) Ökosystemforschung. Springer,
Berlin Heidelberg New York, pp 61−78
Byrd GT, Sage RF, Brown RH (1992) A comparison of dark respiration between C_3 and C_4
plants. Plant Physiol 100:191−198
Caldwell MM (1977) The effects of solar UV-B radiation (280−315 nm) on higher plants:
Implications of stratospheric ozone reduction. In: Castellani A (ed) Research in photo-
biology. Plenum Press, New York, pp 597−607
Caldwell MM (1988) Plant root systems and competition. In: Greuter W, Zimmer B (eds)
Proc XIV. International Botanical Congress. Koeltz, Königstein, pp 385−404
Caldwell MM, White RS, Moore RT, Camp LB (1977) Carbon balance, productivity and
water use of cold-winter desert shrub communities dominated by C_3 and C_4 species.
Oecologia 29:275−300
Caldwell MM, Robberecht R, Flint SD (1983) Internal filters: prospects for UV-acclimation
in higher plants. Physiol Plant 58:445−450
Cape JN, Vogt TC (1991) Application of Härtel's turbidity test across Europe. Can J For
Res 21:1423−1429
Cappelletti C (1961) Ricerche sulla permeabilitá delle cuticole alle radiazioni ultraviolette.
I. Piante di duna. Acc Naz Lincei Cl Fis Mat Nat 30:331−342
Caprio JM, Hopp RJ, Williams JS (1974) Computer mapping in phenological analysis. In:
Lieth H (ed) Phenology and seasonality modeling. Ecological Studies 8. Springer, Berlin
Heidelberg New York, pp 77−82

Carpenter StB, Smith ND (1981) A comparative study of leaf thickness among southern Appalachian hardwoods. Can J Bot 59:1393–1396

Carter DL (1982) Salinity and plant productivity. In: Rechcigl M (ed) CRC Handbook of agricultural productivity, vol I: Plant productivity. CRC Press, Boca Raton, pp 117–133

Carter TR, Porter JH, Parry ML (1992) Some implications of climatic change for agriculture in Europe. J Exp Bot 43:1159–1167

Čatský J, Tichá I (1980) Ontogenetic changes in the internal limitations to bean-leaf photosynthesis. 5. Photosynthetic and photorespiration rates and conductances for CO_2 transfer as affected by irradiance. Photosynthetica 14:392–400

Čermák J, Jeník J, Kučera J, Žídék V (1984) Xylem water flow in a crack willow tree (*Salix fragilis* L.) in relation to diurnal changes of environment. Oecologia 64:145–151

Čermák J, Cienciala E, Kučera J, Hällgren JE (1992) Radial velocity profiles of water flow in trunks of Norway spruce and oak and the response of spruce to severing. Tree Physiol 10:367–380

Cernusca A (1977) Alpine Umweltprobleme. Ergebnisse des Forschungsprojekts Achenkirch. In: Beiträge zur Umweltgestaltung A 62. Schmidt, Berlin

Ceulemans R, Saugier B (1991) Photosynthesis. In: Raghavendra AS (ed) Physiology of trees. Wiley, London, pp 21–50

Ceulemans R, Impens I, Hebrant F, Moermans R (1980) Evaluation of field productivity for several poplar clones based on their gas exchange variables determined under laboratory conditions. Photosynthetica 14:355–362

Chapin FSt (1989) The cost of tundra plant structures: evaluation of concepts and currencies. Am Nat 133:1–19

Chapin FSt, Shaver GR (1985) Arctic. In: Chabot BF, Mooney HA (eds) Physiological ecology of North American plant communities. Chapman & Hall, New York, pp 16–40

Chapman SB (1976) Methods in plant ecology. Blackwell, Oxford

Chartier Ph, Bethenod O (1977) La productivité primaire à l'echelle de la feuille. In: Moyse A (ed) Les processus de la production végétale primaire. Gauthier-Villars, Paris, pp 77–112

Chen HH, Li PH (1980) Characteristics of cold acclimation and deacclimation in tuber-bearing *Solanum* species. Plant Physiol 65:1146–1148

Chmora SN (1993) Dykhanie listev na svetu. In: Tselniker YuL (ed) Rost i gazoobmen CO_2 u lesnykh derevev. Nauka, Moskva, pp 105–129

Cichan MA (1986) Conductance in the wood of selected carboniferous plants. Paleobiology 12:302–310

Cienciala E, Lindroth A, Čermák J, Hällgren JE, Kučera J (1992) Assessment of transpiration estimates for *Picea abies* trees during a growing season. Trees 6:121–127

Clark J (1961) Photosynthesis and respiration in white spruce and balsam fir. State Univ Coll For Syracuse, New York

Claussen W, Biller E (1977) Die Bedeutung der Saccharose- und Stärkegehalte der Blätter für die Regulierung der Netto-Photosyntheseraten. Z Pflanzenphysiol 81:189–198

Cleland RE (1986) The role of hormones in wall loosening and plant growth. Aust J Plant Physiol 13:93–103

Cole DW, Rapp M (1980) Elemental cycling in forest ecosystems. In: Reichle DE (ed) Dynamic properties of forest ecosystems. Cambridge University Press, Cambridge, pp 341–409

Coleman JS, McConnaughay KDM, Bazzaz FA (1993) Elevated CO_2 and plant nitrogen-use: is reduced tissue nitrogen concentration size-dependent? Oecologia 93:195–200

Coley PD (1986) Costs and benefits of defense by tannins in a neotropical tree. Oecologia 70:238–241

Collins RP, Jones MB (1985) The influence of climatic factors on the distribution of C_4-species in Europe. Vegetatio 64:121–129

Cooper JP (1977) Photosynthetic efficiency of maize compared with other field crops. Ann Appl Biol 87:237–242

Cowan IR (1965) Transport of water in the soil-plant-atmosphere system. J Appl Ecol 2:221–239

Cowan IR, Lange OL, Green TGA (1992) Carbon-dioxide exchange in lichens: determination of transport and carboxylation characteristics. Planta 187:282–294

Cox GW, Atkins MD (1979) Agricultural ecology – an analysis of world food production systems. Freeman, San Francisco

Crawford RMM (1977) Tolerance of anoxie and ethanol metabolism in germinating seeds. New Phytol 79:511–517

Crawford RMM (1989) Studies in plant survival. Ecological case histories of plant adaptation to adversity. Blackwell, Oxford

Cronan CS, Reiners WA (1983) Canopy processing of acidic precipitation by coniferous and hardwood forests in New England. Oecologia 59:216–223

Cumming JR, Taylor GJ (1990) Mechanisms of metal tolerance in plants: physiological adaptations for exclusion of metal ions from the cytoplasm. In: Alscher RG, Cumming JR (eds) Stress responses in plants: adaptation. Wiley-Liss, New York, pp 329–356

Darrall NM, Jäger HJ (1984) Biochemical diagnostic tests for the effect of air pollution on plants. In: Koziol MJ, Whatley FR (eds) Gaseous air pollutants and plant metabolism. Butterworths, London, pp 333–349

Däßler HG (1991) Einfluß von Luftverunreinigungen auf die Vegetation. Ursachen –Wirkungen – Gegenmaßnahmen, 4. Aufl. Fischer, Jena

Dawson TE (1993) Hydraulic lift and water use by plants: implications for water balance, performance and plant-plant interactions. Oecologia 95:565–574

Day TA (1993) Relating UV-B radiation screening effectiveness of foliage to absorbing-compound concentration and anatomical characteristics in a diverse group of plants. Oecologia 95:542–550

Day TA, Vogelmann TC, DeLucia EH (1992) Are some plant life forms more effective than others in screening out ultraviolet-B radiation? Oecologia 92:513–519

Deevey ES (1970) Mineral cycles. Sci Am 223:148–158

De Filippis LF, Hampp R, Ziegler H (1981) The effect of sublethal concentration of zinc, cadmium and mercury on *Euglena*. Adenylates and energy charge. Z Pflanzenphysiol 103:1–7

Demmig B, Björkman O (1987) Comparison of the effect of excessive light on chlorophyll fluorescence (77K) and photon yield of O_2 evolution in leaves of higher plants. Planta 171:171–184

Demmig-Adams B, Adams III WW (1992) Photoprotection and other responses of plants to high light stress. Annu Rev Plant Physiol Plant Mol Biol 43:599–626

Diamantoglou S, Rhizopoulou S, Kull U (1989) Energy content, storage substances, and construction and maintenance costs of mediterranean deciduous leaves. Oecologia 81:528–533

Dickson RE (1991) Assimilate distribution and storage. In: Raghavendra AS (ed) Physiology of trees. Wiley, New York, pp 51–85

Doley D (1981) Tropical and subtropical forests and woodlands. In: Kozlowski TT (ed) Water deficits and plant growth, Vol VI: Woody plant communities. Academic Press, New York, pp 209–323

Doley D, Grieve BJ (1966) Measurement of sap flow in a eucalypt by thermoelectric methods. Aust For Res 2:3–27

Downs RJ, Hellmers H (1975) Environment and the experimental control of plant growth. Academic Press, London

Downs MR, Nadelhoffer KJ, Melillo JM, Aber JD (1993) Foliar and fine root nitrate reductase activity in seedlings of four forest tree species in relation to nitrogen availability. Trees 7:233–236

Dufrêne E, Saugier B (1993) Gas exchange of oil palm in relation to light, vapour pressure deficit, temperature and leaf age. Funct Ecol 7:97–104

Duvigneaud P (1967) Ecosystèmes et biosphère. Min Educ Nat Cult, Brüssel

Duvigneaud P, Denaeyer-De Smet S (1970) Biological cycling of minerals in temperate deciduous forests. In: Reichle DE (ed) Analysis of temperate forest ecosystems. Ecological studies 1. Springer, Berlin Heidelberg New York, pp 199–225

Duvigneaud P, Denaeyer-De Smet S (1973) Considérations sur l'écologie de la nutrition minérale des tapis végétaux naturels. Oecol Plant 8:219–246

Duvigneaud P, Denaeyer-De Smet S, Ambroes P, Timperman J, Marbaise JL (1969) Recherche sur l'écosystème forêt. B. La chênaie mélangé calcicole de Virelles-Blaimont. Bull Soc R Bot Belge 102:317–327

Dye PJ, Olbrich BW (1993) Estimating transpiration from 6-year-old *Eucalyptus grandis* trees: development of a canopy conductance model and comparison with independent sap flux measurements. Plant Cell Environ 16:45–53

Dykyjová D, Kvet J (1978) Pond littoral ecosystems. Structure and functioning. Springer, Berlin Heidelberg New York

Eagles CF, Wilson D (1982) Photosynthetic efficiency and plant productivity. In: Rechcigl M (ed) Handbook of agricultural productivity, vol I. CRC Press, Boca Raton, pp 213–247

Earnshaw MJ, Carver KA, Gunn TC, Kerenga K, Harvey V, Griffiths H, Broadmeadow MSJ (1990) Photosynthetic pathway, chilling tolerance and cell sap osmotic potential values of grasses along an altitudinal gradient in Papua New Guinea. Oecologia 84:280–288

Eber W (1991) Morphology in modern ecological research. In: Esser G, Overdieck D (eds) Modern ecology: basic and applied aspects. Elsevier, Amsterdam, pp 3–20

Eberhardt E (1955) Der Atmungsverlauf alternder Blätter und reifender Früchte. Planta 45:57–68

Edwards PJ (1982) Studies of mineral cycling in a montane rain forest in New Guinea. V. Rates of cycling in throughfall and litter fall. J Ecol 70:807–827

Egunjobi JK (1974) Dry matter, nitrogen and mineral element distribution in an unburnt savanna during the year. Oecol Plant 9:1–10

Ehleringer JA (1979) Photosynthesis and photorespiration: biochemistry, physiology, and ecological implications. HortScience 14:217–222

Eliáš P, Kratochvilová I, Janouš D, Marek M, Masarovičová E (1989) Stand microclimate and physiological activity of tree leaves in an oak-hornbeam forest. I. Stand microclimate. Trees 4:227–233

Ellenberg H (1939) Über Zusammensetzung, Standort und Stoffproduktion bodenfeuchter Eichen- und Buchen-Mischwaldgesellschäften Nordwest-Deutschlands. Mitt Florist Soziol Arbeitsgem Niedersachsen 5:3–135

Ellenberg H (1977) Stickstoff als Standortsfaktor, insbesondere für mitteleuropäische Pflanzengesellschaften. Oecol Plant 12:1–22

Ellenberg H (1986) Vegetation Mitteleuropas mit den Alpen, 4. Aufl. Ulmer, Stuttgart

Emmingham WH, Waring RH (1977) An index of photosynthesis for comparing forest sites in western Oregon. Can J For Res 7:165–174

Engel L, Fock H, Schnarrenberger C (1986) CO_2 and H_2O gas exchange of the high alpine plant *Oxyria digyna* (L.) Hill. 2. Response to high irradiance stress and supraoptimal leaf temperatures. Photosynthetica 20:304–314

Enríquez S, Duarte CM, Sand-Jensen K (1993) Patterns in decomposition rates among photosynthetic organisms: the importance of detritus $C:N:P$ content. Oecologia 94:457–471

Epstein E (1972) Mineral nutrition of plants. Wiley, New York

Epstein E (1994) The anomaly of silicon in plant biology. Proc Natl Acad Sci USA 91:11–17

Ernst WHO (1976) Physiological and biochemical aspects of metal tolerance. In: Mansfield IA (ed) Effects of air pollutants on plants. Cambridge University Press, Cambridge, pp 115–133

Ernst WHO (1990) Mine vegetation in Europe. In: Shaw AJ (ed) Heavy metal tolerance in plants: evolutionary aspects. CRC Press, Boca Raton, pp 21–37

Ernst WHO, Joosse van Damme ENG (1983) Umweltbelastung durch Mineralstoffe – biologische Effekte. Fischer, Stuttgart

Esau K (1953) Plant anatomy. Wiley, New York

Esser G (1987) Sensitivity of global carbon pools and fluxes to human and potential climatic impacts. Tellus 39B:245–260

Esterbauer H, Grill D, Zotter M (1978) Peroxidase in Nadeln von *Picea abies* (L.) Karst. Biochem Physiol Pflanz 172:155–159

Evans JR (1983) Nitrogen and photosynthesis in the flag leaf of wheat (*Triticum aestivum* L.). Plant Physiol 72:297–302

Evans JR (1989) Photosynthesis and nitrogen relationships in leaves of C_3 plants. Oecologia 78:9–19

Evans LT (1973) The effect of light on plant growth, development and yield. In: Slatyer RO (ed) Plant response to climatic factors. UNESCO, Paris, pp 21–35

Evenari M (1984) Seed physiology: from ovule to maturing seed. Bot Rev 50:143–170

Evenari M, Schulze ED, Kappen L, Buschbom U, Lange OL (1975) Adaptive mechanisms in desert plants. In: Vernberg FJ (ed) Physiological adaptation to the environment. Intext Educational Publishers, New York, pp 111–130

Ewers FW, Zimmermann MH (1984) The hydraulic architecture of balsam fir (*Abies balsamea*). Physiol Plant 60:453–458

Farrar JF, Williams ML (1991) The effects of increased atmospheric carbon dioxide and temperature on carbon partitioning, source-sink relations and respiration. Plant Cell Environ 14:819–830

Field ChB (1991) Ecological scaling of carbon gain to stress and resource availability. In: Mooney HA, Winner WE, Pell EJ (eds) Response of plants to multiple stresses. Academic Press, San Diego

Finck A (1969) Pflanzenernährung in Stichworten. F. Hirtl, Kiel

Flowers TJ (1985) Physiology of halophytes. Plant Soil 89:41–56

Flowers TJ, Troke PF, Yeo AR (1977) The mechanism of salt tolerance in halophytes. Annu Rev Plant Physiol 28:89–121

Francois LE, Clark RA (1978) Salt tolerance of ornamental shrubs, trees, and iceplant. J Am Soc Hortic Sci 103:280–283

Franz H (1979) Ökologie der Hochgebirge. Ulmer, Stuttgart

Freedman B (1989) Environmental ecology. The impacts of pollution and other stresses on ecosystem structure and function. Academic Press, San Diego

Freer-Smith PH (1984) Response of six broad-leaved trees during long-term exposure to SO_2 and NO_2. New Phytol 97:49–61

Frey-Wyssling A (1949) Stoffwechsel der Pflanzen, 2. Aufl. Büchergilde Gutenberg, Zürich

Fritts HC (1976) Tree rings and climate. Academic Press, New York

Fujikawa S, Miura K (1986) Freezing tolerance in edible mushrooms. Jpn J Freez Dry 32:14–17

Gaff DF (1980) Protoplasmic tolerance of extreme water stress. In: Turner NC, Kramer PJ (eds) Adaptation of plants to water and high temperature stress. Wiley, New York, pp 207–230

Garbutt K, Bazzaz FA (1984) The effects of elevated CO_2 on plants. III. Flower, fruit and seed production and abortion. New Phytologist 98:433–446

Garsed SG (1985) SO_2 uptake and transport. In: Winner WE, Mooney HA, Goldstein RA (eds) Sulfur dioxide and vegetation. Physiology, ecology and policy issues. Stanford University Press, Stanford, pp 75–95

Gartner BL (1991) Stem hydraulic properties of vines vs. shrubs of western poison oak, *Toxicodendron diversilobum*. Oecologia 87:180–189

Gates DM (1965) Energy, plants and ecology. Ecology 46:1–14

Gäumann E (1935) Der Stoffhaushalt der Buche (*Fagus sylvatica*) im Laufe eines Jahres. Ber Schweiz Bot Ges 44:157–334

Gausman HW (1985) Plant leaf optical properties in visible and near-infrared light. In: Grad Studies, Texas Tech Univ No 29. Texas Tech Press, Lubbock, pp 78

Gausman HW, Allen WA (1973) Optical parameters of leaves of 30 plant species. Plant Physiol 52:57–62

Gebauer G, Rehder H, Wollenweber B (1988) Nitrate, nitrate reduction and organic nitrogen in plants from different ecological and taxonomic groups of Central Europe. Oecologia 75:371–385

Geiger R (1961) Das Klima der bodennahen Luftschicht, 4. Aufl. Vieweg, Braunschweig

Gershenzon J (1984) Changes in the levels of plant secondary metabolites under water and nutrient stress. In: Timmermann BN, Steelink C, Loewus FA (eds) Phytochemical adaptations to stress. Plenum Press, New York, pp 273–320

Gerwick BC, Williams-III GJ, Uribe EG (1977) Effects of temperature on the Hill reaction and photophosphorylation in isolated cactus chloroplasts. Plant Physiol 60:430–432

Gerwick C (1982) Response of terrestrial plants to mineral nutrients. In: Mitsui A, Black CC (eds) CRC Handbook of biosolar resources, vol 1. CRC Press, Boca Raton, pp 213–222

Geurten I (1950) Untersuchungen über den Gaswechsel von Baumrinden. Forstwiss Centralbl 69:704–743

Gianinazzi S, Gianinazzi-Pearson V (1988) Mycorrhizae: a plant's health insurance. Chim Oggi 10:56–58

Giaquinta RT, Geiger DR (1973) Mechanism of inhibition of translocation by localized chilling. Plant Physiol 51:372–377

Gimingham CH (1972) Ecology of heathlands. Chapman & Hall, London

Glatzel G (1983) Mineral nutrition and water relations of hemiparasitic mistletoes: a question of partitioning. Experiments with *Loranthus europaeus* on *Quercus petraea* and *Quercus robur.* Oecologia 56:193–201

Glatzel G, Kazda M, Grill D, Halbwachs G, Katzensteiner K (1987) Ernährungsstörungen bei Fichte als Komplexwirkung von Nadelschäden und erhöhter Stickstoffdeposition – ein Wirkungsmechanismus des Waldsterbens? Allg Forst-Jagdztg 158:91–97

Goldstein G, Rada F, Azocar A (1985) Cold hardiness and supercooling along an altitudinal gradient in Andean giant rosette species. Oecologia 68:147–152

Golley FB, McGinnis JT, Clements RG, Child GI, Duever MJ (1975) Mineral cycling in a tropical moist forest ecosystem. University of Georgia Press, Athens

Goryshina TK (1969) Rannevesennie efemeroidy lesostepnyck dubrav. Izdatel'stvo Leningradskogo Universiteta Leningrad

Goryshina TK (1980) Structural and functional features of the leaf assimilatory apparatus in plants of a forest-steppe oakwood. I. Leaf plastid apparatus in plants of various forest strata. Acta Oecol 1:47–54

Goryshina TK (1989) Fotosinteticheskii apparat rastenii i usloviya sredy. Izdatel'stvo Leningr Univ, Leningrad

Gottstein D, Gross D (1992) Phytoalexins of woody plants. Trees 6:55–68

Goudriaan J, van Laar HH (1994) Modelling potential crop growth processes. Kluwer Acad. Publ., Dordrecht

Grabherr G, Cernusca A (1977) Influence of radiation, wind, and temperature on the CO_2 gas exchange of the alpine dwarf shrub community *Loiseleurietum cetrariosum.* Photosynthetica 11:22–28

Grace J (1977) Plant response to wind. Academic Press, London

Granhall U, Lid-Torsvik V (1975) Nitrogen fixation by bacteria and free-living blue-green algae in tundra areas. In: Wiegolaski FE, Kallio P, Rosswall T (eds) Fennoscandian tundra ecosystems. 1. Plants and microorganisms. Springer, Berlin Heidelberg New York, pp 305–315

Gratani L (1993) Response to microclimate of morphological leaf attributes, photosynthetic and water relations of evergreen sclerophyllous shrub species. Photosynthetica 29:573–582

Graumlich LJ, Brubaker LB (1986) Reconstruction of annual temperature (1590–1979) for Longmire, Washington, derived from tree rings. Quat Res 25:223–234

Griffiths H, Smith JAC, Lüttge U, Popp M, Cram WJ, Diaz M, Lee HSJ, Medina E, Schäfer C, Stimmel KH (1989) Ecophysiology of xerophytic and halophytic vegetation of a coastal alluvial plain in northern Venezuela. IV. *Tillandsia flexuosa* Sw. and *Schomburgkia humboldtiana* Reichb., epiphytic CAM plants. New Phytol 111:273–282

Griffiths H (1991) Applications of stable isotope technology in physiological ecology. Funct Ecol 5:254–269

Grill D, Guttenberger H, Zellnig G, Bermadinger E (1989) Reactions of plant cells on air pollution. Phyton 29:277–290

Grill E (1989) Phytochelatins in plants. In: Hamer DH, Winge DR (eds) Metal ion homeostasis. Molecular biology and chemistry. Liss, New York, pp 283–300

Grime JP, Hodgson JG (1969) Ecological aspects of the mineral nutrition of plants. In: Rorison IH (ed) Ecological aspects of the mineral nutrition of plants. Blackwell, Oxford, pp 67–99

Grin AM (1972) Wasserhaushalt der russischen Ebene. Umschau 72:551–554

Guderian R, Tingey DT, Rabe R (1985) Effects of photochemical oxidants on plants. In: Guderian R (ed) Air pollution by photochemical oxidants. Formation, transport, control and effects on plants. Springer, Berlin Heidelberg New York, pp 127–296

Gulmon SL, Mooney HA (1986) Costs of defense on plant productivity. In: Givnish TJ (ed) On the economy of plant form and function. Cambridge University Press, Cambridge, pp 681–698

Haberlandt G (1924) Physiologische Pflanzenanatomie, 6. Aufl. Engelmann, Leipzig

Häckel H (1990) Meteorologie, 2. Aufl. Ulmer, Stuttgart

Hagen C, Bornman JF, Braune W (1992) Reversible lowering of modulated chlorophyll fluorescence after saturating flashes in *Haematococcus lacustris* (Volvocales) at room temperature. Physiol Plant 86:593–599

Hager A (1967) Untersuchungen über die lichtinduzierten reversiblen Xanthophyllumwandlungen an *Chlorella* und *Spinacia*. Planta 74:148–172

Hager A (1975) Die reversiblen, lichtabhängigen Xanthophyllumwandlungen im Chloroplasten. Ber Dtsch Bot Ges 88:27–44

Hager A (1980) The reversible, light-induced conversions of xanthophylls in the chloroplast. In: Czygan FC (ed) Pigments in plants, 2nd edn. Fischer, Stuttgart, pp 57–79

Hagihara A, Hozumi K (1991) Respiration. In: Raghavendra AS (ed) Physiology of trees. J Wiley, New York, pp 87–110

Halbwachs G, Wimmer R (1987) Holzanatomische Aspekte bei der Einwirkung von Immissionen auf Bäume. In: Rossmanith HP (ed) Waldschäden – Holzwirtschaft. Österr Agrarverlag, Wien, pp 133–147

Hall AE (1979) A model of leaf photosynthesis and respiration for predicting carbon dioxide assimilation in different environments. Oecologia 143:299–316

Hall DO, Scurlock JMO, Bolhàr-Nordenkampf HR, Leegood RC, Long SP (1993) Photosynthesis and production in a changing environment: a field and laboratory manual. Chapman & Hall, London

Hänninen H (1990) Modelling bud dormancy release in trees from cool and temperate regions. Acta For Fenn 213:1–47

Harborne JB (1988) Introduction to ecological biochemistry, 3rd edn. Academic Press, London

Harper JL (1977) Population biology of plants. Academic Press, London

Harper JL (1989) The value of a leaf. Oecologia 80:53–58

Härtel O (1972) Langjährige Meßreihen mit dem Trübungstext – Ergebnisse und Folgerungen. Oecologia 9:103–111

Härtel O (1976) Wie lassen sich Pflanzenschäden definieren? Umschau 76:347–350

Hartmann G, Nienhaus F, Butin H (1988) Farbatlas der Waldschäden. Diagnose von Baumkrankheiten. Ulmer, Stuttgart

Hartsema AM, Luyten I, Blaauw AH (1930) The optimal temperatures from flower formation to flowering (Rapid flowering of Darwin tulips II). Verh K Akad Wet Amsterdam 2. Sect, XXVII, 1. Med 30:37–45

Hartung W, Davies WJ (1991) Drought-induced changes in physiology and ABA. In: Davies WJ, Jones HG (eds) Abscisic acid. Physiology and biochemistry. Bios Sci Publ, Oxford

Hattersley PW (1983) The distribution of C_3 and C_4 grasses in Australia in relation to climate. Oecologia 57:113–128

Havaux M (1992) Stress tolerance of photosystem II in vivo. Antagonistic effects of water, heat, and photoinhibition stresses. Plant Physiol 100:424–432

Heide OM (1993) Daylength and thermal time responses of budburst during dormancy release in some northern deciduous trees. Physiol Plant 88:531–540

Henson IE, Turner NC (1991) Stomatal responses to abscisic acid in three lupin species. New Phytol 117:529–534

Henssen A, Jahns HM (1991) Lichens, 3rd edn. Thieme, Stuttgart

Hesketh J, Baker D (1967) Light and carbon assimilation by plant communities. Crop Sci 7:285–293

Heß D (1991) Pflanzenphysiologie, 9. Aufl. Ulmer, Stuttgart

Hettelingh JP, Downing RJ, De Smet PAM (1991) Mapping critical loads for Europe. (CCE Techn Rep 1). Natl Inst Publ Health and Environm Protection, Bilthoven, The Netherlands

Heun AM, Gorham J, Lüttge R, Wyn-Jones G (1981) Changes of water-relation characteristics and levels of organic cytoplasmic solutes during salinity induced transition of *Mesembryanthemum crystallinum* from C_3 photosynthesis to crassulacean acid metabolism. Oecologia 50:66–72

Hincha DK (1989) Low concentrations of trehalose protect isolated thylakoids against mechanical freeze-thaw damage. Biochim Biophys Acta 987:231–234

Hincha DK, Heber U, Schmitt JM (1989) Freezing ruptures thylakoid membranes in leaves, and rupture can be prevented in vitro by cryoprotective proteins. Plant Physiol Biochem 27:795–801

Hinckley TM, Lassoie JP, Running SW (1978) Temporal and spatial variations in the water status of forest trees. For Sci Monogr 20:1–72

Hintikka V (1963) Über das Großklima einiger Pflanzenareale in zwei Klimakoordinatensystemen dargestellt. Ann Bot Soc Vanamo 35:1–64

Hiroi T, Monsi M (1966) Dry matter economy of *Helianthus annuus* communities grown at varying densities on light intensities. J Fac Sci Tokyo 9:241–285

Hock B, Elstner E (1988) Pflanzentoxikologie. Der Einfluß von Schadstoffen und Schadwirkungen auf Pflanzen. 2. Aufl. Bibliographisches Inst, Mannheim

Hoffmann G (1972) Wachstumsrhythmus der Wurzeln und Sproßachsen von Forstgehölzen. Flora 161:303–319

Hoflacher H, Bauer H (1982) Light acclimation in leaves of the juvenile and adult life phases of ivy (*Hedera helix*). Physiol Plant 56:177–182

Höfler K (1920) Ein Schema für die osmotische Leistung der Pflanzenzelle. Ber Dtsch Bot Ges 38:288–298

Hook DD, Scholtens JR (1978) Adaptations and flood tolerance of tree species. In: Hook DD, Crawford RMM (eds) Plant life in anaerobic environments. Ann Arbor Sci Publ, Ann Arbor, pp 299–331

Horak O, Kinzel H (1971) Typen des Mineralstoffwechsels bei den höheren Pflanzen. Österr Bot Z 119:475–495

Horn R, Schulze ED, Hantschel R (1989) Nutrient balance and element cycling in healthy and declining Norway spruce stands. In: Schulze ED, Lange OL, Oren R (eds) Forest decline and air pollution. Springer, Berlin Heidelberg New York, pp 444–455

Horsman DC, Wellburn AR (1976) Guide to the metabolic and biochemical effects of air pollutants on higher plants. In: Mansfield TA (ed) Effects of air pollutants on plants. Cambridge University Press, Cambridge, pp 185–199

Howe HF, Westley LC (1986) Ecology of pollination and seed dispersal. In: Crawley MJ (ed) Plant ecology. Blackwell, London, pp 185–215

Hsiao ThC (1973) Plant responses to water stress. Annu Rev Plant Physiol 24:519–570

Hsiao ThC, O'Toole JC, Yambao EB, Turner NC (1984) Influence of osmotic adjustment on leaf rolling and tissue death in rice (*Oryza sativa* L.). Plant Physiol 75:338–341

Huber B (1956) Allgemeine Grundlagen der Wasserleitung. In: Ruhland W (ed) Handbuch der Pflanzenphysiologie, Bd 3. Springer, Berlin Heidelberg New York, pp 509–513

Huettl RF, Mueller-Dombois D (1993) Forest decline in the Atlantic and Pacific region. Springer, Berlin Heidelberg New York

Hunt R, Hand DW, Hannah MA, Neal AM (1991) Response to CO_2 enrichment in 27 herbaceous species. Funct Ecol 5:410–421

Idso SB, Kimball BA (1992) Seasonal fine-root biomass development of sour orange trees grown in atmospheres of ambient and elevated CO_2. Plant Cell Environ 15:337–341

IGBP (1992) The international geosphere-biosphere programme: a study of global change. Rep No 12. IGBP, Stockholm

Ihne E (1905) Phänologische Karte des Frühlingseinzugs in Mitteleuropa. Petermanns Geogr Mitt 5:97–108

Iljin WS (1940) Boden und Pflanze. II. Physiologie und Biochemie der Kalk- und Kieselpflanzen. Rozpravy Ruské Véd. spol. badat. v Praze 10:75–115

Incoll LD, Long SP, Ashmore MR (1977) SI units in publications in plant science. Curr Adv Plant Sci 28:331–343

Isermann K (1980) Recent findings in plant nutrition and new developments in fertilizer research. Plant Res Dev 12:1–48

Isermann K (1983) Bewertung natürlicher und anthropogener Stoffeinträge über die Atmosphäre als Standortfaktoren im Hinblick auf die Versauerung land- und forstwirtschaftlich genutzter Böden. VDI-Ber 500:307–335

Iversen J (1944) Viscum, Hedera and Ilex as climate indicators. Geol Foren Stockh Forh 66:463–483

Iwaki H, Midorikawa B (1968) Principles for estimating root production in herbaceous perennials. In: Ghilarov MS, Kovda VA, Novichkova-Ivanova LN, Rodin LE, Svishnikova VM (eds) Methods of productivity studies in root system and rhizosphere organisms. Nauka, Leningrad, pp 72–78

Jäger HJ (1982) Biochemical indication of an effect of air pollution on plants. In: Steubing L, Jäger HJ (eds) Monitoring of air pollutants by plants. Junk, The Hague, pp 99–107

Jäger HJ, Bender J, Weigel HJ (1989) Stand der Diskussion über Richtwerte für Schadstoffkonzentrationen in der Luft. Angew Bot 63:559–575

Janiesch P (1978) Ökophysiologische Untersuchungen von Erlenbruchwäldern. I. Die edaphischen Faktoren. Oecol Plant 13:43–57

Jarvis PG (1981) Stomatal conductance, gaseous exchange and transpiration. In: Grace J, Ford ED, Jarvis PG (eds) Plants in their atmospheric environment. Blackwell, Oxford, pp 175–204

Jarvis PG (1989) Atmospheric carbon dioxide and forests. Philos Trans R Soc Lond B 324:369–392

Jarvis PG, Leverenz JW (1983) Productivity of temperate, deciduous and evergreen forests. In: Lange OL, Nobel PS, Osmond CB, Ziegler H (eds) Encyclopedia of plant physiology, vol 12D. Springer, Berlin Heidelberg New York, pp 233–280

Jeffrey DW (1987) Soil-plant relationships. An ecological approach. Croom Helm, London

Jerlov NG (1976) Marine optics. Elsevier, Amsterdam

Jeschke WD, Atkins CA, Pate JS (1985) Ion circulation via phloem and xylem between root and shoot of nodulated white lupin. J Plant Physiol 117:319–330

Johnson DW, Cole DW, Bledsoe CS, Cromack K, Edmonds RL, Gessel SP, Grier CC, Richards BN, Vogt KA (1982) Nutrient cycling in forests of the Pacific Northwest. In: Edmonds RL (ed) Analysis of coniferous forest ecosystems in the western United States. Hutchinson Ross, Stroudsburg, pp 186–232

Johnson PhL, Atwood DM (1970) Aerial sensing and photographic study of the El Verde rain forest. In: Odum HT, Pigeon RF (eds) A tropical rain forest. Book 1, B-5. USAEC Techn Inf Ctr, Oak Ridge

Jordan CF, Kline JR (1977) Transpiration of trees in a tropical rainforest. J Appl Ecol 14:853–860

Jordano P (1992) Fruits and frugivory. In: Fenner M (ed) Seeds. The ecology of regeneration in plant communities. CAB International, Wallingford, pp 105–156

Kairiukshtis LA (1967) Ratsionalnoe ispolzovanie solnechnoi energii kak faktor povysheniya produktivnosti listvenno-elovykh nasazhenii. In: Tselniker JuL (ed) Svetovoi rezhim fotosintez i produktivnost lesa. Nauka, Moskva, pp 151–166

Kaiser WM (1982) Correlation between changes in photosynthetic activity and changes in total protoplast volume in leaf tissue from hygro-, meso- and xerophytes under osmotic stress. Planta 154:538–545

Kaiser WM, Kaiser G, Martinoia E, Heber U (1988) Salt toxicity and mineral deficiency in plants: cytoplasmic ion homeostasis, a necessity for growth and survival under stress. In: Kleinkauf H, Döhren RV, Jaenicke L (eds) The roots of modern biochemistry. De Gruyter, Berlin, pp 722–733

Kallio P, Veum AK (1975) Analysis of precipitation at Fennoscandian tundra sites. In: Wielgolaski FE, Kallio P, Rosswall T (eds) Fennoscandian tundra ecosystems. 1. Plants and microorganisms. Springer, Berlin Heidelberg New York, pp 333–338

Kandler O (1992) The German forest decline situation: a complex disease or a complex of diseases. In: Manion PD, Lachance D (eds) Forest decline concepts. APS Press, St. Paul, pp 59–84

Kappen L (1965) Untersuchungen über die Widerstandsfähigkeit der Gametophyten einheimischer Polypodiaceen gegenüber Frost, Hitze und Trockenheit. Flora 156:101–116

Kappen L (1981) Ecological significance of resistance to high temperature. In: Lange OL, Nobel PS, Osmond CB, Ziegler H (eds) Encyclopedia of plant physiology 12 A. Springer, Berlin Heidelberg New York, pp 439–474

Kappen L (1993) Plant activity under snow and ice, with particular reference to lichens. Arctic 46:297–302

Kappen L, Lange OL, Schulze ED, Evenari M, Buschbom U (1976) Distribution pattern of water relations and net photosynthesis of *Hammada scoparia* (Pomel) Iljin in a desert environment. Oecologia 23:323–334

Kärenlampi L, Tammisola J, Hurme H (1975) Weight increase of some lichens as related to carbon dioxide exchange and thallus moisture. In: Wielgolaski FE, Kallio P, Rosswall T (eds) Fennoscandian tundra ecosystems. 1. Plants and microorganisms. Springer, Berlin Heidelberg New York, pp 135–137

Karow AM, Webb WR (1965) Tissue freezing – a theory for injury and survival. Cryobiology 2:99–108

Kaufmann MR (1977) Soil temperature and drying cycle effects on water relations of *Pinus radiata*. Can J Bot 55:2413–2418

Kausch W (1955) Saugkraft und Wassernachleitung im Boden als physiologische Faktoren, unter besonderer Berücksichtigung des Tensiometers. Planta 45:217–265

Kausch W (1959) Der Einfluß von edaphischen und klimatischen Faktoren auf die Ausbildung des Wurzelwerkes der Pflanzen, unter besonderer Berücksichtigung einiger algerischer Wüstenpflanzen. Habilitationsschrift, Darmstadt

Kausch W (1968) Das Wurzelwerk der Pflanzen als Organ für die Wasseraufnahme. Umschau 2:38–44

Kawano S, Masuda J (1980) The productive and reproductive biology of flowering plants. VII. Resource allocation and reproductive capacity in wild populations of *Helionopsis orientalis* (Thunb.) C. Tanaka (Liliaceae). Oecologia 45:307–317

Keck RW, Boyer JS (1974) Chloroplast response to low leaf water potentials. III. Differing inhibition of electron transport and photophosphorylation. Plant Physiol 53:474–479

Keller Th (1975) Zur Phytotoxizität von Fluorimmissionen für Holzarten. Mitt Eidg Anst Forstl Versw 51:305–331

Keller Th (1980) The effect of a continuous springtime fumigation with SO_2 on CO_2 uptake and structure of the annual ring in spruce. Can J For Res 10:1–6

Keller Th (1982) Zum Nachweis einer Umweltbelastung durch Luftverunreinigungen. Schweiz Z Forstwes 133:873–884

Kelliher FM, Leuning R, Schulze ED (1993) Evaporation and canopy characteristics of coniferous forests and grasslands. Oecologia 95:153–163

Kennedy RA, Rumpho ME, Fox ThC (1992) Anaerobic metabolism in plants. Plant Physiol 100:1–6

Kerner Fv (1888) Untersuchungen über die Schneegrenze im mittleren Inntal. Denkschr K Österr Akad Wiss Math-Naturw Cl 54, II:1–62

Kershaw KA (1985) Physiological ecology of lichens. Cambridge University Press, Cambridge

Kiendl J (1953) Zum Wasserhaushalt des *Phragmitetum communis* und des *Glycerietum aquaticae*. Ber Dtsch Bot Ges 66:246–263

Kimura M (1969) Ecological and physiological studies on the vegetation of Mt. Shimagare. VII. Analysis of production processes of a young *Abies* stand based on the carbohydrate economy. Bot Mag Tokyo 82:6–19

Kinzel H (1982) Pflanzenökologie und Mineralstoffwechsel. Ulmer, Stuttgart

474   References in Tables and Figures

Kira T, Owaga H, Yoda K, Ogino K (1964) Primary production by a tropical rain forest of southern Thailand. Bot Mag Tokyo 77:428–429

Kira T, Shinozaki K, Hozumi K (1969) Structure of forest canopies as related to their primary productivity. Plant Cell Physiol 10:129–142

Kislyuk IM (1964) Issledovanie povrezhdayushchego leistviya okhlazhdeniya na kletki listev rastenii, chuvstvitelnykh k kholodu. Nauka, Moskau, Leningrad

Kislyuk IM, Alexandrov VYa, Denko EI, Feldman NL, Kamentseva IE, Lutova MI, Shukhtina HG, Vaskovsky MD (1977) Thermostability of cells and temperature conditions of species life. Phytotron Newsl 15:59–64

Kline JR, Martin JR, Jordan CF, Koranda JJ (1970) Measurement of transpiration in tropical trees with tritiated water. Ecology 51:1068–1073

Klinge H (1976) Bilanzierung von Hauptnährstoffen im Ökosystem tropischer Regenwald (Manaus), vorläufige Daten. Biogeographica 7:59–77

Kluge M, Ting IP (1978) Crassulaceen acid metabolism. Analysis of an ecological adaptation. Springer, Berlin Heidelberg New York

Koike T (1990) Autumn coloring, photosynthetic performance and leaf development of deciduous broad-leaved trees in relation to forest succession. Tree Physiol 7:21–32

Konings H (1990) Physiological and morphological differences between plants with a high NAR or a high LAR as related to environmental conditions. In: Lambers H, Cambridge ML, Konings H, Pons TL (eds) Causes and consequences of variation in growth rate and productivity of higher plants. SPB Academic Publishing, The Hague, pp 101–123

Körner Ch (1982) CO_2 exchange in the alpine sedge *Carex curvula* as influenced by canopy structure, light and temperature. Oecologia 53:98–104

Körner Ch (1989) The nutritional status of plants from high altitudes. A worldwide comparison. Oecologia 81:379–391

Körner Ch, Arnone III JA (1992) Responses to elevated carbon dioxide in artificial tropical ecosystems. Science 257:1672–1675

Körner Ch, Renhardt U (1987) Dry matter partitioning and root length/leaf area ratios in herbaceous perennial plants with diverse altitudinal distribution. Oecologia 74:411–418

Körner Ch, Pelaez Menendez-Riedl S, John PCL (1989) Why are bonsai plants small? A consideration of cell size. Aust J Plant Physiol 16:443–448

Köstler JN, Brückner E, Bibelriether H (1968) Die Wurzeln der Waldbäume. Parey, Hamburg

Kowalski St (1987) Mycotrophy of trees in converted stands remaining under strong pressure of industrial pollution. Angew Bot 61:65–83

Koziol MJ, Whatley FR (1984) Gaseous air pollutants and plant metabolism. Butterworths, London

Kozlowski TT (1971) Growth and development of trees. I: Seed germination, ontogeny, and shoot growth. Academic Press, New York, pp 443

Kozlowski TT (1992) Carbohydrate sources and sinks in woody plants. Bot Rev 58:107–222

Kozlowski TT, Kramer PJ, Pallardy StG (1991) The physiological ecology of woody plants. Academic Press, San Diego

Kramer D, Römheld V, Landsberg E, Marschner H (1980) Induction of transfer cell formation by iron deficiency in the root epidermis of *Helianthus annuus* L. Planta 147:325–339

Kramer PJ (1949) Plant and soil water relationship. McGraw-Hill, New York

Kramer PJ, Kozlowski TT (1979) Physiology of trees, 2nd edn. McGraw-Hill, New York

Kreeb K (1965) Die ökologische Bedeutung der Bodenversalzung. Angew Bot 39:1–15

Kreeb K (1974a) Ökophysiologie der Pflanzen. Fischer, Jena

Kreeb K (1974b) Pflanzen an Salzstandorten. Naturwissenschaften 61:337–343

Krizek DT (1982) Guidelines for measuring and reporting environmental conditions in controlled-environment studies. Physiol Plant 56:231–235

Kronenberg GHM, Kendrick RE (1986) The physiology of action. In: Kendrick RE, Kronenberg GHM (eds) Photomorphogenesis in plants. Nijhoff, Dordrecht, pp 99–114

Krüssmann G (1970) Taschenbuch der Gehölzverwendung, 2. Aufl. Parey, Berlin

Kubín St (1985) Definition, Bewertung und Messung der photosynthetisch aktiven Strahlung. Gartenbauwissenschaft 50:120–128

Kull U, Herbig A, Frei O (1992) Construction and economy of plant stems as revealed by use of the Bic-method. Ann Bot 69:327–334

Künstle E, Mitscherlich G (1977) Photosynthese, Transpiration und Atmung in einem Mischbestand im Schwarzwald. Teil IV: Bilanz. Allg Forst-Jagdztg 148:227–239

Kuntze H, Niemann J, Roeschmann G, Schwerdtfeger G (1988) Bodenkunde, 4. Aufl. Ulmer, Stuttgart

Kuroiwa S (1969) Total photosynthesis of a foliage in relation to inclination of leaves. In: Productivity of photosynthetic systems, models and methods. IBP/PP, Trebon, pp 58–63

Kursar TA, Coley PD (1992) Delayed development of the photosynthetic apparatus in tropical rain forest species. Funct Ecol 6:411–422

Kutík J, Zima J, Šesták Z, Volfová A (1988) Ontogenetic changes in the internal limitations to bean-leaf photosynthesis. 10. Chloroplast ultrastructure in primary and first trifoliate leaves. Photosynthetica 22:511–515

Kutschera L (1960) Wurzelatlas mitteleuropäischer Ackerunkräuter und Kulturpflanzen. Deutscher Landwirtschaftsverlag, Frankfurt

Kutschera L, Lichtenegger E (1992) Wurzelatlas mitteleuropäischer Grünlandpflanzen. Bd II: Pteridophyta und Dicotyledonae (Magnoliopsida). Teil 1: Morphologie, Anatomie, Ökologie, Verbreitung, Soziologie, Wirtschaft. Fischer, Stuttgart

Kuttler W (1984) Spurenstoffe in der Atmosphäre – ihre Verteilung und regionale Ablagerung. Geodynamik 5:29–76

Kuttler W (1991) Transfer mechanisms and deposition rates of atmospheric pollutants. In: Esser G, Overdieck D (eds) Modern ecology: basic and applied aspects. Elsevier, Amsterdam, pp 509–538

Kyparissis A, Manetas Y (1993) Seasonal leaf dimorphism in a semi-deciduous mediterranean shrub: ecophysiological comparisons between winter and summer leaves. Acta Oecol 14:23–32

Kyriakopoulos E, Larcher W (1976) Saugspannungsdiagramme für austrocknende Blätter von *Quercus ilex* L. Z Pflanzenphysiol 77:268–271

Laatsch W (1954) Dynamik der mitteleuropäischen Mineralböden. Steinkopf, Dresden

Lachaud S (1989) Participation of auxin and abscisic acid in the regulation of seasonal variations in cambial activity and xylogenesis. Trees 3:125–137

Ladefoged K (1963) Transpiration of forest trees in closed stands. Physiol Plant 16:378–414

Lahmann E (1990) Luftverunreinigung – Luftreinhaltung. Parey, Berlin

Landolt E (1971) Ökologische Differenzierungsmuster bei Artengruppen im Gebiet der Schweizerflora. Boissiera 19:129–148

Lang ARG, Klepper B, Cumming MJ (1969) Leaf water balance during oscillation of stomatal aperture. Plant Physiol 44:826–830

Lang GA, Early JD, Martin GC, Darnell RL (1987) Endo-, para- and ecodormancy: physiological terminology and classification for dormancy research. HortSci 22:371–377

Lang M, Lichtenthaler HK (1991) Changes in the blue-green and red fluorescence-emission spectra of beech leaves during the autumnal chlorophyll breakdown. J Plant Physiol 138:550–553

Lange OL (1959) Untersuchungen über den Wärmehaushalt und Hitzeresistenz mauretanischer Wüsten- und Savannenpflanzen. Flora 147:595–651

Lange OL (1967) Investigations on the variability of heat resistance in plants. In: Troshin AS (ed) The cell and environmental temperature. Pergamon Press, London, pp 131–141

Lange OL, Schulze ED, Koch W (1970) Experimentell-ökologische Untersuchungen an Flechten der Negev-Wüste. II. CO_2-Gaswechsel und Wasserhaushalt von *Ramalina maciformis* (Del.) Bory am natürlichen Standort während der sommerlichen Trockenperiode. Flora 159:38–62

Lange OL, Schulze ED, Kappen L, Buschbom U, Evenari M (1975) Photosynthesis of desert plants as influenced by internal and external factors. In: Gates DM, Schmerl RB (eds) Perspectives of biophysical ecology. Springer, Berlin Heidelberg New York, pp 121–143

Lange OL, Beyschlag W, Tenhunen JD (1987) Control of leaf carbon assimilation – input of chemical energy into ecosystems. In: Schulze ED, Zwölfer H (eds) Potentials and limitations of ecosystem analysis. Springer, Berlin Heidelberg New York, pp 149–163

Larcher W (1961) Jahresgang des Assimilations- und Respirationsvermögens von *Olea europaea* L. ssp. *sativa* Hoff. et Link., *Quercus ilex* L. und *Quercus pubescens* Willd. aus dem nördlichen Gardaseegebiet. Planta 56:575–606

Larcher W (1963 a) Orientierende Untersuchung über das Verhältnis von CO_2-Aufnahme zu Transpiration bei fortschreitender Bodenaustrocknung. Planta 60:339–343

Larcher W (1963 b) Zur Frage des Zusammenhanges zwischen Austrocknungsresistenz und Frosthärte bei Immergrünen. Protoplasma 57:569–587

Larcher W (1963 c) Zur spätwinterlichen Erschwerung der Wasserbilanz von Holzpflanzen an der Waldgrenze. Ber Naturwiss Med Ver Innsb 53:125–137

Larcher W (1965) The influence of water stress on the relationship between CO_2-uptake and transpiration. In: Slavík B (ed) Water stress in plants. Academia, Prag, pp 184–194

Larcher W (1970) Kälteresistenz und Überwinterungsvermögen mediterraner Holzpflanzen. Oecol Plant 5:267–286

Larcher W (1973 a) Limiting temperatures for life functions in plants. In: Precht H, Christophersen J, Hensel H, Larcher W (eds) Temperature and life, 2nd edn. Springer, Berlin Heidelberg New York

Larcher W (1973 b) Ökologie der Pflanzen, 1. Aufl. Ulmer, Stuttgart

Larcher W (1977) Ergebnisse des IBP-Projektes Zwergstrauchheide Patscherkofel. Sitzungsber Österr Akad Wiss Math-Naturwiss Kl I 186:301–371

Larcher W (1980) Klimastreß im Gebirge – Adaptationstraining und Selektionsfilter für Pflanzen. In: Rheinisch-Westf Akad Wiss N 291. Westdeutscher Verlag, Leverkusen, pp 49–88

Larcher W (1981) Effects of low temperature stress and frost injury on plant productivity. In: Johnson CB (ed) Physiological processes limiting plant productivity. Butterworths, London, pp 253–269

Larcher W (1985) Kälte und Frost. In: Sorauer P (ed) Handbuch der Pflanzenkrankheiten, 7. Aufl, Bd 1/5. Parey, Berlin, pp 107–326

Larcher W (1987) Streß bei Pflanzen. Naturwissenschaften 74:158–167

Larcher W, Bauer H (1981) Ecological significance of resistance to low temperatures. In: Lange OL, Nobel PS, Osmond CB, Ziegler H (eds) Encyclopedia of plant physiology 12 A. Springer, Berlin Heidelberg New York, pp 403–437

Larcher W, Bodner M (1980) Dosisletalität-Nomogramm zur Charakterisierung der Erkältungsempfindlichkeit tropischer Pflanzen. Angew Bot 54:273–278

Larcher W, Mair B (1969) Die Temperaturresistenz als ökophysiologisches Konstitutionsmerkmal. 1. *Quercus ilex* und andere Eichenarten des Mittelmeergebietes. Oecol Plant 4:347–376

Larcher W, Neuner G (1989) Cold-induced sudden reversible lowering of in vivo chlorophyll fluorescence after saturating light pulses. A sensitive marker for chilling susceptibility. Plant Physiol 89:740–742

Larcher W, Thomaser-Thin W (1988) Seasonal changes in energy content and storage patterns of mediterranean sclerophylls in a northernmost habitat. Acta Oecol 9:271–283

Larcher W, Wagner J (1983) Ökologischer Zeigerwert und physiologische Konstitution von *Sempervivum montanum.* Verh Ges Ökol 11:253–264

Larcher W, Holzner M, Pichler J (1989) Temperaturresistenz inneralpiner Steppengräser. Flora 183:115–131

Larcher W, Wagner J, Thammathaworn A (1990) Effects of superimposed temperature stress on in vivo chlorophyll fluorescence of *Vigna unguiculata* under saline stress. J Plant Physiol 136:92–102

Lassoie JP, Hinckley TM, Grier ChC (1985) Coniferous forests of the Pacific Northwest. In: Chabot BF, Mooney HA (eds) Physiological ecology of North American plant communities. Chapman & Hall, New York, pp 126–161

Läuchli A (1976) Symplasmic transport and ion release to the xylem. In: Wardlaw IF, Passioura JB (eds) Transport and transfer processes in plants. Academic Press, New York, pp 101–112

Lauer MJ, Pallardy StG, Blevins DG, Randall DD (1989) Whole leaf carbon exchange characteristics of phosphate-deficient soybeans (*Glycine max* L.). Plant Physiol 91:848–854

Lauscher F (1981) Säkulare Schwankungen der Dezennienmittel und extreme Jahreswerte der Temperatur in allen Erdteilen. Analysen mit Hilfe der World Weather Records. Arb Zentralanstalt Met Geodyn, H 48 Publ 252

Lawlor DW (1993) Photosynthesis: metabolism, control and physiology. 2nd ed. Longman, London

Lee DW (1986) Unusual strategies of light absorption in rain-forest herbs. In: Givnish ThJ (ed) On the economy of plant form and function. Cambridge University Press, Cambridge, pp 105–131

Lee JA, Stewart GR (1978) Ecological aspects of nitrogen metabolism. Adv Bot Res 6:1–43

Legge AH, Krupa SV (eds) (1990) Acidic deposition: sulphur and nitrogen oxides. Lewis, New York

Lendzian KJ (1987) Aufnahme und zellphysiologische Wirkungen von Luftschadstoffen. Naturwissenschaften 74:282–288

Lendzian KJ, Unsworth MH (1983) Ecophysiological effects of atmospheric pollutants. In: Lange OL, Nobel PS, Osmond CB, Ziegler H (eds) Encyclopedia of plant physiology, vol 12D. Springer, Berlin Heidelberg New York, pp 465–502

Lerch G (1991) Pflanzenökologie, 5. Aufl. Akademie Verlag, Berlin

Levitt J (1980a) Responses of plants to environmental stresses. I. Chilling, freezing and high temperature stresses, 2nd edn. Academic Press, New York

Levitt J (1980b) Responses of plants to environmental stresses. II. Water, radiation, salt, and other stresses. Academic Press, New York

Lichtenthaler HK (1988) In vivo chlorophyll fluorescence as a tool for stress detection in plants. In: Lichtenthaler HK (ed) Applications of chlorophyll fluorescence. Kluwer, Dordrecht, pp 129–142

Lichtenthaler HK, Buschmann C, Döll M, Fietz HJ, Bach T, Kozel U, Meier D, Rahmsdorf U (1981) Photosynthetic activity, chloroplast ultrastructure, and leaf characteristics of high-light and low-light plants and of sun and shade leaves. Photosynth Res 2:115–141

Liese W, Schneider M, Eckstein D (1975) Histometrische Untersuchungen am Holz einer rauchgeschädigten Fichte. Eur J For Pathol 5:152–161

Lieth H (1962) Die Stoffproduktion der Pflanzendecke. Fischer, Stuttgart

Lieth H (1970) Phenology in productivity studies. In: Reichle DE (ed) Analysis of temperate forest ecosystems. Springer, Berlin Heidelberg New York, pp 29–46

Lieth H (1972) Über die Primärproduktion der Pflanzendecke der Erde. Angew Bot 46:1–37

Lieth H (1974) Phenology and seasonality modeling. Springer, Berlin Heidelberg New York

Lieth H (1975a) Modeling the primary productivity of the world. In: Lieth H, Whittaker RH (eds) Primary productivity of the biosphere. Springer, Berlin Heidelberg New York, pp 237–283

Lieth H (1975b) Measurement of caloric values. In: Lieth H, Whittaker R (eds) Primary production of the biosphere. Springer, Berlin Heidelberg New York, pp 119–129

Lieth H, Markert BA (1988) Aufstellung und Auswertung ökosystemarer Element-Konzentrations-Kataster. Springer, Berlin Heidelberg New York

Likens GE, Bormann FH, Pierce RS, Eaton JS, Johnson NM (1977) Biogeochemistry of a forested ecosystem. Springer, Berlin Heidelberg New York

Lin SY, Dence CW (1992) Methods in lignin chemistry. Springer, Berlin Heidelberg New York

LoGullo MA, Salleo S (1988) Different strategies of drought resistance in three mediterranean sclerophyllous trees growing in the same environmental conditions. New Phytol 108:267–276

Longman KA, Jeník J (1987) Tropical forest and its environment, 2nd edn. Longman, London

Loomis RS, Gerakis PA (1975) Productivity of agricultural ecosystems. In: Cooper JP (ed) Photosynthesis and productivity in different environments. Cambridge University Press, Cambridge

Lösch R (1984) Species-specific responses to temperature in acid metabolism and gas exchange performance of Macaronesian Sempervivoideae. In: Margaris NS, Arianoustou-Farragitaki M, Oechel WC (eds) Being alive on land. Junk, The Hague, pp 117–126

Lösch R (1987) Die Produktionsphysiologie von *Aeonium gorgoneum* und anderer nicht-kanarischer Aeonien (Phanerogamae: Crassulaceae). Cour Forschungsinst Senckenb 95:201–209

Lösch R (1990) Water relations of Canarian laurel forest trees. In: Analysis of water transport in plants and cavitation of xylem conduits. Int Workshop, 29–31 May 1990, Vallombrosa/Firenze

Lösch R, Kappen L (1983) Die Temperaturresistenz makaronesischer Sempervivoideae. Verh Ges Ökol 10:521–528

Lossaint P, Rapp M (1971) Répartition de la matière organique, productivité et cycles des éléments minéraux dans des écosystèmes de climat méditerranéen. In: Duvigneaud P (ed) Productivity of forest ecosystems. UNESCO, Paris, pp 597–617

Lossaint P, Rapp M (1978) La forêt méditerranéenne de chênes verts. In: Lamotte M, Bourliere C (eds) Problèmes d'écologie. Ecosystèmes terrestres. Masson, Paris, pp 129–185

Lucas WJ (1987) Functional aspects of cells in root apices. In: Gregory PJ, Lake JV, Rose DA (eds) Root development and function. Cambridge University Press, Cambridge, pp 27–52

Ludlow MM (1989) Strategies of response to water stress. In: Kreeb KH, Richter H, Hinckley TM (eds) Structural and functional responses to environmental stresses. SPB Acad Publ, The Hague, pp 269–281

Ludlow MM, Wilson GL (1971) Photosynthesis of tropical pasture plants. II. Temperature and illuminance. Austr J Biol Sci 24:1065–1075

Lüning K (1984) Temperature tolerance and biogeography of seaweeds: the marine algal flora of Helgoland (North Sea) as an example. Helgol Meeresunters 38:305–317

Lüning K (1990) Seaweeds. Their environment, biography, and ecophysiology. Wiley, New York

Lüttge U (1973) Stofftransport der Pflanzen. Springer, Berlin Heidelberg New York

Lüttge U, Sellner M, Schnabl H, Zimmermann U (1984) Ökotoxikologische Bewertung von Umweltchemikalien und Entwicklung von biologischen Testsystemen aufgrund der Eigenschaften pflanzlicher Membranen. GIT-Suppl 4:36–42

Luyten I, Versluys MC, Blaauw AH (1932) The optimal temperatures from flower formation to flowering for *Hyacinthus orientalis*. Verh K Akad Wet Amsterdam 2 Sect XXIX, 5 Med 36:57–64

Lyons JM (1973) Chilling injury in plants. Annu Rev Plant Physiol 24:445–466

Lyons JM, Raison JK (1970) Oxidative activity of mitochondria isolated from plant tissues sensitive and resistant to chilling injury. Plant Physiol 45:386–389

Lyr H, Fiedler HJ, Tranquillini W (eds) (1992) Physiologie und Ökologie der Gehölze. Fischer, Jena

Maier R (1979) Zur Bioindikation von Bleiwirkungen in Pflanzen über Enzyme. Verh Ges Ökol 7:315–322

Malkina IS, Tselniker JuL (1990) Sezonnaya dinamika summarnogo dykhaniya i dykhaniyu podderzhaniya u stvolov lesnykh dereviev. Bot Zh 75:1138–1144

Marek M (1988) Photosynthetic characteristics of *Ailanthus* leaves. Photosynthetica 22:179–183

Mar-Möller C, Müller D, Nielsen J (1954) Graphic presentation of dry matter production of European beech. Forstl Forsøgsvaes Dan 21:327–335

Marschner H (1985) Nährstoffdynamik in der Rhizosphäre. Ber Dtsch Bot Ges 98:291–309

Marschner H (1986) Mineral nutrition of higher plants. Academic Press, London

Marshall C, Watson MA (1992) Ecological and physiological aspects of reproductive allocation. In: Marshall C, Grace J (eds) Fruit and seed production. Cambridge University Press, Cambridge, pp 173–202

Martinoia E (1992) Transport processes in vacuoles of higher plants. Bot Acta 105:232–234

Masing V, Svirezev YM, Löffler H, Patten BC (1990) Wetlands in the biosphere. In: Patten BC (ed) Wetlands and shallow continental water bodies, vol 1. SPB Acad Publ, The Hague, pp 313–344

Matthysse AG, Scott TK (1984) Functions of hormones at the whole plant level of organization. In: Scott TK (ed) Encyclopedia of plant physiology, vol 10. Springer, Berlin Heidelberg New York, pp 219–243

Matile Ph (1991) Vom Ergrünen und Vergilben der Blätter. Veröff Naturf Ges Zürich, vol 136/5, Zürich

Matile Ph, Altenburger R (1988) Rhythms of fragrance emission in flowers. Planta 174:242–247

Matile Ph, Düggelin T, Schellenberg M, Rentsch D, Bortlik K, Peisker C, Thomas H (1989) How and why is chlorophyll broken down in senescent leaves? Plant Physiol Biochem 27:595–604

Mauseth JD (1988) Plant anatomy. Cummings, Menlo Park

McCree KJ (1981) Photosynthetically active radiation. In: Lange OL, Nobel PS, Osmond CB, Ziegler H (eds) Encyclopedia of plant physiology, vol 12A. Springer, Berlin Heidelberg New York, pp 41–55

McManmon M, Crawford RMM (1971) A metabolic theory of flooding tolerance: the significance of enzyme distribution and behaviour. New Physiol 70:299–306

McWilliam JR, Ferrar PJ (1974) Photosynthetic adaptation of higher plants to thermal stress. In: Bieleski RL, Ferguson AR, Cresswell MM (eds) Mechanisms of regulation of plant growth. Royal Soc N Zealand, Bull 12, Wellington, pp 467–476

Mengel K (1984) Ernährung und Stoffwechsel der Pflanze, 6. Aufl. Fischer, Jena

Mengel K, Kirkby EA (1982) Principles of plant nutrition, 3rd edn. Intern Potash Inst, Bern

Merino J, Field C, Mooney HA (1984) Construction and maintenance costs of mediterranean-climate evergreen and deciduous leaves. Acta Oecol 5:211–229

Meyer FM (1978) Bäume in der Stadt. Ulmer, Stuttgart

Meyer N (1968) Histochemische Untersuchungen über jahreszeitliche Veränderungen der Fermentaktivität und des Stärkegehaltes in Trieben einiger *Prunus*-Arten. Flora A 159:215–232

Michael G (1967) Über die Beanspruchung des Wasserhaushaltes einiger immergrüner Gehölze im Mittelgebirge im Zusammenhang mit dem Frosttrocknisproblem. Arch Forstwes 16:1015–1032

Michalowski CB, Olsen SW, Piepenbrock M, Schmitt JM, Bohnert HJ (1989) Time course of mRNA induction elicited by salt stress in the common ice plant (*Mesembryanthemum crystallinum*). Plant Physiol 89:811–816

Milburn JA (1979) Water flow in plants. Longman, London

Milthorpe FL, Moorby J (1979) An introduction to crop physiology, 2nd edn. Cambridge University Press, Cambridge

Mitrakos K (1981) Temperature germination responses in three mediterranean evergreen sclerophylls. In: Margaris NS, Mooney HA (eds) Components of productivity of mediterranean-climate regions. Basic and applied aspects. Junk, Den Haag, pp 277–279

Mitscherlich G (1970) Wald, Wachstum und Umwelt. Eine Einführung in die ökologischen Grundlagen des Waldwachstums. Bd I: Form und Wachstum von Baum und Bestand. Sauerländer, Frankfurt

Mitscherlich G (1971) Wald, Wachstum und Umwelt. Eine Einführung in die ökologischen Grundlagen des Waldwachstums. Bd II: Waldklima und Wasserhaushalt. Sauerländer, Frankfurt

Mizutani J (1989) Plant allelochemicals and their roles. In: Chou CH, Waller GR (eds) Phytochemical ecology – allelochemicals, mycotoxins and insect pheromones and allomomes. Academia Sinica Monograph, Inst Bot Ser Taipei, pp 155–165

Molchanov AA (1971) Cycles of atmospheric precipitation in different types of forests of natural zones of the USSR. In: Duvigneaud P (ed) Productivity of forest ecosystems. UNESCO, Paris, pp 49–68

Monsi M, Saeki T (1953) Über den Lichtfaktor in den Pflanzengesellschaften, seine Bedeutung für die Stoffproduktion. Jpn J Bot 14:22–52

Monteith JL (1973) Principles of environmental physics. Edward Arnolds, London

Mooney HA (1972) The carbon balance of plants. Annu Rev Ecol Syst 3:315–346

Moor H (1964) Die Gefrier-Fixation lebender Zellen und ihre Anwendung in der Elektronen-mikroskopie. Z Zellforsch 62:546–580

Mori S, Hagihara A (1991) Root respiration in *Chamaecyparis obtusa* trees. Tree Physiol 8:217–225

Morozov VL, Belaya GA (1988) Ekologiya dalnevostochnogo krupnotravya. Nauka, Moskva

Morse SR (1990) Water balance in *Hemizonia luzulifolia:* the role of extracellular polysaccharides. Plant Cell Environ 13:39–48

Moser W, Brzoska W, Zachhuber K, Larcher W (1977) Ergebnisse des IBP-Projekts Hoher Nebelkogel 3184 m. Sitzungsber Österr Akad Wiss Math-Naturwiss Kl I 186:386–419

Müller E, Stierlin HR (1990) Sanasilva Kronenbilder, 2. Aufl. Eidgenössische Forschungsanstalt für Wald, Schnee und Landschaft, Birmensdorf

Müllerstael H (1968) Untersuchungen über den Gaswechsel zweijähriger Holzpflanzen bei fortschreitender Bodenaustrocknung. Beitr Biol Pflanze 44:319–341

Müller-Stoll WR (1935) Ökologische Untersuchungen an Xerothermpflanzen des Kraichgaus. Z Bot 29:161–253

Nadelhoffer KJ, Giblin AE, Shaver GR, Linkins AE (1992) Microbial processes and plant nutrient availability in arctic soils. In: Chapin FSt, Jefferies RL, Reynolds JF, Shaver GR, Svoboda J (eds) Arctic ecosystems in a changing climate. An ecophysiological perspective. Academic Press, San Diego, pp 281–300

Nakhutsrishvili GS (1974) Ekologija vysokogornych rastenii i fitozenozov zentralnogo Kavkaza. Ritmika razvitiya, fotosintez, ekobiomorfy. Mezniereba, Tbilisi

Nakhutsrishvili GS, Gamtsemlidze ZG (1984) Zhizne rastenii v ekstremalnykh usloviyakh vysokogornii. Nauka, Leningrad

Napp-Zinn K (1988) Anatomie des Blattes. II. Blattanatomie der Angiospermen. B: Experimentelle und ökologische Anatomie des Angiospermenblattes. 2. Lieferung. In: Linsbauer K (ed) Handbuch der Pflanzenanatomie, Bd VII/2 B. Bornträger, Berlin

Natr L (1975) Influence of mineral nutrition on photosynthesis and the use of assimilates. In: Cooper JP (ed) Photosynthesis and productivity in different environments. Cambridge University Press, Cambridge, pp 537–555

Neale PJ (1987) Algal photoinhibition and photosynthesis in the aquatic environment. In: Kyle DJ, Osmond CB, Arntzen CJ (eds) Photoinhibition. Elsevier, Amsterdam, pp 36–64

Neales TF (1975) The gas exchange patterns of CAM plants. In: Marcelle R (ed) Environmental and biological control of photosynthesis. Junk, Den Haag, pp 299–310

Negisi K (1966) Photosynthesis, respiration and growth in 1 year old seedlings of *Pinus densiflora*, *Cryptomeria japonica* and *Chamaecyparis obtusa*. Bull Tokyo Univ For 62:1–115

Negisi K (1974) Respiration rates in relation to diameter and age in stem of branch sections of young *Pinus densiflora* trees. Bull Tokyo Univ For 66:209–222

Nelson ND (1984) Woody plants are not inherently low in photosynthetic capacity. Photosynthetica 18:600–605

Ni BR, Pallardy StG (1991) Response of gas exchange to water stress in seedlings of woody angiosperms. Tree Physiol 8:1–9

Nicolas ME, Simpson RJ, Lambers H, Dalling MJ (1985) Effects of drought on partitioning of nitrogen in two wheat varieties differing in drought tolerance. Ann Bot 55:743–754

Nimz H (1974) Das Lignin der Buche – Entwurf eines Konstitutionsschemas. Angew Chem 86:336–344

Nobel PS (1977) Water relations and photosynthesis of a barrel cactus, *Ferocactus acanthodes*, in the Colorado Desert. Oecologia 27:117–133

Nobel PS (1988) Environmental biology of agaves and cacti. Cambridge University Press, Cambridge

Nobel PS (1991a) Physicochemical and environmental plant physiology. Academic Press, New York, pp 635

Nobel PS (1991 b) Environmental productivity indices and productivity for *Opuntia ficus-indica* under current and elevated atmospheric CO_2 levels. Plant Cell Environ 14:637–646

Nobel PS (1991 c) Achievable productivities of certain CAM plants: basis for high values compared with C_3 and C_4 plants. New Phytol 119:183–205

Nobel PS, Jordan PW (1983) Transpiration stream of desert species: resistances and capacitances for a C_3, a C_4, and a CAM plant. J Exp Bot 34:1379–1391

Nobel PS, Cavelier J, Andrade JL (1992a) Mucilage in cacti: Its apoplastic capacitance, associated solutes, and influence on tissue water relations. J Exp Bot 43:641–648

Nobel PS, García-Moya E, Quero E (1992b) High annual productivity of certain agaves and cacti under cultivation. Plant Cell Environ 15:329–335

Nobel W, Mayer Th, Kohler A (1983) Submerse Wasserpflanzen als Testorganismen für Belastungsstoffe. Z Wasser Abwasser Forsch 16:87–90

Nultsch W (1991) Allgemeine Botanik, 9. Aufl. Thieme, Stuttgart

Obeso JR (1993) Does defoliation affect reproductive output in herbaceous perennials and woody plants in different ways? Funct Ecol 7:150–155

Odum EP (1971) Fundamentals of ecology, 3rd edn. Saunders, Philadelphia

Odum HT, Pigeon RF (1970) A tropical rain forest. A study of irradiation and ecology at El Verde/Puerto Rico. Office Inf Serv, US Atomic Energy Comm, Oak Ridge, Tennessee

Oehlkers F (1956) Das Leben der Gewächse. Ein Lehrbuch der Botanik. Springer, Berlin Heidelberg New York

Ogino K, Ninomiya I, Yoshikawa K (1986) Sap flow rate of several tree species in a tropical rain forest in West Sumatra. In: Hotta M (ed) Diversity and dynamics of plant life in Sumatra. Report and Coll. Papers, Part 1. Kyoto University, Sumatra Nature Study, pp 1–9

Ohga N, Ikushima I (1970) Measurement of CO_2 and O_2 contents in a soil. JIBP Level III, Report for 1969, Tokyo, pp 40–43

Okano K, Ito O, Takeba G, Shimizu A, Totsuka T (1984) Effects of NO_2 and O_3 alone or in combination on kidney bean plants. V. ^{13}C-assimilate partitioning as affected by NO_2 and/or O_3. In: Studies on effects of air pollutant mixtures on plants, No 66, Part 2. Res Rep Natl Inst Environm, Tsukuba, pp 49–57

Olsen RA, Bennett JH, Blume D, Brown JC (1981) Chemical aspects of the Fe stress response mechanism in tomatoes. J Plant Nutr 3:905–921

Omasa K, Shimazaki KI, Aiga I, Larcher W, Onoe M (1987) Image analysis of chlorophyll fluorescence transients for diagnosing the photosynthetic system of attached leaves. Plant Physiol 84:748–752

Ondok JP, Pokorný J, Květ J (1984) Model of diurnal changes in oxygen, carbon dioxide and bicarbonate concentrations in a stand *Elodea canadensis* Michx. Aquat Bot 19:293–305

Ormrod DP (1978) Pollution in horticulture. Elsevier, Amsterdam

Ortiz-Lopez A, Ort DR, Boyer JS (1991) Photophosphorylation in attached leaves of *Helianthus annuus* at low water potentials. Plant Physiol 96:1018–1025

Osmond CB (1978) Crassulacean acid metabolism: a curiosity in context. Annu Rev Plant Physiol 29:379–414

Osmond CB, Björkman O, Anderson DJ (1980) Physiological processes in plant ecology. Toward a synthesis with *Atriplex*. Springer, Berlin Heidelberg New York

Osmond CB, Winter K, Ziegler H (1982) Functional significance of different pathways of CO_2 fixation in photosynthesis. In: Lange OL, Nobel PS, Osmond CB, Ziegler H (eds) Encyclopedia of plant physiology 12 B. Springer, Berlin Heidelberg New York, pp 479–547

O'Toole JC, Cruz RT, Singh TN (1979) Leaf rolling and transpiration. Plant Sci Lett 16:111–114

Ott J (1988) Meereskunde. Ulmer, Stuttgart

Overdieck D (1993) Elevated CO_2 and the mineral content of herbaceous and woody plants. Vegetatio 104/105:403–411

Ovington JD (1954) A comparison of rainfall in different woodlands. Forestry 27:41–53

Paembonan SA, Hagihara A, Hozumi K (1992) Long-term respiration in relation to growth and maintenance processes of the aboveground parts of a hinoki forest tree. Tree Physiol 10:101–110

Paine RT (1971) The measurement and application of the calorie to ecological problems. Annu Rev Ecol Syst 2:145–164

Parker J (1952) Desiccation in conifer leaves: anatomical changes and determination of the lethal level. Bot Gaz 114:189–198

Parlange J-Y, Waggoner PE (1970) Stomatal dimensions and resistance to diffusion. Plant Physiol 46:337–342

Parthier B (1991) Jasmonates, new regulators of plant growth and development: many facts and few hypotheses on their actions. Bot Acta 104:446–454

Pate JS (1976) Nutrient mobilization and cycling: case studies for carbon and nitrogen in organs of a legume. In: Wardlaw IF, Passioura JB (eds) Transport and transfer processes in plants. Academic Press, New York, pp 447–462

Patterson DT, Duke SO (1979) Effect of growth irradiance on the maximum photosynthetic capacity of water hyacinth [*Eichhornia crassipes* (Mart.) Solms]. Plant Cell Physiol 20:177–184

Pearcy RW, Ehleringer J, Mooney HA, Rundel PW (1989) Plant physiological ecology. Field methods and instrumentation. Chapman & Hall, London New York

Pereira JS (1994) Gas exchange and growth. In: Schulze ED, Caldwell MM (eds) Ecophysiology of photosynthesis. Springer, Berlin Heidelberg New York, pp 147–181

Pilon-Smits EAH, 't Hart H, Maas JW, Meesterburrie JAN, Kreuler R, van Brederode J (1991) The evolution of crassulacean acid metabolism in *Aeonium* inferred from carbon isotope composition and enzyme activities. Oecologia 91:548–553

Pipp E, Larcher W (1987) Energiegehalte pflanzlicher Substanz: II. Ergebnisse der Datenverarbeitung. Sitzungsber Österr Akad Wiss Math-Naturwiss Kl I 196:249–310

Pisek A (1958) Versuche zur Frostresistenzprüfung von Rinde, Winterknospen und Blüten einiger Arten von Obstgehölzen. Gartenbauwissenschaft 23:54–74

Pisek A, Cartellieri E (1941) Der Wasserverbrauch einiger Pflanzenvereine. Jahrb Wiss Bot 90:256–291

Pisek A, Larcher W (1954) Zusammenhang zwischen Austrocknungsresistenz und Frosthärte bei Immergrünen. Protoplasma 44:30–46

Pisek A, Schiessl R (1947) Die Temperaturbeeinflußbarkeit der Frosthärte von Nadelhölzern und Zwergsträuchern an der alpinen Waldgrenze. Ber Naturwiss Med Ver Innsb 47:33–52

Pisek A, Tranquillini W (1951) Transpiration und Wasserhaushalt der Fichte (*Picea excelsa*) bei zunehmender Luft- und Bodentrockenheit. Physiol Plant 4:1–27

Pisek A, Winkler E (1958) Assimilationsvermögen und Respiration der Fichte (*Picea excelsa* Link.) in verschiedener Höhenlage und der Zirbe (*Pinus cembra* L.) an der alpinen Waldgrenze. Planta 51:518–543

Pisek A, Knapp H, Ditterstorfer J (1970) Maximale Öffnungsweite und Bau der Stomata mit Angaben über ihre Größe und Zahl. Flora 159:459–479

Pitman MG, Lüttge U (1983) The ionic environment and plant ionic relations. In: Lange OL, Nobel PS, Osmond CB, Ziegler H (eds) Encyclopedia of plant physiology, vol 12 C. Springer, Berlin Heidelberg New York, pp 5–34

Pokorný J, Ondok JP (1982) Photosynthesis and primary production in submerged macrophyte stands. In: Gopal B, Turner RE, Wetzel RG, Whigham DF (eds) Wetlands: ecology and management. Nat Inst Ecology, Jaipur, pp 207–214

Pokorný J, Ondok JP (1991) Macrophyte photosynthesis and aquatic environment. Academica, Praha, pp 1–117

Polster H (1967) Wasserhaushalt. In: Lyr H, Polster H, Fiedler HJ (eds) Gehölzphysiologie. Fischer, Jena

Polster H, Fuchs S (1963) Winterassimilation und -Atmung der Kiefer (*Pinus silvestris* L.) im mitteldeutschen Binnenlandklima. Arch Forstwes 12:1011–1024

Polster H, Neuwirth G (1958) Assimilationsökologische Studien an einem fünfjährigen Pappelbestand. Arch Forstwes 7:749–875

Popp M (1984) Chemical composition of Australian mangroves. II. Low molecular weight carbohydrates. Z Pflanzenphysiol 113:411–421

Popp M, Albert R (1980) Freie Aminosäuren und Stickstoffgehalt in Halophyten des Neusiedlersee-Gebietes. Flora 170:229–239

Popp M, Larher F, Weigel P (1984) Chemical composition of Australian mangroves. III. Free amino acids, total methylated onium compounds and total nitrogen. Z Pflanzenphysiol 114:15–25

Pospišilová J (1975) Development of water stress in kale leaves of different insertion levels. Biol Plant 17:392–399

Pospišilová J, Šantrůček J (1994) Stomatal patchiness. Biol Plant 36:481–519

Powles StB (1984) Photoinhibition of photosynthesis induced by visible light. Annu Rev Plant Physiol 35:15–44

Pyke DA (1989) Limited resources and reproductive constraints in annuals. Funct Ecol 3:221–228

Quarrie SA, Jones HG (1977) Effects of abscisic acid and water stress on development and morphology of wheat. J Exp Bot 28:192–203

Rabe R (1990) Bioindikation von Luftverunreinigungen. In: Kreeb KH (ed) Methoden zur Pflanzenökologie und Bioindikation. Fischer, Jena, pp 275–301

Rabotnov TA (1978) On coenopopulations of plants reproducing by seeds. In: Structure and functioning of plant populations. Verh K Ned Akad Wetensch Afd Natuurkd, 70:1–26

Rabotnov TA (1995) Phytozönologie. Ulmer, Stuttgart

Rambal S (1992) *Quercus ilex* facing water stress: a functional equilibrium hypothesis. Vegetatio 99–100:147–153

Ramus J, Rosenberg G (1980) Diurnal photosynthetic performance of seaweeds measured under natural conditions. Mar Biol 56:21–28

Rapp M (1969) Production de litière et apport au sol d'éléments minéraux dans deux écosystèmes méditerranéens: la forêt de *Quercus ilex* L. et la garrigue de *Quercus coccifera* L. Oecol Plant 4:377–410

Rapp M (1971) Cycle de la matière organique et des éléments minéraux dans quelques écosystèmes méditerranéens. In: IBP: Ecologie du Sol. Centre Nat de la Recherche Scientifique 40, Paris, pp 19–184

Raunkiaer C (1910) Statistik der Lebensformen als Grundlage für die biologische Pflanzengeographie. Beih Biol Centralbl 27:171–206 d

Raven JA (1977) The evolution of vascular land plants in relation to supracellular transport processes. Adv Bot Res 5:154–240

Rawson HM, Turner NC, Begg JE (1978) Agronomic and physiological responses of soybean and sorghum crops to water deficits. IV. Photosynthesis, transpiration and water use efficiency of leaves. Aust J Plant Physiol 5:195–209

Rehder H, Schäfer A (1978) Nutrient turnover studies in alpine ecosystems. IV. Communities of the Central Alps and comparative survey. Oecologia 34:309–327

Reich PB (1987) Quantifying plant response to ozone: a unifying theory. Tree Physiol 3:63–91

Reich PB, Uhl C, Walters MB, Ellsworth DS (1991) Leaf lifespan as a determinant of leaf structure and function among 23 Amazonian tree species. Oecologia 86:16–24

Rennenberg H (1984) The fate of excess sulfur in higher plants. Annu Rev Plant Physiol 35:121–153

Retter W (1965) Untersuchungen zur Assimilationsökologie und Temperaturresistenz des Buchenlaubes. Dissertation, Innsbruck

Richards JH, Caldwell MM (1987) Hydraulic lift: substantial nocturnal water transport between soil layers by *Artemisia tridentata* roots. Oecologia 73:486–489

Richards PW (1979) The tropical rain forest. An ecological study. Cambridge University Press, Cambridge

Richter H (1972) Wie entstehen Saugspannungsgradienten in Bäumen? Ber Dtsch Bot Ges 85:341–351

Richter H (1976) The water status in the plant – experimental evidence. In: Lange OL, Kappen L, Schulze ED (eds) Water and plant life. Springer, Berlin Heidelberg New York, pp 42–58

Richter H (1978) A diagram for the description of water relations in plant cells and organs. J Exp Bot 29:1197–1203

Richter H, Kikuta SB (1989) Osmotic and elastic components of turgor adjustment in leaves under stress. In: Kreeb KH, Richter H, Hinckley TM (eds) Structural and functional responses to environmental stresses. Academic Publishing, The Hague, pp 129–137

Ried A (1953) Photosynthese und Atmung bei xerostabilen und xerolabilen Krustenflechten in der Nachwirkung vorausgegangener Entquellungen. Planta 41:436–438

Ristic Z, Cass AA (1991) Morphological characteristics of leaf epidermal cells in lines of maize that differ in endogenous levels of abscisic aci and drough resistance. Bot Gazette 152:439–445

Robberecht R, Caldwell MM (1978) Leaf epidermal transmittance of ultraviolet radiation and its implications for plant sensitivity to ultraviolet radiation induced injury. Oecologia 32:277–287

Roden JS, Pearcy RW (1993) The effect of flutter on the temperature of poplar leaves and its implications for carbon gain. Plant Cell Environ 16:571–577

Rodin L, Bazilevich NI (1967) Production and mineral cycling in terrestrial vegetation. Oliver & Boyd, Edinburgh

Roeckner E (1992) Past, present and future levels of greenhouse gases in the atmosphere and model projections of related climatic changes. J Exp Bot 43:1097–1109

Roller M (1963) Durchschnittswerte phänologischer Phasen aus dem Zeitraum 1946 bis 1960 für 103 Orte Österreichs. Wetter Leben 15:1–12

Römheld V, Kramer D (1983) Relationship between proton efflux and rhizodermal transfer cells induced by iron deficiency. Z Pflanzenphysiol 113:73–83

Rook DA (1969) The influence of growing temperature on photosynthesis and respiration of *Pinus radiata* seedlings. N Z J Bot 7:43–55

Rorison IH (1969) Ecological interferences from laboratory experiments on mineral nutrition. In: Rorison IH (ed) Ecological aspects of the mineral nutrition of plants. Blackwell, Oxford, pp 155–175

Rorison IH, Sutton F (1976) Climate, topography and germination. In: Evans GC, Bainbridge R, Rackham O (eds) Light as an ecological factor: II. Blackwell, Oxford, pp 361–383

Rosenberg NJ (1974) Microclimate: the biological environment. Wiley, New York

Ross J (1981) The radiation regime and architecture of plant stands. Junk, The Hague

Rosswall T, Flower-Ellis JGK, Johansson LG, Jonsson S, Rydén BE, Sonesson M (1975) Stordalen (Abisko), Sweden. In: Rosswall T, Heal OW (eds) Structure and function of tundra ecosystems. Ecological Bulletin 20, Stockholm, Natl Sci Foundation Editorial Service, pp 265–294

Rouschal E (1938) Zur Ökologie der Macchien. Jahrb Wiss Bot 87:436–523

Rozema J, Van Diggelen J (1991) A comparative study of growth and photosynthesis of four halophytes in response to salinity. Acta Oecol 12:673–681

Rozema J, Bijwaard P, Prast G, Broekman R (1985) Ecophysiological adaptations of coastal halophytes from foredunes and salt marshes. Vegetatio 62:499–521

Ruhland W (1915) Untersuchungen über die Hautdrüsen der Plumbaginaceen. Ein Beitrag zur Biologie der Halophyten. Jahrb Wiss Bot 55:409–498

Rundel PW (1978) The ecological role of secondary lichen substances. Biochem Syst Ecol 6:157–170

Rundel PW, Lange OL (1980) Water relations and photosynthetic response of a desert moss. Flora 169:329–335

Ruthsatz B, Hofmann U (1984) Die Verbreitung von C_4-Pflanzen in den semiariden Anden NW-Argentiniens mit einem Beitrag zur Blattanatomie ausgewählter Beispiele. Phytocoenologia 12:219–249

Rychnovská M (1965) A contribution to the ecology of the genus *Stipa*. II. Water relations of plants and habitat on the hill of Krizová hora near the town of Moravsky Krumlov. Preslia 37:42–52

Rychnovská M (1979) Bandania ekosystemow lakowych w Czechoslowacji. (Grassland ecosystem research in Czechoslovakia). Wiad Ekol 25:29–39

Rychnovská M (ed) (1993a) Structure and functioning of seminatural meadows. Elsevier, Amsterdam

Rychnovská M (1993b) Temperate semi-natural grasslands of Eurasia. In: Coupland RT (ed) Natural grasslands. Ecosystems of the world, vol 8b. Elsevier, Amsterdam, pp 125–166

Rychnovská M, Čermák J, Šmid P (1980) Water output in a stand of *Phragmites communis* Trin. Acta Sci Nat Acad Sci Bohemostov Brno 14:3–30

Sage RF, Pearcy RW (1987) The nitrogen use efficiency of C_3 and C_4 plants. II. Leaf nitrogen effects on the gas exchange characteristics of *Chenopodium album* (L.) and *Amaranthus retroflexus* (L.). Plant Physiol 84:959–963

Saglio PH, Rancillac M, Bruzan F, Pradet A (1984) Critical oxygen pressure for growth and respiration of excised and intact roots. Plant Physiol 76:151–154

Sakai A (1983) Comparative study on freezing resistance of conifers with special reference to cold adaptation and its evolutive aspects. Can J Bot 9:2323–2332

Sakai A, Larcher W (1987) Frost survival of plants. Responses and adaptation to freezing stress. Springer, Berlin Heidelberg New York

Sale PJM (1974) Productivity of vegetable crops in a region of high solar input. Aust J Plant Physiol 1:283–296

Salisbury EI (1916) The oak-hornbeam woods of Hertfordshire. J Ecol 4:83–117

Salisbury FB (1982) Photoperiodism. Hort Rev 4:66–105

Salisbury FB (1985) Plant adaptations to the light environment. In: Kaurin A, Junttila O, Nilsen J (eds) Plant production in the North. Norwegian University Press, Tromsø, pp 43–61

Salisbury FB (1991) Système internationale: the use of SI units in plant physiology. J Plant Physiol 131:1–7

Salisbury FB, Ross CW (1992) Plant physiology, 4th edn. Wadsworth, Belmont

Salisbury FB, Spomer GG (1964) Leaf temperatures of alpine plants in the field. Planta 60:497–505

Sato T (1982) Phenology and wintering capacity of sporophytes and gametophytes of ferns native to northern Japan. Oecologia 55:53–61

Saure MC (1985) Dormancy release in deciduous fruit trees. Hort Rev 7:239–300

Savage MJ (1979) Use of the international system of units in the plant sciences. HortSci 14:492–495

Schennikow AP (1932) Phänologische Spektra von Pflanzengesellschaften. In: Abderhalden E (ed) Handbuch der biologischen Arbeitsmethoden, Bd II/6. Springer, Berlin Heidelberg New York, pp 251–266

Schlee D (1992) Ökologische Biochemie, 2. Aufl. Springer, Berlin Heidelberg New York

Schmidt JE, Kaiser WM (1987) Response of the succulent leaves of *Peperomia magnoliaefolia* to dehydration. Plant Physiol 83:190–194

Schmidt L (1977) Phytomassevorrat und Nettoprimärproduktivität alpiner Zwergstrauchbestände. Oecol Plant 12:195–213

Schmidt W, Neubauer Ch, Kolbowski J, Schreiber U, Urbach W (1990) Comparison of effects of air pollutants (SO_2, O_3, NO_2) on intact leaves by measurements of chlorophyll fluorescence and P_{700} absorbance changes. Photosynth Res 25:241–248

Schmidt-Vogt H (1989) Die Fichte. Bd II/2: Krankheiten, Schäden, Fichtensterben. Parey, Hamburg

Schnelle F (1955) Pflanzenphänologie. Akademische Verlagsgesellschaft, Leipzig

Schnelle F (1986) Ergebnisse aus den Internationalen Phänologischen Gärten in Europa – Mittel 1973–1982. Wetter Leben 38:5–17

Schnock G (1971) Le bilan de l'eau dans l'écosystème forêt. Application à une chênaie mélangée de haute Belgique. In: Duvigneaud P (ed) Productivity of forest ecosystems. UNESCO, Paris, pp 41–42

Schreiber U, Berry JA (1977) Heat-induced changes of chlorophyll fluorescence in intact leaves correlated with damage of the photosynthetic apparatus. Planta 136:233–238

Schroeder D (1969) Bodenkunde in Stichworten. Hirt, Kiel

Schubert R (ed) (1991) Bioindikation in terrestrischen Ökosystemen, 2. Aufl. Fischer, Stuttgart

Schulte PJ, Hinckley TM (1985) A comparison of pressure-volume curve data analysis techniques. J Exp Bot 36:1590–1602

Schultz J (1988) Die Ökozonen der Erde. Ulmer, Stuttgart

Schulz JP (1960) Ecological studies on rain forest in northern Suriname. North Holland, Amsterdam

Schulze E, Stix E (1990) Beurteilung phytotoxischer Immissionen, für die noch keine Luftqualitätskriterien festgelegt sind. Angew Bot 64:225–235

Schulze ED (1970) Der CO_2-Gaswechsel der Buche (*Fagus silvatica* L.) in Abhängigkeit von den Klimafaktoren im Freiland. Flora 159:177–232

Schulze ED (1982) Plant life forms and their carbon, water and nutrient relations. In: Lange OL, Nobel PS, Osmond CB, Ziegler H (eds) Encyclopedia of plant physiology 12 B. Springer, Berlin Heidelberg New York, pp 615–676

Schulze ED (1986) Carbon dioxide and water vapor exchange in response to drought in the atmosphere and in the soil. Annu Rev Plant Physiol 37:247–274

Schulze ED, Hall AE (1982) Stomatal responses, water loss and CO_2 assimilation rates of plants in contrasting environments. In: Lange OL, Nobel PS, Osmond CB, Ziegler H (eds) Encyclopedia of plant physiology 12 B. Springer, Berlin Heidelberg New York, pp 181–230

Schulze ED, Lange OL, Koch W (1972) Ökophysiologische Untersuchungen an Wild- und Kulturpflanzen der Negev-Wüste. III. Tagesverläufe der Nettophotosynthese und Transpiration am Ende der Trockenzeit. Oecologia 9:317–340

Schulze ED, Cermák J, Matyssek R, Penka M, Zimmermann R, Vasícek F, Gries W, Kucera J (1985) Canopy transpiration and water fluxes in the xylem of the trunk of *Larix* and *Picea* trees – a comparison of xylem flow porometer and cuvette measurements. Oecologia 66:475–483

Schweingruber FH (1983) Der Jahrring. Standort, Methodik, Zeit und Klima in der Dendrochronologie. Haupt, Bern

Schweingruber FH, Kontic R, Winkler-Seifert A (1983) Eine jahrringanalytische Studie zum Nadelbaumsterben in der Schweiz. Ber Eidg Anst Forstl Versuchsw Nr 253, Birmensdorf

Seeley EJ, Kammereck R (1977) Carbon flux in apple trees: the effects of temperature and light intensity on photosynthetic rates. J Am Soc Hort Sci 102:731–733

Seeley S (1990) Hormonal transduction of environmental stresses. HortSci 25:1369–1376

Selye H (1936) A syndrome produced by diverse nocuous agents. Nature 138:32

Semikhatova OA (1974) Energetika dykhaniya rastenii pri povyshennoi temperature. Nauka, Leningrad

Semikhatova OA, Gerasimenko TV, Ivanova TI (1992) Photosynthesis, respiration, and growth of plants in the Soviet Arctic. In: Chapin FSt, Jefferies RL, Reynolds JF, Shaver GR, Svoboda J (eds) Arctic ecosystems in a changing climate. An ecophysiological perspective. Academic Press, San Diego, pp 169–192

Sen DN (1982) Environment and plant life in Indian desert. Geobios, Jodhpur

Senft WH (1978) Dependence of light-saturated rates of algal photosynthesis on intracellular concentrations of phosphorus. Limnol Oceanogr 23:709–718

Serre F (1976a) Les rapports de la croissance et du climat chez le Pin d'Alep (*Pinus halepensis* Mill.). I. Méthodes utilisées. L'activité cambiale et le climat. Oecol Plant 11:143–171

Serre F (1976b) Les rapports de la croissance et du climat chez le Pin d'Alep (*Pinus halepensis* Mill.). II. L'allongement des pousses et des aiguilles, et le climat. Discussion generale. Oecol Plant 11:201–224

Šesták Z (1985) Photosynthesis during leaf development. Junk, Dordrecht

Šesták Z, Čatský J, Jarvis PG (1971) Plant photosynthetic production. Manual of methods. Junk, Den Haag

Shantz HL (1927) Drought resistance and soil moisture. Ecology 8:145–157

Shmueli E (1960) Chilling and frost damage in banana leaves. Bull Res Counc Isr 8 d:225–288

Shtěstěnko AP (1969) Osobeniosti stroeniya podzhemnykh organov rastenii predelnykh vysot proisrastaniya na Pamire. In: Problemy botaniki XI. Nauka, Leningrad, pp 284

Sikorska E, Kacperska A (1982) Freezing-induced membrane alterations: injury or adaptation? In: Li Ph, Sakai A (eds) Plant cold hardiness and freezing stress. Academic Press, New York, pp 261–272

Siminovitch D (1981) Common and disparate elements in the processes of adaptation of herbaceous and woody plants to freezing – a perspective. Cryobiology 18:166–185

Sinclair R (1983) Water relations of tropical epiphytes. II. Performance during droughting. J Exp Bot 34:1664–1675

Sirén G, Sivertsson E (1976) Överlevelse och produktion hos snabbväxande *Salix*- och *Populus*-kloner för skogsindustri och energiproduktion. Dept Reforest Stockholm Res, Note No 83, Stockholm

Slavík B (1974) Methods of studying plant water relations. Springer, Berlin Heidelberg New York

Slavíková J (1965) Die maximale Wurzelsaugkraft als ökologischer Faktor. Preslia (Praha) 37:419–428

Smidt St, Gabler K, Puxbaum H (1990) Die zeitliche und vertikale Zunahme der Ozonkonzentrationen. Österr Forstztg 247. Folge 7:58–60

Smillie RM, Nott R (1979) Heat injury in leaves of alpine, temperate and tropical plants. Aust J Plant Physiol 6:135–141

Smith PF (1962) Mineral analysis of plant tissues. Annu Rev Plant Physiol 13:81–108

Smith S, Weyers JDB, Berry WG (1989) Variation in stomatal characteristics over the lower surface of *Commelina communis* leaves. Plant Cell Environ 12:653–659

Sobrado MA (1986) Aspects of tissue water relations and seasonal changes of leaf water potential components of evergreen and deciduous species coexisting in tropical dry forests. Oecologia 68:413–416

Sobrado MA (1991) Cost-benefit relationships in deciduous and evergreen leaves of tropical dry forest species. Funct Ecol 5:608–616

Sonesson M (ed) (1980) Ecology of a subarctic mire. In: Ecological Bulletins 30. NFR Edit Serv, Stockholm

Sperry JS, Sullivan JEM (1992) Xylem embolism in response to freeze-thaw cycles and water stress in ring-porous, diffuse-porous, and conifer species. Plant Physiol 100:605–613

Sprugel DG (1990) Components of woody-tissue respiration in young *Abies amabilis* (Dougl.) Forbes trees. Trees 4:88–98

Stadler J, Gebauer G (1992) Nitrate reduction and nitrate content in ash trees (*Fraxinus excelsior* L.): distribution between compartments, site comparison and seasonal variation. Trees 6:236–240

Stålfelt MG (1973) Der Gasaustausch der Moose. Planta 27:30–60

Stanhill G (1970) The water flux in temperate forests: precipitation and evapotranspiration. In: Reichle DE (ed) Analysis of temperate forest ecosystems. Springer, Berlin Heidelberg New York, pp 247–256

Steenbjerg F (1951) Yield curves and chemical plant analyses. Plant Soil 3:97–109

Steenbjerg F, Jakobsen ST (1963) Plant nutrition and yield curves. Soil Sci 95:69–88

Steiner M (1934) Zur Ökologie der Salzmarschen der nordöstlichen Vereinigten Staaten von Nordamerika. Jahrb Wiss Bot 81:94–202

Steinhauser F, Eckel O, Lauscher F (1960) Klimatographie von Österreich. Springer, Wien

Stetter KO, Fiala G, Huber G, Huber R, Segerer A (1990) Hyperthermophilic microorganisms. FEMS Microbiol Rev 75:117–124

Steubing L (1976) Niedere und Höhere Pflanzen als Indikatoren für Immissionsbelastungen. Landschaft Stadt 8:97–144

Steubing L, Dapper H (1964) Der Kreislauf des Chlorids im Meso-Ökosystem einer binnenländischen Salzwiese. Ber Dtsch Bot Ges 8:97–144

Steubing L, Haneke J, Biermann J, Gnittke J (1989) Urangehalte in Pflanzen, Bodenwasser- und Bodenproben im Anomaliengebiet um Aigendorf. Angew Bot 63:361–374

Stocker O (1947) Probleme der pflanzlichen Dürreresistenz. Naturwissenschaften 34:362–371

Stocker O (1952) Grundriß der Botanik. Springer, Berlin Heidelberg New York

Stocker O (1956a) Die Abhängigkeit der Transpiration von den Umweltfaktoren. In: Ruhland W (ed) Handbuch der Pflanzenphysiologie, Bd 3. Springer, Berlin Heidelberg New York, pp 436–488

Stocker O (1956b) Wasseraufnahme und Wasserspeicherung bei Thallophyten. In: Ruhland W (ed) Handbuch der Pflanzenphysiologie, Bd 3. Springer, Berlin Heidelberg New York, pp 160–172

Stocker O (1970) Der Wasser- und Photosynthese-Haushalt von Wüstenpflanzen der mauretanischen Sahara: I. Regengrüne und immergrüne Bäume. Flora 159:539–572

Stocker O (1971) Der Wasser- und Photosynthese-Haushalt von Wüstenpflanzen der mauretanischen Sahara: II. Wechselgrüne, Rutenzweig- und stammsukkulente Bäume. Flora 160:445–494

Stocker O (1972) Der Wasser- und Photosynthese-Haushalt von Wüstenpflanzen der mauretanischen Sahara. III. Kleinsträucher, Stauden und Gräser. Flora 161:46–110

Stocker O (1974a) Der Wasser- und Photosynthesehaushalt von Wüstenpflanzen der südalgerischen Sahara. I. Standorte und Versuchspflanzen. Flora 163:46–88

Stocker O (1974b) Der Wasser- und Photosynthese-Haushalt von Wüstenpflanzen der südalgerischen Sahara. III. Jahresgang und Konstitutionstypen. Flora 163:480–529

Stowe LG, Teeri JA (1978) The geographic distribution of C_4 species of the Dicotyledonae in relation to climate. Am Nat 112:609–623

Stoy V (1965) Photosynthesis, respiration and carbohydrate accumulation in spring wheat in relation to yield. Physiol Plant 4:1–125

Strain BR, Cure JD (1985) Direct effects of increasing carbon dioxide on vegetation. US Dept Energy, Carbon Dioxide Res Div, DOE/ER0238, Washington

Streit L, Feller U (1982) Changing activities of nitrogen-assimilating enzymes during growth and senescence of dwarf beans (*Phaseolus vulgaris* L.). Z Pflanzenphysiol 108:273–281

Strugger S (1938) Die lumineszenzmikroskopische Analyse des Transpirationsstromes in Parenchymen. I. Mitteilung: Die Methode und die ersten Beobachtungen. Flora 133:56–68

Stulen I (1986) Interactions between nitrogen and carbon metabolism in a whole plant context. In: Lambers H, Neeteson JJ, Stulen I (eds) Fundamental, ecological and agricultural aspects of nitrogen metabolism in higher plants. Nijhoff, Dordrecht, pp 261–278

Sucoff E (1975) Effect of deicing salts on woody vegetation along Minnesota roads. In: Techn Bull 303, Forestry Ser 20. Agricultural Experiment Station, Minnesota, pp 3–49

Sugimoto K (1973) Studies on transpiration and water requirement of *indica* and *japonica* rice plants. Jpn J Trop Agricult 16:260–264

Sveshnikova VM (1979) Dominanty kasakhstanskikh stepei. Nauka, Leningrad

Svoboda J (1977) Ecology and primary production of raised beach communities, Truelove Lowland. In: Bliss LC (ed) Truelove Lowland, Devon Island, Canada: a high arctic ecosystem. University of Alberta Press, Edmonton, pp 185–216

Tagawa H (1979) An investigation of initial regeneration in an evergreen broadleaved forest. II. Seedfall, seedling production, survival and age distribution of seedlings. Bull Yokohama Phytosoc Soc Jpn 16:379–391

Tardieu F, Zhang J, Gowing DJG (1993) Stomatal control by both [ABA] in the xylem sap and leaf water status: a test of a model for droughted or ABA-fed field-grown maize. Plant Cell Environ 16:413–420

Teeri JA, Stowe LG (1976) Climatic patterns and the distribution of C_4 grasses in North America. Oecologia 23:1–12

Tenhunen JD, Reynolds JF, Lange OL, Dougherty RL, Harley PC, Kummerow J, Rambal S (1989) Quinta: a physiologically based growth simulator for drought-adapted woody plant species. In: Pereira JS, Landsberg JJ (eds) Biomass production by fast-growing trees. Kluwer, Dordrecht, pp 135–168

Terashima I (1992) Anatomy of non-uniform leaf photosynthesis. Photosynth Res 31:195–212

Terashima I, Saeki T (1985) A new model for leaf photosynthesis incorporating the gradients of light environment and of photosynthetic properties of chloroplasts within a leaf. Ann Bot 56:489–499

Terjung WH, Louie SSF, O'Rourke PA (1976) Toward an energy budget model of photosynthesis predicting world productivity. Vegetatio 32:31−53

Tesche M (1989) Umweltstreß. In: Schmidt-Vogt H (ed) Die Fichte, Bd II/2. Parey, Berlin, pp 346−384

Thimijan RW, Heins RD (1983) Photometric, radiometric, and quantum light units of measure: a review of procedures for interconversion. HortScience 18:818−822

Thomas RB, Reid CD, Ybema R, Strain BR (1993) Growth and maintenance components of leaf respiration of cotton grown in elevated carbon dioxide partial pressure. Plant Cell Environ 16:539−546

Thomas W (1927) Nitrogenous metabolism of *Pyrus malus*. III. The partition of nitrogen in leaves, one- and two-year branch growth and non-bearing spurs throughout a year's cycle. Plant Physiol 2:109−137

Tichá I, Čatský J, Hodánová D, Pospíšilová, Kaše M, Šesták Z (1985) Gas exchange and dry matter accumulation during leaf development. In: Šesták Z (ed) Photosynthesis during leaf development. Academia, Praha, pp 157−216

Tikhomirov BA (1963) Ocherki po biologii rastenii Arktiki. Izdat Akad Nauk, Moskva

Tieszen LL, Senyimba MM, Imbamba SK, Troughton JH (1979) The distribution of C_3 and C_4 grasses and carbon isotope discrimination along an altitudinal and moisture gradient in Kenya. Oecologia 37:337−350

Tietz D, Tietz A (1982) Streß im Pflanzenreich. Biol unserer Zeit 12:113−119

Tomlinson PB (1990) The structural biology of palms. Clarendon Press, Oxford

Tomsett AB, Thurman DA (1988) Molecular biology of metal tolerances of plants. Plant Cell Environ 11:383−394

Tranquillini W (1959) Die Stoffproduktion der Zirbe an der Waldgrenze während eines Jahres. 1. Standortsklima und CO_2-Assimilation. Planta 54:107−129

Tranquillini W (1979) Physiological ecology of the alpine timberline. Springer, Berlin Heidelberg New York

Tranquillini W (1982) Frost-drought and its ecological significance. In: Lange OL, Nobel PS, Osmond CB, Ziegler H (eds) Encyclopedia of plant physiology, vol 12B. Springer, Berlin Heidelberg New York, pp 379−400

Treshow M, Anderson FK (1991) Plant stress from air pollution. Wiley, Chichester

Tretiach M (1993) Photosynthesis and transpiration of evergreen mediterranean and deciduous trees in an ecotone during a growing season. Acta Oecol 14:341−360

Troll C (1955) Der jahreszeitliche Ablauf des Naturgeschehens in den verschiedenen Klimagürteln der Erde. Stud Gen 8:713−733

Troll C (1960) Die Physiognomik der Gewächse als Ausdruck der ökologischen Lebensbedingungen. Wiss Abh Dtsch Geographentag 1959, Wiesbaden, pp 97−122

Troll C (1964) Karte der Jahreszeitklimate der Erde. Erdkunde 18:5−28

Tselniker (Zelniker) YuL (1968) Die Lichtdurchlässigkeit der Baumkronen und der Zustand des Jungholzes unter dem Bestandesschirm. Dtsch Akad Landwirtsch Wiss Berlin, Tagungsber 100:189−198

Tselniker YuL (1978) Fiziologicheskie osnovy tenevynoslivosti drevesnykh rastenii. Nauka, Moskva

Tselniker YuL (ed) (1993) Rost i gazoobmen CO_2 u lesnykh derevev. Nauka, Moskva

Tsutsumi T (1987) The nitrogen cycle in a forest. Mem Coll Agric, Kyoto Univ 130:1−16

Tuba Z, Lichtenthaler HK, Maroti I, Csintalan Z (1993) Resynthesis of thylakoids and functional chloroplasts in the desiccated leaves of the poikilochlorophyllous plant *Xerophyta scabrida* upon rehydration. J Plant Physiol 142:742−748

Tuba Z, Lichtenthaler HK, Csintalan Z, Nagy Z, Szente K (1994) Reconstitution of chlorophylls and photosynthetic CO_2 assimilation upon rehydration of the desiccated poikilochlorophyllous plant *Xerophyta scabrida* (Pax) Th. Dur. et Schinz. Planta 192:414−420

Tuhkanen S (1980) Climatic parameters and indices in plant geography. Acta Phytogeogr Suec 67:1−105

Tumanov II (1927) Ungenügende Wasserversorgung und das Welken der Pflanzen als Mittel zur Erhöhung ihrer Dürreresistenz. Planta 3:391−480

Turner H (1970) Grundzüge der Hochgebirgsklimatologie. In: Die Welt der Alpen. Pinguin, Innsbruck, pp 170–182

Turner NC (1979) Drought resistance and adaptation to water deficits in crop plants. In: Mussell H, Staples RC (eds) Stress physiology in crop plants. Wiley, New York, pp 343–372

Turner NC (1986) Adaptation to water deficits: a changing perspective. Aust J Plant Physiol 13:175–190

Turner NC, Kramer PJ (1980) Adaptation of plants to water and high temperature stress. Wiley, New York

Tyler G (1990) Bryophytes and heavy metals: a literature review. Bot J Linn Soc 104:231–253

Tyree MT, Ewers FW (1991) The hydraulic architecture of trees and other woody plants. New Phytol 119:345–360

Tyree MT, Sperry JS (1989) Vulnerability of xylem to cavitation and embolism. Annu Rev Plant Physiol Mol Biol 40:19–38

Tyurina MM, Gogoleva GA (1975) Rol pokoya i temperatury v morozostoikosti yabloni. 12. Int Bot Congr, Abstr II, Leningrad, 488 pp

Uchijima Z, Seino H (1987) Distribution maps of net primary productivity of natural vegetation and related climatic elements on continents. Natl Inst Agro-Environ Sci, Yatabe

UNESCO (1979) Carte de la répartition mondiale des regions arides. In: Notes téchniques MAB7. UNESCO, Paris

Van Hove LWA, Van Kooten O, Adema EH, Vredenberg WJ, Pieters GA (1989) Physiological effects of long-term exposure to low and moderate concentrations of atmospheric NH_3 on poplar leaves. Plant Cell Environ 12:899–908

Van Valen L (1975) Life, death, and energy of a tree. Biotropica 7:260–269

Vannier G (1994) Perspective I. In: Block W, Vannier G: What is ecophysiology? Two perspectives. Acta Oecol 15:5–12

Vareschi V (1951) Zur Frage der Oberflächenentwicklung von Pflanzengesellschaften der Alpen und der Subtropen. Planta 40:1–35

Vareschi V (1953) Sobre las superficies de assimilación do sociedades vegetales de Cordilleras tropicales y extratropicales. Bol Soc Venez Cienc Nat 14:121–173

Vareschi V (1960) Effectos del viento en los Llanos, durante la epoca de sequía. Bol Soc Venez Cienc Nat 96:118–127

Vareschi V (1962) La quema como factor ecologico en los Llanos. Bol Soc Venez Cienc Nat 23:9–31

Vareschi V, Huber O (1971) La radiación solar y las estaciones anuales de los Llanos de Venezuela. Bol Soc Venz Cienc Nat 29:50–135

Vernon DM, Ostrem JA, Bohnert HJ (1993) Stress perception and response in a facultative halophyte: the regulation of salinity-induced genes in Mesembryanthemum crystallinum. Plant Cell Environ 16:437–444

Vins B (1962) Pouziti letokruhovych analyz k prukazu kourovych skod (Verwendung von Jahrringanalysen zum Nachweis von Rauchschäden). Lesnictví 8:263–280

Vogel JC, Fuls A, Danin A (1986) Geographical and environmental distribution of C_3 and C_4 grasses in the Sinai, Negev, and Judean deserts. Oecologia 70:258–265

Vogt K (1991) Carbon budgets of temperate forest ecosystems. Tree Physiol 9:69–86

Voznesenskii VL (1977) Fotosintez pustynnykh rastenii. Nauka, Leningrad

Wagner J, Tengg G (1993) Phänoembryologie der Hochgebirgspflanzen Saxifraga oppositifolia und Cerastium uniflorum. Flora 188:203–212

Waisel Y (1972) Biology of halophytes. Academic Press, New York

Walker RB (1991) Measuring mineral nutrient utilization. In: Lassoie JP, Hinckley TM (eds) Techniques and approaches in forest tree ecophysiology. CRC Press, Boca Raton, pp 183–206

Walker RB, Gessel StP (1990) Mineral deficiency symptoms in Pacific Northwest conifers. West J Appl For 5:96–98

Wallace T (1951) The diagnosis of mineral deficiencies in plants by visual symptoms. Her Majesty's Stationary Office, London

Wallace W, Pate JS (1967) Nitrate assimilation in higher plants with special reference to the cocklebur (*Xanthium pennsylvanicum* Wallr.). Ann Bot 31:213–228

Walter H (1931) Die Hydratur der Pflanze und ihre physiologisch-ökologische Bedeutung. Fischer, Jena

Walter H (1960) Einführung in die Phytologie, III/1. Standortslehre, 2. Aufl. Ulmer, Stuttgart

Walter H (1967) Die physiologischen Voraussetzungen für den Übergang der autotrophen Pflanzen vom Leben im Wasser zum Landleben. Z Pflanzenphysiol 56:170–185

Walter H (1968) Die Vegetation der Erde in ökophysiologischer Betrachtung, vol 2. Die gemäßigten und arktischen Zonen. Fischer, Jena

Walter H, Breckle SW (1986) Ökologie der Erde 3 – Spezielle Ökologie der Gemäßigten und Arktischen Zonen Euro-Nordasiens. Fischer, Stuttgart

Walter H, Breckle SW (1991) Ökologie der Erde 1 – Ökologische Grundlagen in globaler Sicht, 2. Aufl. Fischer, Stuttgart

Walter H, Kreeb K (1970) Die Hydratation und Hydratur des Protoplasmas der Pflanzen und ihre ökophysiologische Bedeutung. Springer, Wien New York

Walter H, Lieth H (1967) Klimadiagramm-Weltatlas. Fischer, Jena

Walter H, Steiner M (1936) Die Ökologie der Ostafrikanischen Mangroven. Z Bot 30:65–193

Walters MB, Kruger EL, Reich PB (1993) Relative growth rate in relation to physiological and morphological traits for northern hardwood tree seedlings: species, light environment and ontogenetic considerations. Oecologia 96:219–231

Wardlaw IF (1974) Temperature control of translocation. In: Bieleski RL, Ferguson AR, Cresswell MM (eds) Mechanisms of regulation of plant growth. Wellington, New Zealand, pp 533–538

Wardlaw IF (1990) The control of carbon partitioning in plants. New Phytol 116:341–381

Wardlaw IF, Passioura JB (1976) Transport and transfer processes in plants. Academic Press, New York, pp 381–391

Wareing PF, Phillips IDJ (1981) The control of growth and differentiation in plants, 2nd edn. Pergamon Press, Oxford

Wargo PhM (1979) Starch storage and radial growth in woody roots of sugar maple. Can J For Res 9:49–56

Wartinger A, Heilmeier H, Hartung W, Schulze ED (1990) Daily and seasonal courses of leaf conductance and abscisic acid in the xylem sap of almond trees (*Prunus dulcis*) under desert conditions. New Phytol 116:561–587

Waughman GJ (1977) The effect of temperature on nitrogenase activity. J Exp Bot 28:949–960

Webb T (1986) Is vegetation in equilibrium with climate? How to interpret late-Quaternary pollen data. Vegetatio 67:75–91

Webb T, Bartlein PJ (1992) Global changes during the last 3 million years: climatic controls and biotic responses. Annu Rev Ecol Syst 23:141–173

Weidner M, Ziemens C (1975) Preadaptation of protein synthesis in wheat seedlings to high temperature. Plant Physiol 56:590–594

Weigel HJ, Halbwachs G, Jäger HJ (1989) The effects of air pollutants on forest trees from a plant physiological view. Z Pflanzenkr Pflanzenschutz 96:203–217

Weigel HJ, Mejer GJ, Jäger HJ (1992) Auswirkungen von Klimaänderungen auf die Landwirtschaft: open-top Kammern zur Untersuchung von Langzeitwirkungen erhöhter CO_2-Konzentrationen auf landwirtschaftliche Pflanzen. Angew Bot 66:135–142

Weiglin Ch, Winter E (1991) Leaf structures of xerohalophytes from an East Jordanian salt pan. Flora 185:405–424

Wellburn AR (1985) SO_2 effects on stromal and thylakoid function. In: Winner WE, Mooney HA, Goldstein RA (eds) Sulfur dioxide and vegetation. Physiology, ecology and policy issues. Stanford University Press, Stanford, pp 133–147

Wellmann E (1983) UV-radiation in photomorphogenesis. In: Shropshire W, Mohr H (eds) Encyclopedia of plant physiology, vol 16B. Springer, Berlin Heidelberg New York, pp 745–756

Werner D (1992a) Physiology of nitrogen-fixing legume nodules: Compartments and functions. In: Stacey G, Burris RH, Evans HJ (eds) Biological nitrogen fixation. Chapman & Hall, New York, pp 399–431

Werner D (1992b) Symbiosis of plants and microbes. Chapman & Hall, London

Whittaker H, Likens GE (1975) The biosphere and man. In: Lieth H, Whittaker R (eds) Primary productivity of the biosphere. Springer, Berlin Heidelberg New York, pp 305–328

Wiebe HH, Brown RW, Daniel TW, Campbell E (1970) Water potential measurements in trees. BioScience 1970:225–226

Williams K, Percival F, Merino J, Mooney HA (1987) Estimation of tissue construction cost from heat of combustion and organic nitrogen content. Plant Cell Environ 10:725–734

Williamson P, Platt T (1991) Ocean biogeochemistry and air-sea CO_2 exchange. IGBP Newsl 6

Wilson ChCh (1948) The effect of some environmental factors on the movements of guard cells. Plant Physiol 23:5–35

Wimmer R, Halbwachs G (1992) Holzbiologische Untersuchungen an fluorgeschädigten Kiefern. Holz Roh Werkstoff 50:261–267

Winner WE, Mooney HA (1980) Ecology of SO_2 resistance: III. Metabolic changes of C_3 and C_4 Atriplex species due to SO_2 fumigations. Oecologia 46:49–54

Winter K (1985) Crassulacean acid metabolism. In: Barber J, Baker NR (eds) Photosynthetic mechanisms and the environment. Elsevier, Dordrecht, pp 329–387

Winter K, Lüttge U (1976) Balance between C_3 and CAM pathway of photosynthesis. In: Lange OL, Kappen L, Schulze ED (eds) Water and plant life. Springer, Berlin Heidelberg New York, pp 323–334

Wirth V, Türk R (1975) Über die SO_2-Resistenz von Flechten und die mit ihr interferierenden Faktoren. Verh Ges Ökol 74:173–179

WMO (1990) Global climate change. World Meteorologcal Organization Secr, Geneva

Wolverton BC, McDonald RC (1979) Upgrading facultative wastewater lagoons with vascular aquatic plants. J Water Pollut Control Fed 51:305–313

Woodward FI, Smith TM (1994) Predictions and measurements of the maximum photoosynthetic rate, A_{max}, at the global scale. In: Schulze ED, Caldwell MM (eds) Ecophysiology of photosynthesis. Ecol Stud 100. Springer, Berlin, pp 491–509

Woodward FI, Thompson GB, McKee IF (1991) The effects of elevated concentrations of carbon dioxide on individual plants, populations, communities and ecosystems. Ann Bot (Suppl 1) 67:23–38

Woodward RG (1976) Photosynthesis and expansion of leaves of soybean grown in two environments. Photosynthetica 10:274–279

Wyn-Jones RG, Gorham J (1983) Osmoregulation: In: Lange OL, Nobel PS, Osmond CB, Ziegler H (eds) Encyclopedia of plant physiology, vol 12C. Springer, Berlin Heidelberg New York, pp 35–58

Wyn-Jones RG, Pritchard J (1989) Stresses, membranes and cell wall. In: Jones HG, Flowers TJ, Jones MB (eds) Plants under stress. Cambridge University Press, Cambridge, pp 95–114

Wyn-Jones RG, Storey R (1981) Betaines. In: Paleg LG, Aspinall D (eds) Physiology and biochemistry of drought resistance in plants. Academic Press, Sydney, pp 171–204

Yoda K (1967) Comparative ecological studies on three main types of forest vegetation in Thailand. III. Community respiration. Nature Life Southeast Asia 5:83–148

Yoda K, Sato H (1975) Daily fluctuation of trunk diameter in tropical rain forest trees. Jpn J Ecol 25:47–48

Yoshie F (1989) Heat resistance of the temperate plants with different life-forms and from different microhabitats. Bull Assoc Nat Sci Senshu Univ 20:75–87

Yoshikawa K, Ogino K, Maiyus M (1986) Some aspects of sap flow rate of tree species in a tropical rain forest in West Sumatra. In: Hotta M (ed) Diversity and dynamics of plant life in Sumatra. Report and Coll Papers, Part 1. Kyoto University, Sumatra Nature Study, pp 45–59

Zeller O (1958) Über die Jahresrhythmik in der Entwicklung der Blütenknospen einiger Obstsorten. Gartenbauwissenschaft 23:167–181

Zhang J, Davies WJ (1990) Changes in the concentration of ABA in xylem sap as a function of changing soil water status can account for changes in leaf conductance and growth. Plant Cell Environ 13:277–285

Zimmermann MH (1983) Xylem structure and the ascent of sap. Springer, Berlin Heidelberg New York

Zimmermann MH, Brown CL (1974) Trees, structure and function, 2nd edn. Springer, Berlin Heidelberg New York

Subject Index